Biomechanics

Thoroughly revised and updated for the second edition, this comprehensive textbook integrates basic and advanced concepts of mechanics with numerical methods and biomedical applications. Coverage is expanded to include a complete introduction to vector and tensor calculus, and new or fully updated chapters on biological materials and continuum mechanics, motion, deformation and rotation, and the constitutive modelling of solids and fluids. Topics such as kinematics, equilibrium, and stresses and strains are also included, as well as the mechanical behaviour of fibres and the analysis of one-dimensional continuous elastic media. Numerical solution procedures based on the finite element method are presented, with accompanying MATLAB-based software and dozens of new biomedical engineering examples and exercises allowing readers to practise and improve their skills. Solutions for instructors are also available online. This is the definitive guide for both undergraduate and graduate students taking courses in biomechanics.

Cees Oomens is Full Professor in Biomechanics of Soft Tissues at the Eindhoven University of Technology.

Marcel Brekelmans is an Associate Professor in Continuum Mechanics at the Eindhoven University of Technology.

Sandra Loerakker is Assistant Professor in Modelling in Mechanobiology at the Eindhoven University of Technology.

Frank Baaijens is Full Professor in Soft Tissue Biomechanics and Tissue Engineering at the Eindhoven University of Technology. He is currently the University's Rector Magnificus.

CAMBRIDGE TEXTS IN BIOMEDICAL ENGINEERING

Series Editors
W. Mark Saltzman *Yale University*
Shu Chien *University of California, San Diego*

Series Advisors
Jerry Collins *Alabama A & M University*
Robert Malkin *Duke University*
Kathy Ferrara *University of California, Davis*
Nicholas Peppas *University of Texas, Austin*
Roger Kamm *Massachusetts Institute of Technology*
Masaaki Sato *Tohoku University, Japan*
Christine Schmidt *University of Florida*
George Truskey *Duke University*
Douglas Lauffenburger *Massachusetts Institute of Technology*

Cambridge Texts in Biomedical Engineering provide a forum for high-quality textbooks targeted at undergraduate and graduate courses in biomedical engineering. They cover a broad range of biomedical engineering topics from introductory texts to advanced topics, including biomechanics, physiology, biomedical instrumentation, imaging, signals and systems, cell engineering, and bioinformatics, as well as other relevant subjects, with a blending of theory and practice. While aiming primarily at biomedical engineering students, this series is also suitable for courses in broader disciplines in engineering, the life sciences and medicine.

"The increased number of exercises and examples used to bring the lectures alive and to illustrate the theory in biomedical applications make this second edition of the book *'Biomechanics: Concepts and Computation'* definitely the reference to teach classical concepts of mechanics and computational modelling techniques for biomedical engineers at Bachelor level. The authors from Eindhoven University of Technology belong to one of the most prestigious Departments of Biomedical Engineering around the world, with a well-recognized expertise in Soft Tissue Biomechanics and Tissue Engineering. I have no hesitation in recommending that book that should be a prerequisite for any student studying biomechanics."

Yohan Payan, *Director of Research at Centre National de la Recherche Scientifique (CNRS), Université Grenoble Alpes*

"A comprehensive textbook for learning all important concepts of biomechanics and their possible applications in sports and medicine. Students will enjoy the opportunity of learning computational modeling in biomechanics from scratch, needing only basic mathematical background. Instructors will appreciate the endless source of problems all resulting from successful experiences of teaching in the authors' career. Definitely recommended in every library."

Stéphane Avril, *École des Mines, St Étienne*

"'*Biomechanics: Concepts and Computation*' remains one of the strongest textbooks ever written in the field of biomechanical education. The theory in the book is thorough and rigorous, and is extremely well illustrated with numerous excellent exercises. I find the chapters describing numerical implementation and finite element formulations especially useful for translating the theory of tissue mechanics to bioengineering practice. I am using this book routinely in my undergraduate and graduate courses and will continue to do so with this second edition."

Amit Gefen, *Tel Aviv University*

Biomechanics

Concepts and Computation

Second Edition

Cees Oomens
Eindhoven University of Technology

Marcel Brekelmans
Eindhoven University of Technology

Sandra Loerakker
Eindhoven University of Technology

Frank Baaijens
Eindhoven University of Technology

CAMBRIDGE
UNIVERSITY PRESS

CAMBRIDGE
UNIVERSITY PRESS

University Printing House, Cambridge CB2 8BS, United Kingdom

One Liberty Plaza, 20th Floor, New York, NY 10006, USA

477 Williamstown Road, Port Melbourne, VIC 3207, Australia

314–321, 3rd Floor, Plot 3, Splendor Forum, Jasola District Centre, New Delhi – 110025, India

79 Anson Road, #06–04/06, Singapore 079906

Cambridge University Press is part of the University of Cambridge.

It furthers the University's mission by disseminating knowledge in the pursuit of education, learning, and research at the highest international levels of excellence.

www.cambridge.org
Information on this title: www.cambridge.org/9781107163720
DOI: 10.1017/9781316681633

First published 2010
Second edition 2018

A catalogue record for this publication is available from the British Library.

Library of Congress Cataloging-in-Publication Data
Names: Oomens, C. W. J., author. | Brekelmans, Marcel, author. | Loerakker, Sandra, author. | Baaijens, Franciscus Petrus Thomas, author.
Title: Biomechanics : concepts and computation / Cees Oomens (Eindhoven University of Technology), Marcel Brekelmans (Eindhoven University of Technology), Sandra Loerakker (Eindhoven University of Technology), Frank Baaijens (Eindhoven University of Technology).
Other titles: Cambridge texts in biomedical engineering.
Description: Second edition. | Cambridge, United Kingdom; New York, NY: Cambridge University Press, 2017. | Series: Cambridge texts in biomedical engineering
Identifiers: LCCN 2017026498| ISBN 9781107163720 | ISBN 1107163722
Subjects: LCSH: Biomechanics.
Classification: LCC QH513 .O56 2017 | DDC 571.4/3–dc23
LC record available at https://lccn.loc.gov/2017026498

ISBN 978-1-107-16372-0 Hardback

Additional resources for this publication at www.cambridge.org/Oomens

Contents

About the Cover — *page* xiii
Preface to the First Edition — xv
Preface to the Second Edition — xvii

1 Vector and Tensor Calculus — 1
1.1 Introduction — 1
1.2 Definition of a Vector — 1
1.3 Vector Operations — 1
1.4 Decomposition of a Vector with Respect to a Basis — 5
1.5 Some Mathematical Preliminaries on Second-Order Tensors — 10
Exercises — 13

2 The Concepts of Force and Moment — 16
2.1 Introduction — 16
2.2 Definition of a Force Vector — 16
2.3 Newton's Laws — 18
2.4 Vector Operations on the Force Vector — 19
2.5 Force Decomposition — 20
2.6 Drawing Convention — 24
2.7 The Concept of Moment — 25
2.8 Definition of the Moment Vector — 26
2.9 The Two-Dimensional Case — 30
2.10 Drawing Convention for Moments in Three Dimensions — 33
Exercises — 34

3 Static Equilibrium — 39
3.1 Introduction — 39
3.2 Static Equilibrium Conditions — 39
3.3 Free Body Diagram — 42
Exercises — 51

4 The Mechanical Behaviour of Fibres 56

4.1 Introduction 56
4.2 Elastic Fibres in One Dimension 56
4.3 A Simple One-Dimensional Model of a Skeletal Muscle 59
4.4 Elastic Fibres in Three Dimensions 62
4.5 Small Fibre Stretches 69
Exercises 73

5 Fibres: Time-Dependent Behaviour 79

5.1 Introduction 79
5.2 Viscous Behaviour 81
5.2.1 Small Stretches: Linearization 84
5.3 Linear Visco-Elastic Behaviour 85
5.3.1 Superposition and Proportionality 85
5.3.2 Generalization for an Arbitrary Load History 88
5.3.3 Visco-Elastic Models Based on Springs and Dashpots: Maxwell Model 92
5.3.4 Visco-Elastic Models Based on Springs and Dashpots: Kelvin–Voigt Model 96
5.4 Harmonic Excitation of Visco-Elastic Materials 97
5.4.1 The Storage and the Loss Modulus 97
5.4.2 The Complex Modulus 99
5.4.3 The Standard Linear Model 101
5.5 Appendix: Laplace and Fourier Transforms 106
Exercises 108

6 Analysis of a One-Dimensional Continuous Elastic Medium 116

6.1 Introduction 116
6.2 Equilibrium in a Subsection of a Slender Structure 116
6.3 Stress and Strain 118
6.4 Elastic Stress–Strain Relation 121
6.5 Deformation of an Inhomogeneous Bar 122
Exercises 129

7 Biological Materials and Continuum Mechanics 133

7.1 Introduction 133
7.2 Orientation in Space 134
7.3 Mass within the Volume *V* 138
7.4 Scalar Fields 141
7.5 Vector Fields 144

7.6 Rigid Body Rotation 149
Exercises 151

8 Stress in Three-Dimensional Continuous Media 155
8.1 Stress Vector 155
8.2 From Stress to Force 156
8.3 Equilibrium 157
8.4 Stress Tensor 164
8.5 Principal Stresses and Principal Stress Directions 172
8.6 Mohr's Circles for the Stress State 175
8.7 Hydrostatic Pressure and Deviatoric Stress 176
8.8 Equivalent Stress 177
Exercises 178

9 Motion: Time as an Extra Dimension 183
9.1 Introduction 183
9.2 Geometrical Description of the Material Configuration 183
9.3 Lagrangian and Eulerian Descriptions 185
9.4 The Relation between the Material and Spatial Time Derivatives 188
9.5 The Displacement Vector 190
9.6 The Gradient Operator 192
9.7 Extra Rigid Body Displacement 196
9.8 Fluid Flow 198
Exercises 199

10 Deformation and Rotation, Deformation Rate and Spin 204
10.1 Introduction 204
10.2 A Material Line Segment in the Reference and Current Configurations 204
10.3 The Stretch Ratio and Rotation 210
10.4 Strain Measures and Strain Tensors and Matrices 214
10.5 The Volume Change Factor 219
10.6 Deformation Rate and Rotation Velocity 219
Exercises 222

11 Local Balance of Mass, Momentum and Energy 227
11.1 Introduction 227
11.2 The Local Balance of Mass 227
11.3 The Local Balance of Momentum 228

11.4 The Local Balance of Mechanical Power 230
11.5 Lagrangian and Eulerian Descriptions of the Balance Equations 231
Exercises 233

12 Constitutive Modelling of Solids and Fluids 235
12.1 Introduction 235
12.2 Elastic Behaviour at Small Deformations and Rotations 236
12.3 The Stored Internal Energy 242
12.4 Elastic Behaviour at Large Deformations and/or Large Rotations 244
12.4.1 Material Frame Indifference 244
12.4.2 Strain Energy Function 250
12.4.3 The Incompressible Neo-Hookean Model 252
12.4.4 The Incompressible Mooney–Rivlin Model 255
12.4.5 Compressible Neo-Hookean Elastic Solid 256
12.5 Constitutive Modelling of Viscous Fluids 261
12.6 Newtonian Fluids 262
12.7 Non-Newtonian Fluids 263
12.8 Diffusion and Filtration 264
Exercises 264

13 Solution Strategies for Solid and Fluid Mechanics Problems 270
13.1 Introduction 270
13.2 Solution Strategies for Deforming Solids 270
13.2.1 General Formulation for Solid Mechanics Problems 271
13.2.2 Geometrical Linearity 272
13.2.3 Linear Elasticity Theory, Dynamic 273
13.2.4 Linear Elasticity Theory, Static 273
13.2.5 Linear Plane Stress Theory, Static 274
13.2.6 Boundary Conditions 278
13.3 Solution Strategies for Viscous Fluids 280
13.3.1 General Equations for Viscous Flow 281
13.3.2 The Equations for a Newtonian Fluid 282
13.3.3 Stationary Flow of an Incompressible Newtonian Fluid 282
13.3.4 Boundary Conditions 283
13.3.5 Elementary Analytical Solutions 283
13.4 Diffusion and Filtration 285
Exercises 287

14 Solution of the One-Dimensional Diffusion Equation by Means of the Finite Element Method 292
14.1 Introduction 292

14.2 The Diffusion Equation 293
14.3 Method of Weighted Residuals and Weak Form 295
14.4 Polynomial Interpolation 297
14.5 Galerkin Approximation 300
14.6 Solution of the Discrete Set of Equations 307
14.7 Isoparametric Elements and Numerical Integration 308
14.8 Basic Structure of a Finite Element Program 312
Exercises 319

15 Solution of the One-Dimensional Convection–Diffusion Equation by Means of the Finite Element Method 327
15.1 Introduction 327
15.2 The Convection–Diffusion Equation 327
15.3 Temporal Discretization 330
15.4 Spatial Discretization 333
Exercises 338

16 Solution of the Three-Dimensional Convection–Diffusion Equation by Means of the Finite Element Method 342
16.1 Introduction 342
16.2 Diffusion Equation 343
16.3 Divergence Theorem and Integration by Parts 344
16.4 Weak Form 345
16.5 Galerkin Discretization 345
16.6 Convection–Diffusion Equation 348
16.7 Isoparametric Elements and Numerical Integration 349
16.8 Example 353
Exercises 356

17 Shape Functions and Numerical Integration 363
17.1 Introduction 363
17.2 Isoparametric, Bi-Linear Quadrilateral Element 365
17.3 Linear Triangular Element 367
17.4 Lagrangian and Serendipity Elements 370
17.4.1 Lagrangian Elements 371
17.4.2 Serendipity Elements 373
17.5 Numerical Integration 373
Exercises 377

18 Infinitesimal Strain Elasticity Problems 382
18.1 Introduction 382
18.2 Linear Elasticity 382
18.3 Weak Formulation 384
18.4 Galerkin Discretization 385
18.5 Solution 391
Exercises 394

References 399
Index 401

About the Cover

The cover contains images reflecting biomechanics research topics at the Eindhoven University of Technology. An important aspect of mechanics is experimental work to determine material properties and to validate models. The application field ranges from microscopic structures at the level of cells to larger organs like the heart. The core of biomechanics is constituted by models formulated in terms of partial differential equations and computer models to derive approximate solutions.

- *Main image*: Myogenic precursor cells have the ability to differentiate and fuse to form multinucleated myotubes. This differentiation process can be influenced by means of mechanical as well as biochemical stimuli. To monitor this process of early differentiation, immunohistochemical analyses are performed to provide information concerning morphology and localization of characteristic structural proteins of muscle cells. In the illustration, the sarcomeric proteins actin (red), and myosin (green) are shown. Nuclei are stained blue. Image courtesy of Mrs Marloes Langelaan.
- *Left top*: To study the effect of a mechanical load on the damage evolution of skeletal tissue, an in-vitro model system using tissue engineered muscle was developed. The image shows this muscle construct in a set-up on a confocal microscope. In the device the construct can be mechanically deformed by means of an indentor. Fluorescent identification of both necrotic and apoptotic cells can be established using different staining techniques. Image courtesy of Mrs Debby Gawlitta.
- *Left middle*: A three-dimensional finite element mesh of the human heart ventricles is shown. This mesh is used to solve the equations of motion for the beating heart. The model was used to study the effect of depolarization waves and mechanics in the paced heart. Image courtesy of Mr Roy Kerckhoffs.
- *Left bottom*: The equilibrium equations are derived from Newton's laws and describe (quasi-)static force equilibrium in a three-dimensional continuum. See Eqs. (8.33), (8.34) and (8.35) in the present book.

Preface to the First Edition

In September 1997, an educational programme in Biomedical Engineering, unique in the Netherlands, started at the Eindhoven University of Technology, together with the University of Maastricht, as a logical step after almost two decades of research collaboration between both universities. This development culminated in the foundation of the Department of Biomedical Engineering in April 1999 and the creation of a graduate programme (MSc) in Biomedical Engineering in 2000 and Medical Engineering in 2002.

Already at the start of this educational programme, it was decided that a comprehensive course in biomechanics had to be part of the curriculum and that this course had to start right at the beginning of the Bachelor phase. A search for suitable material for this purpose showed that excellent biomechanics textbooks exist. But many of these books are very specialized to certain aspects of biomechanics. The more general textbooks address mechanical or civil engineers or physicists who wish to specialize in biomechanics, so these books include chapters or sections on biology and physiology. Almost all books that were found are at Masters or post-graduate level, requiring basic to sophisticated knowledge of mechanics and mathematics. At a more fundamental level, only books could be found that were written for mechanical and civil engineers.

We decided to write our own course material for the basic training in mechanics appropriate for our candidate biomedical engineers at Bachelor level, starting with the basic concepts of mechanics and ending with numerical solution procedures, based on the finite element method. The course material assembled in the current book comprises three courses for our biomedical engineering curriculum, distributed over the three years of their Bachelor studies. Chapters 1 to 6 mostly treat the basic concepts of forces, moments and equilibrium in a discrete context in the first year. Chapters 7 to 13 in the second year discuss the basis of continuum mechanics, and Chapters 14 to 18 in the third year are focussed on solving the field equations of mechanics using the finite element method.

What makes this book different from other basic mechanics or biomechanics treatises? Of course, as in standard books, there is the usual attention focussed on kinematics, equilibrium, stresses and strains. But several topics are discussed that are normally not found in one single textbook or only described briefly.

- Much attention is given to large deformations and rotations and non-linear constitutive equations (see Chapters 4, 9 and 10).
- A separate chapter is devoted to one-dimensional visco-elastic behaviour (Chapter 5).
- Special attention is given to long, slender, fibre-like structures (Chapter 4).
- The similarities and differences in describing the behaviour of solids and fluids and aspects of diffusion and filtration are discussed (Chapters 12 to 16).
- Basic concepts of mechanics and numerical solution strategies for partial differential equations are integrated in one single textbook (Chapters 14 to 18).

Because of the usually rather complex geometries (and non-linear aspects) found in biomechanical problems, hardly any relevant analytical solutions can be derived for the field equations, and approximate solutions have to be constructed. It is the opinion of the authors that, at Bachelor level, at least the basis for these numerical techniques has to be addressed.

In Chapters 14 to 18 extensive use is made of a finite element code written in MATLAB by one of the authors, which is especially developed as a tool for students. Applying this code requires that the user has a licence for the use of MATLAB, which can be obtained via MathWorks (www.mathworks.com). The finite element code, which is a set of MATLAB scripts, including manuals, is freely available and can be downloaded from the website: www.tue.nl/biomechanicsbook.

Preface to the Second Edition

Since 2009, when this book was published for the first time, we have been using it in our Biomechanics courses in the educational programme Biomedical Engineering, giving us hands-on experience with the book and the exercises. When we were investigating ideas for a second edition, we found that external reviewers were primarily asking for more examples and exercises, concurring with our own thoughts on the book. Over the years, we have assembled quite a number of examples that were often used to animate the lectures and to illustrate the theory in biomedical applications. At the same time, the number of available exercises increased considerably. Eventually, adding many of these examples and increasing the number of exercises are the most significant changes in the second edition. The major changes in the text are:

- Mathematical preliminaries are now concentrated in Chapter 1 and no longer spread over different chapters.
- At some points in the original text, explanations were terse and too concise for students. Based on our experiences over the past eight years, we have extended the text at a number of points and, most importantly, added the earlier mentioned new examples. The biggest change is in Chapter 12, including a separate section on material frame indifference of constitutive equations and a more extensive treatment of hyperelastic materials.

The objectives of the book did not change. It is still meant to be a basic training in mechanics, appropriate for our candidate biomedical engineers at Bachelor level, starting with the basic concepts of mechanics and ending with numerical solution procedures. This book differs from most books on biomechanics, which are usually aimed at students with already considerable knowledge in continuum mechanics and wishing to enter the field of biomechanics. Consequently, those books pay great attention to biology and physiology. In contrast, we assume that students start at a very basic level in terms of mechanics, but already have substantial physiological and biological background knowledge.

1 Vector and Tensor Calculus

1.1 Introduction

Before we can start with biomechanics, it is necessary to introduce some basic mathematical concepts and to introduce the mathematical notation that will be used throughout the book. The present chapter is aimed at understanding some of the basics of vector calculus, which are necessary to elucidate the concepts of force and momentum that will be treated in the next chapter.

1.2 Definition of a Vector

A **vector** is a mathematical entity having both a magnitude (length or size) and a direction. For a vector $\vec{a}$, it holds (see Fig. 1.1), that:

$$\vec{a} = a\vec{e}. \tag{1.1}$$

The **length** of the vector $\vec{a}$ is denoted by $|\vec{a}|$ and is equal to the length of the arrow. The length is equal to a, when a is positive, and equal to $-a$ when a is negative. The **direction** of $\vec{a}$ is given by the unit vector $\vec{e}$ combined with the sign of a. The unit vector $\vec{e}$ has length 1. The vector $\vec{0}$ has length zero.

1.3 Vector Operations

Multiplication of a vector $\vec{a} = a\vec{e}$ by a positive scalar α yields a vector $\vec{b}$ having the same direction as $\vec{a}$ but a different magnitude $\alpha|\vec{a}|$:

$$\vec{b} = \alpha\vec{a} = \alpha a\vec{e}. \tag{1.2}$$

This makes sense: pulling twice as hard on a wire creates a force in the wire having the same orientation (the direction of the wire does not change), but with a magnitude that is twice as large.

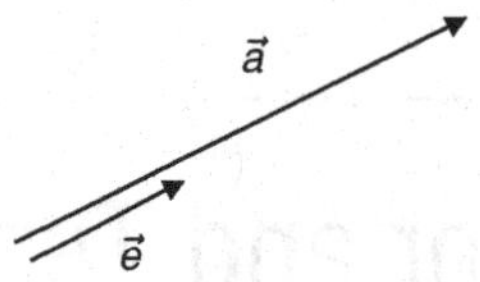

Figure 1.1

The vector $\vec{a} = a\vec{e}$ with $a > 0$.

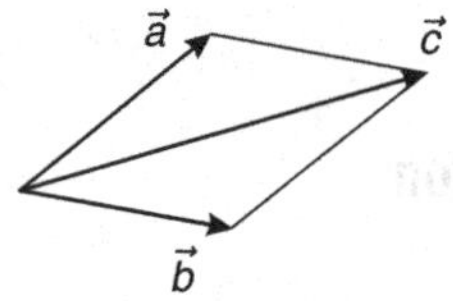

Figure 1.2

Graphical representation of the sum of two vectors: $\vec{c} = \vec{a} + \vec{b}$.

The **sum** of two vectors $\vec{a}$ and $\vec{b}$ is a new vector $\vec{c}$, equal to the diagonal of the parallelogram spanned by $\vec{a}$ and $\vec{b}$ (see Fig. 1.2):

$$\vec{c} = \vec{a} + \vec{b}. \tag{1.3}$$

This may be interpreted as follows. Imagine two thin wires which are attached to a point P. The wires are being pulled at in two different directions according to the vectors $\vec{a}$ and $\vec{b}$. The length of each vector represents the magnitude of the pulling force. The net force vector exerted on the attachment point P is the vector sum of the two vectors $\vec{a}$ and $\vec{b}$. If the wires are aligned with each other and the pulling direction is the same, the resulting force direction clearly coincides with the direction of the two wires, and the length of the resulting force vector is the sum of the two pulling forces. Alternatively, if the two wires are aligned but the pulling forces are in opposite directions and of equal magnitude, the resulting force exerted on point P is the zero vector $\vec{0}$.

The **inner product** or **dot product** of two vectors is a scalar quantity, defined as

$$\vec{a} \cdot \vec{b} = |\vec{a}||\vec{b}| \cos(\phi), \tag{1.4}$$

where ϕ is the smallest angle between $\vec{a}$ and $\vec{b}$ (see Fig. 1.3). The inner product is **commutative**, i.e.

$$\vec{a} \cdot \vec{b} = \vec{b} \cdot \vec{a}. \tag{1.5}$$

The inner product can be used to define the length of a vector, since the inner product of a vector with itself yields ($\phi = 0$):

$$\vec{a} \cdot \vec{a} = |\vec{a}||\vec{a}| \cos(0) = |\vec{a}|^2. \tag{1.6}$$

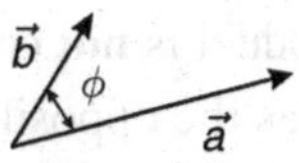

Figure 1.3

Definition of the angle ϕ.

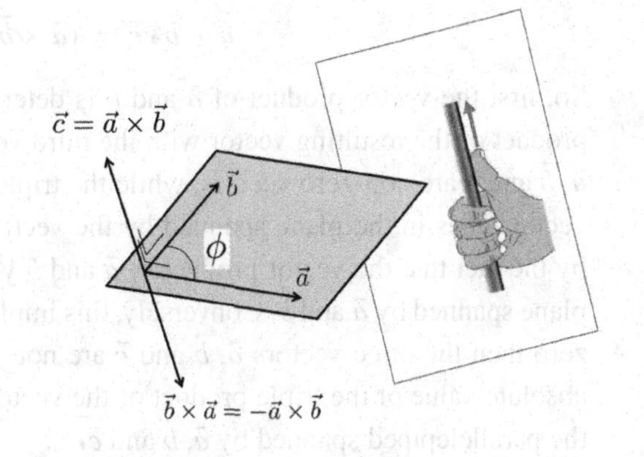

Figure 1.4

Vector product $\vec{c} = \vec{a} \times \vec{b}$. The direction of vector $\vec{c}$ is determined by the corkscrew or right-hand rule.

If two vectors are perpendicular to each other the inner product of these two vectors is equal to zero, since in that case $\phi = \frac{\pi}{2}$:

$$\vec{a} \cdot \vec{b} = 0, \text{ if } \phi = \frac{\pi}{2}. \tag{1.7}$$

The **cross product** or **vector product** of two vectors $\vec{a}$ and $\vec{b}$ yields a new vector $\vec{c}$ that is perpendicular to both $\vec{a}$ and $\vec{b}$ such that $\vec{a}$, $\vec{b}$ and $\vec{c}$ form a right-handed system. The vector $\vec{c}$ is denoted as

$$\vec{c} = \vec{a} \times \vec{b}\,. \tag{1.8}$$

The length of the vector $\vec{c}$ is given by

$$|\vec{c}| = |\vec{a}||\vec{b}| \sin(\phi), \tag{1.9}$$

where ϕ is the smallest angle between $\vec{a}$ and $\vec{b}$. The length of $\vec{c}$ equals the area of the parallelogram spanned by the vectors $\vec{a}$ and $\vec{b}$. The vector system $\vec{a}$, $\vec{b}$ and $\vec{c}$ forms a right-handed system, meaning that if a corkscrew were used rotating from $\vec{a}$ to $\vec{b}$ the corkscrew would move into the direction of $\vec{c}$ (see Fig. 1.4).

The vector product of a vector $\vec{a}$ with itself yields the zero vector, since in that case $\phi = 0$:

$$\vec{a} \times \vec{a} = \vec{0}. \tag{1.10}$$

The vector product is **not** commutative, since the vector product of $\vec{b}$ and $\vec{a}$ yields a vector that has the opposite direction to the vector product of $\vec{a}$ and $\vec{b}$:

$$\vec{a} \times \vec{b} = -\vec{b} \times \vec{a}. \tag{1.11}$$

The **triple product** of three vectors $\vec{a}$, $\vec{b}$ and $\vec{c}$ is a scalar, defined by

$$\vec{a} \times \vec{b} \cdot \vec{c} = (\vec{a} \times \vec{b}) \cdot \vec{c}. \tag{1.12}$$

So, first the vector product of $\vec{a}$ and $\vec{b}$ is determined and subsequently the inner product of the resulting vector with the third vector $\vec{c}$ is taken. If all three vectors $\vec{a}$, $\vec{b}$ and $\vec{c}$ are non-zero vectors, while the triple product is equal to zero, then the vector $\vec{c}$ lies in the plane spanned by the vectors $\vec{a}$ and $\vec{b}$. This can be explained by the fact that the vector product of $\vec{a}$ and $\vec{b}$ yields a vector perpendicular to the plane spanned by $\vec{a}$ and $\vec{b}$. Conversely, this implies that if the triple product is non-zero then the three vectors $\vec{a}$, $\vec{b}$ and $\vec{c}$ are not in the same plane. In that case the absolute value of the triple product of the vectors $\vec{a}$, $\vec{b}$ and $\vec{c}$ equals the volume of the parallelepiped spanned by $\vec{a}$, $\vec{b}$ and $\vec{c}$.

The **dyadic** or **tensor product** of two vectors $\vec{a}$ and $\vec{b}$ defines a linear transformation operator called a **dyad** $\vec{a}\vec{b}$. Application of a dyad $\vec{a}\vec{b}$ to a vector $\vec{p}$ yields a vector into the direction of $\vec{a}$, where $\vec{a}$ is multiplied by the inner product of $\vec{b}$ and $\vec{p}$:

$$\vec{a}\vec{b} \cdot \vec{p} = \vec{a}\,(\vec{b} \cdot \vec{p}). \tag{1.13}$$

So, application of a dyad to a vector transforms this vector into another vector. This transformation is linear, as can be seen from

$$\vec{a}\vec{b} \cdot (\alpha\vec{p} + \beta\vec{q}) = \vec{a}\vec{b} \cdot \alpha\vec{p} + \vec{a}\vec{b} \cdot \beta\vec{q} = \alpha\vec{a}\vec{b} \cdot \vec{p} + \beta\vec{a}\vec{b} \cdot \vec{q}. \tag{1.14}$$

The transpose of a dyad $(\vec{a}\vec{b})^{\mathrm{T}}$ is defined by

$$(\vec{a}\vec{b})^{\mathrm{T}} \cdot \vec{p} = \vec{b}\vec{a} \cdot \vec{p}, \tag{1.15}$$

or simply

$$(\vec{a}\vec{b})^{\mathrm{T}} = \vec{b}\vec{a}. \tag{1.16}$$

An operator $\boldsymbol{A}$ that transforms a vector $\vec{a}$ into another vector $\vec{b}$ according to

$$\vec{b} = \boldsymbol{A} \cdot \vec{a}, \tag{1.17}$$

is called a second-order tensor $\boldsymbol{A}$. This implies that the dyadic product of two vectors is a second-order tensor.

In three-dimensional space, a set of three vectors $\vec{c}_1$, $\vec{c}_2$ and $\vec{c}_3$ is called a **basis** if the triple product of the three vectors is non-zero, hence if all three vectors are non-zero vectors and if they do not lie in the same plane:

$$\vec{c}_1 \times \vec{c}_2 \cdot \vec{c}_3 \neq 0. \tag{1.18}$$

The three vectors $\vec{c}_1$, $\vec{c}_2$ and $\vec{c}_3$ composing the basis are called basis vectors.

If the basis vectors are mutually perpendicular vectors, the basis is called an **orthogonal** basis. If such basis vectors have unit length, then the basis is called **orthonormal**. A **Cartesian basis** is an orthonormal, right-handed basis with basis vectors independent of the location in the three-dimensional space. In the following we will indicate the Cartesian basis vectors with $\vec{e}_x$, $\vec{e}_y$ and $\vec{e}_z$.

1.4 Decomposition of a Vector with Respect to a Basis

As stated above, a Cartesian vector basis is an orthonormal basis. Any vector can be decomposed into the sum of, at most, three vectors parallel to the three basis vectors $\vec{e}_x$, $\vec{e}_y$ and $\vec{e}_z$:

$$\vec{a} = a_x\vec{e}_x + a_y\vec{e}_y + a_z\vec{e}_z. \tag{1.19}$$

The components a_x, a_y and a_z can be found by taking the inner product of the vector $\vec{a}$ with respect to each of the basis vectors:

$$\begin{aligned} a_x &= \vec{a} \cdot \vec{e}_x \\ a_y &= \vec{a} \cdot \vec{e}_y \\ a_z &= \vec{a} \cdot \vec{e}_z, \end{aligned} \tag{1.20}$$

where use is made of the fact that the basis vectors have unit length and are mutually orthogonal, for example:

$$\vec{a} \cdot \vec{e}_x = a_x\vec{e}_x \cdot \vec{e}_x + a_y\vec{e}_y \cdot \vec{e}_x + a_z\vec{e}_z \cdot \vec{e}_x = a_x. \tag{1.21}$$

The components, say a_x, a_y and a_z, of a vector $\vec{a}$ with respect to the Cartesian vector basis, may be collected in a **column**, denoted by $\underset{\sim}{a}$:

$$\underset{\sim}{a} = \begin{bmatrix} a_x \\ a_y \\ a_z \end{bmatrix}. \tag{1.22}$$

So, with respect to a Cartesian vector basis, any vector $\vec{a}$ may be decomposed into components that can be collected in a column:

$$\vec{a} \longleftrightarrow \underset{\sim}{a}\,. \tag{1.23}$$

This 'transformation' is only possible and meaningful if the vector basis with which the components of the column $\underset{\sim}{a}$ are defined has been specified. The choice of a different vector basis leads to a different column representation $\underset{\sim}{a}$ of the vector

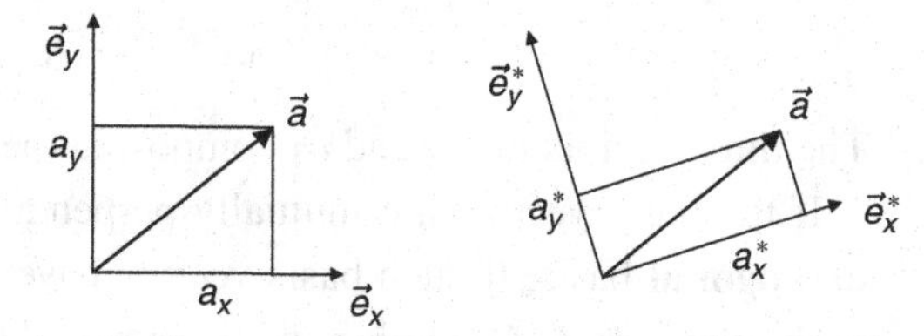

Figure 1.5

Vector $\vec{a}$ with respect to vector bases $\{\vec{e}_x, \vec{e}_y\}$ and $\{\vec{e}_x^*, \vec{e}_y^*\}$.

$\vec{a}$, as illustrated in Fig. 1.5. The vector $\vec{a}$ has two different column representations, $\underset{\sim}{a}$ and $\underset{\sim}{a}^*$, depending on which vector basis is used. If, in a two-dimensional context, $\{\vec{e}_x, \vec{e}_y\}$ is used as a vector basis then

$$\vec{a} \longleftrightarrow \underset{\sim}{a} = \begin{bmatrix} a_x \\ a_y \end{bmatrix}, \tag{1.24}$$

while, if $\{\vec{e}_x{}^*, \vec{e}_y{}^*\}$ is used as a vector basis:

$$\vec{a} \longleftrightarrow \underset{\sim}{a}^* = \begin{bmatrix} a_x^* \\ a_y^* \end{bmatrix}. \tag{1.25}$$

Consequently, with respect to a Cartesian vector basis, vector operations such as multiplication, addition, inner product and dyadic product may be rewritten as 'column' (actually matrix) operations.

Multiplication of a vector $\vec{a} = a_x\vec{e}_x + a_y\vec{e}_y + a_z\vec{e}_z$ with a scalar α yields a new vector, say $\vec{b}$:

$$\begin{aligned}\vec{b} = \alpha\vec{a} &= \alpha(a_x\vec{e}_x + a_y\vec{e}_y + a_z\vec{e}_z) \\ &= \alpha a_x\vec{e}_x + \alpha a_y\vec{e}_y + \alpha a_z\vec{e}_z.\end{aligned} \tag{1.26}$$

So

$$\vec{b} = \alpha\vec{a} \longleftrightarrow \underset{\sim}{b} = \alpha\underset{\sim}{a}. \tag{1.27}$$

The sum of two vectors $\vec{a}$ and $\vec{b}$ leads to

$$\vec{c} = \vec{a} + \vec{b} \longleftrightarrow \underset{\sim}{c} = \underset{\sim}{a} + \underset{\sim}{b}. \tag{1.28}$$

Using the fact that the Cartesian basis vectors have unit length and are mutually orthogonal, the inner product of two vectors $\vec{a}$ and $\vec{b}$ yields a scalar c according to

$$\begin{aligned}c = \vec{a} \cdot \vec{b} &= (a_x\vec{e}_x + a_y\vec{e}_y + a_z\vec{e}_z) \cdot (b_x\vec{e}_x + b_y\vec{e}_y + b_z\vec{e}_z) \\ &= a_xb_x + a_yb_y + a_zb_z.\end{aligned} \tag{1.29}$$

In column notation this result is obtained via

$$c = \underset{\sim}{a}^{\mathrm{T}}\underset{\sim}{b}\,, \tag{1.30}$$

where $\underset{\sim}{a}^{\mathrm{T}}$ denotes the **transpose** of the column $\underset{\sim}{a}$, defined as

$$\underset{\sim}{a}^{\mathrm{T}} = \left[a_x \; a_y \; a_z\right], \tag{1.31}$$

such that:

$$\underset{\sim}{a}^{\mathrm{T}}\underset{\sim}{b} = \left[a_x \; a_y \; a_z\right]\begin{bmatrix} b_x \\ b_y \\ b_z \end{bmatrix} = a_x b_x + a_y b_y + a_z b_z. \tag{1.32}$$

Using the properties of the basis vectors of the Cartesian vector basis:

$$\begin{aligned}
\vec{e}_x \times \vec{e}_x &= \vec{0} \\
\vec{e}_x \times \vec{e}_y &= \vec{e}_z \\
\vec{e}_x \times \vec{e}_z &= -\vec{e}_y \\
\vec{e}_y \times \vec{e}_x &= -\vec{e}_z \\
\vec{e}_y \times \vec{e}_y &= \vec{0} \\
\vec{e}_y \times \vec{e}_z &= \vec{e}_x \\
\vec{e}_z \times \vec{e}_x &= \vec{e}_y \\
\vec{e}_z \times \vec{e}_y &= -\vec{e}_x \\
\vec{e}_z \times \vec{e}_z &= \vec{0},
\end{aligned} \tag{1.33}$$

the vector product of a vector $\vec{a}$ and a vector $\vec{b}$ is directly computed by means of

$$\begin{aligned}
\vec{a} \times \vec{b} &= (a_x\vec{e}_x + a_y\vec{e}_y + a_z\vec{e}_z) \times (b_x\vec{e}_x + b_y\vec{e}_y + b_z\vec{e}_z) \\
&= (a_y\, b_z - a_z\, b_y)\vec{e}_x + (a_z\, b_x - a_x\, b_z)\vec{e}_y + (a_x\, b_y - a_y\, b_x)\vec{e}_z.
\end{aligned} \tag{1.34}$$

If by definition $\vec{c} = \vec{a} \times \vec{b}$, then the associated column $\underset{\sim}{c}$ can be written as:

$$\underset{\sim}{c} = \begin{bmatrix} a_y\, b_z - a_z\, b_y \\ a_z\, b_x - a_x\, b_z \\ a_x\, b_y - a_y\, b_x \end{bmatrix}. \tag{1.35}$$

The dyadic product $\vec{a}\vec{b}$ transforms another vector $\vec{c}$ into a vector $\vec{d}$, according to the definition

$$\vec{d} = \vec{a}\vec{b} \cdot \vec{c} = \boldsymbol{A} \cdot \vec{c}\,, \tag{1.36}$$

with $\boldsymbol{A}$ the second-order tensor equal to the dyadic product $\vec{a}\vec{b}$. In column notation this is equivalent to

$$\underset{\sim}{d} = \underset{\sim}{a}(\underset{\sim}{b}^{\mathrm{T}}\underset{\sim}{c}) = (\underset{\sim}{a}\,\underset{\sim}{b}^{\mathrm{T}})\underset{\sim}{c}\,, \tag{1.37}$$

with $\underset{\sim}{a}\,\underset{\sim}{b}^{\mathrm{T}}$ a (3 × 3) matrix given by

$$\underline{A} = \underset{\sim}{a}\,\underset{\sim}{b}^{\mathrm{T}} = \begin{bmatrix} a_x \\ a_y \\ a_z \end{bmatrix} \begin{bmatrix} b_x & b_y & b_z \end{bmatrix} = \begin{bmatrix} a_x b_x & a_x b_y & a_x b_z \\ a_y b_x & a_y b_y & a_y b_z \\ a_z b_x & a_z b_y & a_z b_z \end{bmatrix}, \tag{1.38}$$

or

$$\underset{\sim}{d} = \underline{A}\,\underset{\sim}{c}. \tag{1.39}$$

In this case $\underline{A}$ is called the matrix representation of the second-order tensor $\boldsymbol{A}$, as the comparison of Eqs. (1.36) and (1.39) reveals.

Example 1.1 Suppose we can write the vectors $\vec{a}$ and $\vec{b}$ as the following linear combination of the Cartesian basis vectors $\vec{e}_x$ and $\vec{e}_y$:

$$\vec{a} = \vec{e}_x + 2\vec{e}_y$$
$$\vec{b} = 2\vec{e}_x + 5\vec{e}_y,$$

and we wish to determine the inner and vector product of both vectors. Then:

$$\begin{aligned} \vec{a} \cdot \vec{b} &= (\vec{e}_x + 2\vec{e}_y) \cdot (2\vec{e}_x + 5\vec{e}_y) \\ &= 2\vec{e}_x \cdot \vec{e}_x + 5\vec{e}_x \cdot \vec{e}_y + 4\vec{e}_y \cdot \vec{e}_x + 10\vec{e}_y \cdot \vec{e}_y \\ &= 2 + 10 \\ &= 12. \end{aligned}$$

When using column notation we can also write:

$$\underset{\sim}{a} = \begin{bmatrix} 1 \\ 2 \\ 0 \end{bmatrix} \qquad \underset{\sim}{b} = \begin{bmatrix} 2 \\ 5 \\ 0 \end{bmatrix},$$

and:

$$\vec{a} \cdot \vec{b} = \underset{\sim}{a}^T \underset{\sim}{b} = [1 \quad 2 \quad 0] \begin{bmatrix} 2 \\ 5 \\ 0 \end{bmatrix} = 12.$$

For the vector product a similar procedure can be used:

$$\begin{aligned} \vec{a} \times \vec{b} &= (\vec{e}_x + 2\vec{e}_y) \times (2\vec{e}_x + 5\vec{e}_y) \\ &= 2\vec{e}_x \times \vec{e}_x + 5\vec{e}_x \times \vec{e}_y + 4\vec{e}_y \times \vec{e}_x + 10\vec{e}_y \times \vec{e}_y \\ &= \vec{0} + 5\vec{e}_z - 4\vec{e}_z + \vec{0} \\ &= \vec{e}_z. \end{aligned}$$

When using column notation for the vector product, Eq. (1.35) has to be used.

Example 1.2 Consider the Cartesian basis $\{\vec{e}_x, \vec{e}_y, \vec{e}_z\}$. We want to know whether the following three vectors given by:

$$\vec{\varepsilon}_1 = 2\vec{e}_x$$
$$\vec{\varepsilon}_2 = \vec{e}_x + 2\vec{e}_y$$
$$\vec{\varepsilon}_3 = \vec{e}_y + 3\vec{e}_z$$

could also be used as a basis. For this purpose we have to determine whether the vectors are independent, and consequently we calculate the triple product:

$$\begin{aligned}(\vec{\varepsilon}_1 \times \vec{\varepsilon}_2) \cdot \vec{\varepsilon}_3 &= \left[2\vec{e}_x \times (\vec{e}_x + 2\vec{e}_y)\right] \cdot (\vec{e}_y + 3\vec{e}_z)\\ &= 4\vec{e}_z \cdot (\vec{e}_y + 3\vec{e}_z) = 12 \neq 0.\end{aligned}$$

This means that the three vectors are independent and might be used as a basis. However, they are not perpendicular and do not have length 1.

Example 1.3 With respect to a Cartesian basis $\{\vec{e}_x, \vec{e}_y, \vec{e}_z\}$ the following vectors are defined:

$$\vec{a} = \vec{e}_x + 2\vec{e}_y$$
$$\vec{b} = 2\vec{e}_x + 5\vec{e}_y$$
$$\vec{c} = 3\vec{e}_x.$$

We want to determine the dyadic products $\boldsymbol{A} = \vec{a}\vec{b}, \boldsymbol{A}^T = \vec{b}\vec{a}$ and the result of $\boldsymbol{A} \cdot \vec{c}$ and $\boldsymbol{A}^T \cdot \vec{c}$. Write

$$\begin{aligned}\boldsymbol{A} = \vec{a}\vec{b} &= (\vec{e}_x + 2\vec{e}_y)(2\vec{e}_x + 5\vec{e}_y)\\ &= 2\vec{e}_x\vec{e}_x + 5\vec{e}_x\vec{e}_y + 4\vec{e}_y\vec{e}_x + 10\vec{e}_y\vec{e}_y,\end{aligned}$$

$$\begin{aligned}\boldsymbol{A}^T = \vec{b}\vec{a} &= (2\vec{e}_x + 5\vec{e}_y)(\vec{e}_x + 2\vec{e}_y)\\ &= 2\vec{e}_x\vec{e}_x + 4\vec{e}_x\vec{e}_y + 5\vec{e}_y\vec{e}_x + 10\vec{e}_y\vec{e}_y\end{aligned}$$

and:

$$\begin{aligned}\boldsymbol{A} \cdot \vec{c} = \vec{a}\vec{b} \cdot \vec{c} &= (2\vec{e}_x\vec{e}_x + 5\vec{e}_x\vec{e}_y + 4\vec{e}_y\vec{e}_x + 10\vec{e}_y\vec{e}_y) \cdot 3\vec{e}_x\\ &= 6\vec{e}_x + 12\vec{e}_y,\end{aligned}$$

$$\begin{aligned}\boldsymbol{A}^T \cdot \vec{c} = \vec{b}\vec{a} \cdot \vec{c} &= (2\vec{e}_x\vec{e}_x + 4\vec{e}_x\vec{e}_y + 5\vec{e}_y\vec{e}_x + 10\vec{e}_y\vec{e}_y) \cdot 3\vec{e}_x\\ &= 6\vec{e}_x + 15\vec{e}_y.\end{aligned}$$

We can also use matrix notation. In that case $\underline{A}$ is the matrix representation of $\boldsymbol{A}$.

$$\underline{A} = \underset{\sim}{a}\underset{\sim}{b}^T = \begin{bmatrix} 1 \\ 2 \\ 0 \end{bmatrix} [2\ 5\ 0] = \begin{bmatrix} 2 & 5 & 0 \\ 4 & 10 & 0 \\ 0 & 0 & 0 \end{bmatrix}$$

and:

$$\underline{A}\,\underset{\sim}{c} = \begin{bmatrix} 2 & 5 & 0 \\ 4 & 10 & 0 \\ 0 & 0 & 0 \end{bmatrix} \begin{bmatrix} 3 \\ 0 \\ 0 \end{bmatrix} = \begin{bmatrix} 6 \\ 12 \\ 0 \end{bmatrix},$$

$$\underline{A}^T\,\underset{\sim}{c} = \begin{bmatrix} 2 & 4 & 0 \\ 5 & 10 & 0 \\ 0 & 0 & 0 \end{bmatrix} \begin{bmatrix} 3 \\ 0 \\ 0 \end{bmatrix} = \begin{bmatrix} 6 \\ 15 \\ 0 \end{bmatrix}.$$

which clearly is not the same as: $\underline{A}\,\underset{\sim}{c}$.

1.5 Some Mathematical Preliminaries on Second-Order Tensors

In this section we will elaborate a bit more about second-order tensors, because they play an important role in continuum mechanics. Remember that every dyadic product of two vectors is a second-order tensor and every second-order tensor can be written as the sum of dyadic products.

An arbitrary second-order tensor $\boldsymbol{M}$ can be written with respect to the Cartesian basis introduced earlier as:

$$\begin{aligned} \boldsymbol{M} = {} & M_{xx}\vec{e}_x\vec{e}_x + M_{xy}\vec{e}_x\vec{e}_y + M_{xz}\vec{e}_x\vec{e}_z \\ & + M_{yx}\vec{e}_y\vec{e}_x + M_{yy}\vec{e}_y\vec{e}_y + M_{yz}\vec{e}_y\vec{e}_z \\ & + M_{zx}\vec{e}_z\vec{e}_x + M_{zy}\vec{e}_z\vec{e}_y + M_{zz}\vec{e}_z\vec{e}_z. \end{aligned} \tag{1.40}$$

The components of the tensor $\boldsymbol{M}$ are stored in the associated matrix $\underline{M}$ defined as

$$\underline{M} = \begin{bmatrix} M_{xx} & M_{xy} & M_{xz} \\ M_{yx} & M_{yy} & M_{yz} \\ M_{zx} & M_{zy} & M_{zz} \end{bmatrix}. \tag{1.41}$$

A tensor identifies a linear transformation. If the vector $\vec{b}$ is the result of the tensor $\boldsymbol{M}$ operating on vector $\vec{a}$, this is written as: $\vec{b} = \boldsymbol{M} \cdot \vec{a}$. In component form, this leads to:

$$\begin{aligned} \vec{b} = {} & (M_{xx}\vec{e}_x\vec{e}_x + M_{xy}\vec{e}_x\vec{e}_y + M_{xz}\vec{e}_x\vec{e}_z \\ & + M_{yx}\vec{e}_y\vec{e}_x + M_{yy}\vec{e}_y\vec{e}_y + M_{yz}\vec{e}_y\vec{e}_z \\ & + M_{zx}\vec{e}_z\vec{e}_x + M_{zy}\vec{e}_z\vec{e}_y + M_{zz}\vec{e}_z\vec{e}_z) \cdot (a_x\vec{e}_x + a_y\vec{e}_y + a_z\vec{e}_z) \end{aligned}$$

$$
\begin{aligned}
&= (M_{xx}a_x + M_{xy}a_y + M_{xz}a_z)\vec{e}_x \\
&+ (M_{yx}a_x + M_{yy}a_y + M_{yz}a_z)\vec{e}_y \\
&+ (M_{zx}a_x + M_{zy}a_y + M_{zz}a_z)\vec{e}_z \\
&= b_x\vec{e}_x + b_y\vec{e}_y + b_z\vec{e}_z \, .
\end{aligned}
\tag{1.42}
$$

Using matrix notation we can write: $\underset{\sim}{b} = \underline{M}\,\underset{\sim}{a}$, in full:

$$
\begin{bmatrix} b_x \\ b_y \\ b_z \end{bmatrix} = \begin{bmatrix} M_{xx} & M_{xy} & M_{xz} \\ M_{yx} & M_{yy} & M_{yz} \\ M_{zx} & M_{zy} & M_{zz} \end{bmatrix} \begin{bmatrix} a_x \\ a_y \\ a_z \end{bmatrix}
= \begin{bmatrix} M_{xx}a_x + M_{xy}a_y + M_{xz}a_z \\ M_{yx}a_x + M_{yy}a_y + M_{yz}a_z \\ M_{zx}a_x + M_{zy}a_y + M_{zz}a_z \end{bmatrix} . \tag{1.43}
$$

Along with the earlier specified matrix $\underline{M}$ the transposed matrix $\underline{M}^{\mathrm{T}}$ is defined according to (taking a mirror image along the principal diagonal):

$$
\underline{M}^{\mathrm{T}} = \begin{bmatrix} M_{xx} & M_{yx} & M_{zx} \\ M_{xy} & M_{yy} & M_{zy} \\ M_{xz} & M_{yz} & M_{zz} \end{bmatrix} . \tag{1.44}
$$

The tensor $\boldsymbol{M}^{\mathrm{T}}$ is associated with the matrix $\underline{M}^{\mathrm{T}}$. Notice that

$$
\begin{aligned}
\underset{\sim}{b} = \underline{M}\,\underset{\sim}{a} \quad &\text{is equivalent to} \quad \underset{\sim}{b}^{\mathrm{T}} = \underset{\sim}{a}^{\mathrm{T}}\underline{M}^{\mathrm{T}}, \\
\vec{b} = \boldsymbol{M}\cdot\vec{a} \quad &\text{is equivalent to} \quad \vec{b} = \vec{a}\cdot\boldsymbol{M}^{\mathrm{T}}.
\end{aligned}
$$

The inverse of the tensor $\boldsymbol{M}$ is denoted by $\boldsymbol{M}^{-1}$. By definition:

$$
\boldsymbol{M}\cdot\boldsymbol{M}^{-1} = \boldsymbol{I}, \tag{1.45}
$$

with $\boldsymbol{I}$ the unit tensor, $\boldsymbol{I} = \vec{e}_x\vec{e}_x + \vec{e}_y\vec{e}_y + \vec{e}_z\vec{e}_z$. The inverse of matrix $\underline{M}$ is denoted by $\underline{M}^{-1}$. By definition:

$$
\underline{M}\,\underline{M}^{-1} = \underline{I}\, , \tag{1.46}
$$

with $\underline{I}$ the unit matrix.

The trace of tensor $\boldsymbol{M}$ (associated matrix $\underline{M}$) is denoted as $\mathrm{tr}(\boldsymbol{M}) = \mathrm{tr}(\underline{M})$ and given by

$$\mathrm{tr}(\boldsymbol{M}) = \mathrm{tr}(\underline{M}) = M_{xx} + M_{yy} + M_{zz}. \tag{1.47}$$

The determinant of the tensor $\boldsymbol{M}$ with matrix representation $\underline{M}$ can be written:

$$\begin{aligned}\det(\boldsymbol{M}) = \det(\underline{M}) = {} & M_{xx}(M_{yy}M_{zz} - M_{yz}M_{zy}) \\ & - M_{xy}(M_{yx}M_{zz} - M_{yz}M_{zx}) \\ & + M_{xz}(M_{yx}M_{zy} - M_{yy}M_{zx}).\end{aligned} \tag{1.48}$$

The deviatoric part of the tensor $\boldsymbol{M}$ is denoted by $\boldsymbol{M}^{\mathrm{d}}$ and defined by

$$\boldsymbol{M}^{\mathrm{d}} = \boldsymbol{M} - \frac{1}{3}\,\mathrm{tr}(\boldsymbol{M})\,\boldsymbol{I}. \tag{1.49}$$

In matrix notation this reads:

$$\underline{M}^{\mathrm{d}} = \underline{M} - \frac{1}{3}\,\mathrm{tr}(\underline{M})\,\underline{I}. \tag{1.50}$$

Let $\boldsymbol{M}$ be an arbitrary symmetric tensor. A non-zero vector $\vec{n}$ is said to be an **eigenvector** of $\boldsymbol{M}$ if a scalar λ exists such that

$$\boldsymbol{M} \cdot \vec{n} = \lambda \vec{n} \qquad \text{or} \qquad (\boldsymbol{M} - \lambda \boldsymbol{I}) \cdot \vec{n} = \vec{0}. \tag{1.51}$$

A non-trivial solution $\vec{n}$ from Eq. (1.51) only exists if

$$\det(\boldsymbol{M} - \lambda \boldsymbol{I}) = 0. \tag{1.52}$$

Using the components of $\underline{M}$ and Eq. (1.48) will lead, after some elaboration, to the following equation:

$$\lambda^3 - I_1\lambda^2 + I_2\lambda - I_3 = 0, \tag{1.53}$$

with

$$\begin{aligned} I_1 &= \mathrm{tr}(\underline{M}) \\ I_2 &= \frac{1}{2}\left[(\mathrm{tr}(\underline{M}))^2 - \mathrm{tr}(\underline{M}\underline{M})\right] \\ I_3 &= \det(\underline{M}). \end{aligned} \tag{1.54}$$

Equation (1.53) is called the **characteristic equation**, and the scalar coefficients I_1, I_2 and I_3 are called the invariants of the symmetric tensor $\boldsymbol{M}$. In tensor form, the invariants can be written as:

$$\begin{aligned} I_1 &= \mathrm{tr}(\boldsymbol{M}) \\ I_2 &= \frac{1}{2}\left[(\mathrm{tr}(\boldsymbol{M}))^2 - \mathrm{tr}(\boldsymbol{M} \cdot \boldsymbol{M})\right] \\ I_3 &= \det(\boldsymbol{M}). \end{aligned} \tag{1.55}$$

Example 1.4 Later in the book, we want to separate shape changes of objects from the volume change. In that case the deviatoric part $\boldsymbol{A}^{\mathrm{d}}$ of a tensor $\boldsymbol{A}$ plays an important role, because this tensor does not change when a tensor of the form $\alpha\boldsymbol{I}$ is added to the original tensor. This can be shown as follows:

$$\begin{aligned}(\boldsymbol{A}+\alpha\boldsymbol{I})^{\mathrm{d}} &= \boldsymbol{A}+\alpha\boldsymbol{I}-\frac{1}{3}\,\mathrm{tr}(\boldsymbol{A}+\alpha\boldsymbol{I})\,\boldsymbol{I}\\ &= \boldsymbol{A}+\alpha\boldsymbol{I}-\frac{1}{3}\,\mathrm{tr}(\boldsymbol{A})\,\boldsymbol{I}-\frac{1}{3}\,\mathrm{tr}(\alpha\boldsymbol{I})\,\boldsymbol{I}\\ &= \boldsymbol{A}+\alpha\boldsymbol{I}-\frac{1}{3}\,\mathrm{tr}(\boldsymbol{A})\,\boldsymbol{I}-\alpha\boldsymbol{I}\\ &= \boldsymbol{A}^{d}\end{aligned}$$

Example 1.5 Consider the tensor $\boldsymbol{A}$ given by:

$$\boldsymbol{A}=\cos(\phi)\vec{e}_x\vec{e}_x-\sin(\phi)\vec{e}_x\vec{e}_y+\sin(\phi)\vec{e}_y\vec{e}_x+\cos(\phi)\vec{e}_y\vec{e}_y+\vec{e}_z\vec{e}_z.$$

Show that, for this tensor, $\boldsymbol{A}^{\mathrm{T}}=\boldsymbol{A}^{-1}$.
The easiest way is to use matrix notation. Then:

$$\begin{aligned}\underline{A}\,\underline{A}^{\mathrm{T}} &= \begin{bmatrix}\cos(\phi) & -\sin(\phi) & 0\\ \sin(\phi) & \cos(\phi) & 0\\ 0 & 0 & 1\end{bmatrix}\begin{bmatrix}\cos(\phi) & \sin(\phi) & 0\\ -\sin(\phi) & \cos(\phi) & 0\\ 0 & 0 & 1\end{bmatrix}\\ &= \begin{bmatrix}\cos^2(\phi)+\sin^2(\phi) & 0 & 0\\ 0 & \cos^2(\phi)+\sin^2(\phi) & 0\\ 0 & 0 & 1\end{bmatrix}\\ &= \begin{bmatrix}1 & 0 & 0\\ 0 & 1 & 0\\ 0 & 0 & 1\end{bmatrix}.\end{aligned}$$

So $\underline{A}\,\underline{A}^{\mathrm{T}}=\underline{I}$ and consequently $\boldsymbol{A}\cdot\boldsymbol{A}^{\mathrm{T}}=\boldsymbol{I}$, which proves that: $\boldsymbol{A}^{\mathrm{T}}=\boldsymbol{A}^{-1}$. This is a property associated with **orthogonal** tensors, which are often used to describe rotations.

Exercises

1.1 The basis $\{\vec{e}_x,\vec{e}_y,\vec{e}_z\}$ has a right-handed orientation and is orthonormal.
(a) Determine $|\vec{e}_i|$ for $i=x,y,z$.
(b) Determine $\vec{e}_i\cdot\vec{e}_j$ for $i,j=x,y,z$.
(c) Determine $\vec{e}_x\cdot\vec{e}_y\times\vec{e}_z$.
(d) Why is: $\vec{e}_x\times\vec{e}_y=\vec{e}_z$?

1.2 Let $\{\vec{e}_x, \vec{e}_y, \vec{e}_z\}$ be an orthonormal vector basis. The force vectors $\vec{F}_x = 3\vec{e}_x + 2\vec{e}_y + \vec{e}_z$ and $\vec{F}_y = -4\vec{e}_x + \vec{e}_y + 4\vec{e}_z$ act on point P. Calculate a vector $\vec{F}_z$ acting on P in such a way that the sum of all force vectors is the zero vector.

1.3 Let $\{\vec{e}_x, \vec{e}_y, \vec{e}_z\}$ be a right-handed and orthonormal vector basis. The following vectors are given: $\vec{a} = 4\vec{e}_z$, $\vec{b} = -3\vec{e}_y + 4\vec{e}_z$ and $\vec{c} = \vec{e}_x + 2\,\vec{e}_z$.

(a) Write the vectors in column notation.

(b) Determine $\vec{a} + \vec{b}$ and $3(\vec{a} + \vec{b} + \vec{c})$.

(c) Determine $\vec{a} \cdot \vec{b}$, $\vec{b} \cdot \vec{a}$, $\vec{a} \times \vec{b}$ and $\vec{b} \times \vec{a}$.

(d) Determine $|\vec{a}|$, $|\vec{b}|$, $|\vec{a} \times \vec{b}|$ and $|\vec{b} \times \vec{a}|$.

(e) Determine the smallest angle between $\vec{a}$ and $\vec{b}$.

(f) Determine a unit normal vector on the plane defined by $\vec{a}$ and $\vec{b}$.

(g) Determine $\vec{a} \times \vec{b} \cdot \vec{c}$ and $\vec{a} \times \vec{c} \cdot \vec{b}$.

(h) Determine $\vec{a}\vec{b} \cdot \vec{c}$, $(\vec{a}\vec{b})^{\mathrm{T}} \cdot \vec{c}$ and $\vec{b}\vec{a} \cdot \vec{c}$.

(i) Do the vectors $\vec{a}$, $\vec{b}$ and $\vec{c}$ form a suitable vector basis? If the answer is yes, do they form an orthogonal basis? If the answer is yes, do they form an orthonormal basis?

1.4 Consider the basis $\{\vec{a}, \vec{b}, \vec{c}\}$ with $\vec{a}, \vec{b}$ and $\vec{c}$ defined as in the previous exercise. The following vectors are given: $\vec{d} = \vec{a} + 2\vec{b}$ and $\vec{e} = 2\vec{a} - 3\vec{c}$.

(a) Determine $\vec{d} + \vec{e}$.

(b) Determine $\vec{d} \cdot \vec{e}$.

1.5 The basis $\{\vec{e}_x, \vec{e}_y, \vec{e}_z\}$ is right-handed and orthonormal. The vectors $\vec{a}_x, \vec{a}_y$ and $\vec{a}_z$ are given by: $\vec{a}_x = 4\vec{e}_x + 3\vec{e}_y$; $\vec{a}_y = 3\vec{e}_x - 4\vec{e}_y$; and $\vec{a}_z = \vec{a}_x \times \vec{a}_y$.

(a) Determine $\vec{a}_z$ expressed in $\vec{e}_x, \vec{e}_y$ and $\vec{e}_z$.

(b) Determine $|\vec{a}_i|$ for $i = x, y, z$.

(c) Determine the volume of the parallelepiped defined by $\vec{a}_x, \vec{a}_y$ and $\vec{a}_z$.

(d) Determine the angle between the lines of action of $\vec{a}_x$ and $\vec{a}_y$.

(e) Determine the vector $\vec{\alpha}_i$ from $\vec{a}_i = |\vec{a}_i|\vec{\alpha}_i$ for $i = x, y, z$. Is $\{\vec{\alpha}_x, \vec{\alpha}_y, \vec{\alpha}_z\}$ a right-handed, orthonormal vector basis?

(f) Consider the vector $\vec{b} = 2\vec{e}_x + 3\vec{e}_y + \vec{e}_z$. Determine the column representation of $\vec{b}$ according to the bases $\{\vec{e}_x, \vec{e}_y, \vec{e}_z\}$, $\{\vec{a}_x, \vec{a}_y, \vec{a}_z\}$ and $\{\vec{\alpha}_x, \vec{\alpha}_y, \vec{\alpha}_z\}$.

(g) Show that: $\vec{a}_x \times \vec{a}_y \cdot \vec{b} = \vec{a}_x \cdot \vec{a}_y \times \vec{b} = \vec{a}_y \cdot \vec{b} \times \vec{a}_x$.

1.6 Assume that $\{\vec{e}_x, \vec{e}_y, \vec{e}_z\}$ is an orthonormal vector basis. The following vectors are defined:

$$\vec{a} = 4\vec{e}_x + 3\vec{e}_y - \vec{e}_z$$
$$\vec{b} = 6\vec{e}_y - \vec{e}_z$$
$$\vec{c} = 8\vec{e}_x - \vec{e}_z\ .$$

Are $\vec{a}, \vec{b}$ and $\vec{c}$ linearly independent? If not, what is the relationship between the vectors?

1.7 The vector basis $\{\vec{e}_x, \vec{e}_y, \vec{e}_z\}$ is orthonormal. Describe how a vector $\vec{a}$ is transformed when the following dyadic products are applied to it:

(a) $\vec{e}_x\vec{e}_x$.

(b) $\vec{e}_x\vec{e}_x + \vec{e}_y\vec{e}_y$.

(c) $\vec{e}_x\vec{e}_x + \vec{e}_y\vec{e}_y + \vec{e}_z\vec{e}_z$.

(d) $\vec{e}_x\vec{e}_y - \vec{e}_y\vec{e}_x + \vec{e}_z\vec{e}_z$.

(e) $\vec{e}_x\vec{e}_x - \vec{e}_y\vec{e}_y + \vec{e}_z\vec{e}_z$.

1.8 Consider the second-order tensor $\boldsymbol{A} = \boldsymbol{I} + 2\vec{e}_x\vec{e}_x + \vec{e}_x\vec{e}_y + \vec{e}_y\vec{e}_x$. Determine $\det(\boldsymbol{A})$.

1.9 Consider the following tensor:

$$\boldsymbol{F} = \boldsymbol{I} + 4\vec{e}_x\vec{e}_x + 2(\vec{e}_x\vec{e}_y + \vec{e}_y\vec{e}_x)$$

and the vector:

$$\vec{b} = 2\vec{e}_x + 2\vec{e}_y + 2\vec{e}_z.$$

It is known that:

$$\boldsymbol{F} \cdot \vec{a} = \vec{b}.$$

Determine the unknown vector $\vec{a}$.

1.10 The following second-order tensors are given:

$$\boldsymbol{A} = \boldsymbol{I} + \vec{e}_x\vec{e}_x + \vec{e}_y\vec{e}_y$$
$$\boldsymbol{B} = 2\vec{e}_x\vec{e}_z.$$

Determine the tensor $\boldsymbol{C} = \frac{1}{2}(\boldsymbol{B} \cdot \boldsymbol{A} + \boldsymbol{A}^T \cdot \boldsymbol{B}^T)$.

2 The Concepts of Force and Moment

2.1 Introduction

We experience the effects of force in everyday life and have an intuitive notion of force. For example, we exert a force on our body when we lift or push an object, while we continuously (fortunately) feel the effect of gravitational forces, for instance while sitting, walking, etc. All parts of the human body in one way or the other are loaded by forces. Our bones provide rigidity to the body and can sustain high loads. The skin is resistant to force: simply pull on the skin to witness this. The cardiovascular system is continuously loaded dynamically owing to the pulsating blood pressure. The bladder is loaded and stretched when it fills up. The intervertebral discs serve as flexible force-transmitting media that give the spine its flexibility. Beside force, we are using levers all the time in our daily life to increase the 'force' that we want to apply to some object, for example by opening doors with the latch, opening a bottle with a bottle-opener. We feel the effect of a lever arm when holding a weight close to our body instead of using a stretched arm. These experiences are the result of the moment that can be exerted by a force. Understanding the impact of force and moment on the human body requires us to formalize the intuitive notion of force and moment. That is the objective of this chapter.

2.2 Definition of a Force Vector

Imagine pulling on a thin wire that is attached to a wall. The pulling force exerted on the point of application is a vector with a physical meaning: it has

- a **length**: the magnitude of the pulling force
- an **orientation** in space: the direction of the wire
- a **line-of-action**, which is the line through the force vector.

The graphical representation of a force vector, denoted by $\vec{F}$, is given in Fig. 2.1. The 'shaft' of the arrow indicates the orientation in space of the force vector. The point of application of the force vector is denoted by the point P.

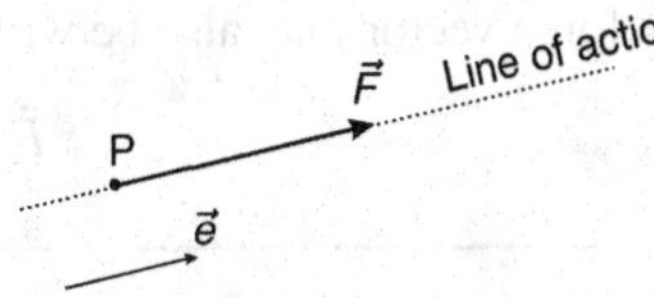

Figure 2.1

The force vector $\vec{F}$ and unit vector $\vec{e}$.

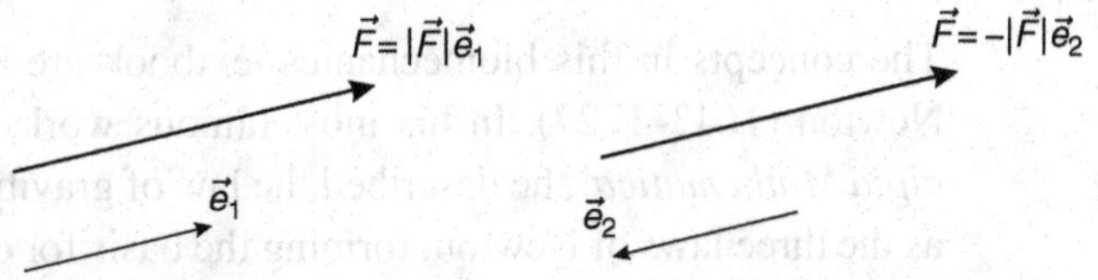

Figure 2.2

Force vector $\vec{F}$ written with respect to $\vec{e}_1$ and written with respect to $\vec{e}_2$.

The magnitude of the force vector is denoted by $|\vec{F}|$. If $\vec{e}$ denotes a unit vector, the force vector may be written as:

$$\vec{F} = F\vec{e}, \tag{2.1}$$

where F may be any rational number (i.e. negative, zero or positive). The absolute value $|F|$ of the number F is equal to the magnitude of force vector:

$$|F| = |\vec{F}|. \tag{2.2}$$

In Fig. 2.2, the force vector $\vec{F}$ is written either with respect to the unit vector $\vec{e}_1$ or with respect to the unit vector $\vec{e}_2$ that has the same working line in space as $\vec{e}_1$ but the opposite direction. Since the unit vector $\vec{e}_1$ has the same direction as the force vector $\vec{F}$:

$$\vec{F} = |\vec{F}|\vec{e}_1 \,. \tag{2.3}$$

In contrast, the unit vector $\vec{e}_2$ has a direction that is opposed to $\vec{F}$; therefore:

$$\vec{F} = -|\vec{F}|\vec{e}_2. \tag{2.4}$$

Example 2.1 Let the force vector $\vec{F}$ be given by:

$$\vec{F} = 2\vec{e}_1.$$

If the unit vectors $\vec{e}_1$ and $\vec{e}_2$ have opposite direction:

$$\vec{e}_2 = -\vec{e}_1,$$

then the force vector may also be written as:

$$\vec{F} = -2\vec{e}_2.$$

2.3 Newton's Laws

The concepts in this biomechanics textbook are based on the work of Sir Isaac Newton (1643–1727). In his most famous work, '*Philosophiae Naturalis Principia Mathematica*', he described the law of gravity and what are currently known as the three laws of Newton, forming the basis for classical mechanics. These laws are:

- Every object in a state of uniform motion tends to remain in that state of motion unless an external force is applied to it. This is often termed simply the 'Law of Inertia'.
- In a one-dimensional context, the second law states that the force F on an object equals the mass m, with SI unit [kg], of the object multiplied by the acceleration a, with dimension [m s^{-2}], of the object:

$$F = ma. \tag{2.5}$$

Consequently, the force F has the dimension [N] (Newton), with:

$$1 \;[\text{N}] = 1 \;[\text{kg m s}^{-2}].$$

This may be generalized to three-dimensional space in a straightforward manner. Let the position of a material particle in space be given by the vector $\vec{x}$. If the particle moves in space, this vector will be a function of the time t, i.e.

$$\vec{x} = \vec{x}(t). \tag{2.6}$$

The velocity $\vec{v}$ of the particle is given by:

$$\vec{v}(t) = \frac{d\vec{x}}{dt}, \tag{2.7}$$

and the acceleration $\vec{a}$ follows from:

$$\vec{a}(t) = \frac{d\vec{v}}{dt} = \frac{d^2\vec{x}}{dt^2}. \tag{2.8}$$

Newton's second law may now be formulated as:

$$\vec{F} = m\vec{a}. \tag{2.9}$$

- The third law states that for every action there is an equal and opposite reaction. This law is exemplified by what happens when we step off a boat onto the bank of a lake: if we move in the direction of the shore, the boat tends to move in the opposite direction.

Example 2.2 Let the position of a particle with mass m for $t \geq 0$ be given by:

$$\vec{x}(t) = \left(1 + \left(\frac{t}{\tau}\right)^2\right)\vec{x}_0,$$

where $\vec{x}_0$ denotes the position of the particle at $t=0$ and τ is a constant, characteristic time. The velocity of this particle is obtained from:

$$\vec{v} = \frac{d\vec{x}}{dt} = \frac{d}{dt}\left((1 + (t/\tau)^2)\vec{x}_0\right) = \frac{d(1 + (t/\tau)^2)}{dt}\vec{x}_0 = (2t/\tau^2)\vec{x}_0,$$

while the acceleration follows from:

$$\vec{a} = \frac{d\vec{v}}{dt} = (2/\tau^2)\vec{x}_0.$$

The force on this particle equals:

$$\vec{F} = (2m/\tau^2)\vec{x}_0.$$

2.4 Vector Operations on the Force Vector

Suppose that a force vector is represented by:

$$\vec{F}_1 = F_1\vec{e}, \tag{2.10}$$

then another force vector, say $\vec{F}_2$, may be obtained by multiplying the force by a factor α, see Fig. 2.3(a):

$$\vec{F}_2 = \alpha F_1\vec{e} = F_2\vec{e}. \tag{2.11}$$

The force vector $\vec{F}_2$ has the same orientation in space as $\vec{F}_1$, but if $\alpha \neq 1$ it will have a different length, and it may have a direction change (if $\alpha < 0$, as shown in Fig. 2.3(a)).

The net result of two force vectors, say $\vec{F}_1$ and $\vec{F}_2$, acting on the same point P is obtained by the vector sum, graphically represented in Fig. 2.3(b):

$$\vec{F}_3 = \vec{F}_1 + \vec{F}_2. \tag{2.12}$$

The vector $\vec{F}_3$ is placed along the diagonal of the parallelogram formed by the vectors $\vec{F}_1$ and $\vec{F}_2$. This implicitly defines the orientation, sense and magnitude of the resulting force vector $\vec{F}_3$.

Clearly, if two force vectors $\vec{F}_1$ and $\vec{F}_2$ are parallel, then the resulting force vector $\vec{F}_3 = \vec{F}_1 + \vec{F}_2$ will be parallel to the vectors $\vec{F}_1$ and $\vec{F}_2$ as well. If $\vec{F}_1 = -\vec{F}_2$,

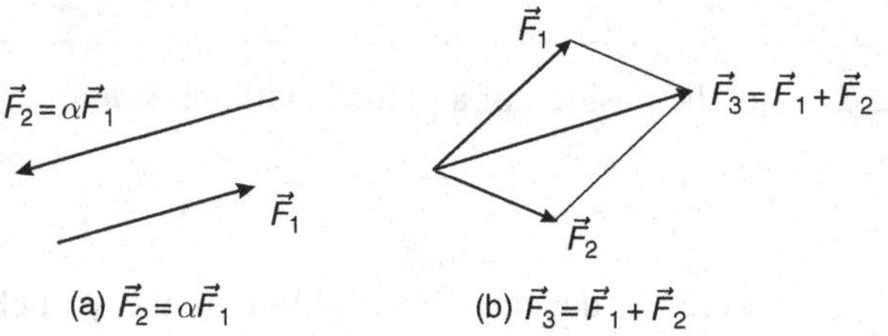

Figure 2.3

Graphical representation of the scalar multiplication of (a) a force vector with $\alpha < 0$ and (b) the sum of two force vectors.

then the addition of these two force vectors yields the so-called zero vector $\vec{0}$, with zero length.

2.5 Force Decomposition

Suppose that a bone is loaded with a force $\vec{F}$ as sketched in Fig. 2.4. The principal axis of the bone has a direction indicated by the unit vector $\vec{e}$. The smallest angle between the force vector $\vec{F}$ and the unit vector $\vec{e}$ is denoted by α. It is useful to know which part of the force $\vec{F}$ acts in the direction of the unit vector $\vec{e}$, indicated by $\vec{F}_t$ (tangential), and which part of the force acts perpendicular to the bone, indicated by the force vector $\vec{F}_n$ (normal). The force vector $\vec{F}$ may, in that case, be written as:

$$\vec{F} = \vec{F}_t + \vec{F}_n. \tag{2.13}$$

To determine the vectors $\vec{F}_t$ and $\vec{F}_n$, vector calculus will be used. The inner product of two vectors, say $\vec{a}$ and $\vec{b}$, is defined as:

$$\vec{a} \cdot \vec{b} = |\vec{a}| \, |\vec{b}| \cos(\alpha), \tag{2.14}$$

where α is the smallest angle between the two vectors $\vec{a}$ and $\vec{b}$; see Fig. 2.5 and Chapter 1 for further details on the properties of the vector inner product. Computation of the inner product requires knowledge of the length of both vectors (i.e. $|\vec{a}|$ and $|\vec{b}|$) and the smallest angle between the two vectors (i.e. α), all physical quantities that can easily be obtained. If the vectors $\vec{a}$ and $\vec{b}$ are perpendicular to each other, hence if $\alpha = \pi/2$, then the inner product equals zero, i.e. $\vec{a} \cdot \vec{b} = 0$. The length of a vector satisfies $|\vec{a}| = \sqrt{\vec{a} \cdot \vec{a}}$.

Now, consider the inner product of an arbitrary vector $\vec{b}$ with a unit vector $\vec{e}$ (i.e. $|\vec{e}| = 1$). Then:

$$\vec{b} \cdot \vec{e} = |\vec{b}| \cos(\alpha). \tag{2.15}$$

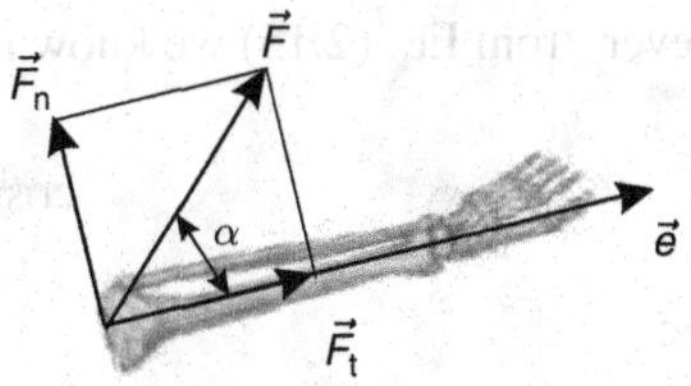

Figure 2.4

Bone loaded by the force vector $\vec{F}$. The orientation of the bone is indicated by the unit vector $\vec{e}$.

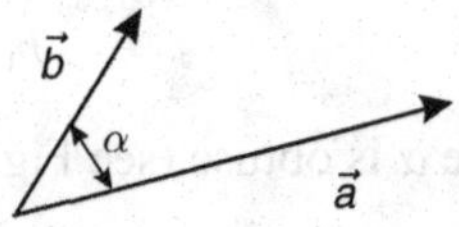

Figure 2.5

Definition of the angle α.

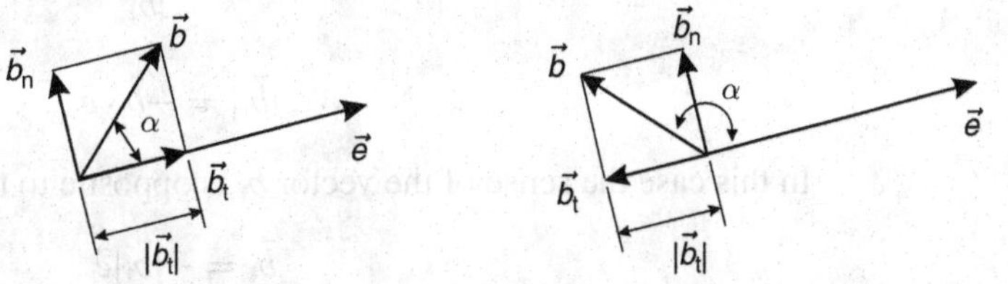

(a) Acute angle α, the length of $\vec{b}_t$ (b) Obtuse angle α, the length of $\vec{b}_t$

Figure 2.6

Vector decomposition in case of (a) an acute and (b) an obtuse angle between the vectors.

Let the vector $\vec{b}$ be written as the sum of a vector parallel to $\vec{e}$, say $\vec{b}_t$, and a vector normal to $\vec{e}$, say $\vec{b}_n$, such that:

$$\vec{b} = \vec{b}_t + \vec{b}_n, \tag{2.16}$$

as depicted in Fig. 2.6(a).

If the angle α between the unit vector $\vec{e}$ and the vector $\vec{b}$ is acute, hence if $\alpha \leq \pi/2$, it is easy to show that this inner product is equal to the length of the vector $\vec{b}_t$, the component of $\vec{b}$ parallel to the unit vector $\vec{e}$; see Fig. 2.6(a). By definition:

$$\cos(\alpha) = \frac{|\vec{b}_t|}{|\vec{b}|}. \tag{2.17}$$

However, from Eq. (2.15) we know that:

$$\cos(\alpha) = \frac{\vec{b} \cdot \vec{e}}{|\vec{b}|}, \tag{2.18}$$

hence:

$$|\vec{b}_t| = \vec{b} \cdot \vec{e}. \tag{2.19}$$

Since the angle α is acute, the vector $\vec{b}_t$ has the same sense as the unit vector $\vec{e}$ such that:

$$\vec{b}_t = |\vec{b}_t|\vec{e} = (\vec{b} \cdot \vec{e})\vec{e}\,. \tag{2.20}$$

If the angle α is obtuse (see Fig. 2.6(b)), hence if $\alpha > \pi/2$, we have:

$$\cos(\pi - \alpha) = -\cos(\alpha) = \frac{|\vec{b}_t|}{|\vec{b}|}. \tag{2.21}$$

With, according to Eq. (2.15), $\cos(\alpha) = \frac{\vec{b} \cdot \vec{e}}{|\vec{b}|}$ this leads to:

$$|\vec{b}_t| = -\vec{b} \cdot \vec{e}. \tag{2.22}$$

In this case the sense of the vector $\vec{b}_t$ is opposite to the unit vector $\vec{e}$, such that:

$$\vec{b}_t = -|\vec{b}_t|\vec{e}\,. \tag{2.23}$$

So, clearly, whether the angle α is acute or obtuse, the vector $\vec{b}_t$ parallel to the unit vector $\vec{e}$ is given by:

$$\vec{b}_t = (\vec{b} \cdot \vec{e})\vec{e}\,. \tag{2.24}$$

Recall that this is only true if $\vec{e}$ has unit length! In conclusion, the inner product of an arbitrary vector $\vec{b}$ with a unit vector $\vec{e}$ defines the magnitude and sense of a vector $\vec{b}_t$ that is parallel to the unit vector $\vec{e}$ such that the original vector $\vec{b}$ may be written as the sum of this parallel vector and a vector normal to the unit vector $\vec{e}$. The vector $\vec{b}_n$ normal to $\vec{e}$ follows automatically from:

$$\vec{b}_n = \vec{b} - \vec{b}_t. \tag{2.25}$$

This implicitly defines the unique decomposition of the vector $\vec{b}$ into a component normal and a component parallel to the unit vector $\vec{e}$.

Based on the considerations above, the force vector $\vec{F}$ in Fig. 2.4 can be decomposed into a component parallel to the bone principal axis $\vec{F}_t$ given by:

$$\vec{F}_t = (\vec{F} \cdot \vec{e})\vec{e}, \tag{2.26}$$

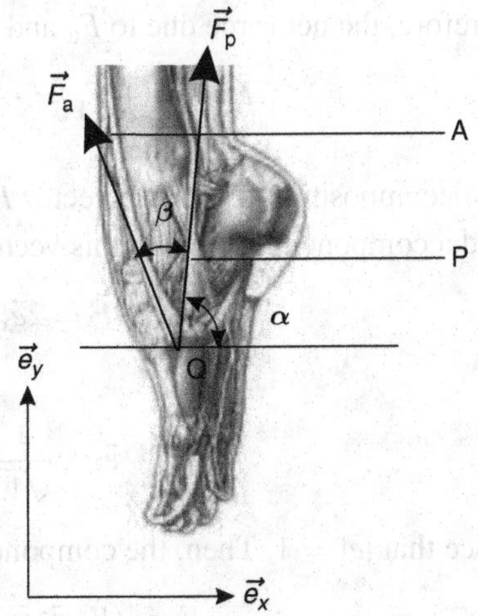

Figure 2.7

Forces of the tendons of the tibialis anterior $\vec{F}_a$ and posterior $\vec{F}_p$, respectively.

where $\vec{e}$ denotes a vector of unit length, and a component normal to the principal axis of the bone:

$$\vec{F}_n = \vec{F} - \vec{F}_t = (\boldsymbol{I} - \vec{e}\,\vec{e}) \cdot \vec{F}. \tag{2.27}$$

Example 2.3 In the foot, the tendons of the tibialis anterior and the tibialis posterior may be identified; see Fig. 2.7. Let the magnitude of the force vectors be given by:

$$F_a = |\vec{F}_a| = 50 \text{ [N]}, \quad F_p = |\vec{F}_p| = 60 \text{ [N]},$$

while the angles α and β are specified by:

$$\alpha = \frac{5\pi}{11}, \quad \beta = \frac{\pi}{6}.$$

What is the net force acting on the attachment point Q of the two muscles on the foot?

First, the force vectors $\vec{F}_a$ and $\vec{F}_p$ are written with respect to the Cartesian coordinate system. Clearly:

$$\begin{aligned}\vec{F}_a &= F_a \left[\cos(\alpha + \beta)\vec{e}_x + \sin(\alpha + \beta)\vec{e}_y\right] \\ &\approx -18.6\vec{e}_x + 46.4\vec{e}_y \ \text{[N]},\end{aligned}$$

and

$$\begin{aligned}\vec{F}_p &= F_p \left[\cos(\alpha)\vec{e}_x + \sin(\alpha)\vec{e}_y\right] \\ &\approx 8.5\vec{e}_x + 59.4\vec{e}_y \ \text{[N]}.\end{aligned}$$

Therefore, the net force due to $\vec{F}_a$ and $\vec{F}_p$ acting on point Q equals:

$$\vec{F} = \vec{F}_a + \vec{F}_p = -10.1\vec{e}_x + 105.8\vec{e}_y \ \text{[N]}.$$

Example 2.4 The decomposition of a force vector $\vec{F}$ into a component parallel to a unit vector $\vec{e}$ and a component normal to this vector is also straightforward. For example, let

$$\vec{F} = 2\vec{e}_x + 6\vec{e}_y \ \text{[N]},$$

and

$$\vec{e} = \frac{1}{\sqrt{13}}(2\vec{e}_x - 3\vec{e}_y).$$

Notice that $|\vec{e}| = 1$. Then, the component of $\vec{F}$ parallel to $\vec{e}$ is obtained from:

$$\begin{aligned}\vec{F}_t &= (\vec{F}\cdot\vec{e})\,\vec{e}\\ &= \underbrace{\left(-\frac{14}{\sqrt{13}}\right)}_{\vec{F}\cdot\vec{e}}\underbrace{\frac{1}{\sqrt{13}}(2\vec{e}_x - 3\vec{e}_y)}_{\vec{e}}\\ &= -\frac{14}{13}(2\vec{e}_x - 3\vec{e}_y) \ \text{[N]}.\end{aligned}$$

2.6 Drawing Convention

Consider two force vectors, $\vec{F}_1$ and $\vec{F}_2$, both parallel to the unit vector $\vec{e}$ as sketched in Fig. 2.8. In this case, the two vectors are identified by numbers F_1 and F_2, rather than by the vector symbols $\vec{F}_1$ and $\vec{F}_2$. These numbers denote the magnitude of the force vector, while the orientation of the arrow denotes the direction of the vector. Consequently, this way of drawing and identifying the vectors implicitly assumes:

$$\vec{F}_1 = F_1\vec{e}\,, \tag{2.28}$$

while:

$$\vec{F}_2 = -F_2\vec{e}\,. \tag{2.29}$$

This drawing convention is generally used in combination with a certain vector basis. In this course, only the Cartesian vector basis is used. In that case, forces acting in the horizontal plane, hence in the $\vec{e}_x$ direction, are frequently identified by H_i (from Horizontal), while forces acting in vertical direction, hence in the $\vec{e}_y$

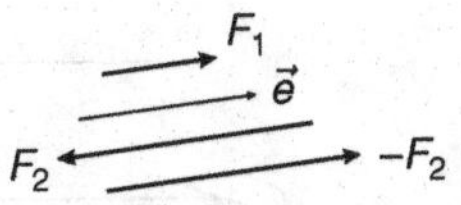

Figure 2.8

Force vectors identified by their magnitude (F_1 and F_2).

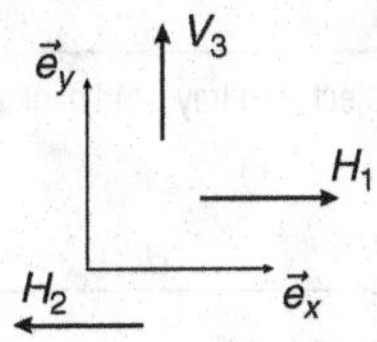

Figure 2.9

Force vectors.

direction, are identified by V_i (from Vertical). For example, the vectors drawn in Fig. 2.9 indicate that:

$$\vec{H}_1 = H_1\vec{e}_x, \quad \vec{H}_2 = -H_2\vec{e}_x, \quad \vec{V}_3 = V_3\vec{e}_y. \tag{2.30}$$

2.7 The Concept of Moment

A simple example of the effect of a moment is experienced when holding a tray with a mass on it that exerts a (gravity) force on the tray: see the schematic drawing in Fig. 2.10(a). This force, which acts at a certain distance d, causes a moment at the position of our hand as is shown in Fig. 2.10(b), where the tray has been removed from the drawing and the resulting load on the hand is indicated by the arrow F, representing the force, and additionally the curved arrow M, representing the moment.

Increasing the distance of the mass with respect to our hand, or increasing the mass, will increase the moment that we experience. In fact, the moment (or torque if you like) that is felt on our hand equals the distance d multiplied by the force due to the mass F:

$$M = dF. \tag{2.31}$$

The moment has a certain orientation in space. Changing the direction of the force F, as visualized in Fig. 2.11(a), will change the orientation of the moment. If the

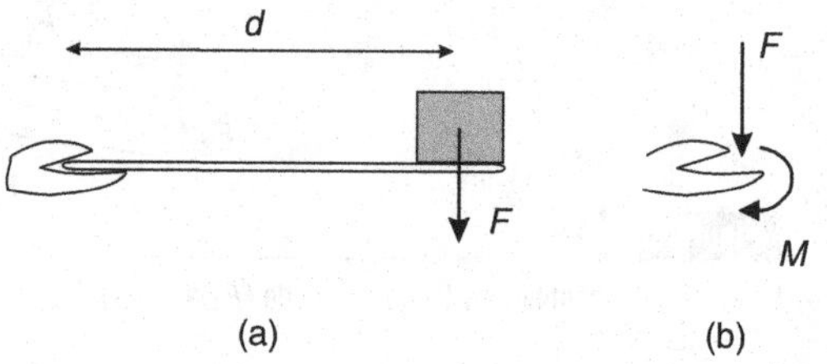

Figure 2.10

(a) Weight of an object on a tray. (b) Loading on the hand.

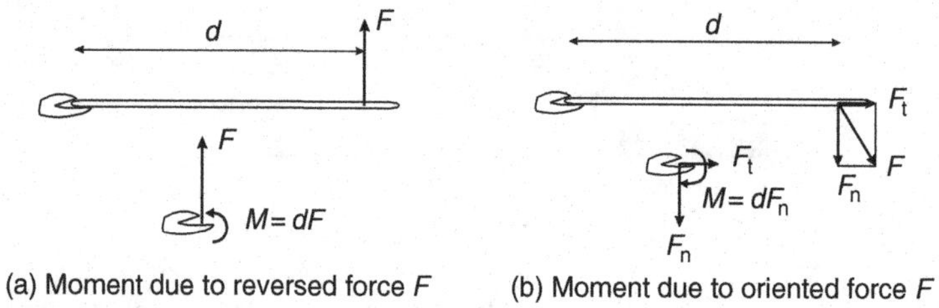

Figure 2.11

Moments due to forces F.

force acts at a certain angle on the tray, as indicated in Fig. 2.11(b), only the force normal to the tray will generate a moment with respect to the hand:

$$M = dF_{\mathrm{n}}. \tag{2.32}$$

In the next section this intuitive notion of moment is formalized.

2.8 Definition of the Moment Vector

A point in space may be identified by its position vector $\vec{x}$: see for instance the three-dimensional example in Fig. 2.12, where $\mathcal{O}$ denotes the location of the origin of the Cartesian vector basis $\{\vec{e}_x, \vec{e}_y, \vec{e}_z\}$.

Assume that a force $\vec{F}$ is applied to a point Q with location $\vec{x}_{\mathrm{Q}}$. The moment vector is defined with respect to a point in space, say P, having location $\vec{x}_{\mathrm{P}}$. The moment exerted by the force $\vec{F}$ with respect to point P is defined as:

$$\vec{M} = (\vec{x}_{\mathrm{Q}} - \vec{x}_{\mathrm{P}}) \times \vec{F} = \vec{d} \times \vec{F}. \tag{2.33}$$

For an interpretation of Eq. (2.33) it is useful to first focus on a two-dimensional configuration. Consider the situation as depicted in Fig. 2.13, where we focus our attention on the plane that is spanned by the vector $\vec{d}$ and the force vector $\vec{F}$. Define

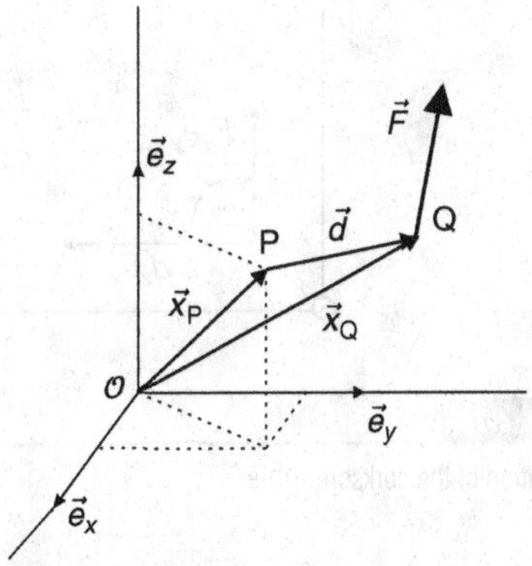

Figure 2.12

Points P and Q in space identified by their position vectors.

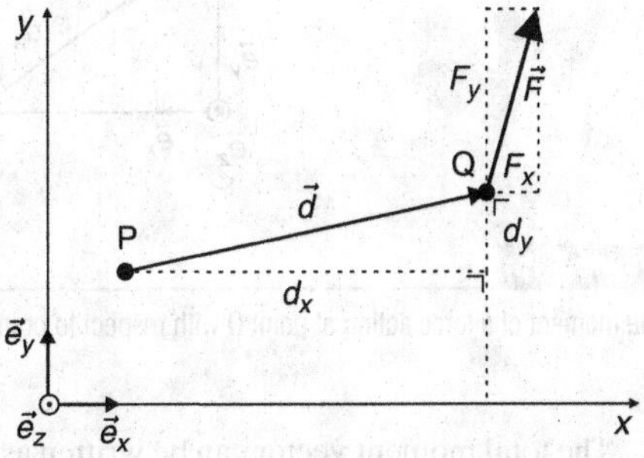

Figure 2.13

The moment of a force acting at point Q with respect to point P.

a Cartesian vector basis $\{\vec{e}_x, \vec{e}_y, \vec{e}_z\}$ with the basis vectors $\vec{e}_x$ and $\vec{e}_y$ in the plane and $\vec{e}_z$ perpendicular to the plane. In this case, vector $\vec{e}_z$ is pointing towards the reader. With respect to this basis, the column representations of the vectors $\vec{d}$ and $\vec{F}$ can be given by:

$$\underset{\sim}{d} = \begin{bmatrix} d_x \\ d_y \\ 0 \end{bmatrix}, \quad \underset{\sim}{F} = \begin{bmatrix} F_x \\ F_y \\ 0 \end{bmatrix}. \tag{2.34}$$

By using Eqs. (2.33) and (1.35) we immediately derive that:

$$\vec{M} = (d_x F_y - d_y F_x)\vec{e}_z. \tag{2.35}$$

From this analysis, several items become clear:

- The moment vector points in a direction perpendicular to the plane that is spanned by the vectors $\vec{d}$ and $\vec{F}$.

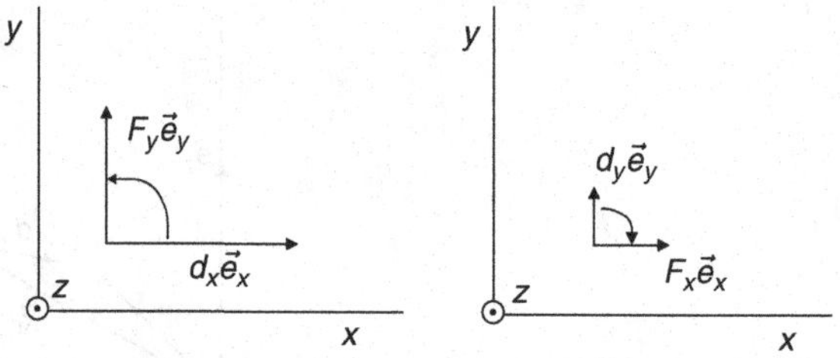

Figure 2.14

Application of the corkscrew rule.

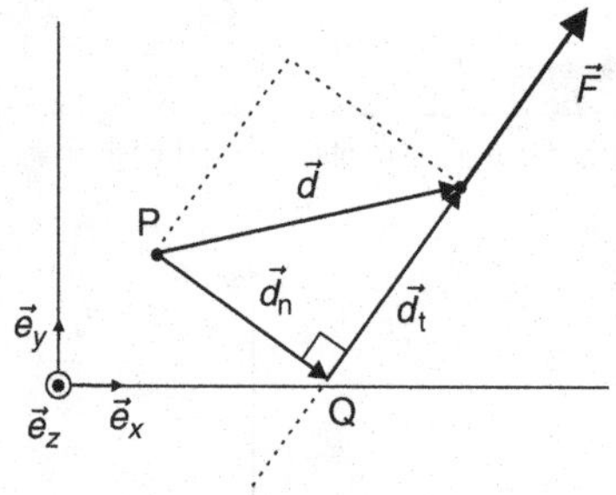

Figure 2.15

The moment of a force acting at point Q with respect to point P.

- The total moment vector can be written as an addition of the moments $\vec{M}_1 = d_x F_y \vec{e}_z$ and $\vec{M}_2 = -d_y F_x \vec{e}_z$. For both composing moments, the force is perpendicular to the working distance of the forces, i.e. $d_x \vec{e}_x$ is perpendicular to $F_y \vec{e}_y$ and $d_y \vec{e}_y$ is perpendicular to $F_x \vec{e}_x$.
- The directions of the composing moments $\vec{M}_1 = d_x F_y \vec{e}_z$ and $\vec{M}_2 = -d_y F_x \vec{e}_z$ follow from the corkscrew rule. To apply this corkscrew rule correctly, place the tails of the two vectors (e.g. $d_x \vec{e}_x$ and $F_y \vec{e}_y$) at the same location in space: see Fig. 2.14. In the case of the combination $d_x \vec{e}_x$ and $F_y \vec{e}_y$, the rotation of the arm to the force is a counterclockwise movement, leading to a vector that points out of the plane, i.e. in the positive $\vec{e}_z$-direction. In the case of the combination $d_y \vec{e}_y$ and $F_x \vec{e}_x$, rotating the arm to the force is a clockwise movement resulting in a moment vector that points into the plane, i.e. in the negative $\vec{e}_z$-direction.

In the definition of the moment vector, the location of the force vector along the line-of-action is not relevant since only the magnitude of the force, the direction and the distance of the point P to the line-of-action are of interest. This is illustrated in Fig. 2.15. We can decompose the vector $\vec{d}$ pointing from the point P to Q into a vector $\vec{d}_n$, perpendicular to the line-of-action of the force $\vec{F}$, and a vector $\vec{d}_t$, parallel to the line-of-action. Then, we can write for the moment $\vec{M}$ of vector $\vec{F}$ with respect to point P:

$$\vec{M} = \vec{d} \times \vec{F} = (\vec{d}_\mathrm{n} + \vec{d}_\mathrm{t}) \times \vec{F} = \vec{d}_\mathrm{n} \times \vec{F}. \tag{2.36}$$

The definition in Eq. (2.33) also assures that the resulting moment is the zero vector if the point P is located on the line-of-action of the force vector (in that case $\vec{d}_\mathrm{n} = \vec{0}$).

In the general three-dimensional case (see Fig. 2.12), the procedure to determine the moment of the force $\vec{F}$ with respect to the point P is similar. The column representations of the vectors $\vec{d}$ and $\vec{F}$ in this case are given by:

$$\underset{\sim}{d} = \begin{bmatrix} d_x \\ d_y \\ d_z \end{bmatrix}, \quad \underset{\sim}{F} = \begin{bmatrix} F_x \\ F_y \\ F_z \end{bmatrix}, \tag{2.37}$$

and the resulting column representation of the moment follows from Eq. (1.35):

$$\underset{\sim}{M} = \begin{bmatrix} d_y F_z - d_z F_y \\ d_z F_x - d_x F_z \\ d_x F_y - d_y F_x \end{bmatrix}. \tag{2.38}$$

Example 2.5 Let the origin of the Cartesian coordinate system be the point with respect to which the moment vector is computed, i.e.

$$\vec{x}_\mathrm{P} = \vec{0} \ \ [\mathrm{m}].$$

The point of application of the force vector $\vec{F}$ is denoted by:

$$\vec{x}_\mathrm{Q} = 2\vec{e}_x + \vec{e}_y \ \ [\mathrm{m}],$$

which means that this point is located in the xy-plane. The force vector is also located in this plane:

$$\vec{F} = 5\vec{e}_y \ \ [\mathrm{N}].$$

The moment of the force $\vec{F}$ with respect to the point P follows from:

$$\begin{aligned} \vec{M} &= (\vec{x}_\mathrm{Q} - \vec{x}_\mathrm{P}) \times \vec{F} \\ &= (2\vec{e}_x + \vec{e}_y) \times 5\vec{e}_y \\ &= 10\underbrace{\vec{e}_x \times \vec{e}_y}_{\vec{e}_z} + 5\underbrace{\vec{e}_y \times \vec{e}_y}_{\vec{0}} \\ &= 10\vec{e}_z \ \ [\mathrm{N\,m}]. \end{aligned}$$

Example 2.6 As before, let:

$$\vec{x}_\mathrm{Q} = 2\vec{e}_x + \vec{e}_y \ \ [\mathrm{m}],$$

and

$$\vec{F} = 5\vec{e}_y \quad [\text{N}],$$

but

$$\vec{x}_\text{P} = 3\vec{e}_z \quad [\text{m}].$$

Then

$$\begin{aligned}\vec{M} &= (\vec{x}_\text{Q} - \vec{x}_\text{P}) \times \vec{F} \\ &= (2\vec{e}_x + \vec{e}_y - 3\vec{e}_z) \times 5\vec{e}_y \\ &= 10\vec{e}_z + 15\vec{e}_x \quad [\text{N m}].\end{aligned}$$

2.9 The Two-Dimensional Case

If all forces act in the same plane, the resulting moment vector with respect to any point in that plane is, by definition, perpendicular to this plane. However, it is common practice in this case to indicate a moment as a curved arrow that shows a clockwise or counterclockwise direction; see Fig. 2.16. Using the notation:

$$\vec{M} = M\vec{e}_z, \tag{2.39}$$

and defining the orientation vector $\vec{e}_z = \vec{e}_x \times \vec{e}_y$ to be pointing out of the plane into the direction of the viewer, a counterclockwise moment corresponds to $M > 0$, while a clockwise moment corresponds to $M < 0$. Figure 2.16(a) shows a two-dimensional body with a force $\vec{F}$ acting on it at point Q. We can define an arbitrary point P in the body. The moment $\vec{M}$ with respect to P as a result of the force $\vec{F}$ will be a vector perpendicular to the plane of drawing. Using the drawing convention as proposed above, Fig. 2.16(a) can be replaced by Fig. 2.16(b). In this case the line-of-action of force $\vec{F}$ is drawn through point P, and the resulting moment is given by a curved arrow in counterclockwise direction. The loading of the body according to Figs 2.16(a) and 2.16(b) is statically, completely equivalent. The same is true for Figs 2.16(c) and 2.16(d) for a clockwise direction of the moment.

Example 2.7 Resulting moment using scalar notation. Following the drawing convention of Section 2.2, the force vectors in Fig. 2.17 are given by:

$$\begin{aligned}\vec{F}_1 &= F_1\vec{e}_x \\ \vec{F}_2 &= -F_2\vec{e}_x \\ \vec{F}_3 &= F_3\vec{e}_x.\end{aligned}$$

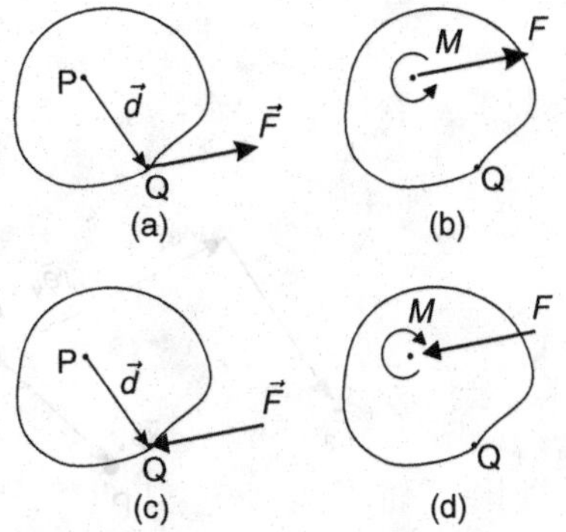

Figure 2.16

Drawing convention for the moment vector for different force vector orientations. Panels (a) and (b) indicate a statically equivalent load for a counterclockwise orientation of the moment vector. Panels (c) and (d) are equivalent for a clockwise orientation of the moment vector.

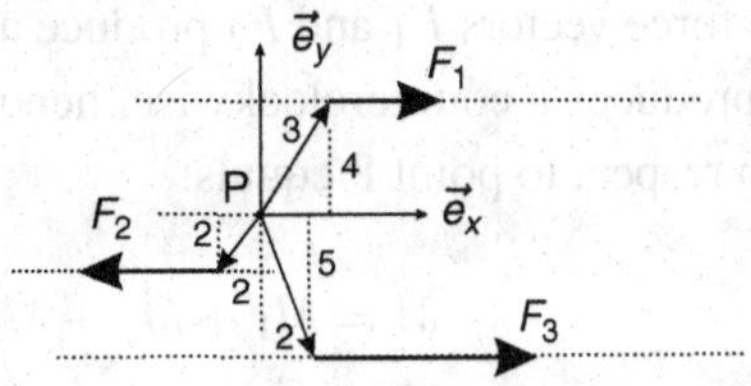

Figure 2.17

Resulting moment in two dimensions (distances in [m]).

Similarly,

$$\vec{d}_1 = 3\ell\vec{e}_x + 4\ell\vec{e}_y$$
$$\vec{d}_2 = -2\ell\vec{e}_x - 2\ell\vec{e}_y$$
$$\vec{d}_3 = 2\ell\vec{e}_x - 5\ell\vec{e}_y.$$

Each of the force vectors $\vec{F}_i$ generates a moment vector with respect to the point P:

$$\vec{M}_i = \vec{d}_i \times \vec{F}_i.$$

Clearly, given the fact that all force vectors are in the plane spanned by the $\vec{e}_x$ and $\vec{e}_y$ vectors, the moment vectors are all in the $\vec{e}_z$ direction:

$$\vec{M}_i = M_i\vec{e}_z.$$

Either using the formal definition of the moment vector or the drawing convention for two-dimensional problems as given above, it follows that:

$$M_1 = -4F_1\ell$$
$$M_2 = -2F_2\ell$$
$$M_3 = 5F_3\ell.$$

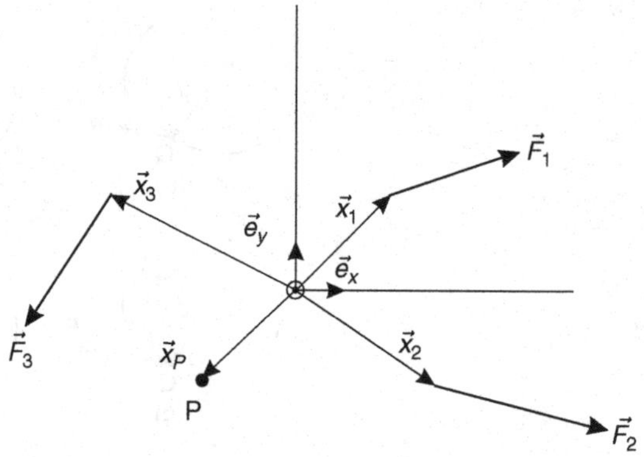

Figure 2.18
Forces and their moment.

The force vectors $\vec{F}_1$ and $\vec{F}_2$ produce a clockwise, hence negative, moment, while $\vec{F}_3$ produces a counterclockwise, hence positive, moment. The resulting moment with respect to point P equals:

$$M = M_1 + M_2 + M_3 = -4F_1 - 2F_2 + 5F_3.$$

Example 2.8 Resulting moment using vector notation. Consider the forces as depicted in Fig. 2.18. The forces are given by:

$$\vec{F}_1 = 3\vec{e}_x + \vec{e}_y \quad [\text{N}]$$
$$\vec{F}_2 = 4\vec{e}_x - \vec{e}_y \quad [\text{N}]$$
$$\vec{F}_3 = -2\vec{e}_x - 3\vec{e}_y \quad [\text{N}],$$

while the points of application are, respectively:

$$\vec{x}_1 = 2\vec{e}_x + 2\vec{e}_y \quad [\text{m}]$$
$$\vec{x}_2 = 3\vec{e}_x - 2\vec{e}_y \quad [\text{m}]$$
$$\vec{x}_3 = -4\vec{e}_x + 2\vec{e}_y \quad [\text{m}].$$

The point P has location:

$$\vec{x}_\text{P} = -2\vec{e}_x - 2\vec{e}_y \quad [\text{m}].$$

The resulting moment of the forces with respect to the point P follows from:

$$\vec{M} = (\vec{x}_1 - \vec{x}_\text{P}) \times \vec{F}_1 + (\vec{x}_2 - \vec{x}_\text{P}) \times \vec{F}_2 + (\vec{x}_3 - \vec{x}_\text{P}) \times \vec{F}_3,$$

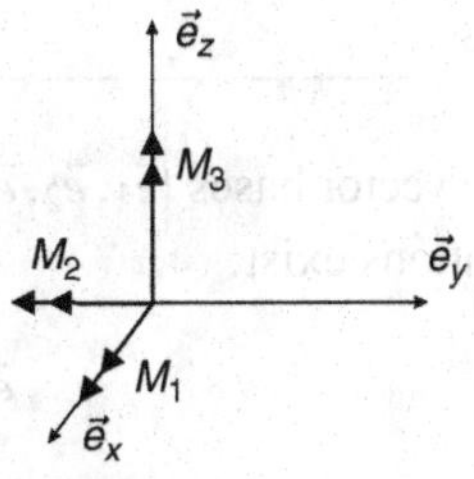

Figure 2.19

Moment vectors identified by means of scalars.

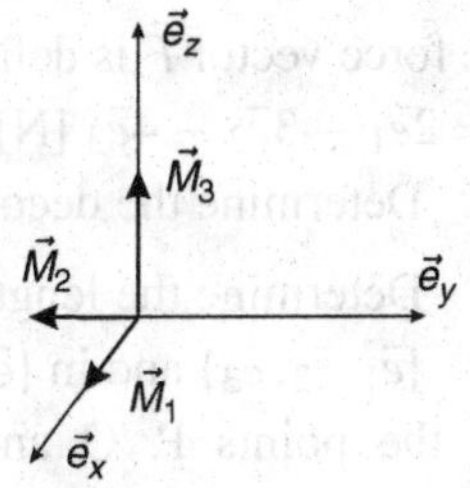

Figure 2.20

Moment vectors identified by means of vectors and by having a single arrowhead.

hence

$$\begin{aligned}\vec{M} &= (4\vec{e}_x + 4\vec{e}_y) \times (3\vec{e}_x + \vec{e}_y) + 5\vec{e}_x \times (4\vec{e}_x - \vec{e}_y) \\ &\quad + (-2\vec{e}_x + 4\vec{e}_y) \times (-2\vec{e}_x - 3\vec{e}_y) \\ &= -8\vec{e}_z - 5\vec{e}_z + 14\vec{e}_z \\ &= \vec{e}_z \quad [\text{N m}].\end{aligned}$$

2.10 Drawing Convention for Moments in Three Dimensions

An arrow drawn with two arrowheads, and identified by a scalar rather than a vector symbol, denotes a moment vector following the right-handed or corkscrew rule. For example, the moment vectors drawn in Fig. 2.19 and identified by the scalars M_1, M_2 and M_3, respectively, correspond to the moment vectors:

$$\vec{M}_1 = M_1\vec{e}_x, \quad \vec{M}_2 = -M_2\vec{e}_y, \quad \vec{M}_3 = M_3\vec{e}_z. \tag{2.40}$$

Notice that the moment vector $\vec{M}_2$ is pointing in the negative y-direction. Alternatively, if in the figure the moment vectors are drawn with a single arrowhead, as in Fig. 2.20, they denote actual vectors and are identified with vector symbols.

Exercises

2.1 The vector bases $\{\vec{e}_1, \vec{e}_2, \vec{e}_3\}$ and $\{\vec{\epsilon}_1, \vec{\epsilon}_2, \vec{\epsilon}_3\}$ are orthonormal. The following relations exist:

$$\vec{\epsilon}_1 = \frac{1}{2}\sqrt{2}\vec{e}_1 + \frac{1}{2}\sqrt{2}\vec{e}_2$$
$$\vec{\epsilon}_2 = -\frac{1}{2}\sqrt{2}\vec{e}_1 + \frac{1}{2}\sqrt{2}\vec{e}_2$$
$$\vec{\epsilon}_3 = \vec{e}_3 \,.$$

The force vector $\vec{F}$ is defined with respect to basis $\{\vec{e}_1, \vec{e}_2, \vec{e}_3\}$ according to: $\vec{F} = 2\vec{e}_1 + 3\vec{e}_2 - 4\vec{e}_3$ [N].

(a) Determine the decomposition of $\vec{F}$ with respect to basis $\{\vec{\epsilon}_1, \vec{\epsilon}_2, \vec{\epsilon}_3\}$.

(b) Determine the length of $\vec{F}$, using the specifications of $\vec{F}$ expressed in $\{\vec{e}_1, \vec{e}_2, \vec{e}_3\}$ and in $\{\vec{\epsilon}_1, \vec{\epsilon}_2, \vec{\epsilon}_3\}$.

2.2 For the points P, Q and R the following location vectors are given, respectively:

$$\vec{x}_{\text{P}} = \vec{e}_x + 2\vec{e}_y \quad [\text{m}]$$
$$\vec{x}_{\text{Q}} = 4\vec{e}_x + 2\vec{e}_y \quad [\text{m}]$$
$$\vec{x}_{\text{R}} = 3\vec{e}_x + e_y \quad [\text{m}].$$

The force vector $\vec{F} = 2\vec{e}_x$ [N] acts on point Q.

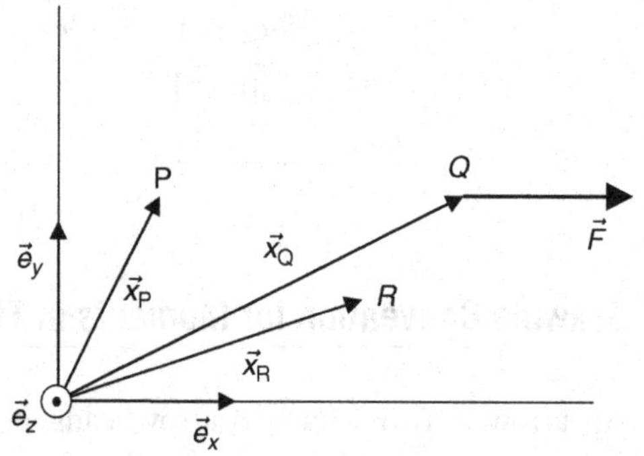

(a) Calculate the moment of the force $\vec{F}$ with respect to point P and with respect to point R.

(b) Write the vectors mentioned above in column notation according to the right-handed orthonormal basis $\{\vec{e}_x, \vec{e}_y, \vec{e}_z\}$ and calculate the moment of the force with respect to the points P and R by using Eq. (1.35).

2.3 Calculate, for each of the situations given below, the resulting moment with respect to point P. We consider a counterclockwise moment positive. F is a positive number.

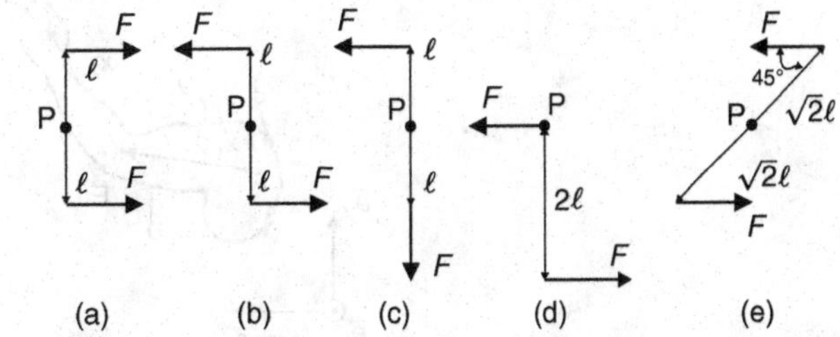

2.4 On an axis, a wheel with radius R is fixed to a smaller wheel with radius r. The forces F and f are tangentially applied to the contours of both wheels (as shown in the figure). Calculate the ratio between the forces F and f in the case where the total moment with respect to the centroid P is zero.

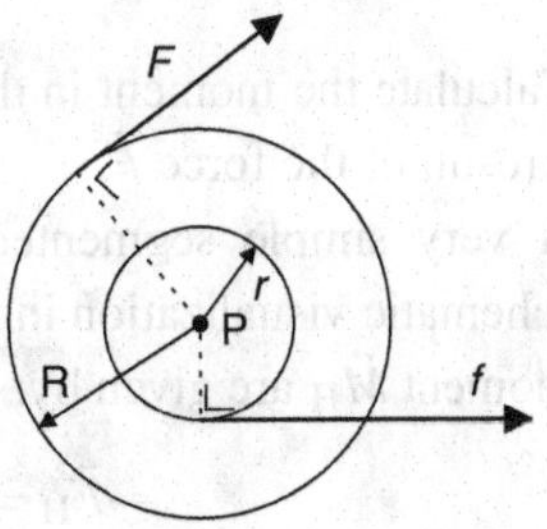

2.5 On the body in the drawing, the moments $\vec{M}_1 = -3\vec{e}_x$ [N m] and $\vec{M}_2 = 4\vec{e}_x$ [N m] and the forces $\vec{F}_1 = -2\vec{e}_z$ [N] and $\vec{F}_2 = \vec{e}_z$ [N] are exerted. The forces are acting at the points $\vec{x}_1 = -2\vec{e}_x + 3\vec{e}_z$ [m] and $\vec{x}_2 = 3\vec{e}_x + \vec{e}_y + 3\vec{e}_z$ [m]. Calculate the resulting moment vector with respect to point P, given by the position vector $\vec{x}_P = 2\vec{e}_x + 4\vec{e}_y + 2\vec{e}_z$ [m].

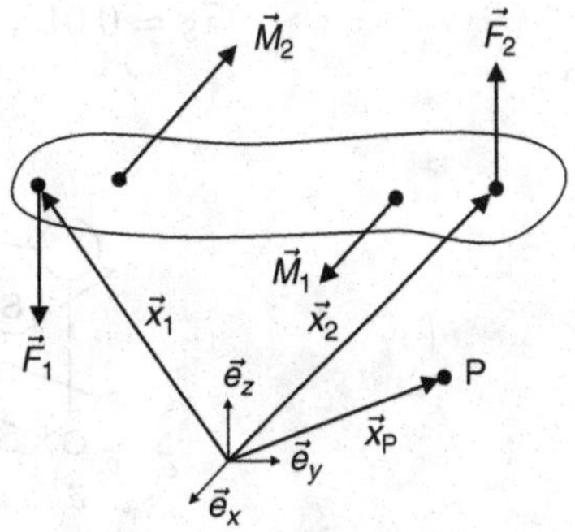

2.6 A person is pulling a rope. This results in a force $\vec{F}$ acting on the hand at point H as depicted in the figure.

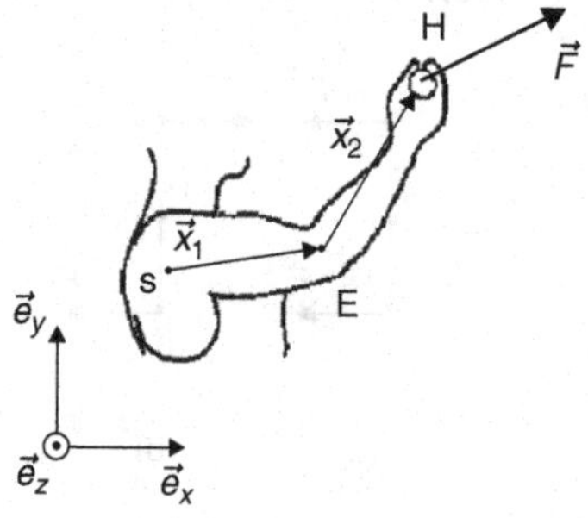

The following vectors are given:

$$\vec{x}_1 = 50\vec{e}_x + 5\vec{e}_y \quad [\text{cm}]$$
$$\vec{x}_2 = 20\vec{e}_x + 30\vec{e}_y \quad [\text{cm}]$$
$$\vec{F} = \vec{e}_x + 2\vec{e}_y \quad [\text{N}].$$

Calculate the moment in the shoulder (point S) and the elbow (point E) as a result of the force $\vec{F}$.

2.7 A very simple segmented model of a sitting person is shown in the schematic visualization in the figure. At the right hand H, a force $\vec{F}_\text{H}$ and a moment $\vec{M}_\text{H}$ are given by:

$$\vec{F}_\text{H} = 30\vec{e}_x + 5\vec{e}_z \quad [\text{N}]$$
$$\vec{M}_\text{H} = 0.02\vec{e}_y + 0.03\vec{e}_z \quad [\text{N m}],$$

related to the Cartesian vector basis $\{\vec{e}_x, \vec{e}_y, \vec{e}_z\}$. In addition, the positions of the hand H and shoulder S with respect to the origin at the lower end of the spine are given by:

$$\vec{x}_\text{H} = 0.05\vec{e}_x - 0.01\vec{e}_y + 0.04\vec{e}_z \quad [\text{m}]$$
$$\vec{x}_\text{S} = 0.01\vec{e}_x - 0.01\vec{e}_y + 0.03\vec{e}_z \quad [\text{m}].$$

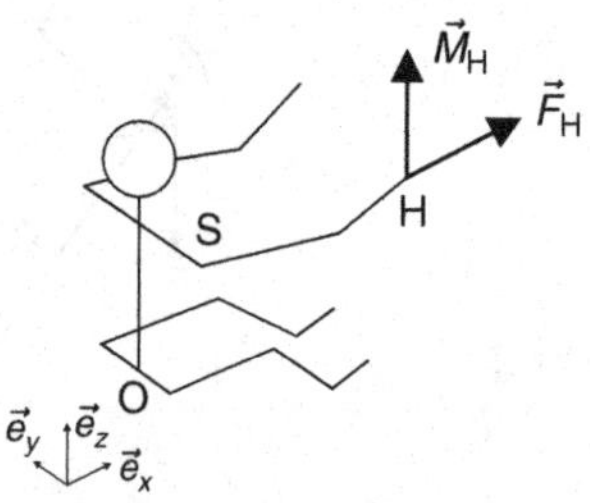

(a) Calculate the moment at the shoulder (point S) as a result of the load at the hand H.

(b) Calculate the moment at the lower end of the spine O as a result of the load at the hand H.

2.8 A person is lifting a dumbbell as shown in the figure. We define a Cartesian basis $\{\vec{e}_x, \vec{e}_y, \vec{e}_z\}$ with the origin in the shoulder of the person. The point U is the centre of gravity of the upper arm, point L is the centre of gravity of the lower arm and point W is the point of application of the combined weight of the dumbbell and the weight of the hand. The positions of these particular points with respect to the shoulder are given by:

$$\vec{x}_{\mathrm{U}} = 5\vec{e}_x - 15\vec{e}_y \quad [\mathrm{cm}]$$
$$\vec{x}_{\mathrm{L}} = 20\vec{e}_x - 20\vec{e}_y \quad [\mathrm{cm}]$$
$$\vec{x}_{\mathrm{W}} = 40\vec{e}_x - 15\vec{e}_y \quad [\mathrm{cm}].$$

The respective forces are:

$$\vec{F}_{\mathrm{U}} = -30\vec{e}_y \quad [\mathrm{N}]$$
$$\vec{F}_{\mathrm{L}} = -20\vec{e}_y \quad [\mathrm{N}]$$
$$\vec{F}_{\mathrm{W}} = -100\vec{e}_y \quad [\mathrm{N}].$$

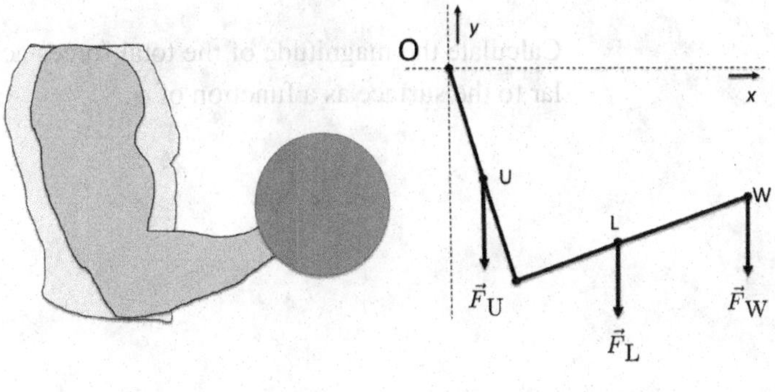

Calculate the resulting moment with respect to the shoulder.

2.9 A cell is firmly attached to a substrate in a discrete number of points with respect to the origin of the cell. At each of these points, forces are acting on the cell, as illustrated in the figure below.
All position and force vectors are known, except for $\vec{F}_5$:

$$\vec{F}_1 = 2\vec{e}_x - 2\vec{e}_y \quad [\mu\mathrm{N}]$$
$$\vec{F}_2 = -3\vec{e}_x - \vec{e}_y \quad [\mu\mathrm{N}]$$
$$\vec{F}_3 = -3\vec{e}_x + \vec{e}_y \quad [\mu\mathrm{N}]$$
$$\vec{F}_4 = \vec{e}_y \quad [\mu\mathrm{N}].$$

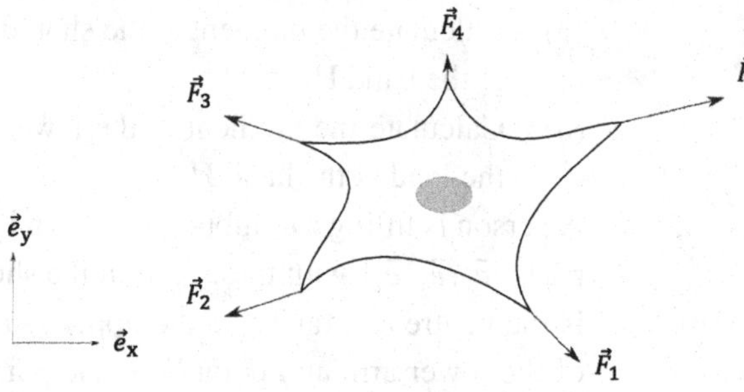

Determine $\vec{F}_5$, given the fact that the cell does not move in this configuration.

2.10 A person is lying on a surface that is positioned at an angle α with respect to the horizontal direction, as shown below. The force that this person is exerting on the surface due to gravity is $\vec{F}_g = -G\vec{e}_y$.

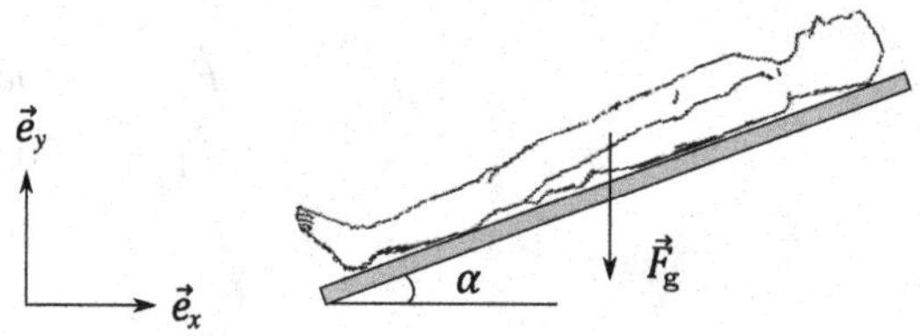

Calculate the magnitude of the total forces acting parallel and perpendicular to the surface as a function of α.

3 Static Equilibrium

3.1 Introduction

According to Newton's law, the acceleration of the centroid of a body multiplied by its mass equals the total force applied to the body, and there will be a spin around the centroid when there is a resulting moment with respect to the centroid. But, in many cases, bodies do not move at all when forces are applied to them. In that case, the bodies are in static equilibrium. A simple example is given in Fig. 3.1. In Fig. 3.1(a) a body is loaded by two forces of equal size but with an opposite direction. The lines-of-action of the two forces coincide and clearly the body is in equilibrium. If the lines-of-action do not coincide, as in Fig. 3.1(b), the forces have a resulting moment and the body will rotate. To enforce static equilibrium, a counteracting moment should be applied to prevent the body rotating, as indicated in Fig. 3.1(c).

3.2 Static Equilibrium Conditions

If a body moves monotonously (no acceleration of the centroid, no rate of rotation around the centroid), the body is in static equilibrium. If the velocities are zero as well, the body is at rest. In both cases, the sum of all forces **and** the sum of all moments (with respect to any point) acting on the body are zero. Suppose that n forces $\vec{F}_i$ $(i = 1, 2, \dots, n)$ are applied to the body. Each of these forces will have a moment M_i with respect to an arbitrary point P. There may be a number of additional moments $\vec{M}_j$ $(j = 1, 2, \dots, m)$ applied to the body. Static equilibrium then requires that

$$\begin{aligned} \sum_{i=1}^{n} \vec{F}_i &= \vec{0} \\ \sum_{i=1}^{n} \vec{M}_i + \sum_{j=1}^{m} \vec{M}_j &= \vec{0}. \end{aligned} \tag{3.1}$$

Simply demanding that the sum of all the forces is equal to zero, to assure equilibrium, is insufficient, since a resulting moment may induce a (rate of)

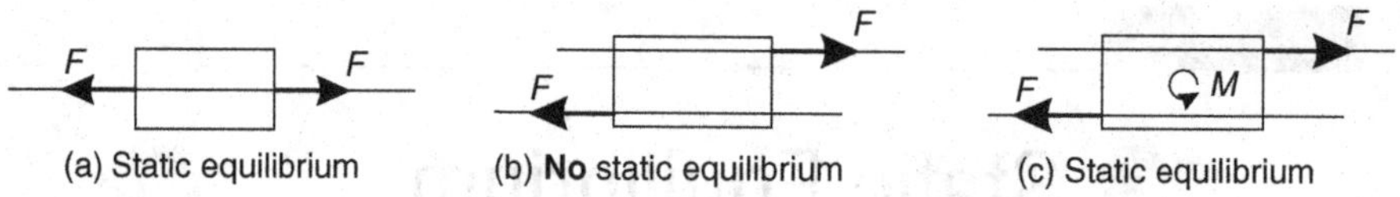

Figure 3.1

Examples of static equilibrium and its violation.

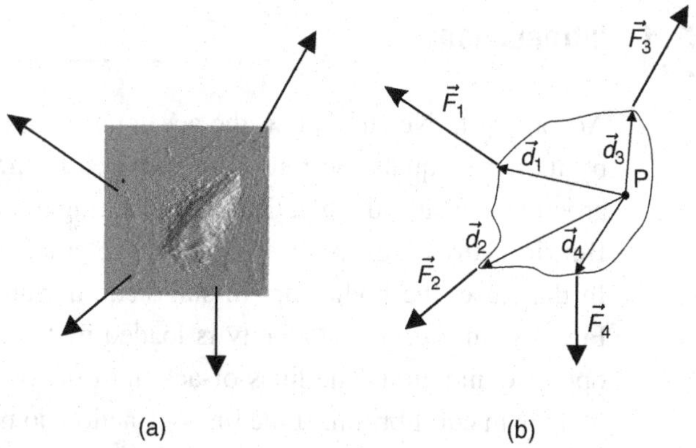

Figure 3.2

An image and a model of a cell.

rotation of the body. Therefore the sum of the moments must vanish as well. The moment vectors associated with each force vector are computed with respect to some point P. However, if the sum of all forces is zero, the sum of the moment vectors should be zero with respect to **any** point P.

Example 3.1 Figure 3.2(a) shows an image of a single cell that was captured by means of an atomic force microscope. Cells attach themselves to the supporting surface at a discrete number of points. The forces acting on these points of the cell are shown as arrows. In Fig. 3.2(b), a model of the cell is given that can be used to examine the equilibrium of forces and moments. In this case, the sum of all the forces should be equal to zero:

$$\sum_{i=1}^{4} \vec{F}_i = \vec{0},$$

while the sum of the moments with respect to the point P (with an arbitrary position $\vec{x}_P$) due to these forces must be zero as well:

$$\sum_{i=1}^{4}(\vec{d}_i \times \vec{F}_i) = \vec{0}.$$

Notice that in this particular case there are no externally applied additional moments.

In the example of Fig. 3.2(b), the moments were determined with respect to point P. If the moment is computed with respect to another point in space, say R, having coordinates $\vec{x}_R = \vec{x}_P + \vec{a}$, then the moment vector with respect to this point R is defined by

$$\begin{aligned}\vec{M} &= \sum_{i=1}^{4}(\vec{d}_i - \vec{a}) \times \vec{F}_i \\ &= \sum_{i=1}^{4}(\vec{d}_i \times \vec{F}_i) - \sum_{i=1}^{4}(\vec{a} \times \vec{F}_i) \\ &= \sum_{i=1}^{4}(\vec{d}_i \times \vec{F}_i) - \vec{a} \times \sum_{i=1}^{4}\vec{F}_i,\end{aligned}$$

which vanishes if the forces are in equilibrium and the sum of the moments with respect to point P equals $\vec{0}$. This implies that any point can be taken to enforce equilibrium of moments. This argument can be generalized in a straightforward manner to any number of forces.

Example 3.2 An example of pure force equilibrium is given in Fig. 3.3. This figure shows an electron micrograph of an actin network supporting the cell membrane. At the intersection point of the network, the molecules are (weakly) cross-linked. Within each of the molecules, a (tensile) force is present, and at the interconnection point force equilibrium must apply. In Fig. 3.3, the forces acting on one of the interconnection points have been sketched, and the sum of these force vectors has to be equal to zero.

With respect to a Cartesian coordinate system, equilibrium requires that the sum of all forces in the x-, y- and z-direction is zero. With the decomposition of a force $\vec{F}$, according to $\vec{F} = F_x\vec{e}_x + F_y\vec{e}_y + F_z\vec{e}_z$, that is the case if:

$$\sum_{i=1}^{n} F_{x,i} = 0$$

$$\sum_{i=1}^{n} F_{y,i} = 0$$

$$\sum_{i=1}^{n} F_{z,i} = 0,$$

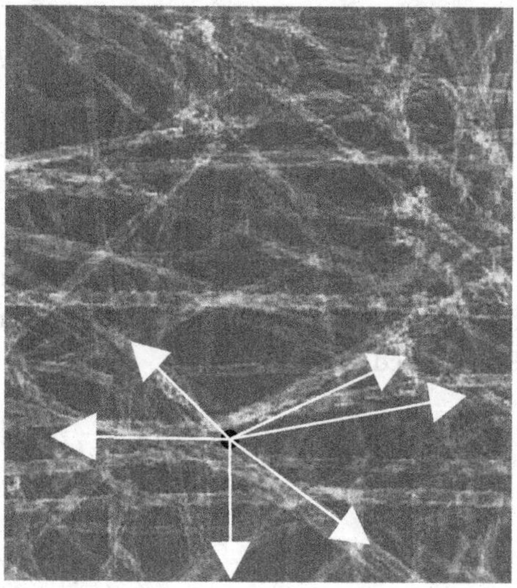

Figure 3.3

Transmission electron microscope (TEM) image of the actin network supporting a cell membrane, with forces acting on an interconnection point.

and the sum of all moments in the x-, y- and z-direction with respect to an arbitrarily selected point P is zero. Choosing point P to coincide with the point of application of the forces immediately reveals that the equilibrium of moments is trivially satisfied.

Equilibrium of forces may also be expressed in column notation, according to:

$$\sum_{i=1}^{n} \underset{\sim}{F}_i = \underset{\sim}{0}.$$

3.3 Free Body Diagram

A free body diagram serves to specify and visualize the complete loading of a body, including the reaction forces and moments acting on the body that is supported in one way or another. The body may be part of a system of bodies and, using the free body diagram, the reaction forces on the body under consideration imposed by the other bodies may be identified. For this purpose the body is isolated from its surroundings, and the proper reaction forces and moments are introduced so as to ensure equilibrium of the body. Clearly, these reaction forces and moments are not known a priori, but the equilibrium conditions may be used to try to compute these unknowns. A distinction must be made between the **statically determinate** and the **statically indeterminate** case.

Requiring force and moment equilibrium provides for a limited number of equations only, and therefore only a limited number of unknowns can be determined. For two-dimensional problems, force equilibrium results in two equations, while the requirement of moment equilibrium supplies only one equation; hence three independent equations can be formulated. Only if the number of unknown reaction loads equals three is the solution of the unknowns possible. Likewise, in the three-dimensional case, imposing force and moment equilibrium generates six independent equations, such that six unknown reactions can be computed. If a free body diagram is drawn and all the reactions can be directly identified from enforcing the equilibrium conditions, this is referred to as the statically determinate case.

If the reactions defined on a free body diagram cannot be calculated by imposing the equilibrium conditions, then this is referred to as the statically **in**determinate case. This is dealt with if more than three forces or moments for two-dimensional problems or more than six forces or moments for three-dimensional problems need to be identified. It should be noted that the equilibrium equations do not suffice if in the two-dimensional case more than one moment is unknown and in the three-dimensional case more than three moments are unknown.

Example 3.3 As a two-dimensional example, consider the body of a single cell, as sketched in Fig. 3.4, that is loaded by a known force F_P, while the body is supported at two points, say A and B. The support is such that at point A only a force in the horizontal direction can be transmitted. This is represented by the rollers, which allow point A to freely move in the vertical direction. At point B, however, forces in both the vertical and horizontal direction can be transmitted from the surroundings to the body, indicated in the figure by a hinge. A free body diagram is sketched in Fig. 3.5. The supports are separated from the body. It is assumed that the supports cannot exert a moment on the body; therefore only reaction forces in the horizontal and vertical direction have been introduced. As a naming convention, all forces in the horizontal direction have been labelled H_α (the subscript α referring to the point of application), while all vertical forces have been labelled V_α. At each of the attachment points, A and B, reaction forces have been introduced on both the body and the support. According to the third law of Newton (see Section 2.3): action = − reaction; forces are defined in the opposite direction with respect to each other, but have equal magnitude. The (three) reaction forces at point A and

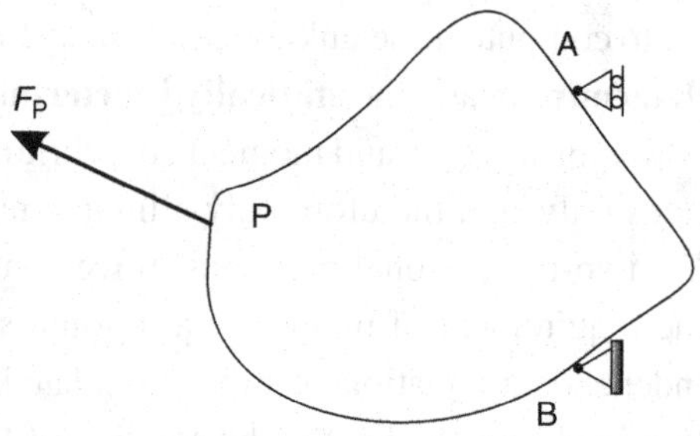

Figure 3.4

A loaded body.

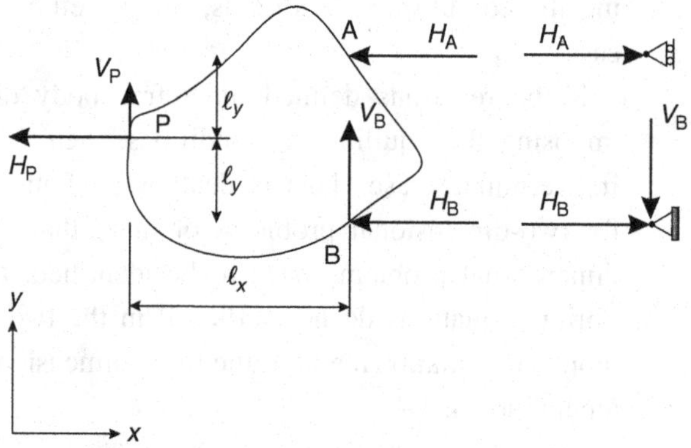

Figure 3.5

Free body diagram of the loaded body.

point B are, for the time being, unknown. They can be calculated by enforcing force and moment equilibrium of the body. Hence, both the sum of all forces in the horizontal direction and the sum of all the forces in the vertical direction acting on the body have to be equal to zero. For this purpose, the load F_P has been decomposed into a horizontal force H_P and a vertical force V_P:

$$- H_P - H_A - H_B = 0$$
$$V_P + V_B = 0.$$

The above expressions represent two equations with three unknowns (H_A, H_B and V_B), being insufficient to determine these. However, the sum of the moments with respect to an arbitrary point has be zero as well. Computing the resulting moment with respect to point A gives:

$$-2\ell_y H_B - \ell_y H_P - \ell_x V_P = 0.$$

This yields **one** additional equation such that the unknown reaction forces may be determined.

The above procedure is not unique in the sense that different points can be used with respect to which the sum of moments should be required to be zero. For instance, rather than using the sum of moments with respect to point A, the sum of moments with respect to point P could have been used.
Suppose that:

$$\ell_x = 2 \quad [\text{m}], \qquad \ell_y = 1 \quad [\text{m}], \qquad H_\text{P} = 20 \quad [\text{N}], \qquad V_\text{P} = 10 \quad [\text{N}].$$

Substitution of these values into the equation above renders:

$$-20 - H_\text{A} - H_\text{B} = 0$$
$$10 + V_\text{B} = 0$$
$$-2H_\text{B} - 20 - 20 = 0.$$

Clearly, from the second equation it follows immediately that $V_\text{B} = -10$ [N], while from the last equation it is clear that $H_\text{B} = -20$ [N], which leaves the first equation to calculate H_A as $H_\text{A} = 0$ [N]. Hence we have the solution:

$$H_\text{A} = 0 \quad [\text{N}], \qquad H_\text{B} = -20 \quad [\text{N}], \qquad V_\text{B} = -10 \quad [\text{N}].$$

Example 3.4 A weightlifter performs a so-called 'deadweight' lift (see Fig. 3.6), and we would like to know the bending moment at the hip of the weightlifter. For this, a line model is made of the body. The weight of the upper body of the lifter is lumped to a vertical force F_B, the weight of the head to a vertical force F_H, the combined weight of the arms and the dumbbells is represented by a vertical force F_W. The distances of the working lines of these forces are shown in the figure.

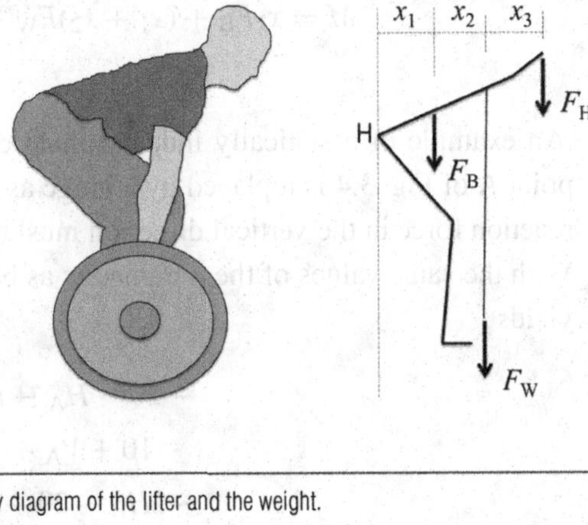

Figure 3.6

Weightlifter and free body diagram of the lifter and the weight.

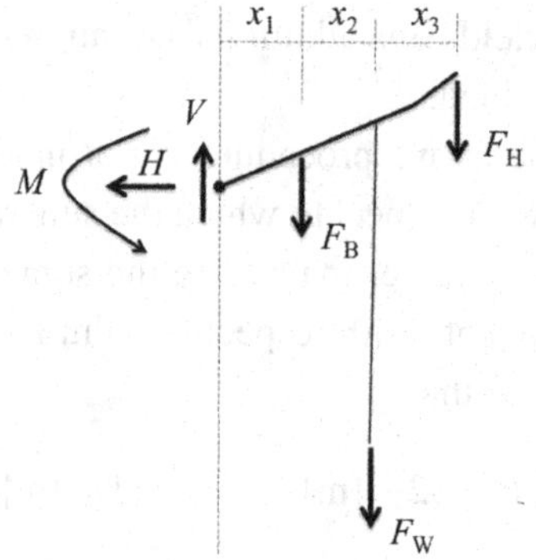

Figure 3.7

Free body diagram to determine the moment in the hip of the weightlifter.

To determine the bending moment in the hip at point H we have to make a new free body diagram with a virtual body cut at the hip. This diagram is shown in Fig. 3.7. Now we have to introduce a horizontal and vertical force and a moment in the hip at the virtual cut through the body. These are the forces and moment that the structures from the lower part of the body are applying to the upper part at the level of the hip.

Horizontal force equilibrium immediately shows that $H = 0$. Equilibrium of the vertical forces leads to:

$$V = F_B + F_H + F_W. \tag{3.1}$$

Note that in this two-dimensional example the directions of the arrows in the figure are important, and the result is the combination of Eq. (3.1) and Fig. 3.7. For the moment M we find:

$$M = x_1 F_B + (x_1 + x_2)F_W + (x_1 + x_2 + x_3)F_H.$$

Example 3.5 An example of a statically indeterminate case appears if the rolling support at point A of Fig. 3.4 is replaced by a hinge as in point B. In that case, an additional reaction force in the vertical direction must be introduced at point A: see Fig. 3.8. With the same values of the parameters as before, force and moment equilibrium yields

$$\begin{aligned} -20 - H_A - H_B &= 0 \\ 10 + V_A + V_B &= 0 \\ -2H_B - 20 - 20 &= 0. \end{aligned}$$

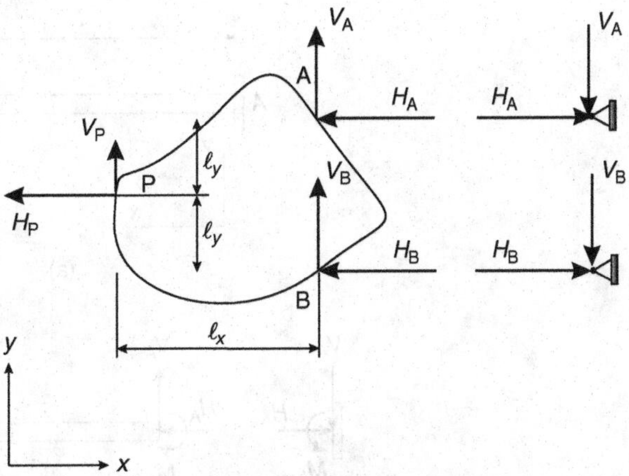

Figure 3.8

Free body diagram of the simple loaded body, statically indeterminate case.

The horizontal reaction forces H_A and H_B can, incidentally, still be calculated, giving, as before:

$$H_A = 0 \quad [N], \qquad H_B = -20 \quad [N].$$

But there is insufficient information to compute V_A and V_B. In fact, there are only three equations to determine the four unknowns (H_A, H_B, V_A and V_B).

The nature of the support defines the possible set of reaction forces that have to be introduced. In the two-dimensional statically indeterminate example given above, the supports are assumed to be hinges or pin-connections. Effectively this means that no moments can be exerted on the support, and only forces in the x- and y-direction have to be introduced.

Example 3.6 An example of a fixed support is given in Fig. 3.9(a), showing a beam that is clamped at one end and loaded at a distance L from this fixation (cantilever beam). The reaction forces and moment can be computed by enforcing force and moment equilibrium. The sum of the forces in the x- and y-direction has to be equal to zero:

$$H_A = 0$$
$$-F + V_A = 0,$$

while the sum of the moments has to be zero, for instance, with respect to point A:

$$M_A - LF = 0.$$

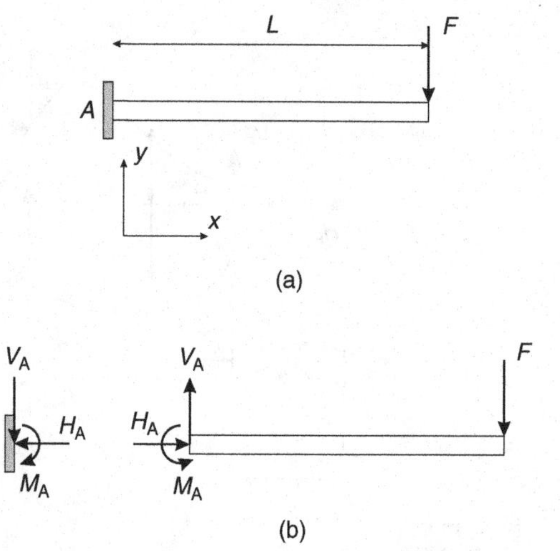

Figure 3.9

Free body diagram of a bar fixed at one end and loaded at the other.

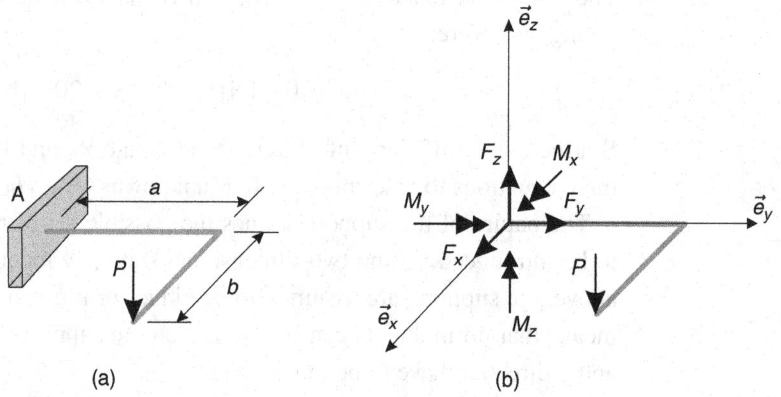

Figure 3.10

A beam construction loaded by a force P and the free body diagram.

Example 3.7 Consider a beam construction, sketched in Fig. 3.10(a), loaded by a force P. The beam is clamped at point A, and we want to determine the reaction loads at point A. First of all a coordinate system is introduced and a free body diagram of the loaded beam construction is drawn, as in Fig. 3.10(b). The applied load is represented by the vector:

$$\vec{P} = -P\vec{e}_z.$$

The reaction force vector on the beam construction at point A is denoted by $\vec{F}$ and is decomposed according to:

$$\vec{F} = F_x\vec{e}_x + F_y\vec{e}_y + F_z\vec{e}_z,$$

while the reaction moment vector at point A is written as

$$\vec{M} = M_x\vec{e}_x + M_y\vec{e}_y + M_z\vec{e}_z.$$

The requirement that the sum of all forces is equal to zero implies that

$$\vec{F} + \vec{P} = \vec{0},$$

and consequently

$$F_x = 0, \quad F_y = 0, \quad F_z - P = 0.$$

The requirement that the sum of all moments with respect to A equals zero leads to:

$$\vec{M} + \vec{d} \times \vec{P} = \vec{0},$$

where the distance vector $\vec{d}$ is given by

$$\vec{d} = b\vec{e}_x + a\vec{e}_y,$$

hence

$$\begin{aligned}\vec{d} \times \vec{P} &= (b\vec{e}_x + a\vec{e}_y) \times (-P\vec{e}_z)\\ &= bP\vec{e}_y - aP\vec{e}_x.\end{aligned}$$

Consequently

$$M_x - aP = 0, \quad M_y + bP = 0, \quad M_z = 0.$$

Example 3.8 Consider the man sketched in Fig. 3.11 who is lifting a weight. We would like to compute the force $\vec{F}_{\mathrm{M}}$ in the muscle connecting the upper arm to the shoulder. A basis $\{\vec{e}_x, \vec{e}_y\}$ is introduced with the origin in the joint, point J. The basis vector $\vec{e}_x$ has the direction of the arm, while basis vector $\vec{e}_y$ is perpendicular to the arm (see figure). The forces $\vec{W} = -W\vec{e}_y$ due to the weight of the arm, and $\vec{W}_0 = -W_0\vec{e}_y$ due to the lifted weight, are both supposed to be known. The reaction force in the joint $\vec{F}_{\mathrm{J}}$ and the force in the muscle $\vec{F}_{\mathrm{M}}$ are both unknown. However, the direction of the force in the muscle is known since this force is oriented with respect to the arm at an angle θ. Consequently

$$\vec{F}_{\mathrm{J}} = F_{\mathrm{J}x}\vec{e}_x + F_{\mathrm{J}y}\vec{e}_y,$$

while the force in the muscle is given by

$$\vec{F}_{\mathrm{M}} = -F_{\mathrm{M}}\cos(\theta)\,\vec{e}_x + F_{\mathrm{M}}\sin(\theta)\,\vec{e}_y.$$

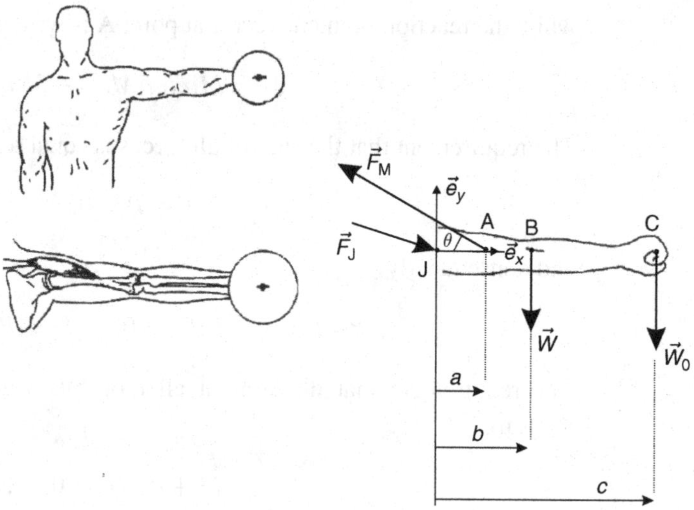

Figure 3.11
Lifting a weight.

Notice that both the x- and y- components of the joint reaction force, F_{Jx} and F_{Jy}, respectively, are unknown (the joint is modelled by a hinge), while for the muscle only the magnitude of the muscle force F_M is unknown.

Application of the force balance in the x- and y-direction yields

$$F_{Jx} - F_M \cos(\theta) = 0$$

and

$$F_{Jy} + F_M \sin(\theta) - W - W_0 = 0.$$

Force equilibrium supplies two equations for three unknowns hence moment equilibrium needs to be enforced as well. With the points A, B and C located at $\vec{x}_A = a\vec{e}_x$, $\vec{x}_B = b\vec{e}_x$ and $\vec{x}_C = c\vec{e}_x$, respectively, moment equilibrium with respect to point J requires:

$$a F_M \sin(\theta) - bW - cW_0 = 0.$$

From this equation it follows that

$$F_M = \frac{bW + cW_0}{a \sin(\theta)}.$$

Hence, from the force balance in the x- and y-directions, the joint reaction forces can be computed immediately.

Suppose that

$$\theta = \frac{\pi}{10}, \quad a = 0.1 \quad [\text{m}], \quad b = 0.25 \quad [\text{m}], \quad c = 0.6 \quad [\text{m}],$$

and

$$W = 5 \quad [\text{N}], \quad W_O = 10 \quad [\text{N}],$$

then

$$F_M = 235 \quad [\text{N}], \quad F_{Jx} = 223 \quad [\text{N}], \quad F_{Jy} = -58 \quad [\text{N}].$$

Exercises

3.1 The position vectors of the points P, Q, R and S are given, respectively, as:

$$\vec{x}_P = \vec{e}_x + 3\vec{e}_y \quad [\text{m}]$$
$$\vec{x}_Q = 4\vec{e}_x + 2\vec{e}_y \quad [\text{m}]$$
$$\vec{x}_R = 3\vec{e}_x + e_y \quad [\text{m}]$$
$$\vec{x}_S = -\vec{e}_x - \vec{e}_y \quad [\text{m}].$$

The force vector $\vec{F}_1 = 2\vec{e}_x$ [N] acts on point Q. The force vector $\vec{F}_2 = -2\vec{e}_x$ [N] acts on point S.

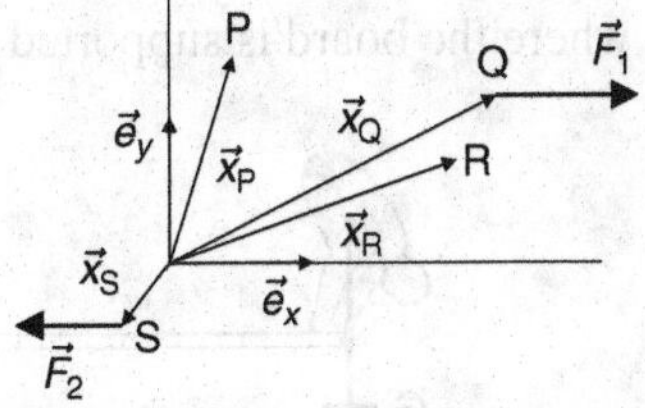

Calculate the resulting moment of the forces $\vec{F}_1$ and $\vec{F}_2$ with respect to the points P and R.

3.2 Determine the reaction forces and moments on the beam construction, experienced at point A due to the fully clamped fixation at point A for both configurations in the figure.

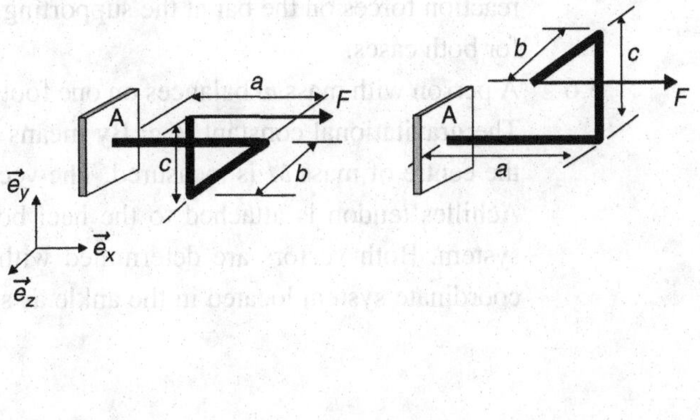

3.3 A person is lying on a board, which is supported at both ends. A vertical reaction force V_A is acting on the board at point A and passes through the origin of the coordinate system. A vertical reaction force V_B is acting at the point located at $\vec{x}_B = \ell\vec{e}_x$, where ℓ is a length. It is known that the weight of the person is F and the weight of the board is P. The vector that determines the centroid C of the person is given by: $\vec{x}_C = \alpha\vec{e}_x + \beta\vec{e}_y$. Determine the reaction forces V_A and V_B as a function of ℓ, α, β, F and P.

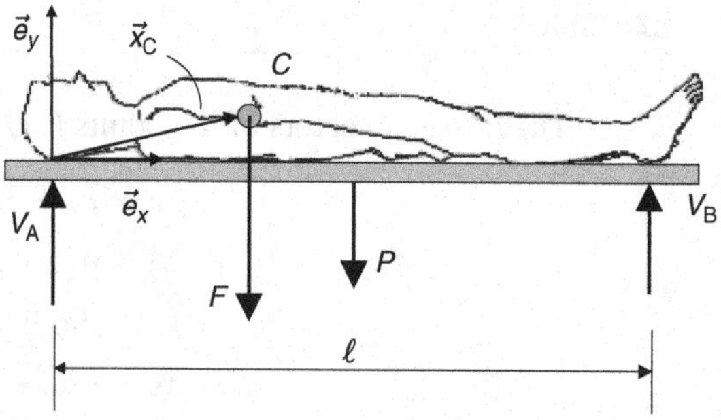

3.4 A swimmer with a weight G is standing at the edge of a diving board. The centre of gravity of the swimmer is just above the edge. The distances a and b are known. Determine the reaction forces on the board at B and C, where the board is supported by rollers and a hinge, respectively.

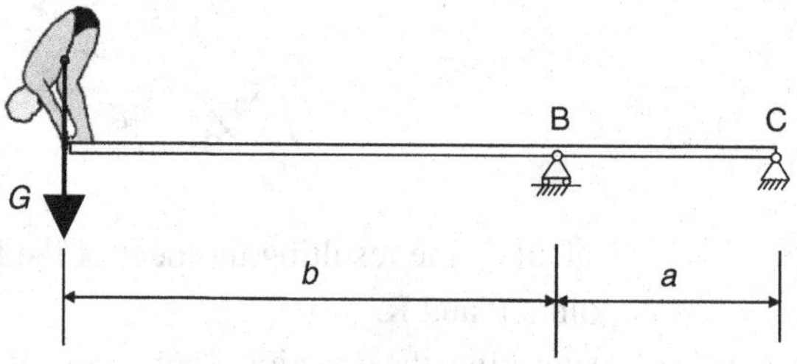

3.5 Wall bars with length ℓ and weight G are placed at an angle α against a wall. Two ways to model the supports are given in the figure. Calculate the reaction forces on the bar at the supporting points as a function of G and α for both cases.

3.6 A person with mass m balances on one foot as sketched in the figure below. The gravitational constant is g. By means of a camera the position $\vec{x}_M$ of the centre of mass M is measured. The vector $\vec{x}_P$ of the point P, where the Achilles tendon is attached to the heel bone, is also measured with this system. Both vectors are determined with respect to the origin of a xy-coordinate system located in the ankle as shown in the figure. In addition,

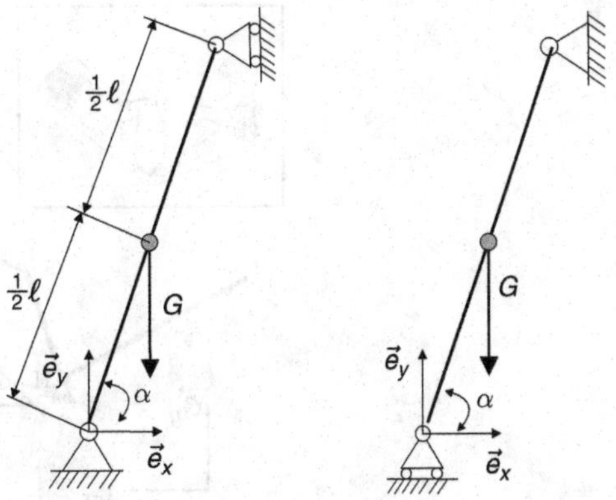

the orientation of the Achilles tendon is measured and represented by the unit vector $\vec{a}$. The results of these measurements are:

$$\vec{a} = \frac{1}{5}(3\vec{e}_x + 4\vec{e}_y)$$
$$\vec{x}_{\mathrm{M}} = 10\vec{e}_x + 100\vec{e}_y \quad [\mathrm{cm}]$$
$$\vec{x}_{\mathrm{P}} = -4\vec{e}_x - 2\vec{e}_y \quad [\mathrm{cm}]\,.$$

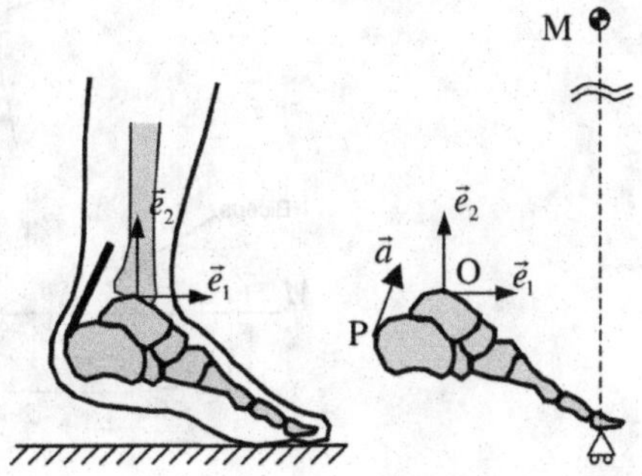

Determine the force in the Achilles tendon, expressed in M and g.

3.7 A girl is doing a workout for the legs by pushing a weight G with her feet up against a slope at 45 degrees to the x-axis. At a certain moment of time the situation can be modelled like in the figure below. The line H–K represents the upper leg, the line K–E the lower leg.
The following position vectors are given:

$$\vec{x}_{\mathrm{H}} = 30\vec{e}_x + 30\vec{e}_y \quad [\mathrm{cm}]$$
$$\vec{x}_{\mathrm{K}} = 50\vec{e}_x + 60\vec{e}_y \quad [\mathrm{cm}]$$
$$\vec{x}_{\mathrm{A}} = 80\vec{e}_x + 80\vec{e}_y \quad [\mathrm{cm}]\,.$$

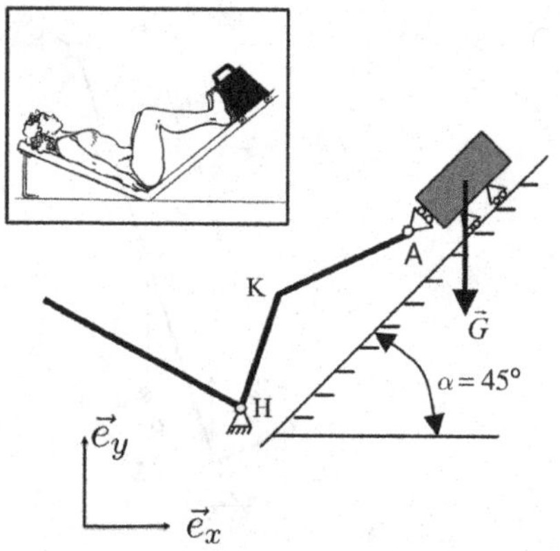

Determine the reaction force $\vec{R}$ that acts upon the upper leg in point H (first define a free body diagram of part H–A that describes all external forces that act upon the leg).

3.8 A person exerts a horizontal force F = 150 [N] (directed to the left) on the test apparatus shown in the drawing.

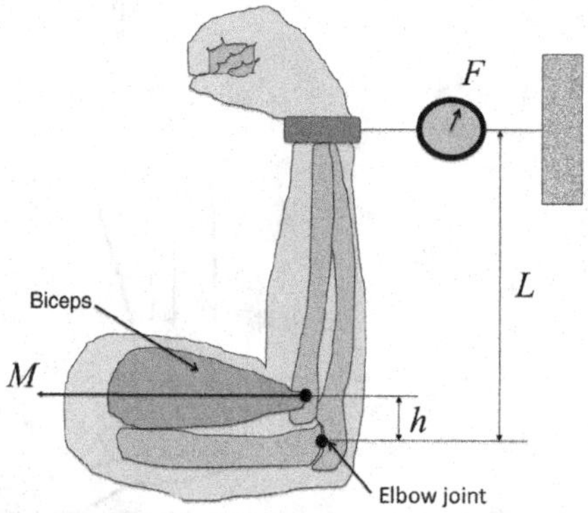

Find the horizontal force M (magnitude and sense) that the biceps exerts on his forearm (L= 0.30 [m], h = 0.040 [m]), assuming that the elbow can be modelled as a perfect joint.

3.9 A sitting person is pulling a rope with a force $\vec{F}$. The working line of the pulling force runs through the point S. On the torso, a gravitational force $\vec{G}$ is also working, which can be assumed to act on point M.

The following points are given, with respect to the basis $\{\vec{e}_x, \vec{e}_y, \vec{e}_z\}$.

$$\begin{aligned} &\text{point O:} \quad \vec{x}_O = 30\vec{e}_x + 30\vec{e}_y \quad [\text{cm}] \\ &\text{point M:} \quad \vec{x}_M = 20\vec{e}_x + 50\vec{e}_y \quad [\text{cm}] \\ &\text{point S:} \quad \vec{x}_S = 10\vec{e}_x + 70\vec{e}_y \quad [\text{cm}] \end{aligned}$$

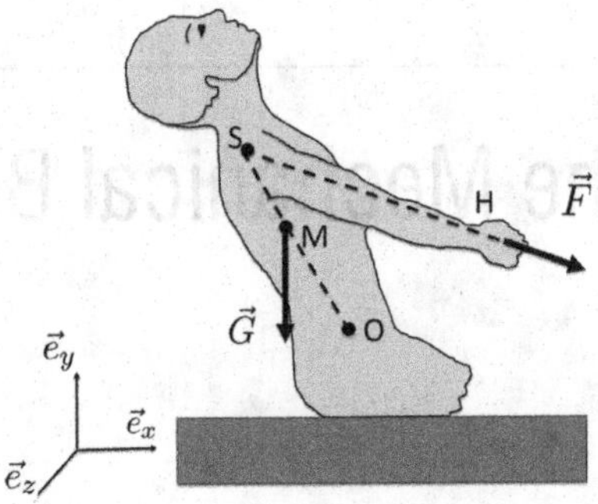

The forces are given by:

$$\vec{F} = F\left(\frac{4}{5}\vec{e}_x - \frac{3}{5}\vec{e}_y\right) \quad [\text{N}]$$

$$\vec{G} = -G\vec{e}_y \quad [\text{N}].$$

Determine the resulting moment of the forces $\vec{F}$ and $\vec{G}$ around the point O.

3.10 A sportsman is doing an exercise that involves holding a dumbbell of 10 [kg] behind his head. The gravitational constant can be assumed to be 10 [m s^{-2}]. With respect to a Cartesian basis, the following vectors are given:

$$\vec{x}_B - \vec{x}_S = -10\,\vec{e}_x + 25\,\vec{e}_y + 10\,\vec{e}_z \quad [\text{cm}]$$

$$\vec{x}_H - \vec{x}_E = -25\,\vec{e}_y \quad [\text{cm}].$$

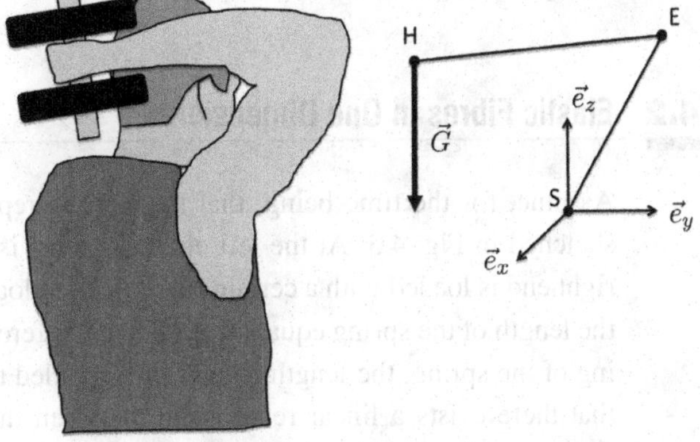

Calculate the force and moment vector that the shoulder has to apply to the arm to balance the weight. In this case we ignore the weight of the arm, so we ask for the 'extra' force and moment as a result of the dumbbell weight.

4 The Mechanical Behaviour of Fibres

4.1 Introduction

Fibres and fibre-like structures play an important role in the mechanical properties of biological tissues. Fibre-like structures may be found in almost all human tissues. A typical example is the fibre reinforcement in a heart valve, Fig. 4.1(a). Another illustration is found in the intervertebral disc as shown in Fig. 4.1(b).

Fibre reinforcement, largely inspired by nature, is frequently used in prosthesis design to optimize mechanical performance. An example is an aortic valve prosthesis, as shown in Fig. 4.2.

Fibres are long slender bodies and, essentially, have a tensile load-bearing capacity along the fibre direction only. The most simple approximation of the often complicated mechanical behaviour of fibres is to assume that they behave elastically. In that case, fibres have much in common with springs. The objective of this chapter is to formulate a relation between the force in the fibre and the change in length of a fibre. Such a relation is called a **constitutive model**.

4.2 Elastic Fibres in One Dimension

Assume, for the time being, that the fibre is represented by a simple spring as sketched in Fig. 4.3. At the left end the spring is attached to the wall, while the right end is loaded with a certain force F. If no load is applied to the spring (fibre) the length of the spring equals ℓ_0, called the reference or initial length. After loading of the spring, the length changes to ℓ, called the current length. It is assumed that there exists a linear relationship between the change in length of the fibre $\ell - \ell_0$ and the applied force:

$$F = a(\ell - \ell_0). \tag{4.1}$$

The constant a reflects the stiffness properties of the spring and can be identified by, for instance, attaching a known weight, i.e. a known force, to the spring in a vertical position and measuring the extension of the spring. However, formulating

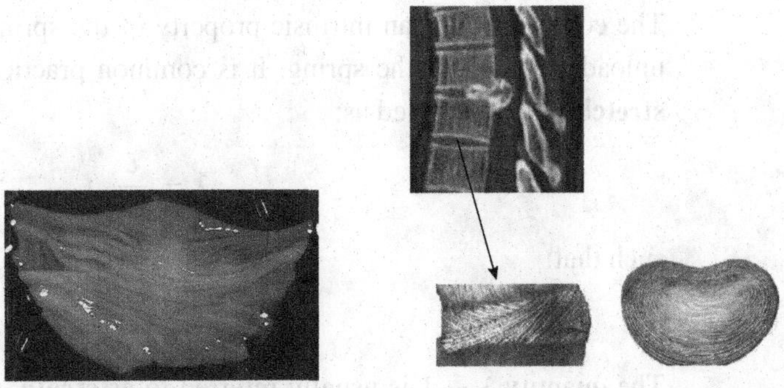

(a) Fibres in a heart valve. Courtesy A. Balguid

(b) Fibres in the intervertebral disc.

Figure 4.1

Examples of fibre structures.

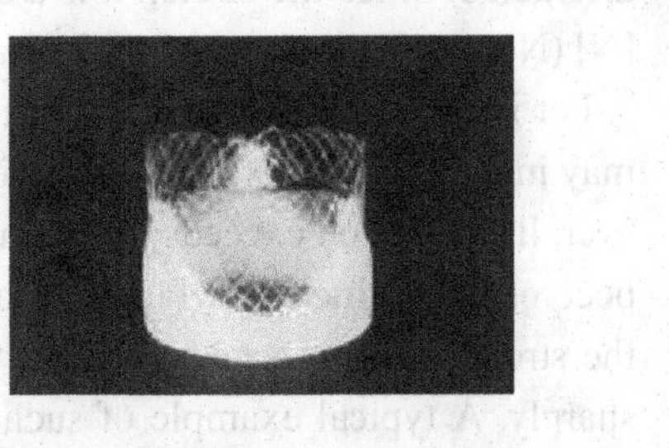

Figure 4.2

Fibre reinforcement in a stented valve prosthesis [5].

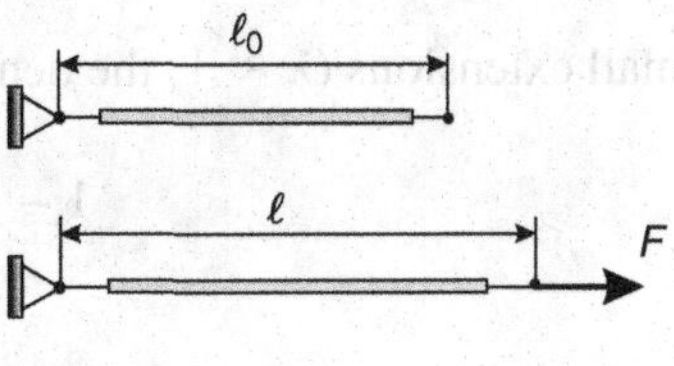

Figure 4.3

Unloaded and loaded spring.

the force–extension relation as in Eq. (4.1) is, although formally correct, not very convenient. If another spring was considered with the same intrinsic properties, but a different initial length, the coefficient a would change as well. Therefore the relation represented by Eq. (4.1) is scaled by the initial, unloaded, length:

$$F = c\frac{\ell - \ell_0}{\ell_0} = c\left(\frac{\ell}{\ell_0} - 1\right). \tag{4.2}$$

The coefficient c is an intrinsic property of the spring that is independent of the unloaded length of the spring. It is common practice to introduce the so-called **stretch** ratio λ, defined as:

$$\lambda = \frac{\ell}{\ell_0}, \tag{4.3}$$

such that:

$$F = c(\lambda - 1). \tag{4.4}$$

The quantity $\lambda - 1$ is usually referred to as **strain**, and it measures the amount of deformation of the spring. Without any elongation of the spring the stretch satisfies $\lambda = 1$, while the strain equals zero. The above force–strain relation represents linear **elastic** behaviour, as depicted in Fig. 4.4(a). If after stretching the fibre returns to its original length, the force equals zero and no energy has been dissipated. Since the stretch λ is a dimensionless quantity and the unit of force is [N] (Newton), the constant c also has unit [N].

For relatively small stretches λ, the actual behaviour of many biological fibres may indeed be approximated by a linear relation between force and stretch. However, if the stretch exceeds a certain value, the force–extension behaviour usually becomes **non-linear**. In fact, in many cases fibres have a **finite extensibility**. If the stretch λ approaches a critical value, say λ_c, the force in the fibre increases sharply. A typical example of such a behaviour is modelled using the following expression for the force–stretch relation:

$$F = \frac{c}{1 - \frac{\lambda - 1}{\lambda_c - 1}}(\lambda - 1). \tag{4.5}$$

For small extensions ($\lambda \approx 1$) the denominator in this expression satisfies

$$1 - \frac{\lambda - 1}{\lambda_c - 1} \approx 1, \tag{4.6}$$

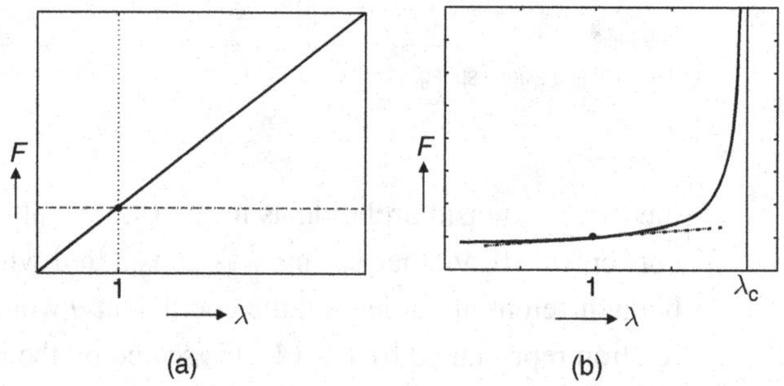

Figure 4.4

Force–stretch relation for linear and non-linear springs.

such that the behaviour is identical to the linear spring. If λ approaches the critical stretch λ_c, the force does indeed increase rapidly with increasing stretch. This is reflected in Fig. 4.4(b). The solid line represents the non-linear, finite extensibility curve according to Eq. (4.5), while the dashed line represents the linear behaviour according to Eq. (4.4).

Example 4.1 Collagen is one of the most abundant proteins in the human body, and is often organised in the form of fibres. Knowing the mechanical properties of collagen, and being able to model this aspect with mechanical models, is important to understand how living tissues respond to mechanical loads. Now suppose that the mechanical behaviour of collagen fibres can be described using the non-linear force–stretch relation given above, and that the critical strain of collagen fibres equals 10%. The local slope of a force–stretch curve represents the apparent stiffness of the material at its current stretch. We can calculate the apparent stiffness at 90% of the critical strain by determining the derivative of the force–stretch curve at that point:

$$\frac{dF}{d\lambda} = \frac{c}{1 - (\lambda - 1)/(\lambda_c - 1)} = \frac{c}{1 - (1.09 - 1)/(1.1 - 1)} = 10c \tag{4.7}$$

This shows that the stiffness of a non-linear fibrous material can vary tremendously, depending on the stretch that is applied to it.

4.3 A Simple One-Dimensional Model of a Skeletal Muscle

The fibres in the skeletal muscle have the unique ability to contract. On a microscopic scale a muscle is composed of contractile myofibrils, which are organised in the form of long fibres. Myofibrils in turn are composed of actin and myosin proteins. The interaction of filaments of these proteins through cross-bridges leads to the contractile properties of the muscle.

The arrangement of these filaments into a sarcomere unit is sketched in Fig. 4.5(a). Upon activation of the muscle, the actin and myosin filaments move with respect to each other, causing the sarcomere to shorten. Upon de-activation of the muscle the actin and myosin filaments return to their original positions owing to the elasticity of the surrounding tissue. In terms of modelling, the change of the sarcomere length implies that the initial, unloaded length of the muscle changes. Let ℓ_0 denote the length of the muscle in the **non-activated** state, while ℓ_c denotes the length of the muscle in the **activated** or **contracted** but **unloaded** state: see

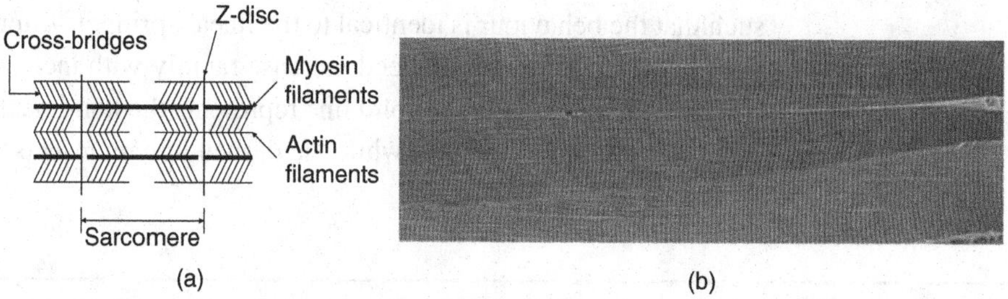

Figure 4.5

(a) Basic structure of a contractile element (sarcomere) of a muscle. (b) Cross section of a muscle; the vertical stripes correspond to Z-discs.

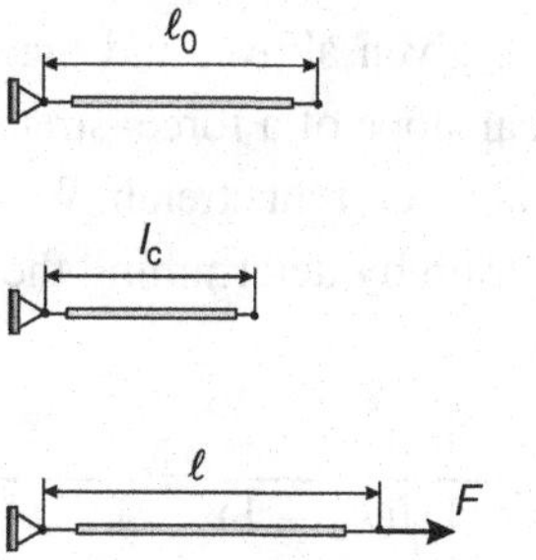

Figure 4.6

Different reference and current lengths of a muscle.

Fig. 4.6. Now, in contrast to a simple elastic spring, the contracted length ℓ_c serves as the reference length, such that the force in the muscle may be expressed as:

$$F = c\left(\frac{\ell}{\ell_c} - 1\right). \tag{4.8}$$

For this it is assumed that, despite the contraction, c does not change. The activated, but unloaded, length ℓ_c of the muscle may be expressed in terms of the non-activated length ℓ_0 using a so-called activation or contraction stretch λ_c defined as:

$$\lambda_c = \frac{\ell_c}{\ell_0}. \tag{4.9}$$

Typically $\lambda_c < 1$ since it represents a contractile action. For simplicity it is assumed that λ_c is known for different degrees of activation of the muscle. Using the activation stretch λ_c, the force–stretch relation for an activated muscle may be rewritten as:

$$F = c\left(\frac{\lambda}{\lambda_c} - 1\right) \quad \text{with } \lambda = \frac{\ell}{\ell_0}. \tag{4.10}$$

Effectively, this expression implies that if the muscle is activated, represented by a certain λ_c, and the muscle is not loaded, hence $F = 0$, the muscle will contract such that

$$\lambda = \lambda_c. \tag{4.11}$$

If, on the other hand, the muscle is activated and forced to have constant length ℓ_0, hence $\lambda = 1$, the force in the muscle equals:

$$F = c\left(\frac{1}{\lambda_c} - 1\right). \tag{4.12}$$

Rather complicated models have been developed to describe the activation of the muscle. A large group of models is based on experimental work by Hill [11], supplying a phenomenological description of the non-linear activated muscle. These models account for the effect of contraction velocity and for the difference in activated and passive state of the muscle. Later, microstructural models were developed, based on the sliding filament theories of Huxley [14]. These models can even account for the calcium activation of the muscle. However, a discussion of these models is beyond the scope of this book.

Example 4.2 In an in-vitro set-up, a skeletal muscle is loaded with an external force P (see Fig. 4.7). The length of the loaded but passive muscle at that time is $\ell = \ell_1$. During the experiment, the muscle is stimulated electrically, and because of this the length reduces to $\frac{3}{4}\ell_1$.

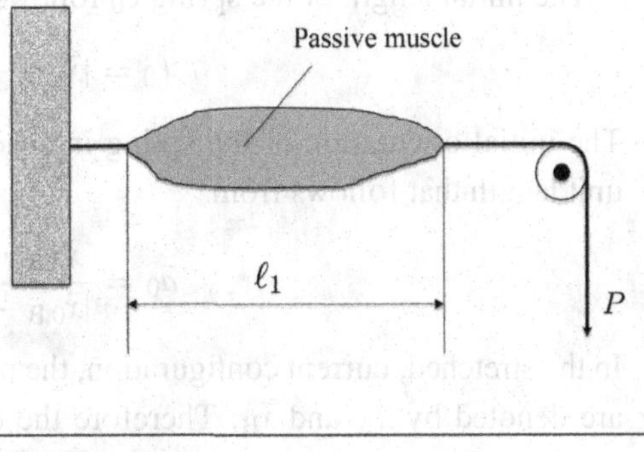

Figure 4.7

Set-up with passive muscle loaded with force P.

In a second experiment, the muscle is again electrically stimulated but the force on the muscle is increased to $2P$, resulting in a length $\ell = \frac{9}{10}\ell_1$. The force extension ratio for the activated muscle is given by:

$$F = c\left(\frac{\ell}{\ell_c} - 1\right).$$

Our goal is to determine the material constant c based on these experiments. For the first experiment, we can write for the activated muscle:

$$P = c\left(\frac{3}{4}\frac{\ell_1}{\ell_c} - 1\right).$$

With this result, we can write ℓ_c as a function of ℓ_1, c and P:

$$\ell_c = \frac{3}{4}\ell_1 \frac{c}{c+P}.$$

The second experiment tells us that:

$$2P = c\left(\frac{9}{10}\frac{\ell_1}{\ell_c} - 1\right) = c\left(\frac{\frac{9}{10}\ell_1}{\frac{3}{4}\ell_1 c/(c+P)} - 1\right).$$

The length ℓ_1 can be eliminated, and this eventually leads to:

$$c = 4P.$$

4.4 Elastic Fibres in Three Dimensions

The above one-dimensional force–extension relation can be generalized to a fibre or spring having an arbitrary position in three-dimensional space. The locations of the end points of the spring, say A and B, in the unstretched, initial configuration are denoted by $\vec{x}_{0,\mathrm{A}}$ and $\vec{x}_{0,\mathrm{B}}$, respectively, see Fig. 4.8.

The initial length of the spring ℓ_0 follows from

$$\ell_0 = |\vec{x}_{0,\mathrm{B}} - \vec{x}_{0,\mathrm{A}}|. \tag{4.13}$$

The initial orientation of the spring in space is denoted by the vector $\vec{a}_0$ having unit length that follows from

$$\vec{a}_0 = \frac{\vec{x}_{0,\mathrm{B}} - \vec{x}_{0,\mathrm{A}}}{|\vec{x}_{0,\mathrm{B}} - \vec{x}_{0,\mathrm{A}}|}. \tag{4.14}$$

In the stretched, current configuration, the positions of the end points of the spring are denoted by $\vec{x}_\mathrm{A}$ and $\vec{x}_\mathrm{B}$. Therefore the current length ℓ of the spring can be computed from

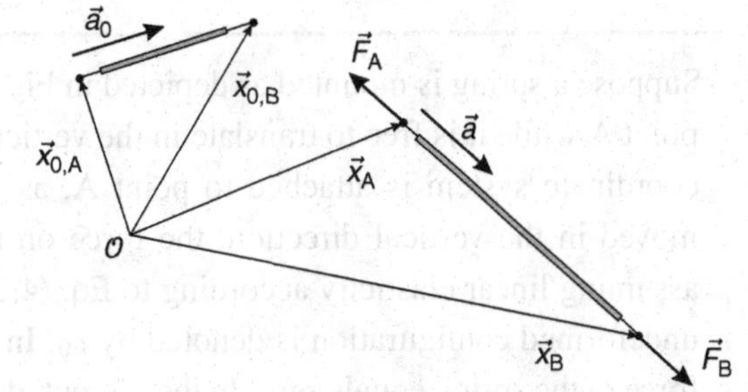

Figure 4.8

A spring in three-dimensional space.

$$\ell = |\vec{x}_B - \vec{x}_A|, \tag{4.15}$$

while the current orientation in space of the spring may be characterized by the vector $\vec{a}$ of unit length:

$$\vec{a} = \frac{\vec{x}_B - \vec{x}_A}{|\vec{x}_B - \vec{x}_A|}. \tag{4.16}$$

Clearly, in analogy with the scalar one-dimensional case, the stretch of the spring λ is defined as

$$\lambda = \frac{\ell}{\ell_0}. \tag{4.17}$$

To cause a stretch of the spring, a force must be applied to the end points A and B. The forces applied on the end points are vectors represented by $\vec{F}_A$ and $\vec{F}_B$. They have equal magnitude but opposite direction:

$$\vec{F}_B = -\vec{F}_A$$

and are parallel to the orientation vector $\vec{a}$:

$$\vec{F}_B = F\vec{a}. \tag{4.18}$$

The scalar F represents the magnitude of the force vector $\vec{F}_B$, and for linearly elastic springs this magnitude follows from the one-dimensional relation Eq. (4.4):

$$F = c(\lambda - 1). \tag{4.19}$$

Therefore, the force vector acting on point B is given by

$$\vec{F}_B = c(\lambda - 1)\vec{a}, \tag{4.20}$$

while the force vector acting on point A is given by

$$\vec{F}_A = -c(\lambda - 1)\vec{a}. \tag{4.21}$$

Example 4.3 Suppose a spring is mounted as depicted in Fig. 4.9. The spring is fixed in space at point A while it is free to translate in the vertical direction at point B. A Cartesian coordinate system is attached to point A, as depicted in Fig. 4.9. If point B is moved in the vertical direction, the force on the spring at point B is computed assuming linear elasticity according to Eq. (4.20). The length of the spring in the undeformed configuration is denoted by ℓ_0. In the undeformed configuration, the force in the spring equals zero. In the current, deformed configuration, the position of point B follows from

$$\vec{x}_{\mathrm{B}} = \ell_0 \vec{e}_x + y \vec{e}_y.$$

Point A is positioned at the origin:

$$\vec{x}_{\mathrm{A}} = \vec{0}.$$

The force vector acting on the spring at point B is written as

$$\vec{F}_{\mathrm{B}} = c(\lambda - 1)\vec{a},$$

while the current length ℓ is written as

$$\ell = |\vec{x}_{\mathrm{B}} - \vec{x}_{\mathrm{A}}| = \sqrt{\ell_0^2 + y^2}.$$

The stretch λ of the spring follows from

$$\lambda = \frac{\ell}{\ell_0} = \frac{\sqrt{\ell_0^2 + y^2}}{\ell_0}.$$

The orientation of the spring as represented by the unit vector $\vec{a}$ is given by

$$\vec{a} = \frac{\vec{x}_{\mathrm{B}} - \vec{x}_{\mathrm{A}}}{|\vec{x}_{\mathrm{B}} - \vec{x}_{\mathrm{A}}|} = \frac{\ell_0 \vec{e}_x + y \vec{e}_y}{\sqrt{\ell_0^2 + y^2}}.$$

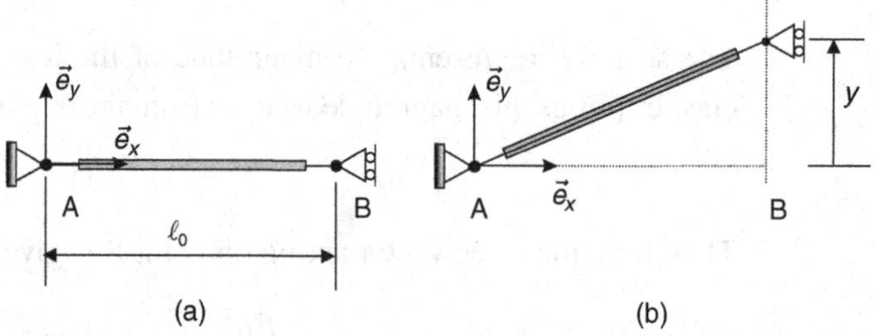

Figure 4.9

Linear elastic spring, fixed at A and free to translate in the $\vec{e}_y$ direction at point B. (a) Undeformed configuration; (b) deformed configuration.

So, in conclusion, the force vector $\vec{F}_B$ applied to the spring at point B equals

$$\vec{F}_B = c \left(\frac{\sqrt{\ell_0^2 + y^2}}{\ell_0} - 1 \right) \frac{\ell_0 \vec{e}_x + y\vec{e}_y}{\sqrt{\ell_0^2 + y^2}}.$$

Given an initial length $\ell_0 = 10$ [cm] and a spring constant $c = 0.5$ [N], the force components of $\vec{F}_B$ in the x- and y-direction (F_x and F_y, respectively) are represented in Fig. 4.10 as a function of the y-location of point B. Notice that both F_x and F_y are non-linear functions of y even though the spring is linearly elastic. The non-linearity stems from the fact that the stretch λ is a non-linear function of y. A non-linear response of this type is called **geometrically non-linear** since it originates from geometrical effects rather than an intrinsic non-linear physical response of the spring.

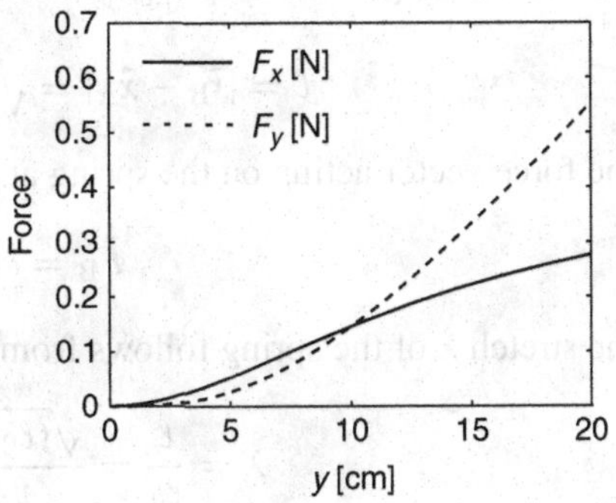

Figure 4.10

Forces in the horizontal and vertical directions exerted on the spring at point B displacing point B in the y-direction.

Example 4.4 Consider a spring, in the undeformed configuration, mounted at an angle α_0 with respect to the x-axis, as depicted in Fig. 4.11. The spring is fixed in space at point A while it is free to translate in the vertical direction at point B. A Cartesian coordinate system is located at point A. If point B is moved in the vertical direction, the force in the spring is computed assuming linear elasticity according to Eq. (4.20). The unstretched length of the spring is denoted by ℓ_0 such that the current position of point B follows from

$$\vec{x}_B = \ell_0 \cos(\alpha_0)\vec{e}_x + y\vec{e}_y.$$

Point A is positioned at the origin:

$$\vec{x}_A = \vec{0}.$$

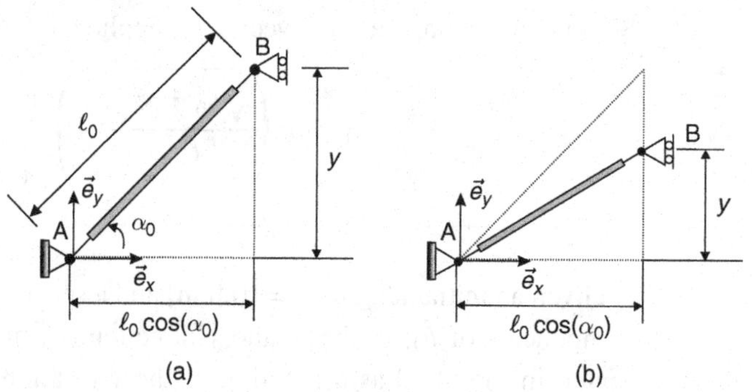

Figure 4.11

(a) Linearly elastic spring in the undeformed configuration oriented at an angle α_0 with respect to $\vec{e}_x$, fixed at A and free to translate in the $\vec{e}_y$-direction at point B. (b) The spring in the deformed configuration.

The current length of the spring satisfies

$$\ell = |\vec{x}_B - \vec{x}_A| = \sqrt{(\ell_0 \cos(\alpha_0))^2 + y^2}.$$

The force vector acting on the spring at point B is written as

$$\vec{F}_B = c(\lambda - 1)\vec{a}.$$

The stretch λ of the spring follows from

$$\lambda = \frac{\ell}{\ell_0} = \frac{\sqrt{(\ell_0 \cos(\alpha_0))^2 + y^2}}{\ell_0}.$$

The orientation of the spring as represented by the unit vector $\vec{a}$ is given by

$$\vec{a} = \frac{\vec{x}_B - \vec{x}_A}{|\vec{x}_B - \vec{x}_A|} = \frac{\ell_0 \cos(\alpha_0)\vec{e}_x + y\vec{e}_y}{\sqrt{(\ell_0 \cos(\alpha_0))^2 + y^2}}.$$

So, in conclusion, the force vector applied to the spring at point B, $\vec{F}_B$, equals

$$\vec{F}_B = c\left(\frac{\sqrt{(\ell_0 \cos(\alpha_0))^2 + y^2}}{\ell_0} - 1\right)\frac{\ell_0 \cos(\alpha_0)\vec{e}_x + y\vec{e}_y}{\sqrt{(\ell_0 \cos(\alpha_0))^2 + y^2}}.$$

Given an initial length $\ell_0 = 1$ [mm], a spring constant $c = 0.5$ [N] and an initial orientation $\alpha_0 = \pi/4$, the force components of $\vec{F}_B$ in the x- and y-directions (F_x and F_y, respectively) are represented in Fig. 4.12 as a function of the y-location of point B. Notice that, as in the previous example, both F_x and F_y are non-linear functions of y, even though the spring is linearly elastic. It is remarkable to see that with decreasing y, starting at the initial position $y_0 = \ell_0 \sin(\alpha_0)$, the magnitude of the force in the y-direction $|F_y|$ first increases and thereafter decreases. This demonstrates a so-called snap-through behaviour. If the translation of point B is

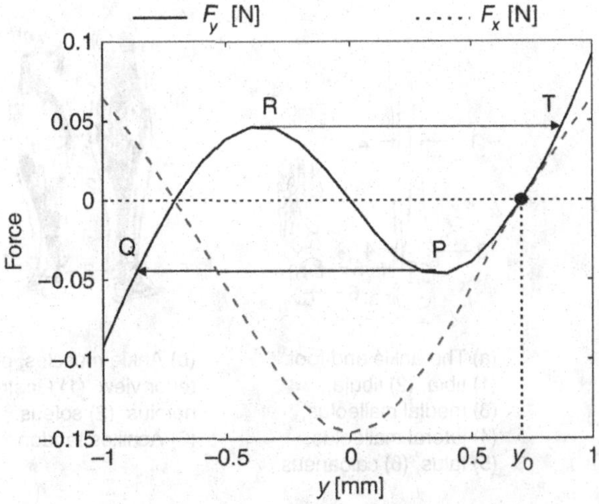

Figure 4.12

Forces in the horizontal and vertical direction exerted on the spring to displace point B in the y-direction. Snap-through behaviour.

driven by an externally applied force, and point P is reached in the force versus y-position curve, the y-coordinate of point B will suddenly move to point Q with equal force magnitude. During the reverse path, a snap through will occur from point R to point T.

Example 4.5 The Achilles tendon is attached to the rear of the ankle (the calcaneus) and is connected to two muscle groups: the gastrocnemius and the soleus, which, in turn, are connected to the tibia; see Fig. 4.13(a). A schematic drawing of this, using a lateral view, is given in Fig. 4.13(c). If the ankle is rotated with respect to the pivot point O, i.e. the origin of the coordinate system, the attachment point A is displaced, causing a length change of the muscle system. The position of the attachment point A is given by

$$\vec{x}_A = -R\sin(\alpha)\,\vec{e}_x - R\cos(\alpha)\,\vec{e}_y,$$

where R is the constant distance of the attachment point A to the pivot point. The angle α is defined in the clockwise direction. The muscles are connected to the tibia at point B, hence $\vec{x}_B = H\vec{e}_y$, with H the distance of point B to the pivot point. The positions in the undeformed, unstretched configuration of these points are

$$\vec{x}_{0,A} = -R\sin(\alpha_0)\,\vec{e}_x - R\cos(\alpha_0)\,\vec{e}_y$$

and

$$\vec{x}_{0,B} = H\vec{e}_y.$$

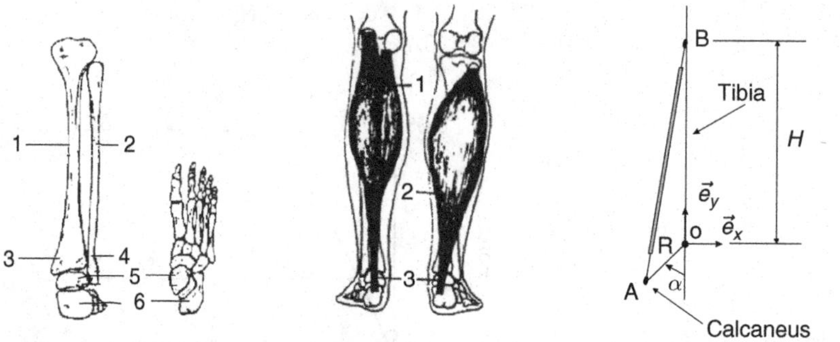

(a) The ankle and foot. (1) tibia, (2) fibula, (3) medial malleolus, (4) lateral malleolus, (5) talus, (6) calcaneus

(b) Ankle muscles, posterior view. (1) Gastrocnemius, (2) soleus, (3) Achilles tendon

(c) Location of muscles

Figure 4.13

Muscle attached to tibia and calcaneus.

Hence, the stretch of the muscles follows from

$$\lambda = \frac{|\vec{x}_{\rm A} - \vec{x}_{\rm B}|}{|\vec{x}_{0,{\rm A}} - \vec{x}_{0,{\rm B}}|} = \frac{\sqrt{(R\sin(\alpha))^2 + (R\cos(\alpha) + H)^2}}{\sqrt{(R\sin(\alpha_0))^2 + (R\cos(\alpha_0) + H)^2}},$$

while the orientation of the muscle is given by

$$\vec{a} = \frac{R\sin(\alpha)\vec{e}_x + (R\cos(\alpha) + H)\vec{e}_y}{\sqrt{(R\sin(\alpha))^2 + (R\cos(\alpha) + H)^2}}.$$

From these results, the force acting on the muscles at point B may be computed: $\vec{F}_{\rm B} = c(\lambda - 1)\vec{a}$. The force components in the x- and y-direction, scaled by the constant c, are depicted in Fig. 4.14 in the case $R = 5$ [cm], $H = 40$ [cm] and an initial angle $\alpha_0 = \pi/4$.

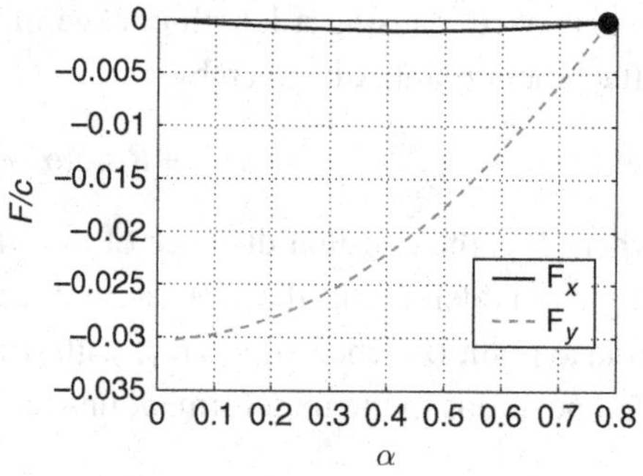

Figure 4.14

Force components in the muscle at B.

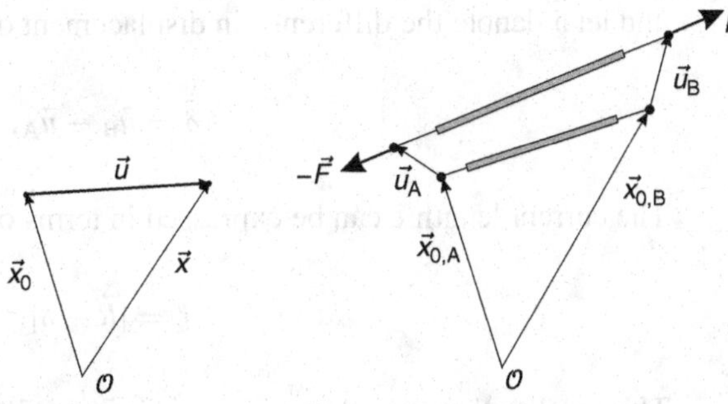

(a) Displacement vector $\vec{u}$ (b) Small extension of the spring

Figure 4.15

Displacement vector and spring extension.

4.5 Small Fibre Stretches

As illustrated by the above example, the finite displacements of the end points of a spring may cause a complicated non-linear response. In the limit of small displacements of the end points, a more manageable relation for the force in the spring results. To arrive at the force versus displacement expression, the concept of displacement first needs to be formalized.

If, as before, the reference position of a certain point is denoted by $\vec{x}_0$, and the current position by $\vec{x}$, then the displacement vector $\vec{u}$ of this point is defined as the difference between the current and initial position of the point; see Fig. 4.15(a):

$$\vec{u} = \vec{x} - \vec{x}_0. \tag{4.22}$$

If a spring, as in Fig. 4.15(b), is loaded by a force vector $\vec{F}$, the end points A and B will displace by the displacement vectors $\vec{u}_A$ and $\vec{u}_B$, respectively. The current position of the end points can be written as

$$\begin{aligned} \vec{x}_A &= \vec{x}_{0,A} + \vec{u}_A \\ \vec{x}_B &= \vec{x}_{0,B} + \vec{u}_B\ . \end{aligned} \tag{4.23}$$

The current length of the spring ℓ may also be expressed in terms of the end-point displacements:

$$\ell = |(\vec{x}_{0,B} + \vec{u}_B) \;-\; (\vec{x}_{0,A} + \vec{u}_A)|. \tag{4.24}$$

Define $\vec{R}$ as the end-to-end vector from point A to point B in the initial, unloaded, configuration, hence

$$\vec{R} = \vec{x}_{0,B} - \vec{x}_{0,A}, \tag{4.25}$$

and let $\vec{\delta}$ denote the difference in displacement of point B and point A:

$$\vec{\delta} = \vec{u}_{\mathrm{B}} - \vec{u}_{\mathrm{A}}. \tag{4.26}$$

The current length ℓ can be expressed in terms of the vectors $\vec{R}$ and $\vec{\delta}$:

$$\ell = |\vec{R} + \vec{\delta}|. \tag{4.27}$$

This implies that (recall that $\vec{a} \cdot \vec{a} = |\vec{a}||\vec{a}| = |\vec{a}|^2$)

$$\begin{aligned} \ell^2 &= (\vec{R} + \vec{\delta}) \cdot (\vec{R} + \vec{\delta}) \\ &= \vec{R} \cdot \vec{R} + 2\vec{R} \cdot \vec{\delta} + \vec{\delta} \cdot \vec{\delta}. \end{aligned} \tag{4.28}$$

If the end-point displacements are sufficiently small, the inner product $\vec{\delta} \cdot \vec{\delta}$ will be small compared with the other inner products in the above expression and may be neglected. Therefore, to a good approximation we have

$$\ell^2 \approx \vec{R} \cdot \vec{R} + 2\vec{R} \cdot \vec{\delta}. \tag{4.29}$$

Consequently, the current length may be written as

$$\ell = \sqrt{\vec{R} \cdot \vec{R} + 2\vec{R} \cdot \vec{\delta}}. \tag{4.30}$$

This may be rewritten in a more convenient form, bearing in mind that each of the inner products yields a scalar:

$$\begin{aligned} \ell &= \sqrt{\vec{R} \cdot \vec{R} \left(1 + 2\frac{\vec{R} \cdot \vec{\delta}}{\vec{R} \cdot \vec{R}}\right)} \\ &= \sqrt{\vec{R} \cdot \vec{R}} \sqrt{1 + 2\frac{\vec{R} \cdot \vec{\delta}}{\vec{R} \cdot \vec{R}}}. \end{aligned}$$

If α is a small number, then $\sqrt{1 + 2\alpha} \approx 1 + \alpha$; hence if $\vec{\delta}$ is sufficiently small the current length ℓ may be approximated by

$$\ell \approx \sqrt{\vec{R} \cdot \vec{R}} \left(1 + \frac{\vec{R} \cdot \vec{\delta}}{\vec{R} \cdot \vec{R}}\right). \tag{4.31}$$

Using

$$\ell_0 = |\vec{R}| = \sqrt{\vec{R} \cdot \vec{R}}, \tag{4.32}$$

this can be rewritten as:

$$\ell = \ell_0 + \frac{\vec{R} \cdot \vec{\delta}}{\ell_0}. \tag{4.33}$$

The stretch λ may now be expressed as:

$$\lambda = \frac{\ell}{\ell_0} = 1 + \frac{\vec{R} \cdot \vec{\delta}}{\ell_0^2}. \tag{4.34}$$

Recall that the force–stretch relation for a spring is given by:

$$\vec{F}_{\mathrm{B}} = c(\lambda - 1)\vec{a}, \tag{4.35}$$

where $\vec{a}$ denotes the vector of unit length pointing from point A to point B and where $\vec{F}_{\mathrm{B}}$ is the force acting on point B.

For sufficiently small displacements of the end points this vector may be approximated by:

$$\vec{a} \approx \frac{\vec{R}}{|\vec{R}|} = \frac{\vec{R}}{\ell_0} = \vec{a}_0. \tag{4.36}$$

Consequently, the following expression of the force vector $\vec{F}_{\mathrm{B}}$ is obtained:

$$\vec{F}_{\mathrm{B}} = c \underbrace{\frac{\vec{R} \cdot \vec{\delta}}{\ell_0^2}}_{\lambda - 1} \underbrace{\frac{\vec{R}}{\ell_0}}_{\vec{a}_0} = c \left(\vec{a}_0 \cdot \frac{\vec{\delta}}{\ell_0} \right) \vec{a}_0, \tag{4.37}$$

or written in terms of the end-point displacements

$$\vec{F}_{\mathrm{B}} = c \, \frac{\vec{a}_0 \cdot (\vec{u}_{\mathrm{B}} - \vec{u}_{\mathrm{A}})}{\ell_0} \vec{a}_0. \tag{4.38}$$

In this expression, three parts may be recognized: first, the stiffness of the spring, c; second the amount of stretch in the direction of the spring (see Fig. 4.16) also called the fibre strain ε:

$$\varepsilon = \lambda - 1 = \frac{\vec{a}_0 \cdot (\vec{u}_{\mathrm{B}} - \vec{u}_{\mathrm{A}})}{\ell_0}; \tag{4.39}$$

and third, the orientation of the spring, represented by the unit vector $\vec{a}_0$.

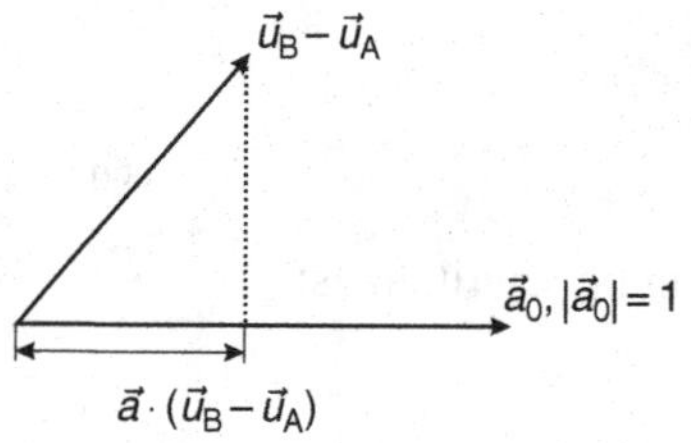

Figure 4.16

Measure of the elongation of the spring.

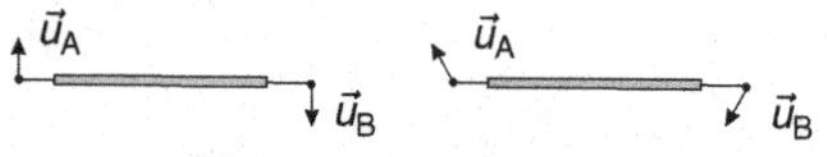

Figure 4.17

Examples of $(\vec{u}_B - \vec{u}_A) \cdot \vec{a}_0 = 0$ leading to $\vec{F} = \vec{0}$.

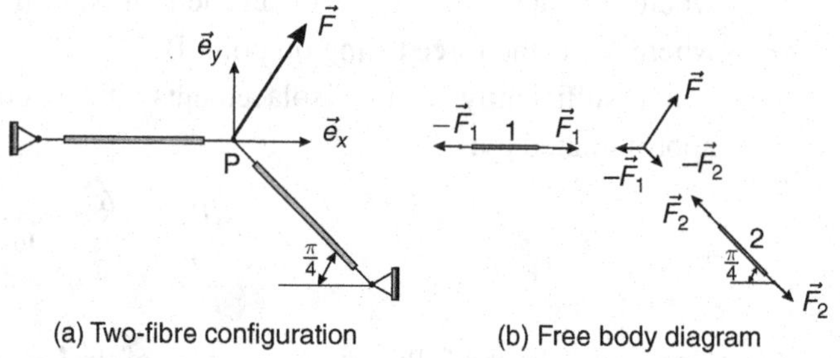

Figure 4.18

Two-fibre configuration and free body diagram of point P.

Example 4.6 An immediate consequence of the linearization process is that, if the displacement difference $\vec{u}_B - \vec{u}_A$ is normal to the fibre axis, i.e. $(\vec{u}_B - \vec{u}_A) \cdot \vec{a}_0 = 0$, the force in the fibre equals zero. Two examples of this are given in Fig. 4.17.

Example 4.7 Suppose that two fibres have been arranged according to Fig. 4.18(a). Both fibres have the same unloaded length ℓ_0 and elastic property c. At point P the two fibres have been connected. A force $\vec{F}$ is applied to this point. The free body diagram with respect to point P is given in Fig. 4.18(b). The orientation of each of the fibres in space is represented by the vector $\vec{a}_i$ $(i = 1, 2)$. These vectors have been chosen such that they point from the supports to point P, hence

$$\vec{a}_1 = \vec{e}_x, \quad \vec{a}_2 = \frac{1}{2}\sqrt{2}(-\vec{e}_x + \vec{e}_y).$$

If $\vec{u}_P = u_x\vec{e}_x + u_y\vec{e}_y$ denotes the displacement of point P, the force vectors $\vec{F}_1$ and $\vec{F}_2$ are described by

$$\vec{F}_1 = \frac{c}{\ell_0}(\vec{a}_1 \cdot \vec{u}_P)\vec{a}_1 = \frac{c}{\ell_0}u_x\vec{e}_x,$$

$$\vec{F}_2 = \frac{c}{\ell_0}(\vec{a}_2 \cdot \vec{u}_P)\vec{a}_2 = \frac{c}{\ell_0}\frac{1}{2}(-u_x + u_y)(-\vec{e}_x + \vec{e}_y).$$

The requirement of force equilibrium at point P implies

$$\vec{F} - \vec{F}_1 - \vec{F}_2 = \vec{0}.$$

With $\vec{F} = F_x\vec{e}_x + F_y\vec{e}_y$, it follows that in the x-direction:

$$F_x + \frac{c}{2\ell_0}(-3u_x + u_y) = 0,$$

while in the y-direction:

$$F_y - \frac{c}{2\ell_0}(-u_x + u_y) = 0.$$

This gives two equations from which the two unknowns (u_x and u_y) can be solved.

Exercises

4.1 The length change of a muscle, with respect to the length ℓ_0 in the relaxed state, can be written as $\delta = \ell - \ell_0$.

(a) Give an expression for the force F in the activated muscle as a function of c, δ, ℓ_0 and ℓ_c.

(b) What is the magnitude of the force F when $\delta = 0$?

4.2 Give a sketch of the force as a function of the active muscle length in the diagram below.

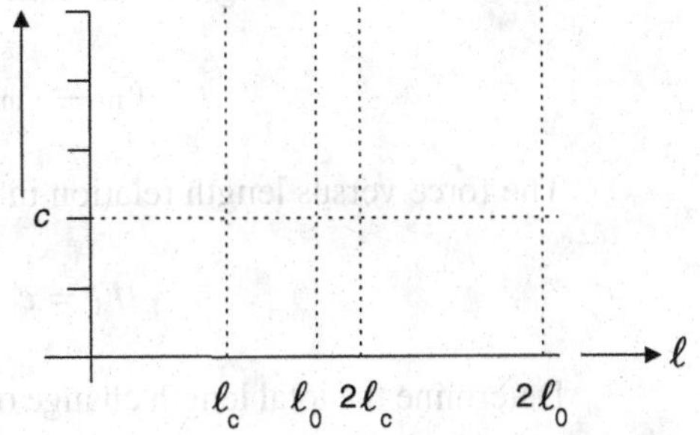

4.3 In reality, the force–length equation for an activated muscle in Eq. (4.10) is only valid in a very limited range of extension ratios. The force that a sarcomere (and thus a skeletal muscle) can exert has a maximum, as depicted in the figure below. Sarcomere lengths at several interesting points in the graph are depicted by Li, with $i = 1, \ldots, 5$.

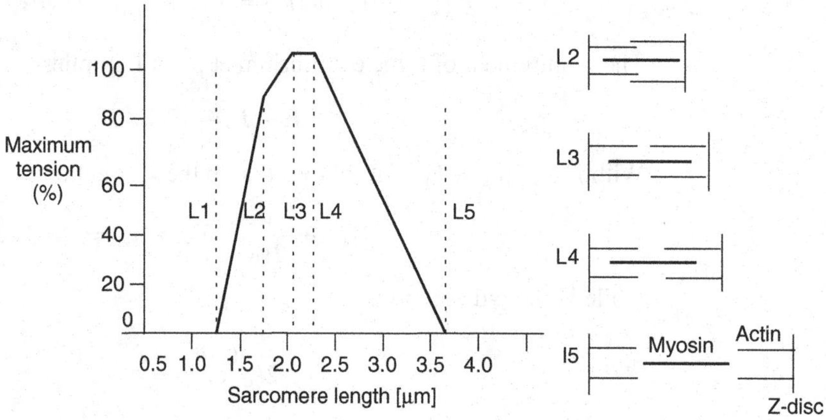

(a) Explain why the force decreases when the sarcomere length exceeds L4.

(b) The force versus length relation between L1 and L2 can be described exactly with Eq. (4.10). Determine the value of c in the case where the maximum force as given in the graph is 100 [N].

4.4 A muscle–tendon complex is loaded with a force F. The combination of the muscle–tendon complex can be schematically depicted as given in the figure.

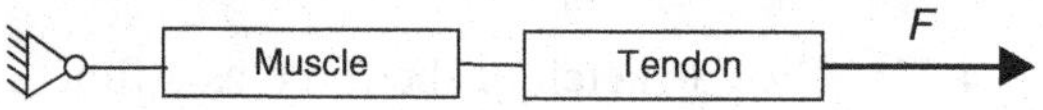

The force versus length relation in the muscle (m) can be written as

$$F_{\mathrm{m}} = c_{\mathrm{m}} \left(\frac{\ell_{\mathrm{m}}}{\ell_{\mathrm{m,c}}} - 1 \right).$$

The force versus length relation in the tendon (t) can be written as

$$F_{\mathrm{t}} = c_{\mathrm{t}} \left(\frac{\ell_{\mathrm{t}}}{\ell_{\mathrm{t,0}}} - 1 \right).$$

Determine the total length change of the muscle–tendon complex as a result of the load F with respect to the unloaded length.

4.5 A fibre is marked on two sides with small dots. The position of these dots is measured in an unloaded reference configuration. From these measurements, it appears that the positions are given as

$$\vec{x}_{0,\mathrm{A}} = -\vec{e}_x + 3\vec{e}_y \ \text{[cm]}$$

$$\vec{x}_{0,\mathrm{B}} = 2\vec{e}_x + 3\vec{e}_y \ \text{[cm]}.$$

The fibre moves, and in the current (deformed) configuration the positions of points A and B are measured again:

$$\vec{x}_\mathrm{A} = 4\vec{e}_x + 2\vec{e}_y \ \text{[cm]}$$

$$\vec{x}_\mathrm{B} = 8\vec{e}_x - \vec{e}_y \ \text{[cm]}.$$

The constant in the force versus extension relation is $c = 300$ [N]. Determine the force vectors $\vec{F}_\mathrm{A}$ and $\vec{F}_\mathrm{B}$ in the deformed configuration.

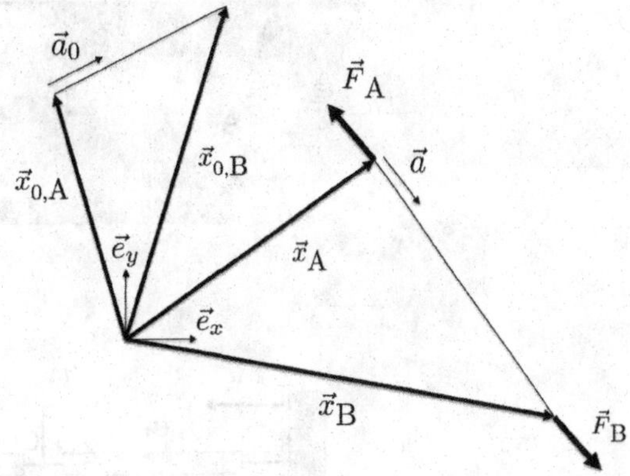

4.6 In the situation that is depicted in the drawing below, only the gluteus medius is active. During adduction of the bone, the femur rotates. The point of application of this muscle on the trochanter B follows a circular path. The point of application on the acetabulum A is given by the vector

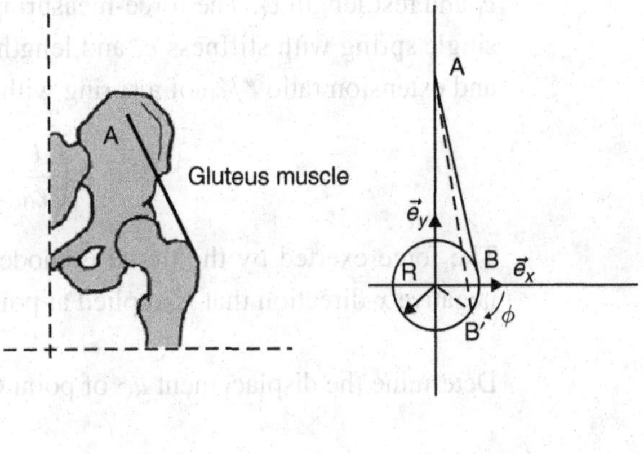

$\vec{x}_A = L\vec{e}_y$. When the adduction angle $\phi = 0$, the length of the muscle is $\ell = \ell_c$. Calculate the force $\vec{F}$ as a function of the adduction angle ϕ.

4.7 A problem in tissue engineering is compaction of the engineered construct. Cells in the construct apply forces to their environment, so a tissue that is grown in an incubator will shrink over time. Especially when biodegradable scaffolds are used, this can have a large impact, because the shape of the final construct is unclear and this influences the functionality. An experiment to measure the (very small) forces exerted by the cells in a construct is shown in the figure. A tissue is fixed to clamps A and B. The clamp B is attached with metal leaf springs to a rigid frame. The bending of the springs can be used to measure the force applied by the cells, and consequently the bending has to be incorporated in the analysis [21].

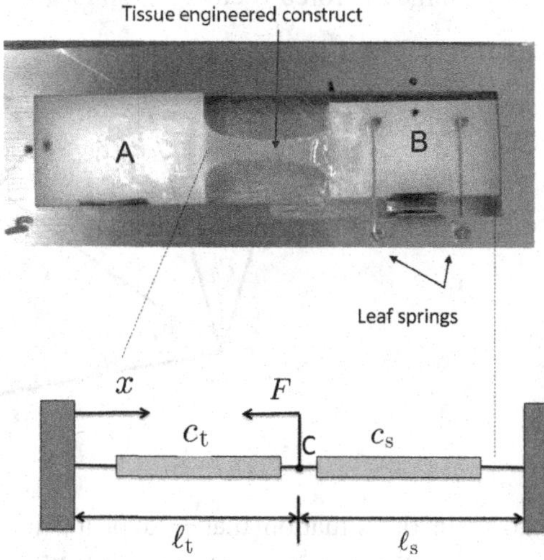

This is done by modelling the complete system with two springs, mutually connected at point C. The tissue is modelled as a spring with stiffness c_t and rest length ℓ_t. The force-measuring leaf springs are lumped into one single spring with stiffness c_s and length ℓ_s. The relation between force F and extension ratio ℓ/ℓ_0 of a spring with stiffness c is given by:

$$F = c\left(\frac{\ell}{\ell_0} - 1\right).$$

The force exerted by the tissue is modelled as a concentrated force F in negative x-direction that is applied at point C.

Determine the displacement u_C of point C as a function of $F, c_t, c_s, \ell_t, \ell_s$.

4.8 In a relaxed state, the end points of a muscle AB are given by the vectors:

$$\vec{x}_A^0 = \vec{e}_x + 3\vec{e}_y \quad [\text{cm}]$$
$$\vec{x}_B^0 = 5\vec{e}_x + 3\vec{e}_y \quad [\text{cm}].$$

After a movement, the end points in the current position of the muscle are given by:

$$\vec{x}_A = \vec{e}_x + 2\vec{e}_y \quad [\text{cm}]$$
$$\vec{x}_B = 5\vec{e}_x - \vec{e}_y \quad [\text{cm}].$$

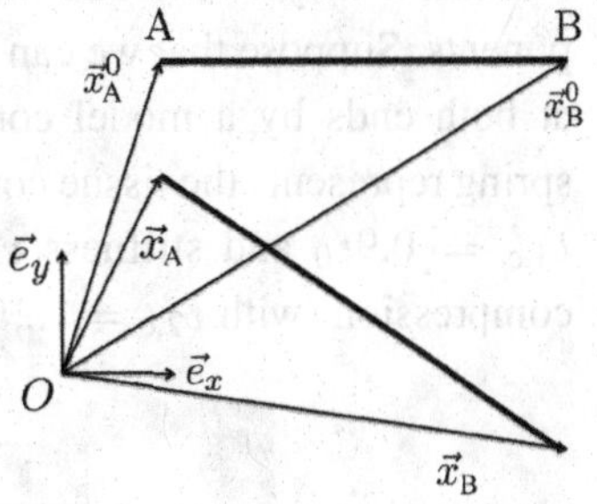

The relation between the passive force F and the current length ℓ of the muscle is given by:

$$F = c\left(\frac{\ell}{\ell_0} - 1\right)$$

with ℓ_0 the unloaded length of the passive muscle. The relation between the active force F and the current length ℓ of the muscle is given by:

$$F = c\left(\frac{\ell}{\ell_c} - 1\right)$$

with ℓ_c the unloaded length of the active muscle.

(a) Determine the force in the muscle in the current position in the case in which the muscle is passive.

(b) Determine the force in the muscle in the current position in the case in which the muscle is active.

4.9 A well-known phenomenon of soft biological tissues is that they contract when they are explanted from the human body, which is also known as prestretch. A researcher identifies the mechanical properties of a piece of material after it is explanted, and finds that it satisfies the following relationship:

$$F = C(\tilde{\lambda} - 1), \tag{E4.1}$$

where C represents the stiffness of the material and $\tilde{\lambda}$ the stretch with respect to the contracted length of the tissue l_c. A disadvantage is that this curve only represents the ex-vivo material behaviour, which may be different from the in-vivo behavior where the unloaded length equals l_0 (with $l_0 > l_c$). Determine the force–stretch behaviour for this material with respect to the in-vivo state (i.e. express the force F in terms of λ, where λ is defined with respect to l_0).

4.10 Cartilage consists of different components, some of which are naturally exposed to extension and others have a tendency to be exposed to compression. This results from the fact that the length l_c to which these components contract if they would be isolated from the cartilage differs between components. Suppose that we can represent a piece of cartilage that is clamped at both ends by a model consisting of two parallel springs, where one spring represents the tissue components that are exposed to extension (with $\ell_{1,c} = 0.9\ell_0$ and stiffness c_1), and the other represents the material in compression (with $\ell_{2,c} = 1.2\ell_0$ and stiffness c_2).

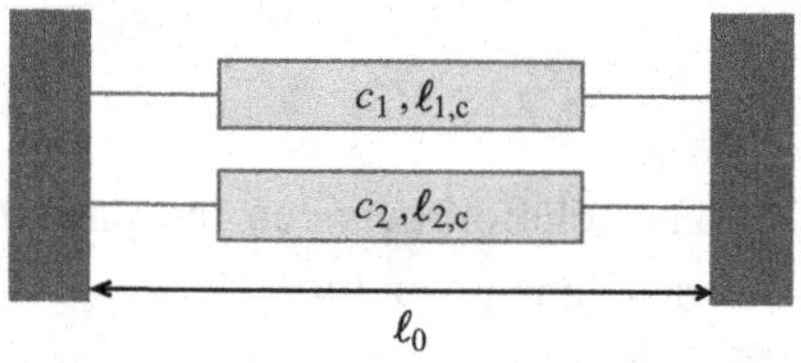

(a) Determine the total force that the tissue is exerting on its surroundings at its original length ℓ_0.

(b) When $c_1 = c_2$, will the tissue expand or contract when the clamps are removed?

5 Fibres: Time-Dependent Behaviour

5.1 Introduction

In the previous chapter on fibres, the material behaviour was constantly considered to be **elastic**, meaning that a unique relation exists between the extensional force and the deformation of the fibre. This implies that the force versus stretch curves for the loading and unloading path are identical. There is no history dependency, and all energy that is stored into the fibre during deformation is regained during the unloading phase. This also implies that the rate of loading or unloading does not affect the force–stretch curves. However, most biological materials do **not** behave elastically.

An example of a loading history and a typical response of a biological material is shown in Figs. 5.1(a) and (b). In Fig. 5.1(a), a deformation history is given that might be used in an experiment to mechanically characterize some material specimen. The specimen is rapidly stretched to a certain value, then the deformation is fixed and after a certain time restored to zero. After a short resting period, the stretch is applied again but to a higher value of the stretch. This deformation cycle is repeated several times. In this case the length change is prescribed and the associated force is measured. Figure 5.1(b) shows the result of such a measurement. When the length of the fibre is kept constant, the force decreases with time. This phenomenon is called **relaxation**. Conversely, if a constant load is applied, the length of the fibre will increase. This is called **creep**.

When the material is subjected to cyclic loading, the force versus stretch relation in the loading process is usually somewhat different from that in the unloading process. This is called **hysteresis** and is demonstrated in Fig. 5.2. The difference in the response paths during loading and unloading implies that energy is dissipated, usually in the form of heat, during the process. Most biological materials show more or less the behaviour given above, which is called **visco-elastic** behaviour. The present chapter discusses how to describe this behaviour mathematically. Pure viscous behaviour, as can be attributed to an ideal fluid, is considered first. The description will be extended to linear visco-elastic behaviour, followed by

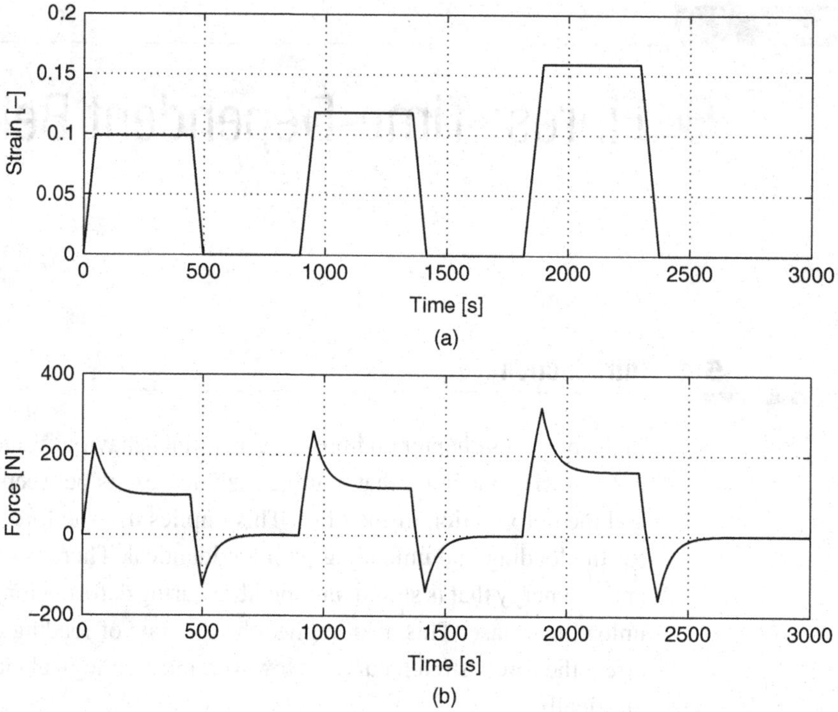

Figure 5.1

Loading history in a relaxation experiment. The deformation of the tissue specimen is prescribed.

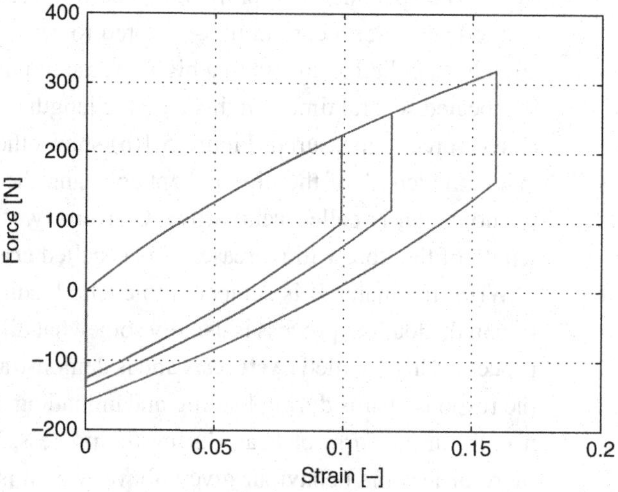

Figure 5.2

Force–strain curve for cyclic loading of a biological material.

a discussion on harmonic excitation, a technique that is often used to determine material properties of visco-elastic materials.

Example 5.1 The mechanical behaviour of axons (long, slender parts of nerve cells that conduct electrical signals to different parts of the human body) is often determined by applying a constant force or deformation to an axon. The displacement of certain locations within the axon can be represented by means of a kymograph, where the change in position of certain points on the axon is followed over time. In the experimentally determined kymograph in Fig. 5.3, which was constructed in response to a constant force application, it is clearly visible that axons demonstrate visco-elastic behaviour. The maximum force is applied almost immediately after the start of the experiment, while the deformation continues to increase with time. If one wants to model the mechanical behaviour of axons (an example is shown at the bottom of Fig. 5.3), then this visco-elastic behaviour should be accounted for, as this material response cannot be captured with purely elastic constitutive laws.

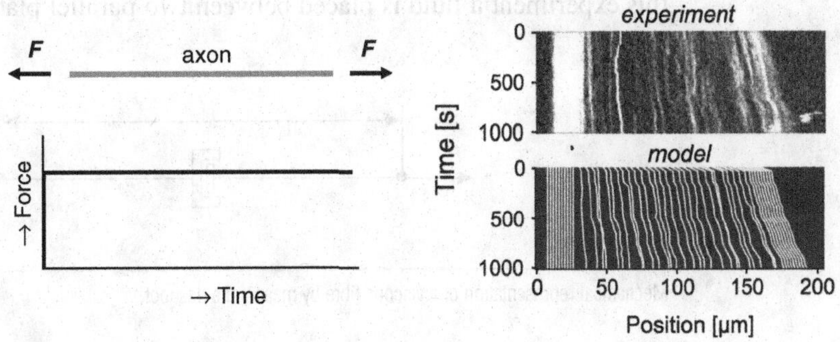

Figure 5.3

Example of a force-driven mechanical test on an axon. The deformation of the axon in the experiment is captured via a kymograph (image courtesy of Kyle Miller, Michigan State University), and can be predicted by a mechanical model (image courtesy of Rijk de Rooij, Stanford University).

5.2 Viscous Behaviour

It is not surprising that biological tissues do not behave purely elastically, since a large percentage of most tissues is water. The behaviour of water can be characterized as 'viscous'. Cast in a one-dimensional format, viscous behaviour during elongation (as in a fibre) may be represented by

$$F = c_\eta \frac{1}{\ell}\frac{d\ell}{dt}, \tag{5.1}$$

where c_η is the damping coefficient in [N s] and $d\ell/dt$ measures the rate of change of the length of the fibre. Mechanically, this force–elongational rate relation may be represented by a dashpot (see Fig. 5.4). Generally:

$$D = \frac{1}{\ell}\frac{d\ell}{dt} \tag{5.2}$$

is called the rate of deformation, which is related to the stretch parameter λ. Recall that

$$\lambda = \frac{\ell}{\ell_0}, \tag{5.3}$$

such that

$$D = \frac{1}{\ell}\frac{d\ell}{dt} = \frac{1}{\lambda}\frac{d\lambda}{dt}. \tag{5.4}$$

Ideally, a fluid stretching experiment should create a deformation pattern as visualized in Fig. 5.5(a). In practice this is impossible, because the fluid has to be spatially fixed and loaded, for instance via end plates, as depicted in Fig. 5.5(b). In this experiment a fluid is placed between two parallel plates at an initial distance

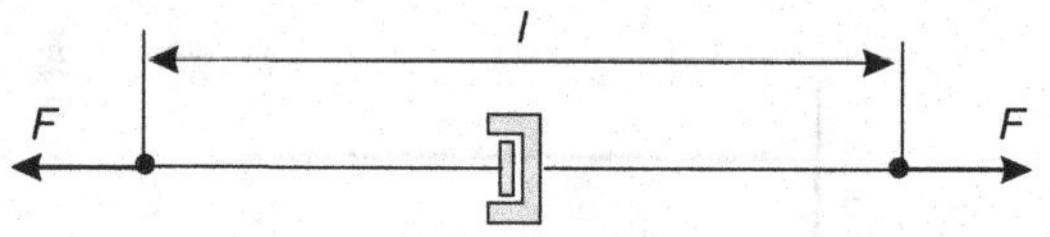

Figure 5.4

Mechanical representation of a viscous fibre by means of a dashpot.

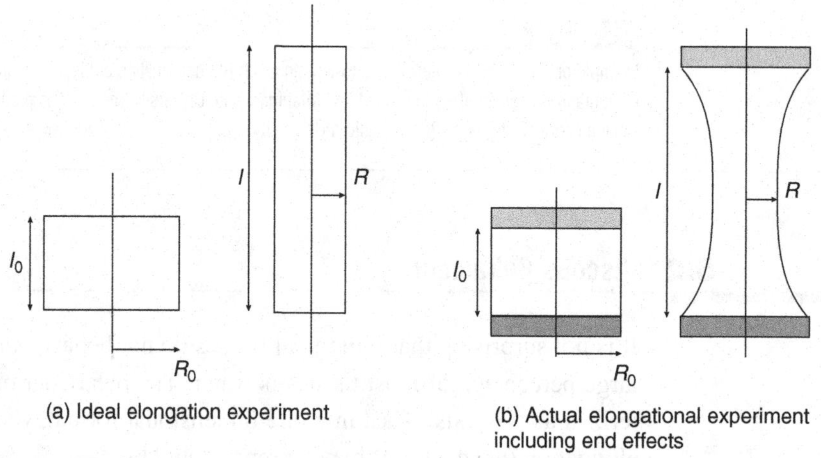

Figure 5.5

Schematic representation of an elongation experiment for fluids.

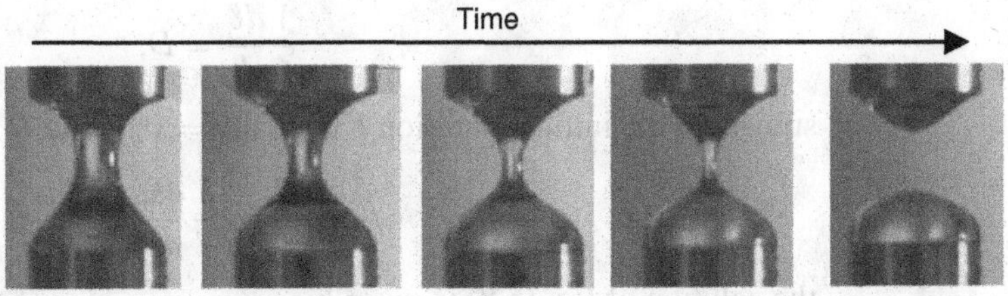

Figure 5.6

Example of uniaxial testing experiment with a fluid.

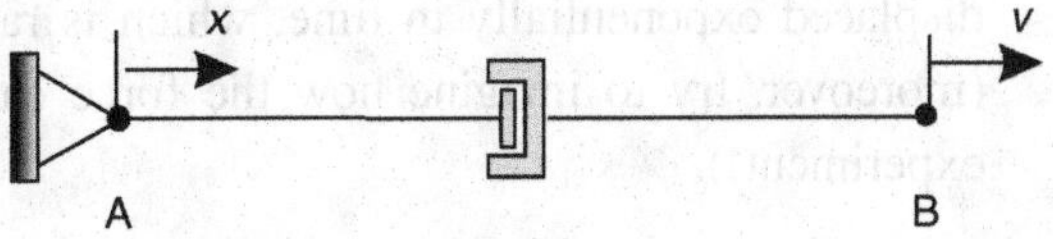

Figure 5.7

Point B is moved with a constant velocity v.

ℓ_0. Next, the end plates are displaced and the force on the end plates is measured. A typical example of a stretched filament is shown in Fig. 5.6. Although this seems to be a simple experiment, it is rather difficult to perform in practice. Figure 5.6 shows an experiment in which the fluid is a little extended initially, after which gravitational sag continues the filament stretching process. Ideally, a filament stretching experiment should be performed at a constant elongational rate. This is not trivial to achieve. For instance, let one end of the dashpot, say point A positioned at the origin (i.e. $x_A = 0$), be fixed in space, while the other end, point B, is displaced with a constant velocity v, as depicted in Fig. 5.7. In that case the position of point B is given by

$$x_B = \ell_0 + vt, \tag{5.5}$$

with ℓ_0 the initial length of the dashpot. The actual length ℓ at time t is given by:

$$\ell = x_B - x_A = \ell_0 + vt. \tag{5.6}$$

Hence, the elongational rate is given by

$$\dot{D} = \frac{1}{\ell}\frac{d\ell}{dt} = \frac{v}{\ell_0 + vt}. \tag{5.7}$$

This shows, that if one end is moved with a constant velocity, the elongational rate decays with increasing time t. Maintaining a constant elongational rate is possible if the velocity of point B is adjusted as a function of time. Indeed, a constant elongational rate $\dot{D}$ implies that the length ℓ must satisfy

$$\frac{1}{\ell}\frac{d\ell}{dt} = D, \tag{5.8}$$

subject to the initial condition $\ell = \ell_0$ at $t = 0$, while D is constant. Since

$$\frac{d\ln(\ell)}{dt} = \frac{1}{\ell}\frac{d\ell}{dt} = D, \tag{5.9}$$

the solution of Eq. (5.8) is given by

$$\ell = \ell_0\, e^{Dt}. \tag{5.10}$$

This means that to maintain a constant elongational rate, point B has to be displaced exponentially in time, which is rather difficult to achieve in practice (moreover, try to imagine how the force can be measured during this type of experiment!).

5.2.1 Small Stretches: Linearization

If u_B and u_A denote the end-point displacements, introduce:

$$\Delta\ell = u_B - u_A. \tag{5.11}$$

The stretch λ may be expressed as

$$\lambda = \frac{\ell_0 + \Delta\ell}{\ell_0}. \tag{5.12}$$

Introducing the strain ε as

$$\varepsilon = \frac{\Delta\ell}{\ell_0}, \tag{5.13}$$

the stretch is written as

$$\lambda = 1 + \varepsilon. \tag{5.14}$$

For sufficiently small strain levels, i.e. $|\varepsilon| \ll 1$, and using the notation $\dot{\varepsilon} = d\varepsilon/dt$, one can write:

$$D = \frac{1}{\lambda}\frac{d\lambda}{dt} = \frac{1}{1+\varepsilon}\dot{\varepsilon} \approx (1-\varepsilon)\dot{\varepsilon} \approx \dot{\varepsilon}. \tag{5.15}$$

Consequently, if $|\varepsilon| \ll 1$ then .

$$D = \frac{1}{\ell}\frac{d\ell}{dt} \approx \dot{\varepsilon} = \frac{1}{\ell_0}\frac{d\ell}{dt}, \tag{5.16}$$

such that Eq. (5.1) reduces to

$$F = c_\eta \dot{\varepsilon}. \tag{5.17}$$

Remark In the literature, the symbol $\dot{\varepsilon}$ is frequently used to denote the elongational rate for large filament stretches instead of D.

Example 5.2 Consider again the experiment depicted in Fig. 5.7, where point B is displaced with a constant velocity, such that $x_B = \ell_0 + vt$. We would like to calculate the strain rate $\dot{\varepsilon}$ for this experiment. In order to do that, we first determine the strain:

$$\varepsilon = \frac{\Delta \ell}{\ell_0} = \frac{vt}{\ell_0}. \tag{5.18}$$

From this, the strain rate can be determined:

$$\dot{\varepsilon} = \frac{v}{\ell_0}. \tag{5.19}$$

Notice that both v and ℓ_0 are constant, which means that the strain rate is equal during the complete experiment, as opposed to the elongation rate D. For small strains (t is small), however, the strain rate $\dot{\varepsilon}$ and elongation rate D are similar.

5.3 Linear Visco-Elastic Behaviour

5.3.1 Superposition and Proportionality

Biological tissues usually demonstrate a combined viscous–elastic behaviour as described in the introduction. In the present section, we assume geometrically and physically linear behaviour of the material. This means that the theory leads to linear relations, expressing the force in terms of the deformation(-rate), and that the constitutive description satisfies two conditions:

- **Superposition** The response to combined loading histories can be described as the summation of the responses to the individual loading histories.
- **Proportionality** When the strain is multiplied by some factor the force is multiplied by the same factor (in fact, proportionality is a consequence of superposition).

To study the effect of these conditions, a unit-step function for the force is introduced, defined as H(t) (Heaviside function):

$$\mathrm{H}(t) = \begin{cases} 0 & \text{if } t < 0 \\ 1 & \text{if } t \geq 0. \end{cases} \tag{5.20}$$

Assume that a unit-step in the force $F(t) = \mathrm{H}(t)$ is applied to a linear visco-elastic material. The response $\varepsilon(t)$ might have an evolution as given in Fig. 5.8. This response denoted by $J(t)$ is called the **creep compliance** or **creep function**. Proportionality means that increasing the force with some factor F_0 leads to a proportional increase in the strain:

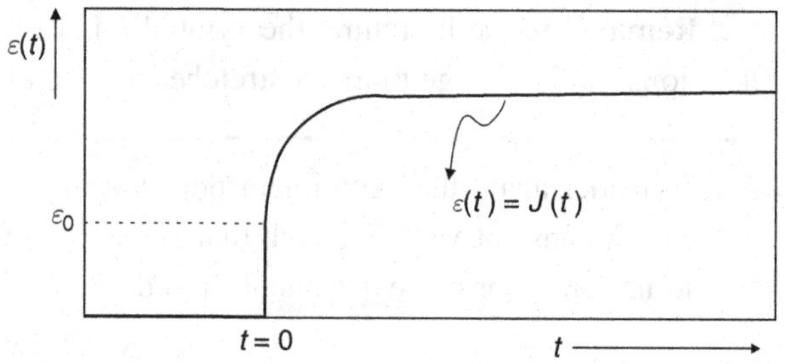

Figure 5.8

Typical example of the strain response after a unit step in the force.

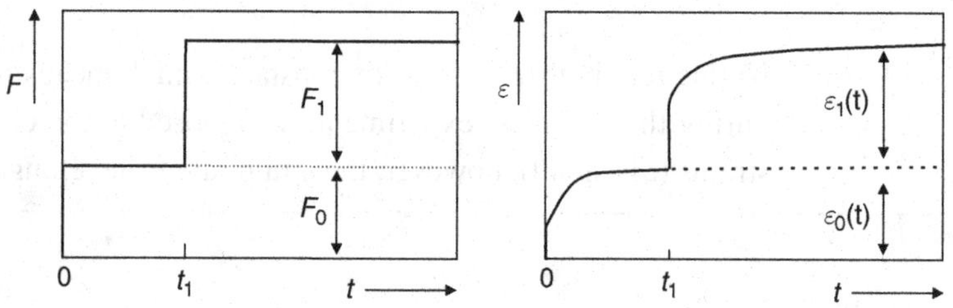

Figure 5.9

Superposition of responses for a linear visco-elastic material.

$$F(t) = \mathrm{H}(t)F_0 \quad \rightarrow \quad \varepsilon(t) = J(t)F_0. \tag{5.21}$$

Superposition implies that applying a load step F_0 at $t = 0$ with response:

$$\varepsilon_0(t) = J(t)F_0, \tag{5.22}$$

followed by a load step F_1 at $t = t_1$ with individual response:

$$\varepsilon_1(t) = J(t - t_1)F_1, \tag{5.23}$$

leads to a total response, which is a summation of the two:

$$\varepsilon(t) = \varepsilon_0(t) + \varepsilon_1(t) = J(t)F_0 + J(t - t_1)F_1. \tag{5.24}$$

This is graphically shown in Fig. 5.9.

Example 5.3 In a laboratory, a tendon with a length of 50 [mm] is mechanically tested. Initially the tendon is in a stationary state (the force in the tendon equals zero and the tendon has a constant length) which is used as the reference state. The mechanical behaviour of the tendon can be described with a linear visco-elastic model, characterized by the creep function $J(t)$ according to:

$$J(t) = J_0(2 - \mathrm{e}^{-t/\tau}),$$

with $\tau = 250$ [s], $J_0 = 0.01$ [N^{-1}] and t expressed in [s]. In the set-up, the tendon is loaded during 500 [s] with a force of 50 [N] and after that unloaded. We want to calculate the strain $\varepsilon(t)$ as a function of time.

For this we can use the superposition and proportionality principle. The load history can be considered to be a superposition of a step load of 50 [N] at $t = 0$ [s] and superposed to that a step load of -50 [N] at $t = 500$ [s]. The response to the first step can be described as:

$$\varepsilon_a(t) = F_0 \, J(t) = 0.5 \left(2 - \mathrm{e}^{-t/250}\right) \qquad \text{for } t \geq 0 \quad [\mathrm{s}].$$

The response to the second step can be described as:

$$\varepsilon_b(t) = 0 \qquad \text{for } t < 500 \quad [\mathrm{s}]$$
$$\varepsilon_b(t) = -F_0 \, J(t - 500) = -0.5 \left(2 - \mathrm{e}^{(t-500)/250}\right) \qquad \text{for } t \geq 500 \quad [\mathrm{s}].$$

The total response for this experiment is:

$$\varepsilon(t) = \varepsilon_a(t) + \varepsilon_b(t) = 0.5 \left(2 - \mathrm{e}^{-t/250}\right) \qquad 0 \leq t < 500 \quad [\mathrm{s}]$$
$$\varepsilon(t) = \varepsilon_a(t) + \varepsilon_b(t) = 0.5 \left(\mathrm{e}^{(t-500)/250)} - \mathrm{e}^{-t/250}\right) \qquad t \geq 500 \quad [\mathrm{s}].$$

Example 5.4 Suppose we do a test with a material and apply a step in the strain ε (for example in a uniaxial stretch test) of 0.01, and we measure the force as a function of time. By dividing the measured force $F(t)$ by the strain, the relaxation function $G(t)$ is found. The result is given in the table below.

Time [s]	$F(t)$ [N]	$G(t)=F(t)/\varepsilon$ [N]
10	2.96	296
20	2.92	292
40	2.85	285
60	2.78	278
80	2.72	272
100	2.67	267
120	2.62	262
140	2.57	257
160	2.52	252
180	2.49	249
200	2.45	245
220	2.42	242
240	2.38	238

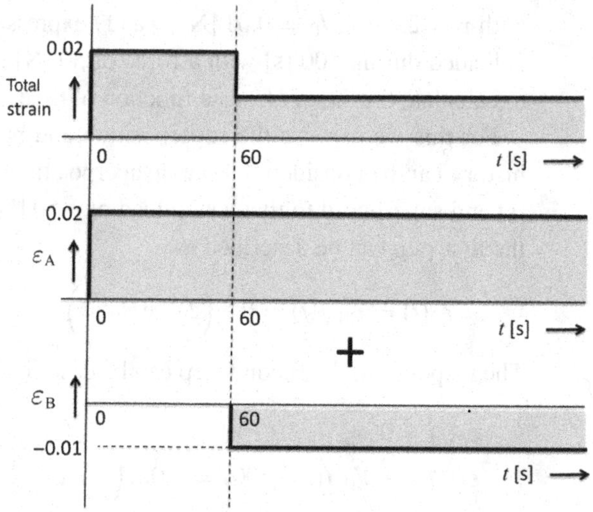

Figure 5.10

Loading history in a relaxation experiment, and the decomposed strains ε_A and ε_B as a function of time.

In a second test, the loading history shown in Fig. 5.10 is applied. At $t = 0$ [s] a step strain of 0.02 is applied. At $t = 60$ [s] the strain is reduced to a level of $\varepsilon(t) = 0.01$. We want to determine the force F at $t = 100$ [s] and at $t = 240$ [s]. For this we use the superposition principle and decompose the strain history by means of a summation of the step strains $\varepsilon_A(t)$ and $\varepsilon_B(t)$ as shown in Fig. 5.10 and use the relaxation function $G(t)$ to derive the force.

To determine the force at $t = 100$ [s] we wil have to add the contributions of the steps ε_A and ε_B at the time since application, so:

$$F(100) = 0.02\,G(100) - 0.01\,G(100 - 60) = 5.34 - 2.85 = 2.49 \quad \text{[N]}.$$

The same can be done for the force at $t = 240$ [s]:

$$F(240) = 0.02\,G(240) - 0.01\,G(240 - 60) = 4.76 - 2.49 = 2.27 \quad \text{[N]}.$$

5.3.2 Generalization for an Arbitrary Load History

Both the superposition and proportionality principles can be used to derive a more general constitutive equation for linear visco-elastic materials. Assume we have an arbitrary excitation as sketched in Fig. 5.11. This excitation can be considered to be built up by an infinite number of small steps in the force.

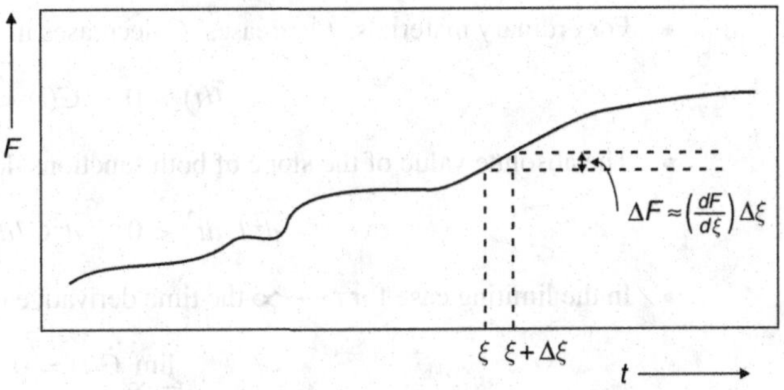

Figure 5.11

An arbitrary force history in a creep test.

The increase ΔF of the force F between time steps $t = \xi$ and $t = \xi + \Delta\xi$ is equal to

$$\Delta F \approx \left(\frac{dF}{d\xi}\right)\Delta\xi = \dot{F}(\xi)\Delta\xi. \tag{5.25}$$

The response at time t as a result of this step at time ξ is given by

$$\Delta\varepsilon(t) = \dot{F}(\xi)\Delta\xi J(t-\xi). \tag{5.26}$$

The time-dependent force $F(t)$ as visualized in Fig. 5.11 can be considered as a composition of sequential small steps. By using the superposition principle, we are allowed to add the responses on all these steps in the force (for each ξ). This will lead to the following integral expression, with all intervals $\Delta\xi$ taken as infinitesimally small:

$$\varepsilon(t) = \int_{\xi=-\infty}^{t} J(t-\xi)\dot{F}(\xi)d\xi. \tag{5.27}$$

This integral was derived first by Boltzmann in 1876.

In the creep experiment, the load is prescribed and the resulting strain is measured. Often, the experimental set-up is designed to prescribe the strain and to measure the associated, required force. If the strain is applied as a step, this is called a relaxation experiment, because after a certain initial increase the force will gradually decrease in time. The same strategy as used to derive Eq. (5.27) can be pursued for an imposed strain history, leading to

$$F(t) = \int_{\xi=-\infty}^{t} G(t-\xi)\dot{\varepsilon}(\xi)d\xi, \tag{5.28}$$

with $G(t)$ the relaxation function.

The functions J and G have some important physical properties:

- For ordinary materials, J increases, G decreases in time (for $t > 0$):

$$\dot{J}(t) > 0 \quad \dot{G}(t) < 0 \tag{5.29}$$

- The absolute value of the slope of both functions decreases:

$$d^2J/dt^2 < 0 \quad d^2G/dt^2 > 0 \tag{5.30}$$

- In the limiting case for $t \to \infty$ the time derivative of $G(t)$ will approach zero:

$$\lim_{t\to\infty} \dot{G}(t) = 0 \tag{5.31}$$

- In the limiting case for $t \to \infty$ the time derivative of $J(t)$ will be greater than or equal to zero:

$$\lim_{t\to\infty} \dot{J}(t) \geq 0 \tag{5.32}$$

It will be clear that there must be a relation between $G(t)$ and $J(t)$, because both functions describe the behaviour of the same material. This relationship can be determined by using Laplace transformation (for a summary of definitions and properties of Laplace transformations, see Appendix 5.5). The Laplace transformation $\hat{x}(s)$ of the time function $x(t)$ is defined as

$$\hat{x}(s) = \int_0^\infty x(t)e^{-st}dt. \tag{5.33}$$

Assuming that the creep and relaxation functions are zero for $t < 0$, the Laplace transforms of Eqs. (5.27) and (5.28) are

$$\hat{\varepsilon}(s) = \hat{J}(s)s\hat{F}(s) \tag{5.34}$$

$$\hat{F}(s) = \hat{G}(s)s\hat{\varepsilon}(s). \tag{5.35}$$

From these equations it is easy to derive that

$$\hat{G}(s)\hat{J}(s) = \frac{1}{s^2}. \tag{5.36}$$

Back transformation leads to

$$\int_{\xi=0}^{t} J(t-\xi)G(\xi)d\xi = t. \tag{5.37}$$

When $G(t)$ or $J(t)$ is known, the material behaviour, at least for one-dimensional tests, is specified.

There are several reasons that a full explicit description of J and G is very difficult. It is not possible to enforce infinitely fast steps in the load, so it is not possible to realize a perfect step. Consequently, it is almost impossible to determine $G(t)$ and $J(t)$ for very small values of t. At the other side of the time domain, the problem is encountered that it is not possible to carry out measurements for an unlimited (infinite) period of time.

Both functions $J(t)$ and $G(t)$ are continuous with respect to time. Often these functions are approximated by discrete spectra. Examples of such spectra are

$$J(t) = J_0 + \sum_{k=1}^{N} f_k(1 - e^{-t/\tau_k}) + t/\eta, \tag{5.38}$$

with J_0, f_k, τ_k $(k = 1, \ldots, N)$ and η material parameters (constants) or

$$G(t) = G_\infty + \sum_{j=1}^{M} g_j\, e^{-t/\tau_j}, \tag{5.39}$$

with G_∞, g_j, τ_j $(j = 1, \ldots, M)$ material constants. These discrete descriptions can be derived from spring–dashpot models, which will be the subject of the next section.

Example 5.5 Suppose that for a material a model is available to describe the relaxation behaviour. The relaxation $G(t)$ is given by:

$$G(t) = G_0 e^{-t/\tau}.$$

At $t = 0^+$ [s] it is derived that $G(0^+) = 2$ [N]. Note: In fact the relaxation function is a discontinuous function at $t = 0$ [s]. We use 0^+ to emphasize that we are infinitesimally close to $t = 0$ [s], but after the step in the strain is applied. It follows that $G_0 = 2$ [N].

At $t = 10^4$ [s] the value of the relaxation function $G(10^4) = 1$ [N]. Thus:

$$G(10^4) = 2e^{-10^4/\tau} = 1$$

leading to: $\tau = 10^4/0.693 = 14{,}430$ [s].

We wish to use the result of the relaxation test to predict the creep behaviour and determine the strain at $t = 1000$ [s] after a load step of 0.1 [N]. For this, we have to determine the creep function $J(t)$ from the relaxation function $G(t)$. Using the Laplace transforms of both functions and Eq. (5.36):

$$\hat{J}(s)\hat{G}(s) = \frac{1}{s^2}.$$

With:

$$\hat{G}(s) = \frac{G_0}{s + 1/\tau},$$

this leads to:

$$\hat{F}(s) = \frac{s + 1/\tau}{G_0 s^2} = \frac{1}{G_0 s} + \frac{1}{G_0 \tau s^2}.$$

Transforming this back to the time domain yields:

$$J(t) = \frac{1}{G_0} + \frac{t}{G_0\tau}.$$

The strain at $t = 1000$ [s] after a step load of 0.1 [N] can now be derived as:

$$\varepsilon(1000) = 0.1J(1000) = 0.1\left(\frac{1}{2} + \frac{1000}{2 \times 14{,}430}\right) = 0.0535.$$

5.3.3 Visco-Elastic Models Based on Springs and Dashpots: Maxwell Model

An alternative way of describing linear visco-elastic materials is by assembling a model using the elastic and viscous components as discussed before. Two examples are given, while only small stretches are considered. In that case, the constitutive models for the elastic spring and viscous dashpot are given by

$$F = c\,\varepsilon, \qquad F = c_\eta \dot{\varepsilon}. \tag{5.40}$$

In the Maxwell model, for the set-up of Fig. 5.12(a), the strain ε is additionally composed of the strain in the spring (ε_s) and the strain in the dashpot (ε_d):

$$\varepsilon = \varepsilon_s + \varepsilon_d, \tag{5.41}$$

implying that

$$\dot{\varepsilon} = \dot{\varepsilon}_s + \dot{\varepsilon}_d. \tag{5.42}$$

The forces in both the spring and the dashpot must be the same, therefore, based on Eq. (5.40), the strain rates in the spring and dashpot satisfy

$$\dot{\varepsilon}_s = \frac{1}{c}\dot{F} \tag{5.43}$$

and

$$\dot{\varepsilon}_d = \frac{F}{c_\eta}. \tag{5.44}$$

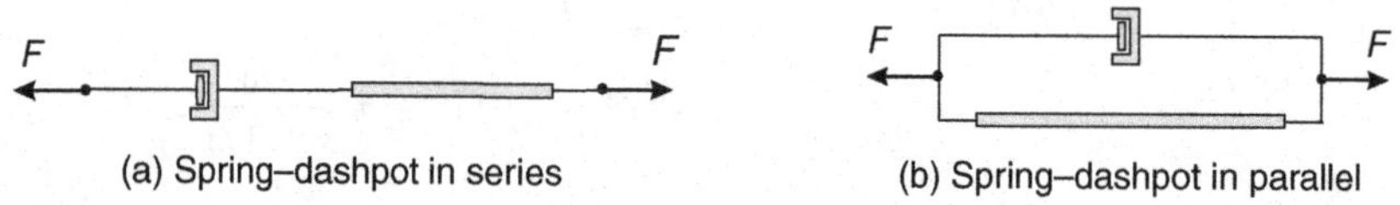

Figure 5.12

A Maxwell (a) and Kelvin–Voigt (b) arrangement of the spring and dashpot.

Equation (5.42) reveals:

$$\dot{\varepsilon} = \frac{1}{c}\dot{F} + \frac{F}{c_\eta}. \tag{5.45}$$

This is rewritten (by multiplication with c) as

$$\dot{F} + \frac{c}{c_\eta}F = c\dot{\varepsilon}, \tag{5.46}$$

and with introduction of the so-called **relaxation time** τ:

$$\tau = \frac{c_\eta}{c}, \tag{5.47}$$

the final expression is obtained:

$$\dot{F} + \frac{1}{\tau}F = c\dot{\varepsilon}. \tag{5.48}$$

This differential equation is subject to the condition that for $t < 0$ the force F and the strain rate $\dot{\varepsilon}$ vanish. To find a solution of this differential equation, the force F is split into a solution F_h of the homogeneous equation:

$$\dot{F}_\mathrm{h} + \frac{1}{\tau}F_\mathrm{h} = 0 \tag{5.49}$$

and one particular solution F_p of the inhomogeneous Eq. (5.48):

$$F = F_\mathrm{h} + F_\mathrm{p}\,. \tag{5.50}$$

The homogeneous solution is of the form:

$$F_\mathrm{h} = c_1 \mathrm{e}^{c_2 t}, \tag{5.51}$$

with c_1 and c_2 constants. Substitution into Eq. (5.49) yields

$$c_1\, c_2\, \mathrm{e}^{c_2 t} + \frac{1}{\tau} c_1\, \mathrm{e}^{c_2 t} = 0 \quad \longrightarrow \quad c_2 = -\frac{1}{\tau} \quad \longrightarrow \quad F_\mathrm{h} = c_1 \mathrm{e}^{-t/\tau}. \tag{5.52}$$

The solution F_p is found by selecting

$$F_\mathrm{p} = C(t)\mathrm{e}^{-t/\tau}, \tag{5.53}$$

with $C(t)$ to be determined. Substitution into Eq. (5.48) yields

$$\underbrace{\frac{dC}{dt}\mathrm{e}^{-t/\tau} - \frac{C}{\tau}\mathrm{e}^{-t/\tau}}_{dF_\mathrm{p}/dt} + \underbrace{\frac{C}{\tau}\mathrm{e}^{-t/\tau}}_{(1/\tau)F_\mathrm{p}} = c\dot{\varepsilon} \quad \longrightarrow \quad \frac{dC}{dt} = c\mathrm{e}^{t/\tau}\dot{\varepsilon}, \tag{5.54}$$

hence

$$C = \int c\mathrm{e}^{\xi/\tau}\dot{\varepsilon}(\xi)d\xi. \tag{5.55}$$

Because the strain rate $\dot{\varepsilon} = 0$ for all $t < 0$, it follows that

$$F_\mathrm{p} = \left(\int_0^t c \mathrm{e}^{\xi/\tau} \dot{\varepsilon}(\xi) d\xi \right) \mathrm{e}^{-t/\tau}. \tag{5.56}$$

Combining Eqs. (5.51) and (5.56), the solution F is given by

$$F = c_1 \mathrm{e}^{-t/\tau} + \left(\int_0^t c \mathrm{e}^{\xi/\tau} \dot{\varepsilon}(\xi) d\xi \right) \mathrm{e}^{-t/\tau}. \tag{5.57}$$

Requiring that for all $t < 0$ the force satisfies $F = 0$ leads to

$$F(t = 0) = c_1, \qquad c_1 = 0. \tag{5.58}$$

Consequently, the solution of the first-order differential equation Eq. (5.48) is given by

$$F(t) = c \int_0^t \mathrm{e}^{-(t-\xi)/\tau} \dot{\varepsilon}(\xi) d\xi. \tag{5.59}$$

It is clear that the integral equation as introduced in the previous section, Eq. (5.28), can be considered as a general solution of a differential equation. In the present case the relaxation spectrum, as defined in Eq. (5.39) is built up by just one single Maxwell element, and in this case: $G(t) = \mathrm{e}^{-t/\tau}$.

To understand the implications of this model, consider a strain history as specified in Fig. 5.13, addressing a spring–dashpot system in which one end point is fixed while the other end point has a prescribed displacement history. The force response is given in Fig. 5.14 in the case of $t^* = 5\tau$. Notice that in this figure the time has been scaled by the relaxation time τ, while the force has been scaled by $c_\eta r$, with r the strain rate; see Fig. 5.13. Two regimes may be distinguished.

(i) For $t < t^*$ the strain proceeds linearly in time, leading to a **constant** strain rate r. In this case the force response is given by (recall that $c_\eta = \tau c$)

$$F = c_\eta r (1 - \mathrm{e}^{-t/\tau}). \tag{5.60}$$

For $t \ll \tau$ it holds that

$$\mathrm{e}^{-t/\tau} \approx 1 - \frac{t}{\tau}, \tag{5.61}$$

such that the force in that case is given by

$$F = c_\eta \frac{t}{\tau} r = c t r. \tag{5.62}$$

So, for relatively small times t, and constant strain rate, the response is dominantly elastic. This is consistent with the spring–dashpot configuration. For small t, only the spring is extended while the dashpot is hardly active. Furthermore, at $t = 0$:

$$\dot{F} = c r, \tag{5.63}$$

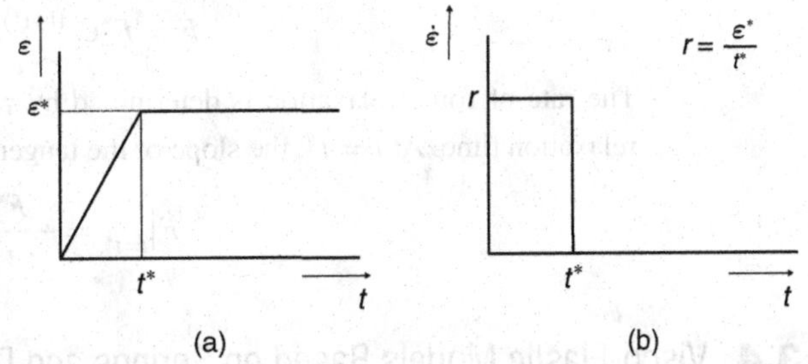

Figure 5.13

Strain (a) and strain rate (b) as a function of time.

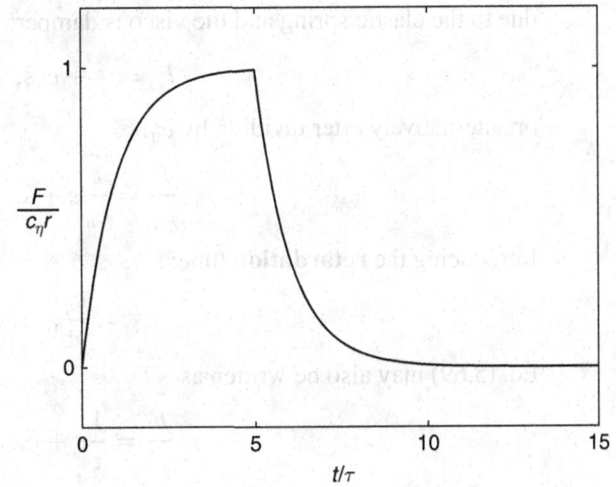

Figure 5.14

Force response of the Maxwell model.

which implies that the line tangent to the force versus time curve should have a slope of cr. For larger values of t, but still smaller than t^* we have

$$e^{-t/\tau} \to 0, \tag{5.64}$$

such that the force is given by

$$F = c_\eta r, \tag{5.65}$$

which is a purely viscous response. In this case the spring has a constant extension, and the force response is dominated by the dashpot. This explains why, at a constant strain rate, the force curve tends towards an asymptote in Fig. 5.14 for sufficiently large t.

(ii) For times $t > t^*$ the strain rate is zero. In that case, the force decreases exponentially in time. This is called **relaxation**. If F^* denotes the force $t = t^*$, the force for $t > t^*$ is given by

$$F = F^* e^{-(t-t^*)/\tau}. \tag{5.66}$$

The rate of force relaxation is determined by τ, which explains why τ is called a relaxation time. At $t = t^*$, the slope of the tangent to the force curve equals

$$\dot{F}\big|_{t=t^*} = -\frac{F^*}{\tau}. \tag{5.67}$$

5.3.4 Visco-Elastic Models Based on Springs and Dashpots: Kelvin–Voigt Model

A second example of combined viscous and elastic behaviour is obtained for the set-up of Fig. 5.12(b). In this case the total force F equals the sum of the forces due to the elastic spring and the viscous damper:

$$F = c\,\varepsilon + c_\eta \dot{\varepsilon}, \tag{5.68}$$

or, alternatively after dividing by c_η:

$$\frac{F}{c_\eta} = \frac{c}{c_\eta}\varepsilon + \dot{\varepsilon}. \tag{5.69}$$

Introducing the **retardation** time:

$$\tau = \frac{c_\eta}{c}, \tag{5.70}$$

Eq. (5.69) may also be written as

$$\frac{F}{c_\eta} = \frac{1}{\tau}\varepsilon + \dot{\varepsilon}. \tag{5.71}$$

The set-up according to Fig. 5.12(b) is known as the Kelvin–Voigt model. In analogy with the Maxwell model, the solution of this differential equation is given by

$$\varepsilon(t) = \frac{1}{c_\eta}\int_0^t e^{-(t-\xi)/\tau} F(\xi) d\xi. \tag{5.72}$$

In the case of a **constant** force F, the strain response is given by

$$\varepsilon(t) = \frac{F}{c}(1 - e^{-t/\tau}). \tag{5.73}$$

This phenomenon of an increasing strain with a constant force (up to a maximum of F/c) is called **creep**. For $t \ll \tau$:

$$\varepsilon \approx \frac{Ft}{c_\eta}, \tag{5.74}$$

corresponding to a viscous response, while for $t \gg \tau$:

$$\varepsilon \approx \frac{F}{c}, \tag{5.75}$$

reflecting a purely elastic response.

5.4 Harmonic Excitation of Visco-Elastic Materials

5.4.1 The Storage and the Loss Modulus

In this section, some methods will be described that can be used to calculate the response of a visco-elastic material for different excitations. The section is aimed at closed-form solutions for the governing equations. Fourier and Laplace transforms and complex function theory are used. First, the methods will be outlined in a general context. After that, an example of a standard linear model will be discussed.

In the previous section, first-order differential equations appeared to describe the behaviour of simple visco-elastic models; however, in general, a linear visco-elastic model is characterized by either a higher-order differential equation (or a set of first-order differential equations):

$$p_0 F + p_1 \frac{dF}{dt} + \cdots + p_M \frac{d^M F}{dt^M} = q_0 \varepsilon + q_1 \frac{d\varepsilon}{dt} + \cdots + q_N \frac{d^N \varepsilon}{dt^N}, \tag{5.76}$$

or an integral equation:

$$\varepsilon(t) = \int_0^t J(t - \xi)\dot{F}(\xi)d\xi \tag{5.77}$$

$$F(t) = \int_0^t G(t - \xi)\dot{\varepsilon}(\xi)d\xi. \tag{5.78}$$

Both types of formulation can be derived from each other.

When a model is used consisting of a number of springs and dashpots, the creep and relaxation functions can be expressed by a series of exponential functions. It is said that the functions form a discrete spectrum. It is possible, and for biological materials sometimes necessary [9], to use continuous functions for $G(t)$ and $J(t)$ thus establishing a more general identification than the differential formulation with limited M and N.

To determine the response to some arbitrary force or strain history, several solution methods are available. In the case of a differential model, a usual way is to determine the homogeneous solution and after that a particular solution. The general solution is the summation of both. This is a method that is applied in the time domain. Another approach is to use Laplace transforms. In this case the differential equation is replaced by an algebraic equation in the Laplace domain which usually is easy to solve. This solution is then transformed back into the time domain (often the harder part). This approach is usually used for functions that are one-sided, meaning that the functions are zero up to a certain time and finite after that time.

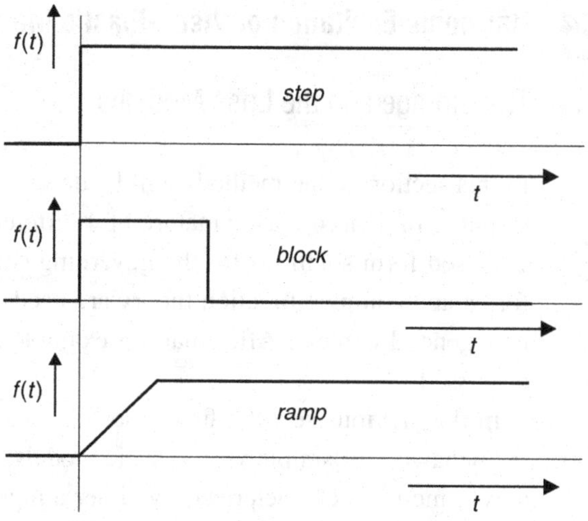

Figure 5.15

Some simple functions that can be used as loading histories and for which closed-form solutions exist for linear visco-elastic models.

The result of the integral formulation can, for the case of discrete spectra, be considered as the general solution of the associated differential equation, and sometimes it is possible to derive closed-form expressions for the integrals. This depends strongly on the spectrum and the load history. If no closed-form solutions are available, numerical methods are necessary to calculate the integrals, which is usually the case for realistic loading histories.

Closed-form solutions can only be generated for simple loading histories. Strictly speaking, the above methods are only applicable for transient signals (zero for $t < 0$ and finite for $t \geq 0$). For examples, see Fig. 5.15.

A frequently used way of excitation in practice is **harmonic excitation**. In that case, the applied loading history has the form of a sine or cosine. Let us assume that the prescribed strain is harmonic according to

$$\varepsilon(t) = \varepsilon_0 \cos(\omega t). \tag{5.79}$$

In case of a linear visco-elastic model the output, i.e. the force, will also be a harmonic:

$$F(t) = F_0 \cos(\omega t + \delta), \tag{5.80}$$

or equivalently:

$$F(t) = F_0 \cos(\delta) \cos(\omega t) \; - \; F_0 \sin(\delta) \sin(\omega t) \,. \tag{5.81}$$

This equation reveals that the force can be decomposed into two terms: the first with amplitude $F_0 \cos(\delta)$ in phase with the applied strain (called the elastic response), the second with amplitude $F_0 \sin(\delta)$, which is 90° out of phase with the applied strain (called the viscous response). Equation (5.81) can also be written as

$$F(t) = \varepsilon_0 E_1 \cos(\omega t) \ - \ \varepsilon_0 E_2 \sin(\omega t), \tag{5.82}$$

with: $E_1 = (F_0/\varepsilon_0)\cos(\delta)$ the **storage modulus** and $E_2 = (F_0/\varepsilon_0)\sin(\delta)$ the **loss modulus**. These names will become clear after considering the amount of energy (per unit length of the sample considered) dissipated during one single loading cycle. The necessary amount of work for such a cycle is

$$\begin{aligned} W &= \int_0^{2\pi/\omega} F\dot{\varepsilon} dt \\ &= -\int_0^{2\pi/\omega} [\varepsilon_0 E_1 \cos(\omega t) \ - \ \varepsilon_0 E_2 \sin(\omega t)]\, \varepsilon_0 \omega \sin(\omega t) dt \\ &= \pi \varepsilon_0^2 E_2. \end{aligned} \tag{5.83}$$

It is clear that part of the work is dissipated as heat. This part, given by Eq. (5.83), is determined by E_2, the loss modulus. During loading the E_1-related part of F also contributes to the stored work, but this energy is released during unloading. That is why E_1 is called the storage modulus.

5.4.2 The Complex Modulus

In literature on visco-elasticity, the **complex modulus** is often used, which is related to the storage and loss modulus. To identify this relation, a more formal way to study harmonic excitation is pursued. In the case of a harmonic signal the Boltzmann integral for the relaxation function can be written as

$$F(t) = \int_{-\infty}^{t} G(t-\xi)\dot{\varepsilon}(\xi) d\xi. \tag{5.84}$$

The domain of the integral in Eq. (5.84) starts at $-\infty$ because it is assumed that at the current time t the harmonic strain is applied for such a long time that all effects from switching on the signal have disappeared (also meaning that using a Laplace transform for these type of signals is not recommended). The time t in the upper boundary of the integral can be removed by substitution of $\xi = t - s$ in Eq. (5.84), thus replacing the integration variable ξ by s:

$$F(t) = -\int_{\infty}^{0} G(s)\dot{\varepsilon}(t-s) ds = \int_{0}^{\infty} G(s)\dot{\varepsilon}(t-s) ds. \tag{5.85}$$

Substitution of (5.79) into this equation leads to

$$F(t) = \varepsilon_0 \cos(\omega t) \left[\omega \int_0^{\infty} G(s) \sin(\omega s) ds \right] - \varepsilon_0 \sin(\omega t) \left[\omega \int_0^{\infty} G(s) \cos(\omega s) ds \right]. \quad (5.86)$$

The terms between brackets [ ] are functions only of the frequency and not of the time. These terms are solely determined by the type of material that is being considered and can be measured. We can write Eq. (5.86) as

$$F(t) = \varepsilon_0 E_1(\omega) \cos(\omega t) - \varepsilon_0 E_2(\omega) \sin(\omega t), \quad (5.87)$$

where the formal definitions for the storage and loss moduli:

$$E_1(\omega) = \omega \int_0^{\infty} G(s) \sin(\omega s) ds \quad (5.88)$$

$$E_2(\omega) = \omega \int_0^{\infty} G(s) \cos(\omega s) ds \quad (5.89)$$

can be recognized.

In the case of harmonic excitation, it is worthwhile to use complex function theory. Instead of (5.79) we write

$$\varepsilon(t) = \mathrm{Re}\{\varepsilon_0 \mathrm{e}^{i\omega t}\}. \quad (5.90)$$

The convolution integral for the force, Eq. (5.84) can be written as

$$F(t) = \int_{\xi=-\infty}^{\infty} G(t-\xi)\dot{\varepsilon}(\xi) d\xi = \int_{s=-\infty}^{\infty} G(s)\dot{\varepsilon}(t-s) ds. \quad (5.91)$$

The upper limit t is replaced by ∞. This is allowed because $G(t)$ is defined such that $G(t-\xi) = 0$ for $\xi > t$. After that, we have substituted $\xi = t - s$. Substitution of (5.90) into (5.91) leads to

$$F(t) = i\omega\varepsilon_0 \mathrm{e}^{i\omega t} \int_{-\infty}^{\infty} G(s) \mathrm{e}^{-i\omega s} ds = i\omega\varepsilon_0 G^*(\omega) \mathrm{e}^{i\omega t}. \quad (5.92)$$

In this equation $G^*(\omega)$ can be recognized as the Fourier transform of $G(t)$ (see Appendix 5.5). It is clear that the force has the same form as the strain, only the force has a complex amplitude. If we define the **complex modulus** $E^*(\omega)$ as

$$E^*(\omega) = i\omega G^*(\omega) = E_1(\omega) + iE_2(\omega), \quad (5.93)$$

substitution of (5.93) into (5.92) gives the real part of $F(t)$:

$$F(t) = \varepsilon_0 E_1 \cos(\omega t) - \varepsilon_0 E_2 \sin(\omega t). \quad (5.94)$$

It is clear again that E_1 and E_2 are the storage and loss moduli. The above expression specifies the form of the force output in the time domain. We can also directly derive the relation between $E(\omega)$ and $G(\omega)$ by using a Fourier transform of Eq. (5.91):

$$F^*(\omega) = G^*(\omega)i\omega\varepsilon^*(\omega) = E^*(\omega)\varepsilon^*(\omega). \tag{5.95}$$

The complex modulus is similar to the transfer function in system theory. The storage modulus is the real part of the transfer function, and the loss modulus is the imaginary part.

When experiments are performed to characterize visco-elastic biological materials, the results are often presented either in the form of the storage and the loss moduli as a function of the excitation frequency, or by using the absolute value of the complex modulus, in combination with the phase shift between input (strain) and output (force), as a function of the frequency. In the case of **linear** visco-elastic behaviour, these properties give a good representation of the material (this has to be tested first). As a second step, often a model is proposed, based on a combination of springs and dashpots, to 'fit' to the given moduli. If this is possible, the material behaviour can be described with a limited number of material parameters (the properties of the springs and dashpots) and all possible selections of properties to describe the material under consideration can be derived from each other. This will be demonstrated in the next subsection for a particular, but frequently used, model, the standard linear model.

5.4.3 The Standard Linear Model

The standard linear model can be represented by one dashpot and two springs, as shown in Fig. 5.16. The upper part is composed of a linear spring; the lower part shows a Maxwell element.

Similar to the procedure used in Section 5.3.3, the total strain of the Maxwell element is considered as an addition of the strain in the dashpot (ε_d) and the strain in the spring ($\varepsilon_s = \varepsilon - \varepsilon_d$). The following relations can be proposed for variables that determine the standard linear model:

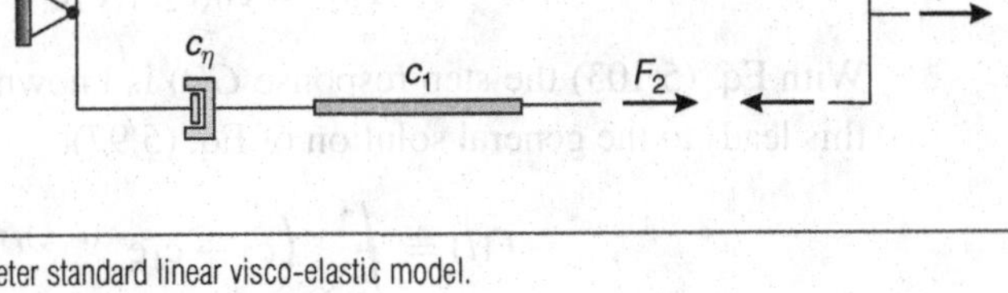

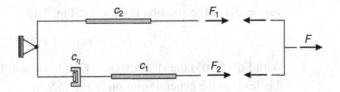

Figure 5.16

The three-parameter standard linear visco-elastic model.

$$\begin{aligned} F &= F_1 + F_2 \\ F_1 &= c_2 \varepsilon \\ F_2 &= c_\eta \dot{\varepsilon_d} \\ F_2 &= c_1(\varepsilon - \varepsilon_d). \end{aligned} \tag{5.96}$$

Elimination of $\dot{\varepsilon_d}$, F_1 and F_2 from this set of equations leads to

$$F + \tau_R \dot{F} = c_2 \varepsilon + (c_1 + c_2)\tau_R \dot{\varepsilon}, \tag{5.97}$$

with $\tau_R = c_\eta / c_1$ the characteristic relaxation time. The force response to a step $\varepsilon(t) = \varepsilon_0 \, H(t)$ in the strain yields

$$F(t) = F(t)_{\text{hom}} + F(t)_{\text{part}} = \alpha e^{-t/\tau_R} + c_2 \varepsilon_0, \tag{5.98}$$

(where 'hom' indicates homogeneous, 'part' particular) with α an integration constant to be determined from the initial conditions. Determining the initial condition at $t = 0$ for this problem is not trivial. It is a jump condition with a discontinuous force F and strain ε at $t = 0$. A way to derive this jump condition is by using the definition of the time derivative:

$$\dot{F} = \lim_{\Delta t \to 0} \frac{F(t + \Delta t) - F(t)}{\Delta t}. \tag{5.99}$$

Let us take two time points a distance Δt apart, one point on the time axis left of $t = 0$, which we call $t = 0^-$, and one point on the right side of $t = 0$, which we call $t = 0^+$. In that case Eq. (5.97) can be written as

$$F(0) + \tau_R \frac{F(0^+) - F(0^-)}{\Delta t} = c_2 \varepsilon(0) + (c_1 + c_2)\tau_R \frac{\varepsilon(0^+) - \varepsilon(0^-)}{\Delta t}, \tag{5.100}$$

or

$$\begin{aligned} &F(0)\Delta t + \tau_R(F(0^+) - F(0^-)) \\ &= c_2 \varepsilon(0)\Delta t + (c_1 + c_2)\tau_R(\varepsilon(0^+) - \varepsilon(0^-)). \end{aligned} \tag{5.101}$$

Because $F(0^-) = 0$ and $\varepsilon(0^-) = 0$, and the terms with Δt vanish when $\Delta t \to 0$, it is found that

$$F(0^+) = (c_1 + c_2)\varepsilon_0, \tag{5.102}$$

so $\alpha = c_1 \varepsilon_0$. The solution is shown in Fig. 5.17:

$$F(t) = \varepsilon_0 (c_2 + c_1 e^{-t/\tau_R}). \tag{5.103}$$

With Eq. (5.103) the step response $G(t)$ is known. Using the Boltzmann integral this leads to the general solution of Eq. (5.97):

$$F(t) = \int_{-\infty}^{t} \left(c_2 + c_1 e^{-(t-\xi)/\tau_R}\right) \dot{\varepsilon}(\xi) d\xi. \tag{5.104}$$

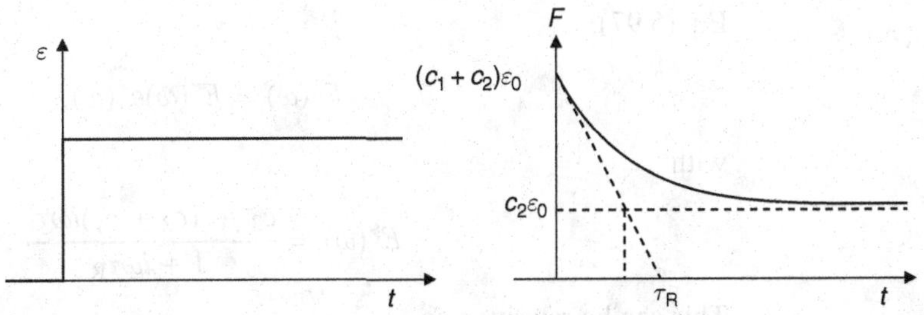

Figure 5.17

Response (right) of the three-parameter model to a step in the strain (left).

There are several ways to determine the creep function. We can solve Eq. (5.97) for a step in the force. This can be done by determining a homogeneous and a particular solution as was done for the relaxation problem. However, it can also be done by means of Laplace transformation of the differential equation. This leads to an algebraic equation that can be solved. The result can be transformed back from the Laplace domain to the time domain.

Instead of again solving the differential equation we can use the relation that exists between the creep function and the relaxation function, Eq. (5.36). A Laplace transformation of $G(t)$ leads to

$$\hat{G}(s) = \frac{c_2}{s} + \frac{c_1}{s + 1/\tau_R} = \frac{c_2(s + 1/\tau_R) + c_1 s}{s(s + 1/\tau_R)} . \tag{5.105}$$

With Eq. (5.36), the Laplace transform of J is found:

$$\hat{J}(s) = \frac{1}{s^2 \hat{G}(s)} = \frac{s + 1/\tau_R}{[(c_1 + c_2)s + c_2/\tau_R]s} . \tag{5.106}$$

$$\hat{J}(s) = -\frac{c_1/c_2}{[(c_1 + c_2)s + c_2/\tau_R]} + \frac{1}{c_2 s} . \tag{5.107}$$

Back transformation leads to

$$J(t) = \frac{1}{c_2}\left(1 - \frac{c_1}{c_2 + c_1} e^{-t/\tau_K}\right), \tag{5.108}$$

with $\tau_K = (c_1 + c_2)\tau_R/c_2$ the characteristic creep time. It is striking that the characteristic creep time is different from the characteristic relaxation time. To be complete, the integral equation for force controlled problems is given:

$$\varepsilon(t) = \int_{-\infty}^{t} \frac{1}{c_2}\left(1 - \frac{c_1}{c_1 + c_2} e^{-(t-\xi)/\tau_K}\right) \dot{F}(\xi) d\xi . \tag{5.109}$$

At this point, it is opportune to mention some terminology from system dynamics. A linear system can be defined by a **transfer function**. For a harmonic excitation the transfer function is found by a Fourier transform of the original differential

Eq. (5.97):

$$F^*(\omega) = E^*(\omega)\varepsilon^*(\omega), \tag{5.110}$$

with

$$E^*(\omega) = \frac{c_2 + (c_2 + c_1)i\omega\tau_R}{1 + i\omega\tau_R}. \tag{5.111}$$

This can be rewritten as

$$E^*(\omega) = c_2 \frac{1 + i\omega\tau_K}{1 + i\omega\tau_R}. \tag{5.112}$$

In system dynamics, it is customary to plot a Bode diagram of these functions. For this we need the absolute value of $E^*(\omega)$:

$$|E^*(\omega)| = c_2 \frac{\sqrt{1 + (\omega\tau_K)^2}}{\sqrt{1 + (\omega\tau_R)^2}}. \tag{5.113}$$

The phase shift $\phi(\omega)$ is

$$\phi(\omega) = \arctan(\omega\tau_K) - \arctan(\omega\tau_R). \tag{5.114}$$

In our case $\tau_K > \tau_R$ because c_1 and c_2 are always positive. The result is given in Fig. 5.18.

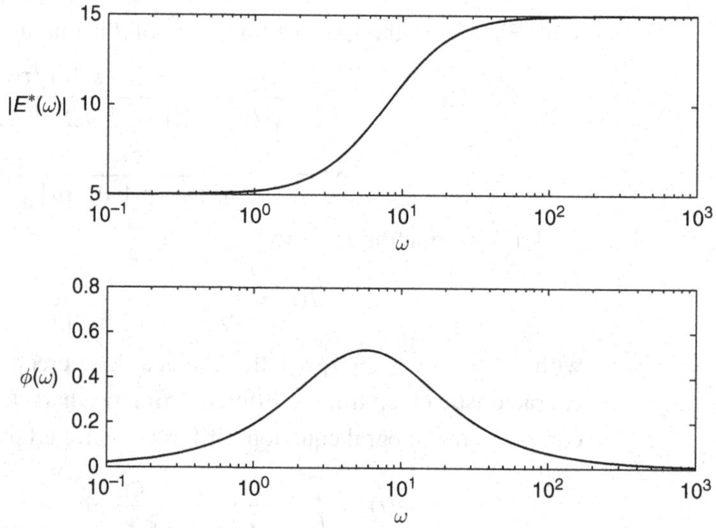

Figure 5.18

The complex modulus and phase shift for a standard linear model with two springs ($c_1 = 10$ [N], $c_2 = 5$ [N]) and one dashpot ($\tau_R = 0.1$ [s]).

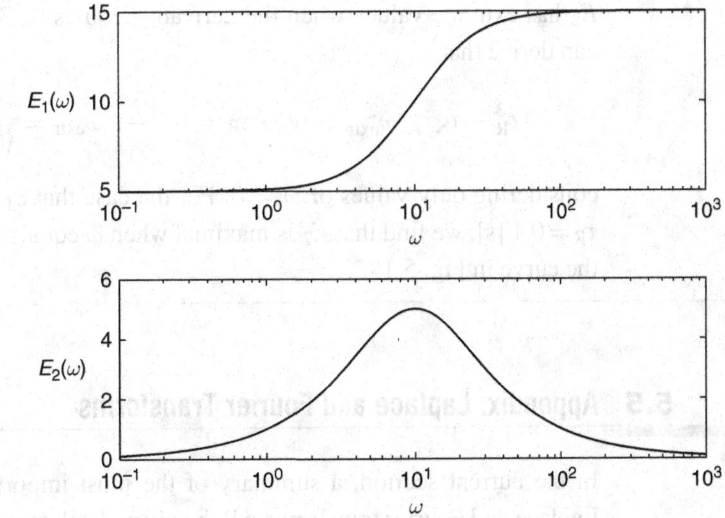

Figure 5.19

The storage and loss moduli for the standard linear model.

The storage and the loss moduli are the real and imaginary part of the complex modulus $E^*(\omega)$:

$$E_1(\omega) = c_2 \frac{1 + \omega^2 \tau_K \tau_R}{1 + \omega^2 \tau_R^2} \tag{5.115}$$

$$E_2(\omega) = c_2 \frac{\omega \tau_K - \omega \tau_R}{1 + \omega^2 \tau_R^2}. \tag{5.116}$$

The result is given in Fig. 5.19. E_1 has a similar shape to $|E^*(\omega)|$, because the asymptotic values for $\omega \to 0$ and $\omega \to \infty$ are the same. The loss modulus E_2 has its maximum at the point where the phase shift is highest. This can be explained. At very high frequencies, the dashpot has an infinite stiffness and the behaviour of the standard linear model is dominated by the two springs. At very low frequencies, the influence of the dashpot is small and the behaviour is dominated by c_2. In these areas the mechanical behaviour is like that of an elastic material.

Example 5.6 For the example given in Fig. 5.19, we would like to calculate for which frequency ω the loss modulus E_2 is maximal. For this, we need to determine the derivative of E_2 with respect to ω:

$$\frac{dE_2}{d\omega} = c_2 \frac{(1 + \omega^2 \tau_R^2)(\tau_K - \tau_R) - (\omega \tau_K - \omega \tau_R)(2\omega \tau_R^2)}{1 + \omega^2 \tau_R^2}, \tag{5.117}$$

$$= c_2 \frac{(\tau_R^3 - \tau_K \tau_R^2)\omega^2 + \tau_K - \tau_R}{1 + \omega^2 \tau_R^2}. \tag{5.118}$$

E_2 has extreme values when the derivative equals 0. For that particular case, we can derive that

$$(\tau_R^3 - \tau_K \tau_R^2)\omega_{extr}^2 + \tau_K - \tau_R = 0 \quad \rightarrow \omega_{extr} = \sqrt{\frac{\tau_R - \tau_K}{\tau_R^3 - \tau_K \tau_R^2}}, \tag{5.119}$$

considering only values of $\omega \geq 0$. For the case that c_1 = 10 [N], c_2 = 5 [N], and τ_R = 0.1 [s], we find that E_2 is maximal when ω equals 10 [s^{-1}], corresponding to the curve in Fig. 5.19.

5.5 Appendix: Laplace and Fourier Transforms

In the current section, a summary of the most important issues with regard to Laplace and Fourier transforms will be given. Both transformations can be applied to differential equations and transform these equations into algebraic equations. In general, the Fourier transform is used for periodic functions; the Laplace transform is used for one-sided functions, meaning that the functions are zero up to a certain time and finite after that time. In terms of visco-elasticity, this means that the Fourier transform is used as a tool to describe harmonic excitation and the Laplace transform is used to describe creep and relaxation.

The **Laplace transform** $\hat{x}(s)$ of a time function $x(t)$ is defined as

$$\hat{x}(s) = \int_0^\infty x(t)e^{-st}dt. \tag{5.120}$$

The most important properties of Laplace transforms are:

- Laplace transform is a linear operation.
- When $x(t)$ is a continuous function, the Laplace transform of the time derivative $\dot{x}(t)$ of $x(t)$ is given by

$$\hat{\dot{x}}(t) = s\hat{x}(s) - x(0), \tag{5.121}$$

  with $x(0)$ the value of the original function $x(t)$ at time $t = 0$.
- Convolution in the time domain is equivalent to a product in the Laplace domain. Using two time functions $x(t)$ and $y(t)$ with Laplace transforms $\hat{x}(s)$ and $\hat{y}(s)$, the following convolution integral $I(t)$ could be defined:

$$I(t) = \int_{\tau=-\infty}^{\infty} x(\tau)y(t-\tau)d\tau. \tag{5.122}$$

  In that case the Laplace transform of this integral can be written as

$$\hat{I}(s) = \hat{x}(s)\,\hat{y}(s). \tag{5.123}$$

- If a function $x(t)$ has a Laplace transform $\hat{x}(s)$, then the Laplace transform of the function $t^n x(t)$, with $n = 1, 2, 3, \ldots$, can be written as

$$\widehat{t^n x(t)} = (-1)^n \frac{d^n \hat{x}(s)}{ds^n}. \tag{5.124}$$

- If a function $x(t)$ has Laplace transform $\hat{x}(s)$, then the Laplace transform of $x(t)/t$, assuming that $\lim_{t \to 0} x(t)/t$ exists, is given by

$$\widehat{(x(t)/t)} = \int_s^\infty x(a) da. \tag{5.125}$$

The **Fourier transform** $x^*(t)$ of time function $x(t)$ is given by

$$x^*(\omega) = \int_{-\infty}^{\infty} x(t) e^{i\omega t} dt, \tag{5.126}$$

with $i = \sqrt{-1}$. The Fourier transform has similar properties to the Laplace transform. The most important properties of Fourier transforms are:

- Fourier transformation is a linear operation.
- When $x(t)$ is a continuous function, the Fourier transform of the time derivative $\dot{x}(t)$ of $x(t)$ is given by

$$\dot{x}^*(t) = i\omega\, x^*(\omega). \tag{5.127}$$

- Convolution in the time domain is equivalent to a product in the Fourier domain. Using two time functions $x(t)$ and $y(t)$ with Fourier transforms $x^*(\omega)$ and $y^*(\omega)$, the following convolution integral $I(t)$ could be defined:

$$I(t) = \int_{-\infty}^{\infty} x(\tau) y(t - \tau) d\tau. \tag{5.128}$$

In that case, the Fourier transform of this integral can be written as

$$I^*(\omega) = x^*(\omega)\, y^*(\omega). \tag{5.129}$$

- If a function $x(t)$ has a Fourier transform $x^*(\omega)$, then the Fourier transform of the function $t^n x(t)$, with $n = 1, 2, 3, \ldots$, can be written as

$$\left(t^n x(t)\right)^* = (i)^n \frac{d^n x^*(\omega)}{d\omega^n}. \tag{5.130}$$

- If a function $x(t)$ has Fourier transform $x^*(\omega)$, then the Fourier transform of $x(t)/t$ is given by

$$(x(t)/t)^* = i\omega \int_\omega^\infty x(a) da. \tag{5.131}$$

Finally, Table 5.1 gives some Laplace and Fourier transforms of often-used functions. In the table, the step function H(t) (Heavyside function) as defined in Eq. (5.20) is used, as well as the delta function $\delta(t)$, defined as

$$\delta(t) = \begin{cases} 0 & \text{if } t \neq 0 \\ \infty & \text{if } t = 0. \end{cases} \tag{5.132}$$

Table 5.1 Time functions with their Fourier and Laplace transforms.

Original function	Fourier transform	Laplace transform
$\delta(t)$	1	1
$\mathrm{H}(t)$		$1/s$
$t\,\mathrm{H}(t)$		$1/s^2$
$e^{-bt}\mathrm{H}(t)$	$1/(i\omega + b)$	$1/(s+b)$
$e^{i\omega_0}$	$\delta(\omega - \omega_0)$	
$\sin(at)\,\mathrm{H}(t)$	$a/(a^2 - \omega^2)$	$a/(s^2 + a^2)$
$\cos(at)\,\mathrm{H}(t)$	$i\omega/(a^2 - \omega^2)$	$s/(s^2 + a^2)$

and

$$\int_{-\infty}^{+\infty} \delta(t)dt = 1. \tag{5.133}$$

Exercises

5.1 A visco-elastic material is described by means of a Maxwell model. The model consists of a linear dashpot, with damping coefficient c_η, in series with a linear spring, with spring constant c (see figure below).

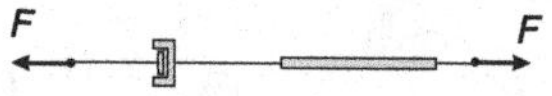

The numerical values for the material properties are:

$$c = 8 \times 10^4 \quad [\mathrm{N}]$$
$$c_\eta = 0.8 \times 10^4 \quad [\mathrm{N\,s}].$$

(a) Derive the differential equation for this model.

(b) Give the response for a unit step in the strain. Make a drawing of the response.

(c) The material is subjected to an harmonic strain excitation with amplitude ε_0 and an angular frequency ω. Give the complex modulus $E^*(\omega)$ and the phase shift $\phi(\omega)$ for this material.

(d) Give the amplitude of the force for the case $\varepsilon_0 = 0.01$ and a frequency $f = 5$ [Hz].

5.2 A material can be characterized with a standard linear model (see figure below). The material is loaded with a step in the force at time $t = 0$. At time $t = t_1$ the material is suddenly unloaded stepwise.

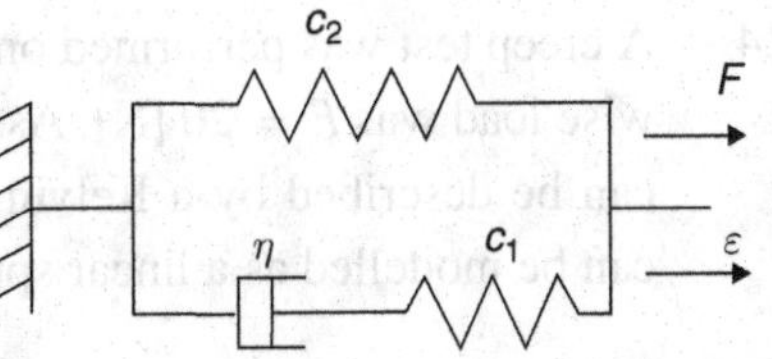

(a) Derive the strain response for this loading history.
(b) Make a drawing of the strain response as a function of time.
(c) Calculate $d\varepsilon/dt$ for $t = 0$.

5.3 In a dynamic experiment, a specimen is loaded with a strain as given in the figure below. The strain is defined by:

$$\varepsilon = 0.01 \ \sin(\omega t),$$

with $\omega = 0.1$ [rad s^{-1}].

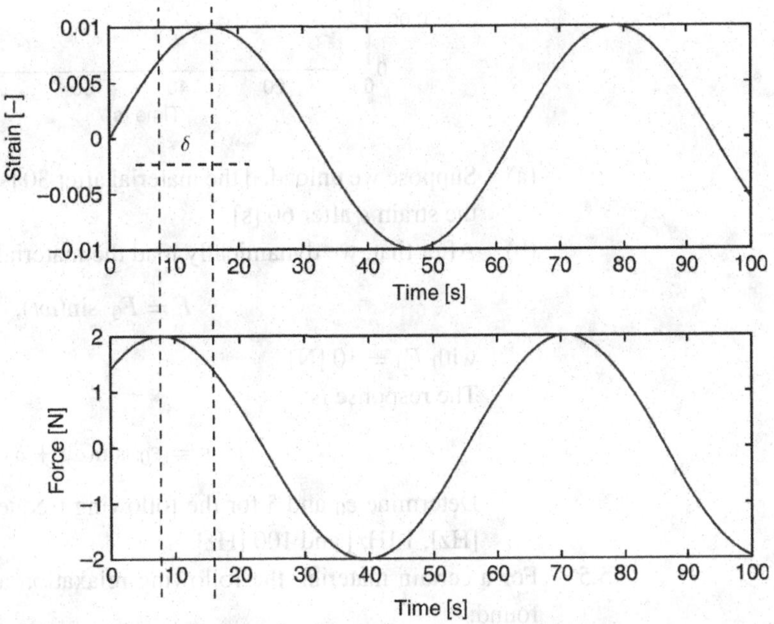

The response is also given in the figure. Let us assume that we are dealing with a material that can be described by a Maxwell model (one spring and dashpot in series).

(a) What is the phase shift if we double the angular frequency of the load?

(b) What will be the amplitude of the force for that frequency?

(c) Now we do a relaxation experiment, with the same material. We apply a step strain of $\varepsilon = 0.02$. What is the force at: $t = 0$ [s], $t = 10$ [s] and $t = 100$ [s]?

5.4 A creep test was performed on some biological material. The applied step-wise load was $F = 20$ [N]. Assume that we are dealing with a material that can be described by a Kelvin–Voigt model. This means that the material can be modelled as a linear spring and dashpot in parallel.

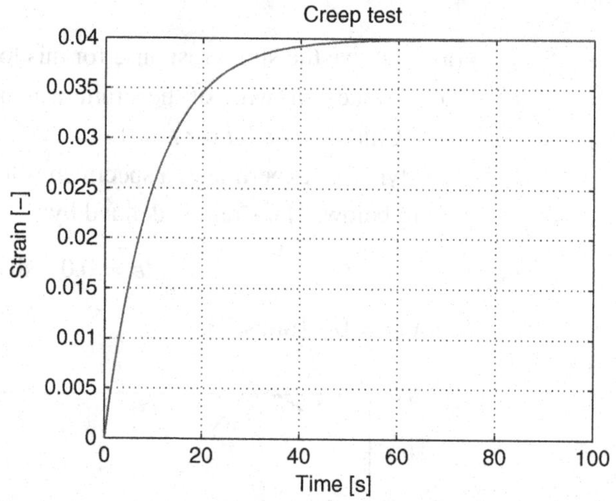

(a) Suppose we unloaded the material after 30 [s]. Determine for this case the strain ε after 60 [s].

(b) After that, we dynamically load the material with a force F:

$$F = F_0 \; \sin(\omega t),$$

with $F_0 = 10$ [N].
The response is:

$$\varepsilon = \varepsilon_0 \; \sin(\omega t + \delta).$$

Determine ε_0 and δ for the following frequencies: $f = 0.01$ [Hz], 0.1 [Hz], 1 [Hz] and 100 [Hz].

5.5 For a certain material, the following relaxation and creep functions were found:

$$G(t) = 1 + \mathrm{e}^{-t} \quad [\mathrm{N}]$$

$$J(t) = 1 - \frac{1}{2}\mathrm{e}^{-t/2} \quad [\mathrm{N}^{-1}].$$

relating force to strain on the material.

(a) Does this material law satisfy Eq. (5.36)?

(b) Test 1: Assume that at $t = 0$ a step in the force is applied of 1 [N]. After 1.5 [s], the step is removed. What is the strain after 3 [s]?

(c) Test 2: At $t = 0$, a step in the strain is applied equal to 0.01. This strain is removed at $t = 1.5$ [s]. What is the value of the force after 3 [s]?

5.6 A piece of tendon material is subjected to the following strain history. At $t = 0$ [s], a step in the strain is applied of 0.01. After 4 [s], the strain is increased by an extra step of 0.01. After 8 [s], the strain is reduced by 0.005. The total strain evolution is given in the figure below.

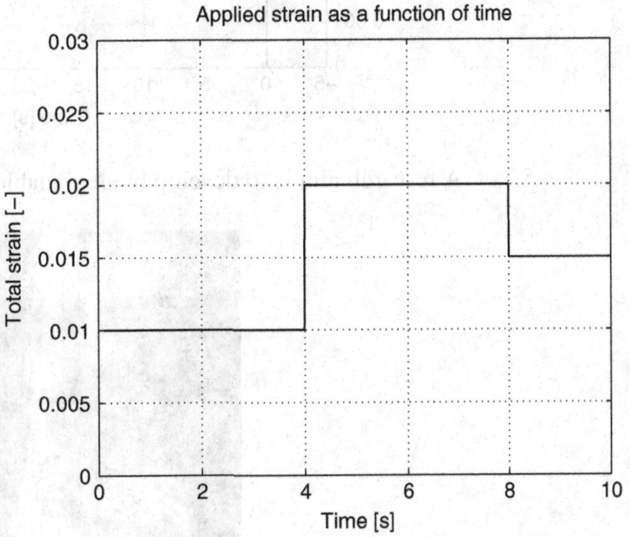

The only information that is available for the material is the result from a creep test as given in the figure on the next page. In the creep test, a tendon specimen was loaded with a mass of 500 [kg]. The gravitational constant is 10 [m s^{-2}].

(a) We try to describe the material with a standard linear model. The relaxation function for this material is:

$$G(t) = c_2 + c_1 e^{-t/\tau_R} .$$

Determine the value of c_1 and c_2 using the information in the figure with the result of the creep test.

(b) Sketch the force response on the strain history as given in the first figure with the applied strain as a function of time.

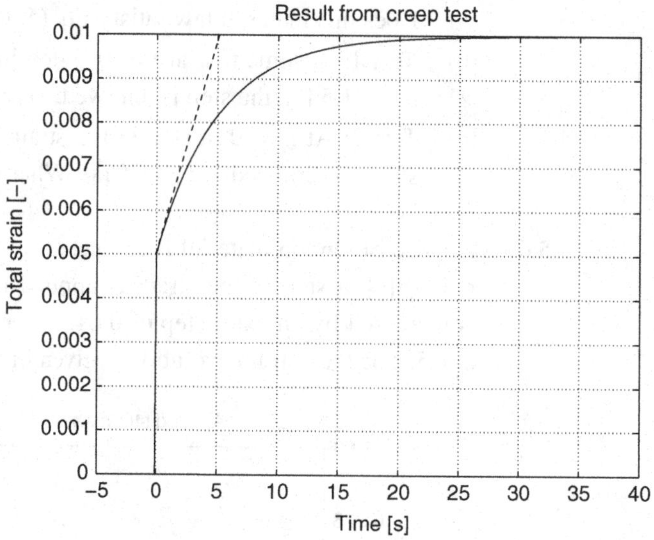

5.7 A research aim is to develop biodegradable bone screws (see figure).

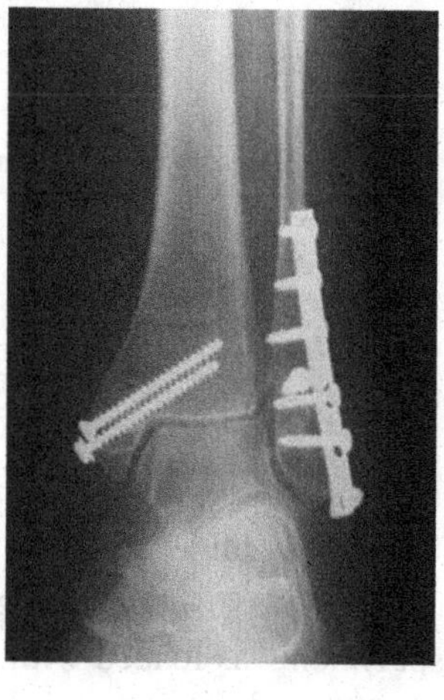

One disadvantage of the application of this type of polymer screws is their visco-elastic material behaviour. In a laboratory test, a researcher fixes two stiff plates (to be considered as rigid bodies) to each other. Initially the screw is in a stress- and strain-free state. The screw is manufactured from a linearly visco-elastic material. The mechanical behaviour of the material can be described by means of the relaxation function $G(t)$:

$$G(t) = 5 \exp\left(-0.07 \left(\frac{t}{t_0}\right)^{0.3}\right),$$

where the time t is expressed in hours and $t_0 = 1$ hour. $G(t)$ is in [10^6 N]. At a certain time the screw is tightened (rapidly) such that the transmitted extensional force equals 1 [kN].

(a) Calculate the axial strain ε_1 in the screw after it is tightened.

(b) Calculate the force F_1 in the screw 24 hours after it is tightened

After 24 hours the screw is tightened again rapidly, such that the extensional force again equals 1 [kN].

(c) Calculate the associated strain ε_2 in the screw

(d) Calculate the force F_2 in the screw again 24 hours later.

5.8 To determine the mechanical properties of biological tissues, a rheometer is often used (see figure). The normal procedure is that a cylindrical tissue specimen is clamped between two plates. Subsequently, the bottom plate is rotated and the moment (torque) that is acting on the fixed top plate is measured. Because the deformation is completely determined by the geometry and not the mechanical behaviour, this is called a viscometric deformation. The rotational angle and the torque are measured, and from this the material behaviour can be derived. For soft biological tissues, sometimes a slightly different procedure is applied. In that case, small rectangular samples are used and these are placed off centre [10] (see the figure). The reason is that often only very small samples can be obtained and by placing them off centre, the torque is higher and the testing machine more accurate. A

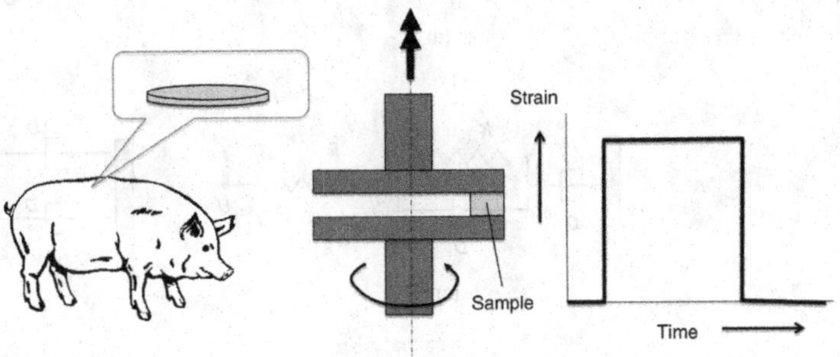

researcher has determined properties of fat in this way. Using the eccentric set-up enables one to calculate the force $F(t)$ that is on the sample, and the problem can be treated as a one-dimensional problem. At small strains, the behaviour can be described with a linear visco-elastic model according to:

$$F(t) = \int_{\xi=-\infty}^{t} G(t-\xi)\dot{\varepsilon}(\xi)\, d\xi,$$

with $G(t)$ the relaxation function, F a force with dimension [N] and $\dot{\varepsilon}$ the strain rate [s^{-1}]. The following relaxation function could be derived from the experiments:

$$G(t) = 10 + 5e^{-t/5} \quad [N].$$

To the fat tissue, a strain is applied as depicted at the right-hand side of the figure. By using a Heavyside function $H(t)$ this can be formulated as:

$$\varepsilon(t) = 0.001[H(t) - H(t - 10)].$$

(a) Give an expression for the force $F(t)$ as a consequence of the applied strain for $t < 10$ [s].

(b) What is the force immediately after the strain is removed at $t = 10^+$ [s]? Is it a compressive or extensional force?

(c) What is the force at $t = 30$ [s]?

5.9 In the figure below four different, linear spring–dashpot models have been drawn that could be used to describe the mechanical behaviour of a biological material; here k and b are material parameters, f and u are force and displacement. The spring and dashpot constants are given in the figure. Each of the models represents some kind of linear visco-elastic behaviour.

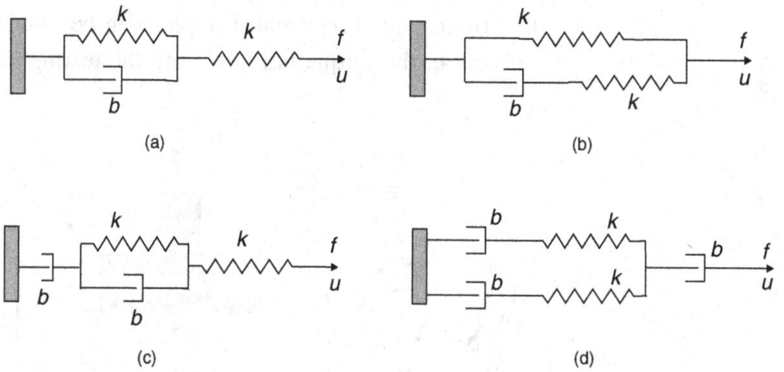

Assume that the length of a tissue specimen is changed by means of a step ε_0 as given in the figure below.

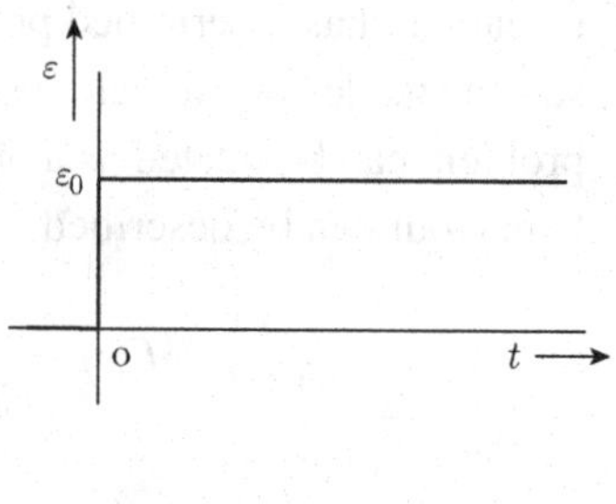

Draw the applied load $f(t)$ as a function of time t for each of the models. Also give relevant values for the initial values of f, asymptotes etc. in the figures.

5.10 The strain $\varepsilon(t)$ of a linear visco-elastic material can be calculated by means of the creep function $J(t)$ provided that the load history $F(t)$ is known. Assume that:

$$J(t) = J_0 + C\left[1 - e^{-t/\tau}\right],$$

with $C = \frac{J_0}{2}$ and J_0 and τ material constants. No deformation or force is applied to the material for $t < 0$. From the time $t = 0$ the material is loaded with a closed-loop loading process specified by:

$$F(t) = F_0\left[\mathrm{H}(t) - 2\mathrm{H}(t - \beta\tau) + \mathrm{H}(t - 2\beta\tau)\right],$$

with H() the Heaviside step function and F_0 a positive constant, which is a measure of the magnitude of the force. The constant β is a measure for the duration of the loading proces.

(a) Give the load as a function of time without using the Heaviside function.

(b) Determine the strain $\varepsilon(t)$ as a function of time and draw a sketch of the result. Note that for $t \gg 2\beta\tau$ the strain is almost reduced to zero again.

(c) Prove that, independent of the numerical values of the relevant parameters, for $t \geq 2\beta\tau$, we always have $\varepsilon(t) < 0$.

6 Analysis of a One-Dimensional Continuous Elastic Medium

6.1 Introduction

In the previous chapters, the global behaviour of fibres was considered, without much attention to the detailed shape of these structures. Only the length change of the fibre played a role in the analysis. In the present chapter we address in a little more detail the deformation of long slender structures. These could be tendons or muscles, but also long bones. The aim is to generalize the concepts introduced in previous chapters for discrete systems (i.e. springs) to continuous systems. To simplify matters, the loading and deformation of a one-dimensional elastic bar is considered.

6.2 Equilibrium in a Subsection of a Slender Structure

Consider a straight bar as visualized in Fig. 6.1(a), loaded by an external force F at $x = L$ and fixed in space at $x = 0$. The figure shows the bar with the x-axis in the longitudinal or axial direction. It is assumed that each cross section initially perpendicular to the axis of the bar remains perpendicular to the axis after loading. In fact, it is assumed that all properties and displacements are a function of the x-coordinate only.

The objective is to compute the displacement of each cross section of the bar due to the loading. The area of the cross section perpendicular to the central axis of the bar as well as the mechanical properties of the bar may be a function of the x-coordinate. Therefore the displacement may be a non-linear function of the axial coordinate.

To compute the displacements of the cross sections, a procedure analogous to the discrete case (elastic springs) is followed. In contrast with the discrete case, the equilibrium conditions are not applied on a global scale but on a local scale. For this purpose, the free body diagram of an arbitrary slice of the bar, for example the grey slice in Fig. 6.1(a), is investigated. This free body diagram is depicted in Fig. 6.1(b). The left side of the slice is located at position x, and the slice has a

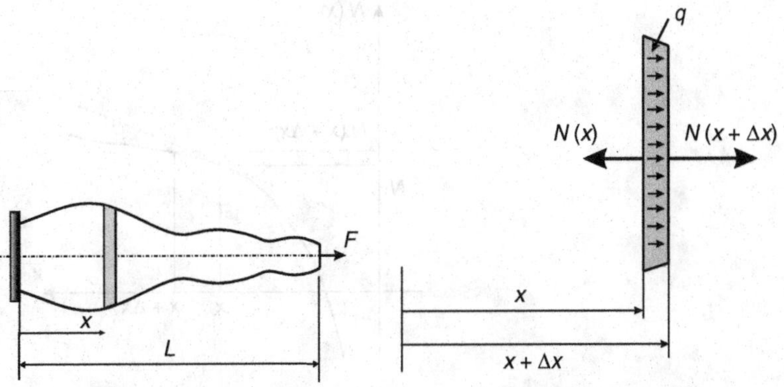

(a) Representation of a bar

(b) Free body diagram of a slice at position x

Figure 6.1

Bar and free body diagram of a slice of the bar.

length Δx. The net force on the left side of the slice equals $N(x)$, while on the right side of the slice a force $N(x+\Delta x)$ is present. The net force on the right side of the bar may be different from the net force on the left side of the slice owing to the presence of a so-called distributed volume force. A volume force Q is a force per unit of volume, and may be due, for instance, to gravity. Integration over the cross section area of the slice yields a load per unit length, called q.

If the distributed load q is assumed constant within the slice of thickness Δx, force equilibrium of the slice implies that

$$N(x) = N(x+\Delta x) + q\Delta x. \tag{6.1}$$

This may also be written as

$$\frac{N(x+\Delta x) - N(x)}{\Delta x} + q = 0. \tag{6.2}$$

If the length of the slice Δx approaches zero, we can write

$$\lim_{\Delta x \to 0} \frac{N(x+\Delta x) - N(x)}{\Delta x} = \frac{dN}{dx}, \tag{6.3}$$

where dN/dx denotes the derivative of $N(x)$ with respect to x. The transition expressed in Eq. (6.3) is illustrated in Fig. 6.2. In this graph, a function $N(x)$ is sketched. The function $N(x)$ is evaluated at x and $x+\Delta x$, while Δx is small. When moving from x to $x+\Delta x$, the function $N(x)$ changes a small amount: from $N(x)$ to $N(x+\Delta x)$. If Δx is sufficiently small, the ratio $\Delta N/\Delta x$ defines the tangent line to the function $N(x)$ at point x, and hence equals the derivative of $N(x)$ with respect to x. Notice that this implies that for sufficiently small Δx:

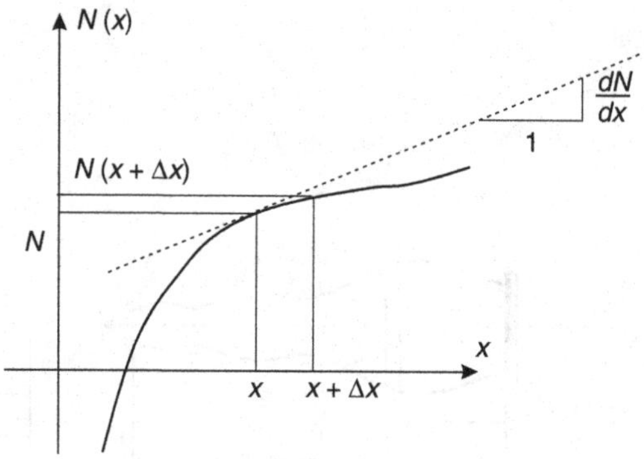

Figure 6.2

First-order derivative of a function $N(x)$.

$$N(x+\Delta x) \approx N(x) + \frac{dN}{dx}\Delta x. \tag{6.4}$$

The result of this transformation is that the force equilibrium relation Eq. (6.2) may be written as

$$\frac{dN}{dx} + q = 0. \tag{6.5}$$

If the load per unit length q equals zero, then the equilibrium equation reduces to

$$\frac{dN}{dx} = 0, \tag{6.6}$$

which means that the force N is constant throughout the bar. It actually has to be equal to the force F applied to the right end of the bar: see Fig. 6.1(a). Consequently, if the slice of Fig. 6.1(b) is considered, the force $N(x)$ equals $N(x+\Delta x)$. In other words, the force in the bar can only be non-constant if q can be neglected.

6.3 Stress and Strain

The equilibrium equation (6.5) derived in the previous section does not give information about the deformation of the bar. For this purpose, a relation between force and strain or strain rate must be defined, similar to the force–strain relation for an elastic spring, discussed in Chapter 4. For continuous media it is more appropriate to formulate a relation between force per unit area (stress) and a deformation measure, such as strain or strain rate. The concepts of stress and strain in continuous media are introduced in this section.

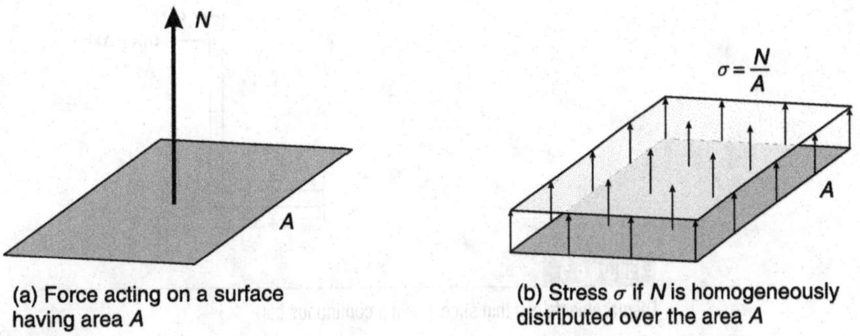

Figure 6.3

Stress σ as a homogeneously distributed force N over an area A.

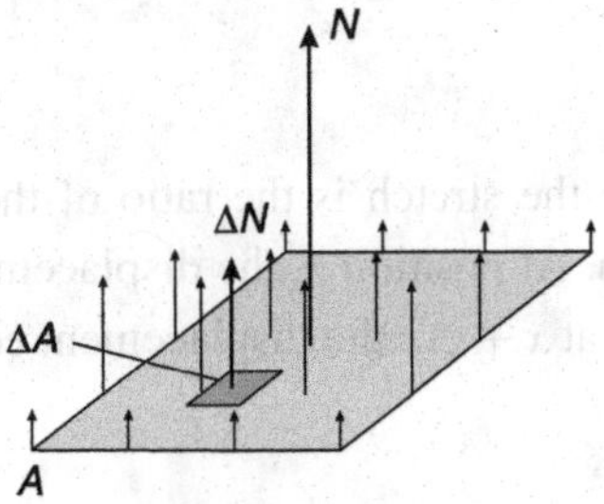

Figure 6.4

Inhomogeneous distribution of the force N.

In the one-dimensional case discussed in this chapter, the force N acting on a cross section of the bar is assumed to be homogeneously distributed over the surface of this cross section. If A denotes the area of the cross section, the **stress** σ is defined as

$$\sigma = \frac{N}{A}. \tag{6.7}$$

So, the stress is a force per unit area. If the force is not homogeneously distributed over a surface (see Fig. 6.4), we must consider a small part of the surface, ΔA. Actually, an infinitesimally small area ΔA is considered. This surface area only carries a part ΔN of the total force N. Then the stress σ is formally defined as

$$\sigma = \lim_{\Delta A \to 0} \frac{\Delta N}{\Delta A}. \tag{6.8}$$

This means that the stress is defined by the ratio of an infinitesimal amount of force ΔN over an infinitesimal amount of area ΔA, while the infinitesimal area ΔA approaches zero. In the following, it is assumed that Eq. (6.7) can be applied.

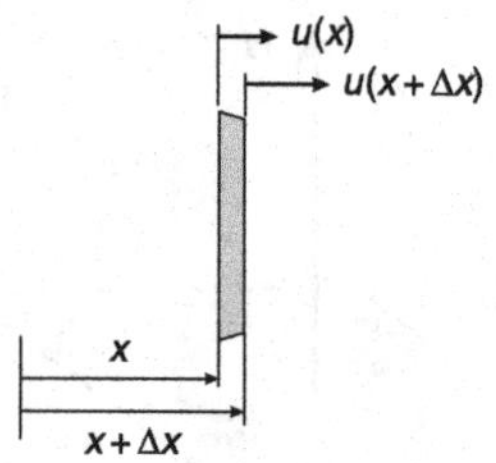

Figure 6.5
Displacements of a thin slice within a continuous bar.

Consider a slice of the bar (shaded) having length Δx, as depicted in Fig. 6.5. The linear **strain** ε is expressed in terms of the stretch λ of the slice by

$$\varepsilon = \lambda - 1, \tag{6.9}$$

where the stretch is the ratio of the deformed length of the slice and the initial length. At position x, the displacement of the cross section of the bar equals $u(x)$, while at $x + \Delta x$ the displacement equals $u(x + \Delta x)$. The initial length of the slice equals:

$$\ell_0 = \Delta x, \tag{6.10}$$

while the current length is given by

$$\ell = \Delta x + u(x + \Delta x) - u(x). \tag{6.11}$$

Therefore, the stretch, that is the ratio of the deformed length to the initial length, is given by:

$$\lambda = \frac{\Delta x + u(x + \Delta x) - u(x)}{\Delta x}. \tag{6.12}$$

Consequently, if the width of the slice Δx approaches zero, the strain is computed from

$$\begin{aligned} \varepsilon &= \lim_{\Delta x \to 0} \frac{\Delta x + u(x + \Delta x) - u(x)}{\Delta x} - 1 \\ &= \lim_{\Delta x \to 0} \frac{u(x + \Delta x) - u(x)}{\Delta x}. \end{aligned} \tag{6.13}$$

Using the definition of the derivative, this yields

$$\varepsilon = \frac{du}{dx}. \tag{6.14}$$

In conclusion, the strain is defined as the derivative of the displacement field u with respect to the coordinate x.

Example 6.1 Consider the shape of a femur, the top part of which is illustrated below. In the one-dimensional situation, we can assume that a vertical force is applied to the bone as a result of body weight and/or movement. The cross-sectional area of the bone in the direction of x varies quite considerably. Consequently, the stress in the femur as a result of the applied force will be inhomogeneous.

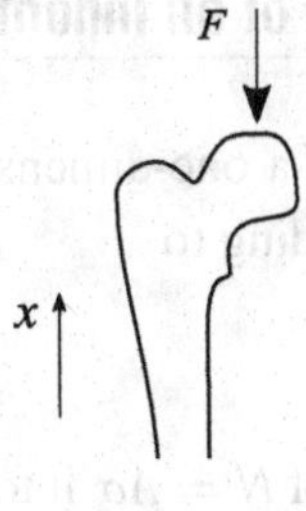

6.4 Elastic Stress–Strain Relation

Recall that the force–strain relation for an elastic spring at small, infinitesimal displacements is given by

$$\vec{F} = c \underbrace{\frac{\vec{a} \cdot (\vec{u}_B - \vec{u}_A)}{l_0}}_{\text{fibre strain}} \vec{a}. \tag{6.15}$$

Here, c represents the stiffness of the spring, while the unit vector $\vec{a}$ denotes the orientation of the spring in space. In analogy with this, the (one-dimensional) stress–strain relation for linearly elastic materials is defined as

$$\sigma = E\varepsilon, \tag{6.16}$$

where E is the so-called **Young's modulus**. Using the definition of the strain in terms of the derivative of the displacement field, this may also be written as

$$\sigma = E\frac{du}{dx}. \tag{6.17}$$

Example 6.2 For a given displacement field $u(x)$, the stress field can be computed. Suppose, for instance, that the Young's modulus is constant and that u is given by a polynomial expression, say:

$$u = a_1 x + 2a_2 x^2 + 5a_3 x^3,$$

with a_1, a_2 and a_3 known coefficients. Then the stress will be

$$\sigma = E\frac{du}{dx} = E(a_1 + 4a_2 x + 15a_3 x^2).$$

6.5 Deformation of an Inhomogeneous Bar

In the case of a one-dimensional bar, the stress at each cross section is uniquely defined according to

$$\sigma = \frac{N}{A}. \tag{6.18}$$

Substitution of $N = A\sigma$ into the equilibrium equation Eq. (6.5) yields

$$\frac{d(A\sigma)}{dx} + q = 0. \tag{6.19}$$

Subsequently, the stress–strain relation Eq. (6.17) is substituted such that the following second-order differential equation in terms of the displacement field $u(x)$ is obtained:

$$\frac{d}{dx}\left(EA\frac{du}{dx}\right) + q = 0. \tag{6.20}$$

In the absence of a force per unit length, $q = 0$, the force in the bar must be constant. The stress σ does not have to be constant, because the cross-sectional area A may be a function of the coordinate x.

Suppose that both the force and the stress are constant (this can only occur if $q = 0$ and A is constant). Then it follows from Eq. (6.20) that

$$EA\frac{du}{dx} = c, \tag{6.21}$$

with c a constant. Nevertheless, the strain $\varepsilon = du/dx$ may be a function of the coordinate x if the Young's modulus is non-constant.

The solution of the differential Eq. (6.20) yields the displacement as a function of x, and once $u(x)$ is known the strain $\varepsilon(x)$ and the stress $\sigma(x)$ in the bar can be retrieved. However, this differential equation can only be solved if two appropriate boundary conditions are specified. Two types of boundary conditions are distinguished. First, there are **essential** boundary conditions, formulated in terms of specified boundary displacements. The displacement $u(x)$ must at least be specified at one end point, and depending on the problem at hand, possibly at two. This is required to uniquely determine $u(x)$ and may be understood as follows. Suppose that $\hat{u}$ satisfies the equilibrium equation Eq. (6.20). Then, if the displacement $u(x)$ is not specified for at least one end point, an arbitrary constant displacement c may be added to $\hat{u}(x)$, while Eq. (6.20) for this modified displacement field $\hat{u}(x) + c$ is still satisfied, since the strain is given by

$$\varepsilon = \frac{d(\hat{u}+c)}{dx} = \frac{d\hat{u}}{dx} + \underbrace{\frac{dc}{dx}}_{=0} = \frac{d\hat{u}}{dx} \tag{6.22}$$

for any constant c. Such a constant c would correspond to a rigid body translation of the bar. So, in conclusion, at least one essential boundary condition **must** be specified.

Second, **natural** boundary conditions, formulated in terms of external boundary loads, may be specified, depending on the problem at hand. In the configuration visualized in Fig. 6.1(a), the bar is loaded by an external load F at the right end of the bar, at $x = L$, with L the length of the bar. At this boundary, the force equals

$$N(x = L) = F = \sigma A. \tag{6.23}$$

Since $\sigma = E\, du/dx$, the natural boundary condition at $x = L$ reads

$$EA\frac{du}{dx} = F. \tag{6.24}$$

Because the equilibrium equation Eq. (6.20) is a second-order differential equation, **two** boundary conditions must be specified: one must be an essential boundary condition (the displacement must be specified at least at one point to avoid rigid body displacement) and the other may be either an essential or natural boundary condition. The combination of the equilibrium equation with appropriate boundary conditions is called a (determinate) **boundary value** problem.

Example 6.3

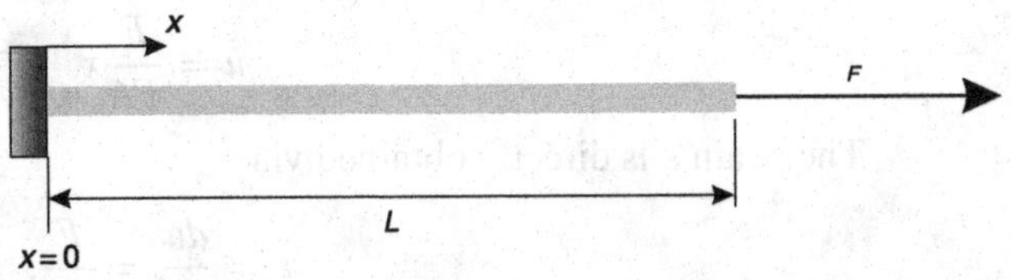

As a first example, the solution of a well-defined boundary value problem for a homogeneous bar without a distributed load is analysed. Consider a bar of length L that has a uniform cross section, hence A is constant, and with constant Young's modulus E. There is no volume load present, hence $q = 0$. At $x = 0$ the displacement is suppressed, hence $u = 0$ at $x = 0$, while at the other end of the bar a force F is applied. Then, the boundary value problem is fully described by the following set of equations:

$$\frac{d}{dx}\left(EA\frac{du}{dx}\right) = 0 \quad \text{for} \quad 0 < x < L$$
$$u = 0 \quad \text{at} \quad x = 0$$
$$EA\frac{du}{dx} = F \quad \text{at} \quad x = L\,.$$

Integrating the equilibrium equation once yields

$$EA\frac{du}{dx} = c,$$

where c denotes an integration constant. This may also be written as

$$\frac{du}{dx} = \frac{c}{EA}.$$

Because both the Young's modulus E and the cross-sectional area A are constant, integration of this relation gives

$$u = \frac{c}{EA}x + d,$$

with d yet another integration constant. So, the solution $u(x)$ is known provided that the integration constants c and d can be determined. For this purpose, the boundary conditions at both ends of the bar are used. First, since at $x = 0$ the displacement $u = 0$, the integration constant d must be zero, hence

$$u = \frac{c}{EA}x.$$

Second, at $x = L$ the force is known, such that

$$F = EA\frac{du}{dx}\bigg|_{x=L} = EA\frac{c}{EA} = c.$$

So the (unique) solution to the boundary value problem reads

$$u = \frac{F}{EA}x.$$

The strain ε is directly obtained via

$$\varepsilon = \frac{du}{dx} = \frac{F}{EA},$$

while the stress σ follows from

$$\sigma = E\varepsilon = \frac{F}{A}$$

as expected.

Example 6.4 Consider, as before, a bar of length L, clamped at one end and loaded by a force F at the other end of the bar. The Young's modulus E is constant throughout the bar, but the cross section varies along the axis of the bar. Let the cross section $A(x)$ be given by

$$A = A_0 \left(1 + \frac{x}{3L}\right),$$

with A_0 a constant, clearly representing the cross-sectional area at $x = 0$.

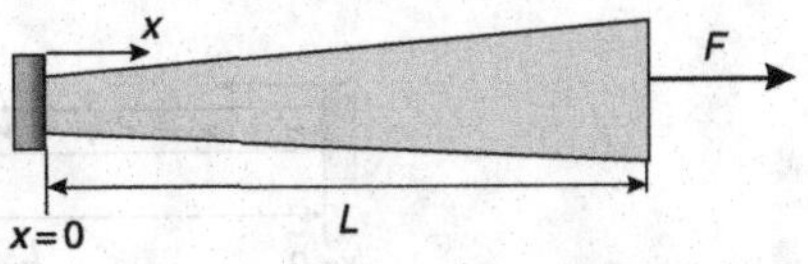

The boundary value problem is defined by the same set of equations as in the previous example:

$$\frac{d}{dx}\left(EA\frac{du}{dx}\right) = 0 \quad \text{for} \quad 0 < x < L$$
$$u = 0 \quad \text{at} \quad x = 0$$
$$EA\frac{du}{dx} = F \quad \text{at} \quad x = L.$$

Integration of the equilibrium equation yields

$$\frac{du}{dx} = \frac{c}{EA} = \frac{c}{EA_0(1 + \frac{x}{3L})},$$

with c an integration constant that needs to be identified. Integration of this result gives

$$u = \frac{3cL}{EA_0} \ln\left(1 + \frac{x}{3L}\right) + d,$$

with d another integration constant. The integration constants c and d can be determined by application of the boundary conditions at $x = 0$ and $x = L$. Since at $x = 0$ the displacement $u = 0$ it follows that

$$u(0) = 0 = \frac{3cL}{EA_0} \ln(1 + 0) + d = d,$$

hence $d = 0$. At $x = L$, the force F is given, such that

$$EA\frac{du}{dx}\bigg|_{x=L} = c = F.$$

Consequently

$$u = \frac{3FL}{EA_0} \ln\left(1 + \frac{x}{3L}\right).$$

The stress σ in the bar is computed directly from

$$\sigma = E\frac{du}{dx} = \frac{F}{A_0(1+\frac{x}{3L})}.$$

Example 6.5 Consider a bar of length L with constant Young's modulus E and cross section A. As before, the bar is fixed at $x = 0$ while at $x = L$ a force F is applied. Furthermore, the bar is loaded by a constant distributed load q.

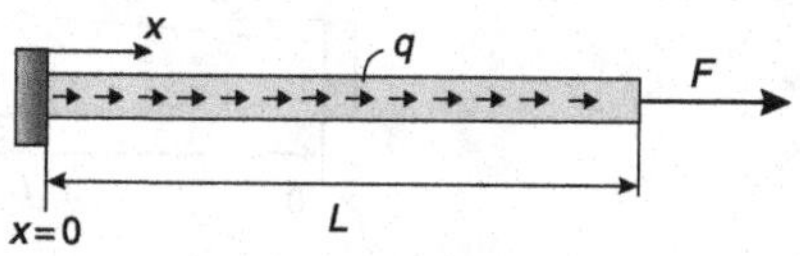

Accordingly, the boundary value problem is given by

$$\frac{d}{dx}\left(EA\frac{du}{dx}\right) = -q \quad \text{for } 0 < x < L$$
$$u = 0 \quad \text{at } x = 0$$
$$EA\frac{du}{dx} = F \quad \text{at } x = L.$$

Integration of the equilibrium equation yields

$$\frac{du}{dx} = -\frac{1}{EA}qx + \frac{c}{EA},$$

where c is an integration constant. Integration of this result yields

$$u = -\frac{1}{2EA}qx^2 + \frac{c}{EA}x + d.$$

Since at $x = 0$ the displacement $u = 0$, it follows immediately that $d = 0$. At $x = L$ the force F is prescribed, hence

$$EA\frac{du}{dx}\bigg|_{x=L} = -qL + c = F,$$

which implies that

$$c = F + qL.$$

Consequently, the displacement u is given by

$$u = -\frac{1}{2EA}qx^2 + \frac{F+qL}{EA}x.$$

Example 6.6 Consider a system of two bars as depicted in Fig. 6.6(a). The Young's modulus and the cross section of each bars are constant and are given by E_1, A_1 and E_2, A_2

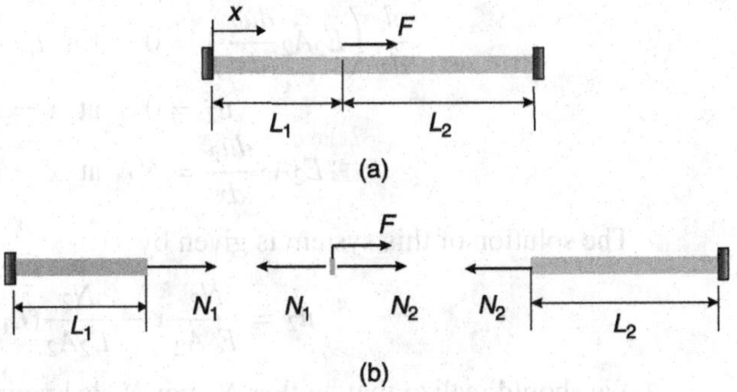

Figure 6.6
A two bar system and associated free body diagram.

for the left and right bar, respectively. The length of the left bar equals L_1 while the right bar has length L_2. The boundary conditions are as depicted in the figure: fixation at both $x = 0$ and $x = L_1 + L_2$. In addition, a concentrated force F is applied at $x = L_1$. Our goal is to determine the displacement of the point $x = L_1$.

To solve this problem, we recognize two difficulties:

- We have to find a way to deal with the concentrated force F.
- The problem is statically indeterminate. Reaction forces cannot be uniquely determined from force equilibrium alone.

The concentrated force has to be incorporated according to the following procedure. Three free body diagrams have to be created by virtually cutting the bar just left of the point where the force is applied and just right of that point. Thus, three free bodies can be distinguished: the left bar, the right bar and a very thin slice around the concentrated force. The free body diagrams are shown in Fig. 6.6(b).

The boundary value problem for the first bar is given by

$$\frac{d}{dx}\left(E_1A_1\frac{du_1}{dx}\right) = 0 \quad \text{for } 0 < x < L_1$$
$$u_1 = 0 \quad \text{at } x = 0$$
$$E_1A_1\frac{du_1}{dx} = N_1 \quad \text{at } x = L_1.$$

Clearly, based on the previous examples, for the first bar the displacement field is given by

$$u_1 = \frac{N_1}{E_1A_1}x.$$

For the second bar, the following boundary value problem holds:

$$\frac{d}{dx}\left(E_2A_2\frac{du_2}{dx}\right)=0 \quad \text{for } L_1<x<L_1+L_2$$
$$u_2=0 \quad \text{at } x=L_1+L_2$$
$$E_2A_2\frac{du_2}{dx}=N_2 \quad \text{at } x=L_1.$$

The solution of this system is given by

$$u_2=\frac{N_2}{E_2A_2}x-\frac{N_2}{E_2A_2}(L_1+L_2).$$

We should realize that neither N_1 nor N_2 is known so far. However, there are two additional equations that have to be satisfied. Force equilibrium of the slice at $x=L_1$ (see Fig. 6.6) requires that:

$$-N_1+N_2+F=0,$$

while the displacement field must be continuous at $x=L_1$: the two bars must remain fitting together, hence:

$$u_1(L_1)=u_2(L_1). \tag{6.25}$$

Based upon the solution for u_1 and u_2, it follows that:

$$u_1(L_1)=\frac{N_1}{E_1A_1}L_1,$$

$$u_2(L_1)=-\frac{N_2}{E_2A_2}L_2.$$

Because the bars must fit together at $x=L_1$ we find

$$N_1=-N_2\frac{E_1A_1}{L_1}\frac{L_2}{E_2A_2}.$$

Use of force equilibrium of the slice yields

$$N_2\frac{E_1A_1}{L_1}\frac{L_2}{E_2A_2}+N_2+F=0,$$

hence

$$N_2=\frac{-F}{1+\frac{E_1A_1}{L_1}\frac{L_2}{E_2A_2}}.$$

Now that N_1 and N_2 have been determined, it is possible to find an expression for the displacement of the bar at point $x=L_1$ as a function of the force F and the material and geometrical properties of both bars:

$$u(L_1)=\frac{FL_1L_2}{E_2A_2L_1+E_1A_1L_2}.$$

Exercises

6.1 A displacement field as a function of the coordinate x is given as: $u(x) = ax^2 + bx + c$, with a, b and c constant coefficients. Determine the strain field as a function of x.

6.2 A strain field as a function of coordinate x is given as $\varepsilon(x) = ax^2 + bx + c$, with a, b and c constant coefficients. Determine the displacement field $u(x)$ satisfying $u(0) = 0$.

6.3 A bar with Young's modulus E and length ℓ is clamped at $x = 0$ and loaded with a force F at $x = \ell$. The cross section is a function of x ($0 \leq x \leq \ell$), according to:

$$A(x) = A_0 \, e^{-\beta x}$$

with A_0 the cross-sectional area at $x = 0$ and with β a positive constant.

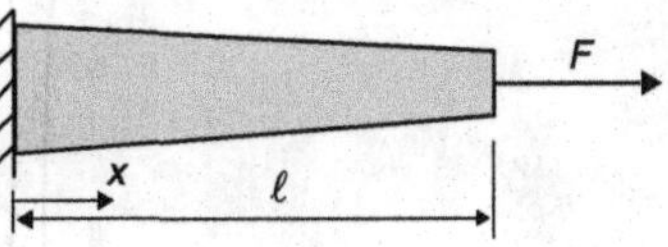

(a) Determine the stress field $\sigma(x)$.
(b) Determine the displacement field $u(x)$.

6.4 A bar with Young's modulus E, cross section A and length ℓ is clamped at $x = 0$. A distributed load $q(x)$ is applied in the x-direction on the bar, according to:

$$q(x) = \alpha \, e^{\beta x}$$

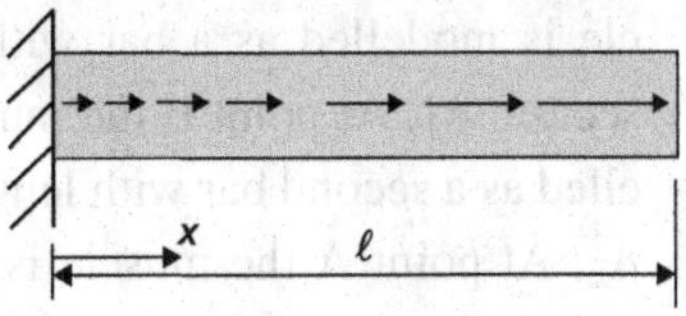

with α and β constant load parameters.
(a) Determine the stress field $\sigma(x)$.
(b) Determine the displacement field $u(x)$.

6.5 A bar with Young's modulus E, density ρ, cross section A and length ℓ is hanging by its own weight. The gravitational acceleration is g. The point where the bar is clamped is located at $x = 0$. Determine the displacement field $u(x)$ and the length change of the bar due to gravity.

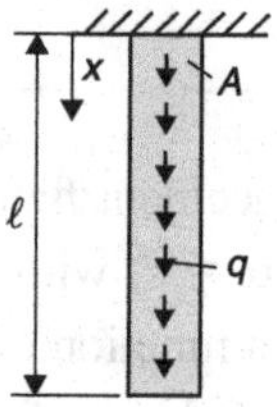

6.6 The central part of a femur is modelled as a straight tube with outer diameter D, inner diameter d and length ℓ. Assume that the cortical bone behaves like a linear elastic material with Young's modulus E. In addition, assume that the bone is loaded with an axial compressive force P.

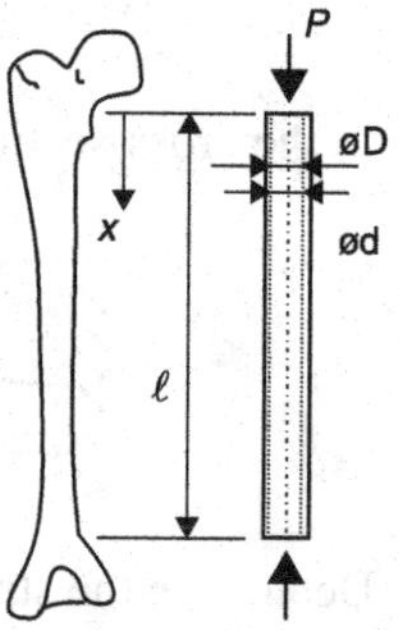

(a) Determine the stress field $\sigma(x)$.

(b) Determine the displacement field $u(x)$. Assume that at $x = \ell$ the displacement $u(\ell)$ satisfies $u(\ell) = 0$.

6.7 Consider a muscle/tendon complex as shown in the figure. To find out how much the tendon and the muscle are extended when the complex as a whole is loaded with a force F, a very crude two bar model can be used. The muscle is modelled as a bar with length ℓ_1, Young's modulus E_1 and cross section A_1. At point B the muscle is attached to the tendon, which is modelled as a second bar with length ℓ_2, Young's modulus E_2 and cross section A_2. At point A the muscle is attached to the bone, which we consider as a rigid fixation. At point C a force F is applied in the direction of the bar (tendon).

(a) Determine the internal force in a cross section between A and B and in a cross section between B and C.

(b) Determine the stress σ in a cross section between A and B and a cross section between B and C.

(c) What happens with the calculated forces and stresses if the Young's moduli of both muscle and tendon are reduced to half of their original value?

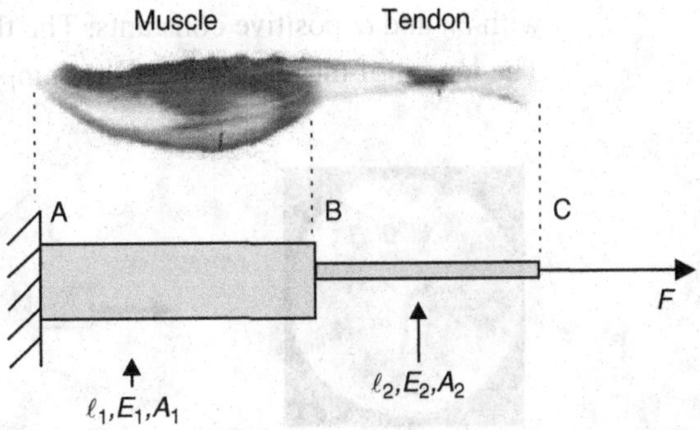

(d) Determine the displacements at point B and C as a result of the applied load F at point C.

6.8 A test is performed to determine the Young's modulus E of a blood vessel (see figure). The vessel has an inner radius r and an outer radius R. It is clamped on the left-hand side at $x = 0$, while the right-hand side at $x = \ell$ is displaced over a distance u_ℓ. The displacement is small compared with the length of the vessel ℓ.

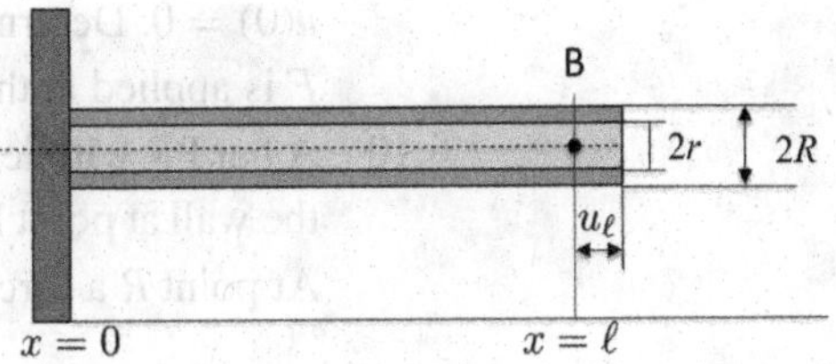

Determine the force F needed to displace the point B by u_ℓ as a function of E, u_ℓ, ℓ, R and r.

6.9 To determine the mechanical properties of a heart valve, a small rectangular piece of tissue is cut out of the valve and clamped in a uniaxial testing machine. After clamping, it can be seen that the test specimen is a little more slender in the middle (point M) than near the clamps (point C; see figure). It is decided to model the geometry of the sample as found after clamping. Because of symmetry only the right half of the sample is analysed. The following equilibrium equation describes the deformation in the x-direction:

$$\frac{d}{dx}\left(EA\frac{du}{dx}\right) = 0.$$

The width $b(x)$ is assumed to be described by:

$$b(x) = b_0\left(1 + \frac{\alpha x}{\ell}\right)$$

with b_0 and α positive constants. The thickness is constant and given by t. The length of the right half of the sample is denoted by ℓ.

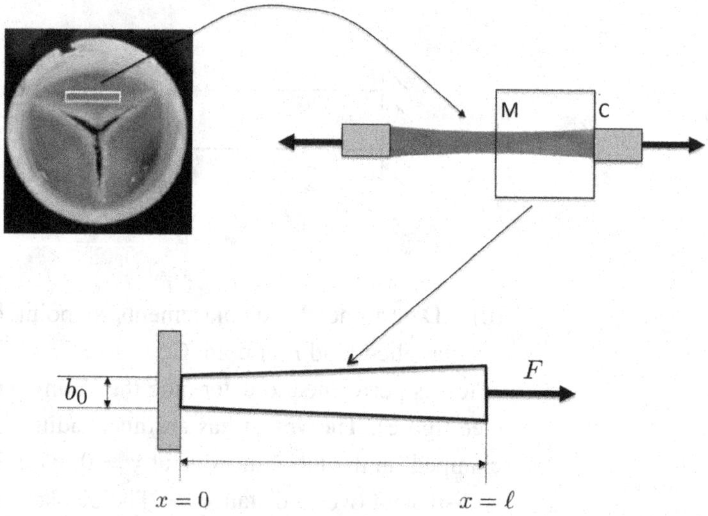

For the point $x = 0$ the displacement $u(0)$ is assumed to be suppressed, $u(0) = 0$. Determine the displacement $u(\ell)$ of the point $x = \ell$ when a force F is applied at that point.

6.10 A bar PR with length 2ℓ, Young's modulus E and cross section A is fixed to the wall at point P. At point Q a force F_1 is acting in the positive x-direction. At point R a force F_2 is acting in the negative x-direction (see figure).

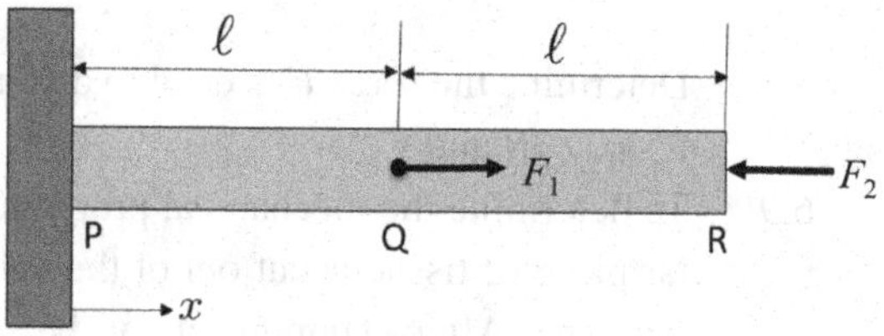

(a) Determine the stress in a cross section between P and Q, and in a cross section between Q and R.

(b) Determine the displacements of the points Q and R.

(c) What is the relation between F_1 and F_2 such that the displacement of point R equals zero? Could you predict that result?

7 Biological Materials and Continuum Mechanics

7.1 Introduction

Up to this point, all treated problems have been in some way one-dimensional. In Chapter 3 we have discussed equilibrium of two- and three-dimensional bodies and in Chapter 4 the fibres were allowed to have some arbitrary orientation in three-dimensional space. But, when deformations were involved, the focus was on fibres and bars, dealing with one-dimensional force–strain relationships. Only one-dimensional equations were solved. In this chapter and the ones that follow, the theory will be extended to the description of three-dimensional bodies, and it is opportune to spend some time looking at the concept of a continuum.

Consider a certain amount of solid and/or fluid material in a three-dimensional space. Although in reality for neighbouring points in space the (physical) character and behaviour of the residing material may be completely different (because of discontinuities at the microscopic level, which become clearer on reducing the scale of observation), it is common practice that a less detailed description (at a macroscopic level) with a more gradual change of physical properties is used. The discontinuous heterogeneous reality is **homogenized** and modelled as a continuum. To make this clearer, consider the bone in Fig. 7.1. Although one might conceive the bone at a macroscopic level, as depicted in Fig. 7.1(a), as a massive structure filling all the volume that it occupies in space, it is clear from Fig. 7.1(b) that at a smaller scale the bone is a discrete structure with open spaces in between (although the spaces can be filled with a softer material or a liquid). At an even smaller scale, at the level of one single trabecula, individual cells can be recognized. This phenomenon is typical for biological materials at any scale, whether it is at the organ level, the tissue level or cellular level, or even smaller. At each level, a very complex structure can be observed, and it is not feasible to take into account every detail of this structure. That is why a homogenization of the structure is necessary, and, depending on the objectives of the modelling (which is the question that has to be answered in the context of the selection of the theoretical model), a certain level of homogenization is chosen. It is always necessary to critically evaluate whether or not the chosen homogenization is allowed and to be

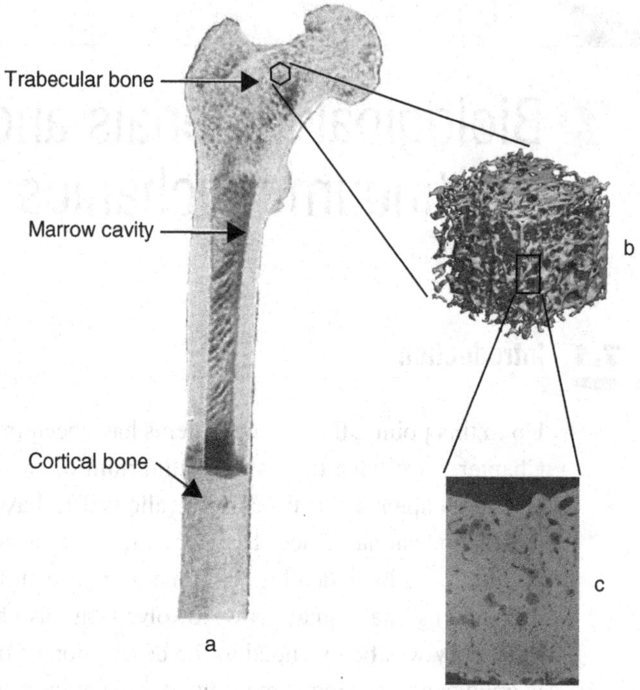

Figure 7.1

Structure of bone at different length scales (images courtesy of Bert van Rietbergen).

aware of the limitations of the model at hand. In Section 7.3 we will discuss when a continuum approach is allowed.

To model a continuum, physical variables are formulated as continuous functions of the position in space. Related to this, some attention will be given in this chapter to the visualization of physical fields. In addition, derivatives of variables will be discussed. The gradient operator, important in all kinds of theoretical derivations, constitutes a central part of this. The chapter ends with a section on the properties of second-order tensors, which form indispensable tools in continuum mechanics.

7.2 Orientation in Space

For an orientation in three-dimensional space, a Cartesian *xyz*-coordinate system is defined with origin O; see Fig. 7.2. The orientation of the coordinate system is laid down by means of the Cartesian vector basis $\{\vec{e}_x, \vec{e}_y, \vec{e}_z\}$, containing mutually perpendicular unit vectors. The position of an arbitrary point P in space can be

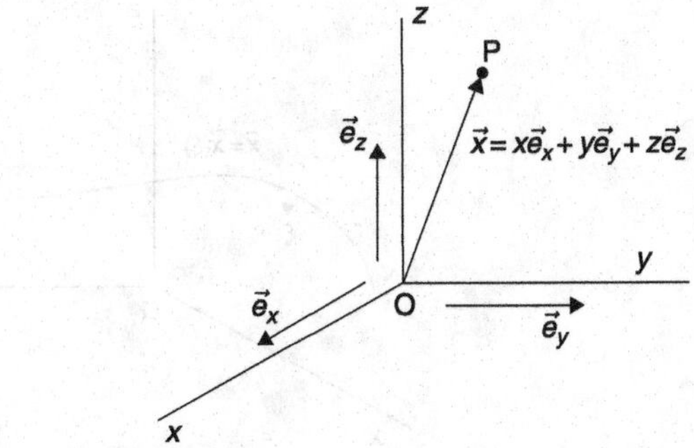

Figure 7.2

Coordinate system and position of a point P in space.

defined with the position vector $\vec{x}$, starting at the origin O and with the end point at P. This position vector can be specified by means of components with respect to the basis, defined earlier:

$$\vec{x} = x\,\vec{e}_x + y\,\vec{e}_y + z\,\vec{e}_z \quad \text{with} \quad \underset{\sim}{x} = \begin{bmatrix} x \\ y \\ z \end{bmatrix} = \begin{bmatrix} \vec{x}\cdot\vec{e}_x \\ \vec{x}\cdot\vec{e}_y \\ \vec{x}\cdot\vec{e}_z \end{bmatrix}. \tag{7.1}$$

The column $\underset{\sim}{x}$ is in fact a representation of the vector $\vec{x}$ with respect to the chosen basis vectors $\vec{e}_x$, $\vec{e}_y$ and $\vec{e}_z$ only. Nevertheless, the position vector at hand will sometimes also be indicated with $\underset{\sim}{x}$. This will not jeopardize uniqueness, because in the present context only one single set of basis vectors will be used.

The geometry of a curve (a set of points joined together) in three-dimensional space can be defined by means of a parameter description $\vec{x} = \vec{x}(\xi)$, in components $\underset{\sim}{x} = \underset{\sim}{x}(\xi)$, where ξ is specified within a certain interval; see Fig. 7.3. The tangent vector, with unit length, at an arbitrary point of this curve is depicted by $\vec{e}$, satisfying

$$\vec{e} = \frac{1}{\ell}\frac{d\vec{x}}{d\xi} \quad \text{with} \quad \ell = \sqrt{\left(\frac{d\vec{x}}{d\xi}\right)\cdot\left(\frac{d\vec{x}}{d\xi}\right)} \tag{7.2}$$

and after a transition to component format:

$$\underset{\sim}{e} = \frac{1}{\ell}\frac{d\underset{\sim}{x}}{d\xi} \quad \text{with} \quad \ell = \sqrt{\left(\frac{d\underset{\sim}{x}}{d\xi}\right)^{\mathrm{T}}\left(\frac{d\underset{\sim}{x}}{d\xi}\right)}. \tag{7.3}$$

The parameter ξ is distance measuring if ℓ equals 1; in that special case the length of an (infinitesimally small) line piece $d\vec{x}$ (with components $d\underset{\sim}{x}$) of the curve is exactly the same as the accompanying change $d\xi$.

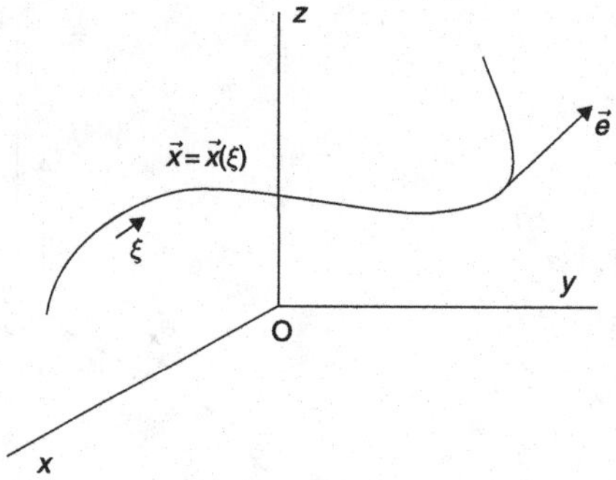

Figure 7.3

A parameterized curve in space.

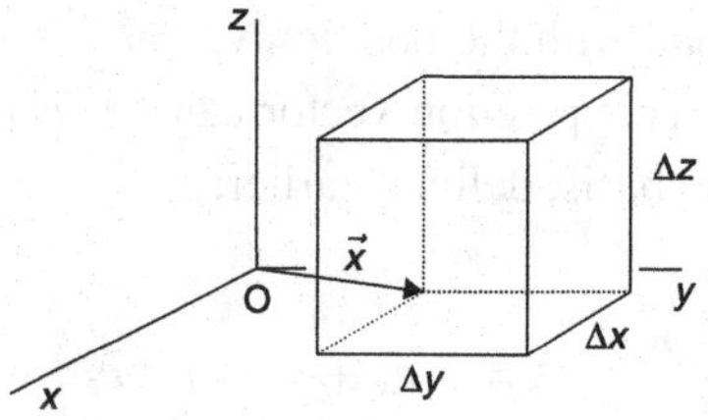

Figure 7.4

A subvolume $\Delta V = \Delta x \Delta y \Delta z$ of V in space.

A bound volume in three-dimensional space is denoted by V. That volume can be regarded as a continuous set of geometric points, of which the position vectors $\vec{x}$ have their end points within V. A subvolume is indicated by ΔV. That subvolume may have the shape of a (rectangular) block, with edges in the same direction as the coordinate axes, $\Delta V = \Delta x \Delta y \Delta z$ (see Fig. 7.4). An infinitesimally small volume element in the shape of a rectangular block is written as $dV = dxdydz$.

Example 7.1 Let us assume that a parameter curve with respect to a Cartesian basis $\{\vec{e}_x, \vec{e}_y\}$ is given by:

$$\vec{x} = \xi \vec{e}_x + \xi^2 \vec{e}_y,$$

then we can find a vector $\vec{a}$ tangent to this curve by:

$$\vec{a} = \frac{d\vec{x}}{d\xi} = \vec{e}_x + 2\xi \vec{e}_y,$$

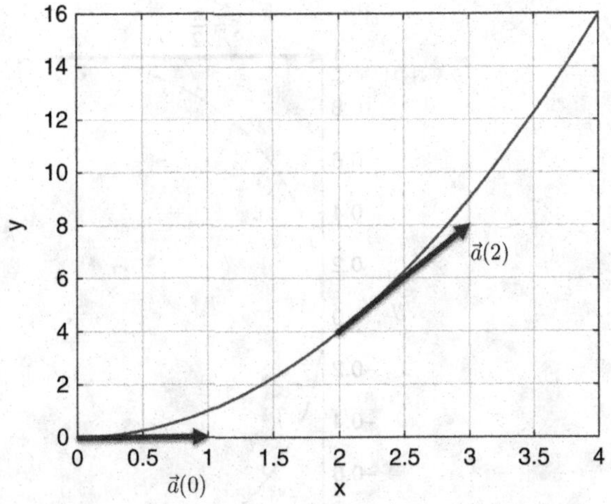

Figure 7.5

A curve in space with tangent vectors.

It is clear that for this curve the length of the vector $\vec{a}(\xi)$ changes with ξ. In Fig. 7.5 the vector $\vec{a}$ is drawn for two values of ξ:

$$\xi = 0 \rightarrow \vec{a}(0) = \frac{d\vec{x}}{d\xi} = \vec{e}_x$$

$$\xi = 2 \rightarrow \vec{a}(2) = \frac{d\vec{x}}{d\xi} = \vec{e}_x + 4\vec{e}_y.$$

In most practical cases we want a tangent vector with unit length, so we use:

$$\vec{e} = \frac{1}{\ell}\frac{d\vec{x}}{d\xi}.$$

In this case: $\ell = \sqrt{1 + 4\xi^2}$ resulting in:

$$\xi = 0 \quad \rightarrow \quad \vec{e}(0) = \vec{e}_x$$

$$\xi = 2 \quad \rightarrow \quad \vec{e}(2) = \frac{1}{\sqrt{17}}(\vec{e}_x + 4\vec{e}_y).$$

Example 7.2 The curve:

$$\vec{x} = \cos(\xi)\,\vec{e}_x + \sin(\xi)\,\vec{e}_y,$$

represents the parameter description of a circle with radius 1. The tangent vector $\vec{a}$ along this curve can be determined according to:

$$\frac{d\vec{x}}{d\xi} = -\sin(\xi)\,\vec{e}_x + \cos(\xi)\,\vec{e}_y$$

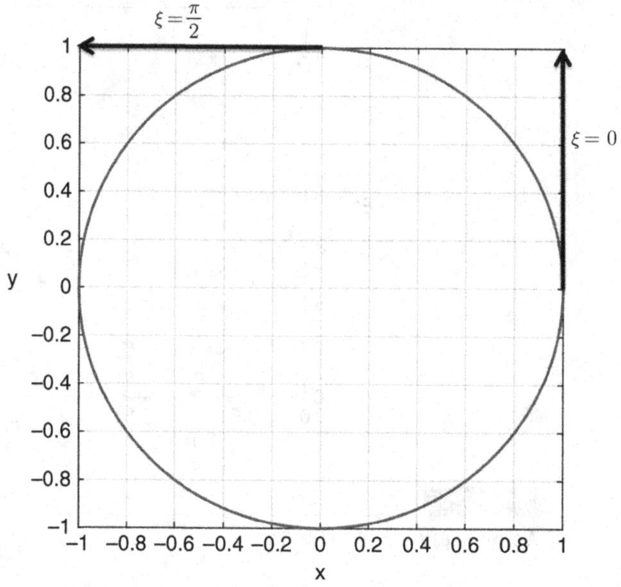

Figure 7.6

Tangent vectors to a circle.

with length:

$$\ell = \sqrt{\sin^2(\xi) + \cos^2(\xi)} = 1.$$

So in this case the length of the tangent vector does not depend on the parameter ξ.

7.3 Mass within the Volume V

Assume that the volume V is filled with a material with a certain mass. Sometimes different fractions of materials can be distinguished, which usually physically interact with each other. In the following we consider one such fraction, which is not necessarily homogeneous when observed in detail (at a 'microscopic' scale). Observation at a microscopic scale, in the current context, means a scale that is much smaller than the global dimensions of the considered volume. To support this visually, one might imagine a kind of base material containing very small substructures that can be visualized under a microscope (individual muscle cells in a skeletal muscle, for example; see Fig. 7.7).

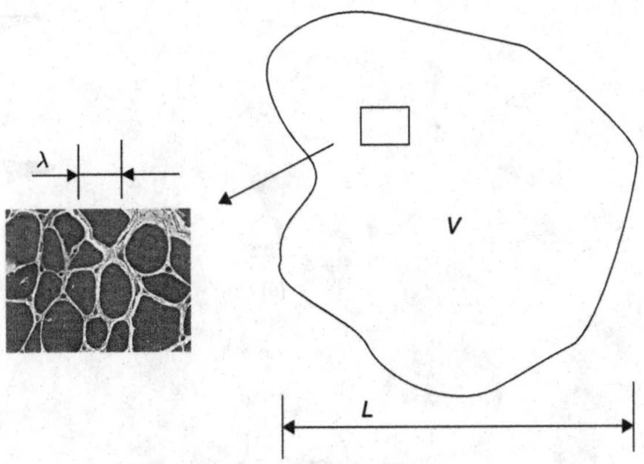

Figure 7.7

Cross section of a heterogeneous structure.

For the substructures (cells in the example), λ may be regarded as a relevant characteristic length scale for the size, as well as for the mutual stacking. The characteristic macroscopic length scale for the volume V is depicted by L. It is assumed that the inequality $L \gg \lambda$ is satisfied, meaning that on a macroscopic scale the heterogeneity at microscopic level is no longer recognizable. Attention is now focussed on physical properties, which are coupled to the material. As an illustration let us take the (mass) density. The local density ρ_ℓ at a spatial point with position vector $\vec{x}$ is defined by

$$\rho_\ell = \frac{dm}{dV} \quad \text{with} \quad dV = dxdydz \to 0, \tag{7.4}$$

where dm is the mass of the considered fraction (the cells) in the volume dV. Ignoring some of the mathematical subtleties, a discontinuous field $\rho_\ell(\vec{x})$ results for the local density, with λ as the normative measure for the mutual distance of the discontinuities. On the microscopic level such a variation can be expected, but on the macroscopic level it is often (not) observable, (often) not interesting and not manageable.

Let us define the (locally averaged) density ρ in the spatial point $\vec{x}$ by

$$\rho = \frac{\Delta m}{\Delta V} \quad \text{with:} \quad \Delta V = \Delta x \Delta y \Delta z \quad \text{and:} \quad \Delta m = \int_{\Delta V} \rho_\ell \, dV. \tag{7.5}$$

In this case it is assumed that the dimensions of ΔV are much larger than λ and at the same time much smaller than L. Provided that it is possible to specify such a subvolume ΔV this results in a continuous density field $\rho(\vec{x})$, from which the

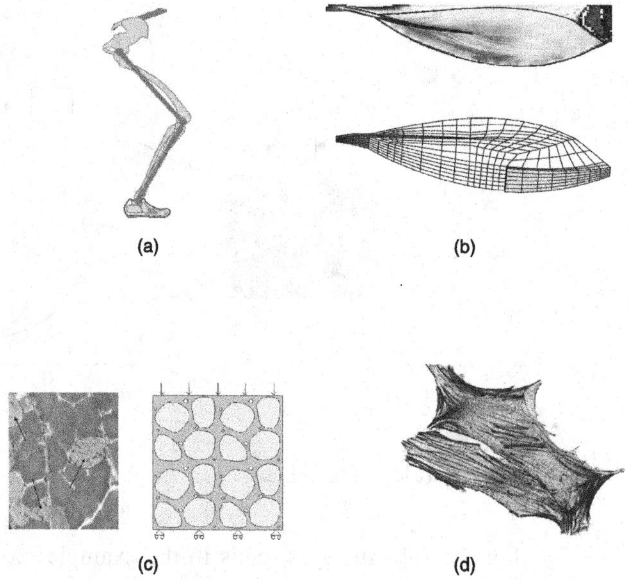

Figure 7.8

Examples of models at different length scales: (a) model of a leg; (b) model of a skeletal muscle; (c) representative volume element of a cross section of a muscle; (d) model of the cytoskeleton of a single cell.

microscopic deviations have disappeared, but which still contains macroscopic variations: the real density field, according to Eq. (7.4), is **homogenized**.

Continuum theory deals with physical properties that, in a way as described above, are made continuous. The results cannot be used on a microscopic level with a length scale λ, but on a much more global level.

In the above reasoning, we talk about two scales: a macroscopic and a microscopic scale. However, in many applications, a number of intermediate steps from large to small can be of interest. This is often seen in technical applications, but is especially of importance in biological materials. In such a case the characteristic measure λ of the components could suddenly become the relevant macroscopic length scale.

Example 7.3 Figure 7.8 illustrates some of the scales that can be distinguished in biomechanical modelling. Figure 7.8(a) could be a model to study a high-jumper. Typically, this kind of model would be used to study the coordination of muscles, describing the human body as a whole and examining how it moves. The macroscopic length scale L would be of the order of 1 [m]. In this case, the user is often interested in stresses that are found in the bones and joints. The length scale λ at which these stresses and strains are studied is of the order of centimetres. Muscles are often described with one-dimensional fibre-like models as discussed in Chapter 4.

Figure 7.8(b) shows a magnetic resonance image of the tibialis anterior of a rat and a schematic of the theoretical finite element model that it was based on to describe its mechanical behaviour ([18]). This was a model designed to study in detail the force transmission through the tibialis anterior. The muscle was modelled as a continuum and the fibre layout was included, as well as a detailed description of the active and passive mechanical behaviour of the muscle cells. For this model, L is of the order of a few centimetres and λ of the order of 100 [μm].

Figure 7.8(c) shows a microscopic cross section of a skeletal muscle together with a so-called representative volume element that is used for a microstructural model of a skeletal muscle. This model was used to study the transport of oxygen from the capillaries to the cells and how the oxygen is distributed over cell and intercellular space. Here the substructures – cells and intercellular spaces – were modelled as a continuum. In this case $L \approx 50$ [μm] and $\lambda \approx 0.1$ [μm]. One could even go another step further in modelling the single cell as shown in Fig. 7.8(d). In this case, the actin skeleton that is shown could be modelled as a fibre-like structure and the cytoplasm of the cell as a continuum, filled with a liquid or a solid ($L \approx 10$ [μm], $\lambda \approx 10$ [nm]).

7.4 Scalar Fields

Let us consider the variation of a scalar physical property of the material (or a fraction of it) in the volume V. An example could be the temperature field: $T = T(\vec{x})$. Visualization of such a field in a three-dimensional configuration could be done by drawing (contour) planes with constant temperature. A clearer picture is obtained when lines of constant temperature (isotherms or contour lines) are drawn on a flat two-dimensional surface of a cross section (see Fig. 7.9). By means of this figure, estimates can be made of the partial derivatives of the temperature with respect to to x (with y and z constant) and y (with x and z constant). For example, at the lower boundary the partial derivatives could be approximated by

$$\frac{\partial T}{\partial x} = 2 \quad \left[{}^\circ\text{C m}^{-1}\right], \quad \frac{\partial T}{\partial y} = 0 \quad \left[{}^\circ\text{C m}^{-1}\right]. \tag{7.6}$$

In a three-dimensional scalar field, the partial derivatives with respect to the coordinates are assembled in a column, i.e. the column with the components of the gradient vector associated with the scalar field. The gradient operator is specified by $\vec{\nabla}$ when vector notation is used and by $\underset{\sim}{\nabla}$ when components are used. In the example of the temperature field, this yields

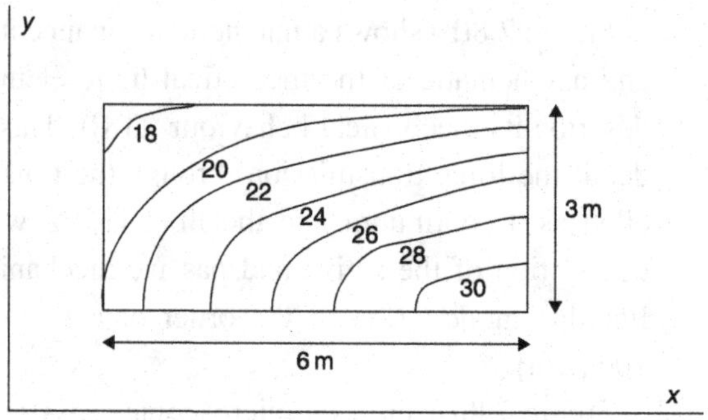

Figure 7.9

Isotherms (°C) in a cross section with constant z.

$$\underset{\sim}{\nabla} T = \begin{bmatrix} \frac{\partial T}{\partial x} \\ \frac{\partial T}{\partial y} \\ \frac{\partial T}{\partial z} \end{bmatrix} \quad \text{with} \quad \underset{\sim}{\nabla} = \begin{bmatrix} \frac{\partial}{\partial x} \\ \frac{\partial}{\partial y} \\ \frac{\partial}{\partial z} \end{bmatrix}, \tag{7.7}$$

and also

$$\vec{\nabla} T = \vec{e}_x \frac{\partial T}{\partial x} + \vec{e}_y \frac{\partial T}{\partial y} + \vec{e}_z \frac{\partial T}{\partial z} \quad \text{with} \quad \vec{\nabla} = \vec{e}_x \frac{\partial}{\partial x} + \vec{e}_y \frac{\partial}{\partial y} + \vec{e}_z \frac{\partial}{\partial z}. \tag{7.8}$$

The gradient of a certain property is often a measure for the intensity of a physical transport phenomenon; the gradient of the temperature, for example, is directly related to the heat flux. Having the gradient of a certain property (T), the derivative of that property along a spatial curve (given in a parameter description; see Fig. 7.3), $\vec{x} = \vec{x}(\xi)$, can be determined (by using the chain rule for differentiation):

$$\begin{aligned} \frac{dT}{d\xi} &= \frac{\partial T}{\partial x}\frac{dx}{d\xi} + \frac{\partial T}{\partial y}\frac{dy}{d\xi} + \frac{\partial T}{\partial z}\frac{dz}{d\xi} \\ &= (\underset{\sim}{\nabla} T)^{\mathrm{T}} \frac{d\underset{\sim}{x}}{d\xi} = (\underset{\sim}{\nabla}^{\mathrm{T}} T)\frac{d\underset{\sim}{x}}{d\xi} \\ &= \left(\frac{d\underset{\sim}{x}}{d\xi}\right)^{\mathrm{T}} \underset{\sim}{\nabla} T. \end{aligned} \tag{7.9}$$

On the right-hand side of the equation the inner product of the (unnormalized) tangent vector to the curve and the gradient vector can be recognized. Equation (7.9) can also be written as

$$\frac{dT}{d\xi} = \frac{d\vec{x}}{d\xi} \cdot \vec{\nabla} T. \tag{7.10}$$

Along a contour line (isotherm) in a two-dimensional cross section, $dT/d\xi = 0$, so the gradient vector $\vec{\nabla}T$ (with components $\underset{\sim}{\nabla}T$) must be perpendicular to the contour line in each arbitrary point. In a three-dimensional configuration, it can be said that the gradient vector of a certain scalar property at a point in the volume V will be perpendicular to the contour plane through that point. The directional derivative dT/de of the temperature (the increase of the temperature per unit length along the curve) in the direction of the unit vector $\vec{e}$, with components $\underset{\sim}{e}$, can be written as

$$\frac{dT}{de} = \underset{\sim}{e}^{\mathrm{T}}\underset{\sim}{\nabla}T \quad \text{and also} \quad \frac{dT}{de} = \vec{e} \cdot \vec{\nabla}T. \tag{7.11}$$

Example 7.4 On a domain in three-dimensional space, given by: $-5\ell \leq x \leq 5\ell$, $-5\ell \leq y \leq 5\ell$, $-5\ell \leq z \leq 5\ell$ with ℓ a constant, a temperature field is given by:

$$T = 100 - \frac{x^2 + 2y^2 + z^2}{\ell^2}.$$

In addition, a curve in space is given by the following parameter description:

$$\vec{x}(\xi) = (1 + 3\xi)\,\vec{e}_x + 2\xi^2\,\vec{e}_z\ .$$

We wish to determine the directional derivative of the temperature T in the point along the curve that is defined by $\xi = 1$. The unit vector $\vec{e}$ tangent to the curve is found by means of:

$$\vec{e} = \frac{1}{\ell}\frac{d\vec{x}}{d\xi} = \frac{1}{\sqrt{9 + 16\xi^2}}(3\vec{e}_x + 4\xi\vec{e}_z).$$

The gradient of T yields:

$$\begin{aligned} \vec{\nabla}T = & \frac{\partial T}{\partial x}\vec{e}_x + \frac{\partial T}{\partial y}\vec{e}_y + \frac{\partial T}{\partial z}\vec{e}_z \\ = & -\frac{2x}{\ell^2}\vec{e}_x - \frac{4y}{\ell^2}\vec{e}_y - \frac{2z}{\ell^2}\vec{e}_z. \end{aligned}$$

By definition:

$$\vec{x} = x\vec{e}_x + y\vec{e}_y + z\vec{e}_z$$

so:

$$x = 1 + 3\xi; \quad y = 0; \quad z = 2\xi^2.$$

For $\xi = 1$ we find:

$$\frac{dT}{de} = \vec{e} \cdot \vec{\nabla}T = -\frac{8}{\ell^2}.$$

Example 7.5 Consider a domain in the form of a cube in three-dimensional space, given by $-\ell \leq x \leq 3\ell$, $-2\ell \leq y \leq 2\ell$ and $-2\ell \leq z \leq 2\ell$ with ℓ a constant and x, y and z the Cartesian coordinates. Within this domain, a concentration field exists with known gradient $\vec{\nabla}c$ which is given by:

$$\vec{\nabla}c = \theta\vec{e}_y \quad \text{with } \theta \ \text{ a constant.}$$

In the domain, a curve is given. The points on the curve satisfy:

$$(x-\ell)^2 + y^2 = \ell^2 \quad \text{and} \quad z = \ell.$$

Calculate the coordinates of the point (or points) on the curve where the derivative of the concentration in the direction of the curve is equal to zero. The curve can be recognized as a circle with radius ℓ and midpoint $(x, y, z) = (\ell, 0, \ell)$, which can be described with the parameter equation:

$$\vec{x}(\xi) = (\ell + \ell\cos(\xi))\,\vec{e}_x + \ell\sin(\xi)\,\vec{e}_y + \ell\vec{e}_z.$$

For the tangent vector along the curve, we find:

$$\vec{e} = -\sin(\xi)\,\vec{e}_x + \cos(\xi)\,\vec{e}_y.$$

The directional derivative of c is zero when:

$$\vec{e}\cdot\vec{\nabla}c = 0 \quad \rightarrow \quad \cos(\xi) = 0,$$

which is the case for:

$$\xi = \frac{\pi}{2} + n\pi \quad \text{with } n = 0, 1, 2, 3, \ldots$$

7.5 Vector Fields

In the previous section, a scalar temperature field was considered. Departing from a given temperature field, we were able to determine the associated (temperature) gradient field. This gradient field can be regarded as a vector field (at every point of the volume V, the components of the associated vector can be determined). In the current chapter, we want to start with a vector field, and use the (momentary) velocity field $\vec{v} = \vec{v}(\vec{x})$ of the material in the volume V as an example. With respect to the Cartesian xyz-coordinate system, the velocity vector $\vec{v}$ and the associated column $\underset{\sim}{v}$ can be written as

$$\vec{v} = v_x\vec{e}_x + v_y\vec{e}_y + v_z\vec{e}_z, \quad \underset{\sim}{v} = \begin{bmatrix} v_x \\ v_y \\ v_z \end{bmatrix}. \tag{7.12}$$

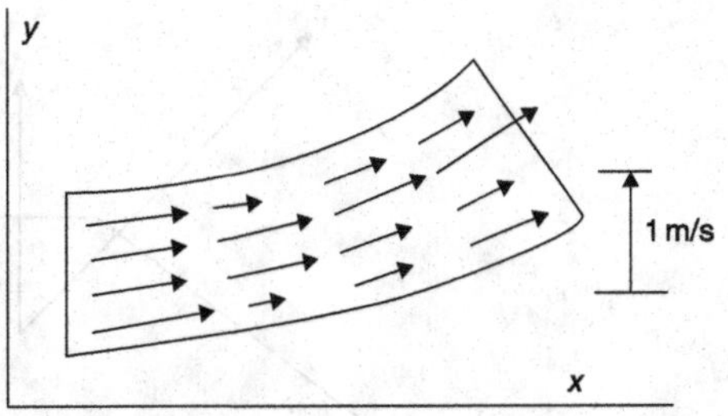

Figure 7.10

Velocity field in a cross section with constant z.

The possibilities for a clear graphical illustration of a vector field (velocity) in a three-dimensional configuration are limited. For a problem with a flat fluid flow pattern, for example:

$$\begin{aligned} v_x &= v_x(x, y) \\ v_y &= v_y(x, y) \\ v_z &= 0, \end{aligned} \tag{7.13}$$

a representation like the one given in Fig. 7.10 can be used. The arrows represent the velocity vector (magnitude and orientation) of the fluid at the tail of the arrow.

At a given velocity $\vec{v}$ in a certain point of V, the velocity component v in the direction of an arbitrary unit vector $\vec{e}$ (components in the column $\underset{\sim}{e}$) can be calculated with

$$v = v(\vec{e}) = \vec{v} \cdot \vec{e} \quad \text{and also} \quad v = v(\underset{\sim}{e}) = \underset{\sim}{v}^{\mathrm{T}} \underset{\sim}{e}. \tag{7.14}$$

In Fig. 7.11, a small plane is drawn at a certain point in V. The plane has an outward (unit) normal $\vec{n}$. The local velocity of the material is specified with $\vec{v}$. Thus, the amount of mass per unit of surface and time flowing out of the plane in the figure is given by

$$\rho\, \vec{v} \cdot \vec{n} = \rho\, \underset{\sim}{v}^{\mathrm{T}} \underset{\sim}{n}. \tag{7.15}$$

This is called the mass flux.

For vector fields, the associated gradient can also be determined. The gradient of a vector field results in a tensor field. For example, for the gradient of the velocity field we find

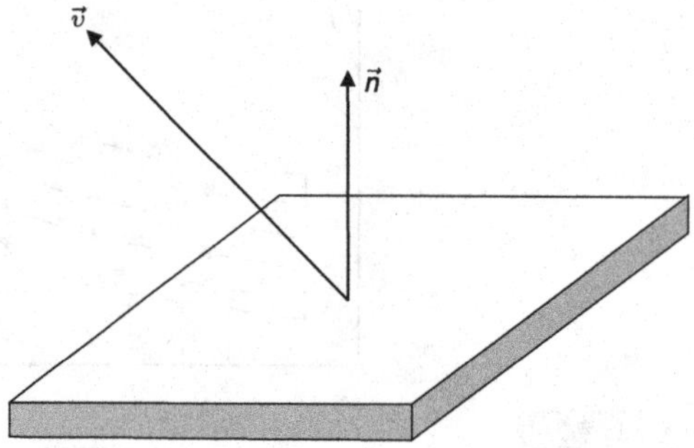

Figure 7.11

Flux of material through a small surface area.

$$
\begin{aligned}
\vec{\nabla}\vec{v} &= \left(\vec{e}_x\frac{\partial}{\partial x} + \vec{e}_y\frac{\partial}{\partial y} + \vec{e}_z\frac{\partial}{\partial z}\right)(v_x\vec{e}_x + v_y\vec{e}_y + v_z\vec{e}_z) \\
&= \frac{\partial v_x}{\partial x}\vec{e}_x\vec{e}_x + \frac{\partial v_y}{\partial x}\vec{e}_x\vec{e}_y + \frac{\partial v_z}{\partial x}\vec{e}_x\vec{e}_z \\
&+ \frac{\partial v_x}{\partial y}\vec{e}_y\vec{e}_x + \frac{\partial v_y}{\partial y}\vec{e}_y\vec{e}_y + \frac{\partial v_z}{\partial y}\vec{e}_y\vec{e}_z \\
&+ \frac{\partial v_x}{\partial z}\vec{e}_z\vec{e}_x + \frac{\partial v_y}{\partial z}\vec{e}_z\vec{e}_y + \frac{\partial v_z}{\partial z}\vec{e}_z\vec{e}_z.
\end{aligned}
\tag{7.16}
$$

The associated matrix representation can be written as

$$
\underset{\sim}{\nabla}\,\underset{\sim}{v}^{\mathrm{T}} = \begin{bmatrix} \frac{\partial v_x}{\partial x} & \frac{\partial v_y}{\partial x} & \frac{\partial v_z}{\partial x} \\ \frac{\partial v_x}{\partial y} & \frac{\partial v_y}{\partial y} & \frac{\partial v_z}{\partial y} \\ \frac{\partial v_x}{\partial z} & \frac{\partial v_y}{\partial z} & \frac{\partial v_z}{\partial z} \end{bmatrix}. \tag{7.17}
$$

It should be noticed that with the current notation an operation of the type $\underset{\sim}{\nabla}\underset{\sim}{v}$ is not allowed.

The transposed form of the tensor $\vec{\nabla}\vec{v}$ is called the velocity gradient tensor $\boldsymbol{L}$:

$$
\boldsymbol{L} = \left(\vec{\nabla}\vec{v}\right)^{\mathrm{T}}. \tag{7.18}
$$

The associated matrix representation reads:

$$
\underset{\sim}{L} = \left(\underset{\sim}{\nabla}\underset{\sim}{v}^{\mathrm{T}}\right)^{\mathrm{T}}. \tag{7.19}
$$

Figure 7.12 gives an illustration of the application of this tensor (matrix). Consider an infinitesimally small material line element $d\vec{x}$ (with components $d\underset{\sim}{x}$) in volume V at time t. If, at the tail of the vector $d\vec{x}$, the velocity is $\vec{v}$, then at the head of the vector $d\vec{x}$ the velocity will be: $\vec{v} + d\vec{x}\cdot\vec{\nabla}\vec{v} = \vec{v} + (\vec{\nabla}\vec{v})^{\mathrm{T}}\cdot d\vec{x} = \vec{v} + \boldsymbol{L}\cdot d\vec{x}$.

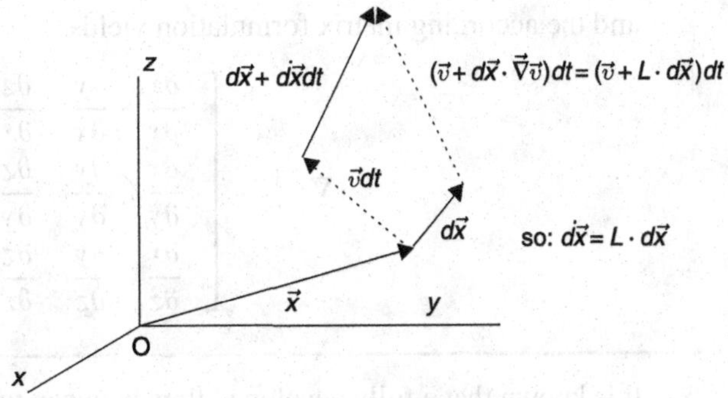

Figure 7.12

Change of a material line element $d\vec{x}$ after a time increment dt.

After an infinitesimally small increase in time dt, the material line element $d\vec{x}$ will change into $d\vec{x} + \dot{d\vec{x}}dt$. By means of the figure it can be proven that:

$$\dot{d\vec{x}} = \boldsymbol{L} \cdot d\vec{x} \quad \text{and also} \quad \dot{d\underset{\sim}{x}} = \underline{L}\, d\underset{\sim}{x}, \tag{7.20}$$

and so the tensor $\boldsymbol{L}$ (with matrix representation $\underline{L}$) is a measure for the current change (per unit of time) of material line elements. This holds for the length as well as for the orientation.

The divergence (an operator that is often used for vector fields) of the velocity, div($\vec{v}$), defined as $\vec{\nabla} \cdot \vec{v}$, can be written as

$$\begin{aligned} \text{div}(\vec{v}) &= \left(\vec{e}_x \frac{\partial}{\partial x} + \vec{e}_y \frac{\partial}{\partial y} + \vec{e}_z \frac{\partial}{\partial z}\right) \cdot (v_x \vec{e}_x + v_y \vec{e}_y + v_z \vec{e}_z) \\ &= \frac{\partial v_x}{\partial x} + \frac{\partial v_y}{\partial y} + \frac{\partial v_z}{\partial z} \\ &= \text{tr}(\boldsymbol{L}), \end{aligned} \tag{7.21}$$

and also

$$\text{div}(\vec{v}) = \frac{\partial v_x}{\partial x} + \frac{\partial v_y}{\partial y} + \frac{\partial v_z}{\partial z} = \text{tr}(\underset{\sim}{\nabla}\, \underset{\sim}{v}^{\text{T}}) = \underset{\sim}{\nabla}^{\text{T}}\, \underset{\sim}{v} = \text{tr}(\underline{L}). \tag{7.22}$$

For the sake of completeness, the following rather trivial result for the gradient $\vec{\nabla}$ applied to the position vector $\vec{x}$ is given:

$$\begin{aligned} \vec{\nabla}\vec{x} = &\frac{\partial x}{\partial x}\vec{e}_x\vec{e}_x + \frac{\partial y}{\partial x}\vec{e}_x\vec{e}_y + \frac{\partial z}{\partial x}\vec{e}_x\vec{e}_z \\ &+ \frac{\partial x}{\partial y}\vec{e}_y\vec{e}_x + \frac{\partial y}{\partial y}\vec{e}_y\vec{e}_y + \frac{\partial z}{\partial y}\vec{e}_y\vec{e}_z \\ &+ \frac{\partial x}{\partial z}\vec{e}_z\vec{e}_x + \frac{\partial y}{\partial z}\vec{e}_z\vec{e}_y + \frac{\partial z}{\partial z}\vec{e}_z\vec{e}_z \\ = &\vec{e}_x\vec{e}_x + \vec{e}_y\vec{e}_y + \vec{e}_z\vec{e}_z = \boldsymbol{I}, \end{aligned} \tag{7.23}$$

and the according matrix formulation yields

$$\underset{\sim}{\nabla}\, \underset{\sim}{x}^{\mathrm{T}} = \begin{bmatrix} \dfrac{\partial x}{\partial x} & \dfrac{\partial y}{\partial x} & \dfrac{\partial z}{\partial x} \\ \dfrac{\partial x}{\partial y} & \dfrac{\partial y}{\partial y} & \dfrac{\partial z}{\partial y} \\ \dfrac{\partial x}{\partial z} & \dfrac{\partial y}{\partial z} & \dfrac{\partial z}{\partial z} \end{bmatrix} = \underline{I}. \qquad (7.24)$$

Example 7.6 It is known that a fully developed flow between two parallel plates has a parabolic velocity profile (see Fig. 7.13). Using the coordinate system as given in the figure, this flow field can be described by:

$$\vec{v} = a\left(1 - \frac{y^2}{h^2}\right)\vec{e}_x,$$

with $2h$ the distance of the plates and with a a positive constant. For this two-dimensional case we can write for the matrix representation $\underline{L}$ of the velocity gradient tensor:

$$\underline{L} = \begin{bmatrix} \frac{\partial v_x}{\partial x} & \frac{\partial v_x}{\partial y} \\ \frac{\partial v_y}{\partial x} & \frac{\partial v_y}{\partial y} \end{bmatrix} = \begin{bmatrix} 0 & -2a\frac{y}{h^2} \\ 0 & 0 \end{bmatrix}$$

and the tensor can be written as:

$$\boldsymbol{L} = -2a\frac{y}{h^2}\vec{e}_x\vec{e}_y.$$

Example 7.7 An experimental set-up that is used to determine the mechanical behaviour of fluids consists of a reservoir containing four rotating cylinders, which rotate with the same velocity, as shown in Fig. 7.14 below. The figure gives a top view of the set-up. The set-up is filled with an incompressible fluid. In the neighbourhood of the origin of the xyz-coordinate system, this leads to a two-dimensional velocity field:

$$\vec{v} = c(-x\vec{e}_x + y\vec{e}_y)$$

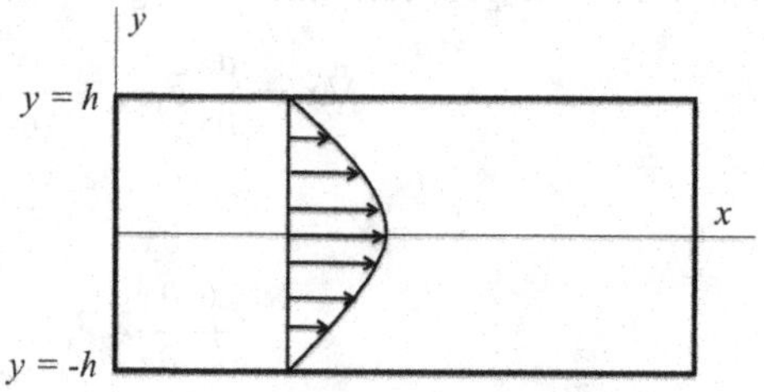

Figure 7.13

Parabolic flow between two plates (Poiseuille flow).

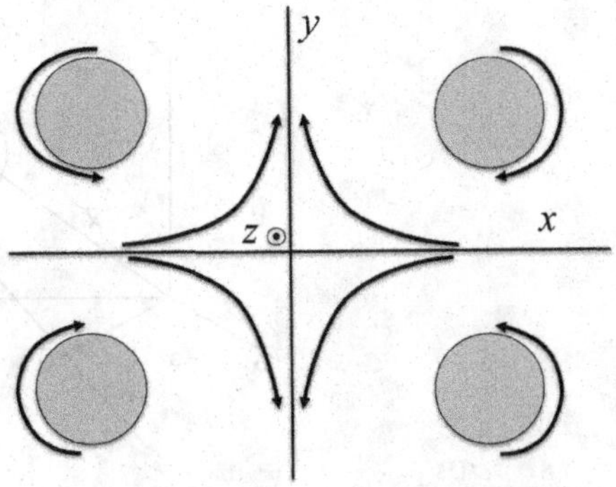

Figure 7.14
Extensional flow.

with c a constant and with $\vec{e}_x$ and $\vec{e}_y$ unit vectors along the x- and y-axis. At the origin itself, the velocity is zero. We consider a small material element of fluid particles at time t with a length ℓ and directed along the y-axis. Our purpose is to calculate the length change $d\ell$ of this material element after an (infinitesimal) time change dt. From Eq. (7.19), it follows for this case:

$$\underline{L} = \begin{bmatrix} -c & 0 \\ 0 & c \end{bmatrix}.$$

The line element under consideration is represented by:

$$d\underset{\sim}{x} = \begin{bmatrix} 0 \\ \ell \end{bmatrix}.$$

By using Eq. (7.20), we find:

$$d\dot{\underset{\sim}{x}} = \underline{L}d\underset{\sim}{x} = \begin{bmatrix} -c & 0 \\ 0 & c \end{bmatrix}\begin{bmatrix} 0 \\ \ell \end{bmatrix} = \begin{bmatrix} 0 \\ c\ell \end{bmatrix}.$$

So after an infinitesimal time step dt, the length change will be: $dx = c\ell dt$.

7.6 Rigid Body Rotation

In the present section, it is assumed that the mass in the rigid volume V rotates around a fixed axis in three-dimensional space. We focus our attention on the velocity field $\vec{v} = \vec{v}(\vec{x})$ and on the velocity gradient tensor $\boldsymbol{L}$ that can be derived from it. Figure 7.15 illustrates the problem being considered. The axis of rotation is defined by means of a point S, fixed in space, with position vector $\vec{x}_S$ and the

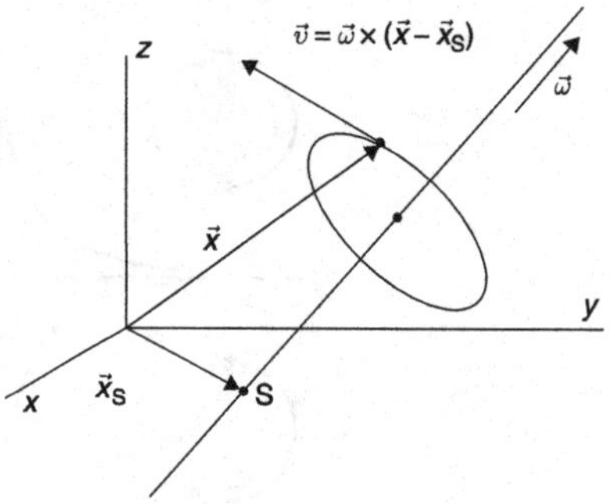

Figure 7.15

Rotation of material.

constant angular velocity vector $\vec{\omega}$. The associated columns with the components with respect to the Cartesian basis can be written as

$$\underset{\sim}{x}_S = \begin{bmatrix} x_S \\ y_S \\ z_S \end{bmatrix}, \quad \underset{\sim}{\omega} = \begin{bmatrix} \omega_x \\ \omega_y \\ \omega_z \end{bmatrix}. \tag{7.25}$$

The velocity vector $\vec{v}$ at a certain point with position vector $\vec{x}$ satisfies

$$\vec{v} = \vec{v}(\vec{x}) = \vec{\omega} \times (\vec{x} - \vec{x}_S). \tag{7.26}$$

It can be derived that the components of the velocity vector $\vec{v}$ and the spatial coordinate vector $\vec{x}$, stored in the columns

$$\underset{\sim}{v} = \begin{bmatrix} v_x \\ v_y \\ v_z \end{bmatrix} \text{ and } \underset{\sim}{x} = \begin{bmatrix} x \\ y \\ z \end{bmatrix}, \tag{7.27}$$

are related according to

$$\underset{\sim}{v} = \underset{\sim}{v}(\underset{\sim}{x}) = \underset{\sim}{\Omega}(\underset{\sim}{x} - \underset{\sim}{x}_S) \text{ with } \underset{\sim}{\Omega} = \begin{bmatrix} 0 & -\omega_z & \omega_y \\ \omega_z & 0 & -\omega_x \\ -\omega_y & \omega_x & 0 \end{bmatrix}. \tag{7.28}$$

The spin matrix $\underset{\sim}{\Omega}$ that is associated with the angular velocity vector $\vec{\omega}$ (with components $\underset{\sim}{\omega}$) is skew symmetric: $\underset{\sim}{\Omega}^T = -\underset{\sim}{\Omega}$. With the uniform rotation as a rigid body considered in this section, we find:

$$\underset{\sim}{L} = \left(\underset{\sim}{\nabla}\, \underset{\sim}{v}^T\right)^T = \left(\left(\underset{\sim}{\nabla}\, (\underset{\sim}{x} - \underset{\sim}{x}_S)^T\right) \underset{\sim}{\Omega}^T\right)^T = \left((\underset{\sim}{I} + \underset{\sim}{O})\underset{\sim}{\Omega}^T\right)^T = \underset{\sim}{\Omega} \tag{7.29}$$

and for the associated velocity gradient tensor $\boldsymbol{L}$:

$$\boldsymbol{L} = \omega_x(\vec{e}_z\vec{e}_y - \vec{e}_y\vec{e}_z) + \omega_y(\vec{e}_x\vec{e}_z - \vec{e}_z\vec{e}_x) + \omega_z(\vec{e}_y\vec{e}_x - \vec{e}_x\vec{e}_y). \tag{7.30}$$

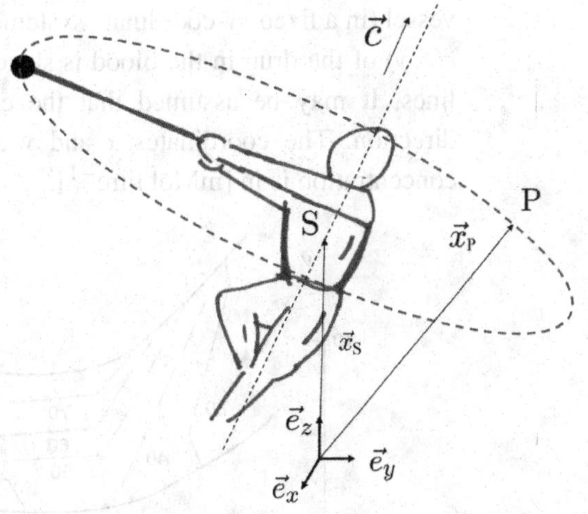

Figure 7.16

Hammer thrower.

Example 7.8 A hammer thrower is swinging a hammer by moving it in a circle around an axis in the direction of vector $\vec{c} = \frac{3}{2}\vec{e}_y + 2\vec{e}_z$ (see Fig. 7.16). He is rotating the hammer with an angular velocity $\omega = 5$ [rad s^{-1}]. The rotation axis passes through the body of the hammer thrower at point $\vec{x}_S = \frac{3}{2}\vec{e}_z$. When he releases the rope to throw the hammer, the hammer is at point P given by: $\vec{x}_P = -2\vec{e}_x + \vec{e}_y + \frac{1}{2}\vec{e}_z$.

We want to calculate the velocity vector $\vec{v}_P$ of the hammer at the point P. For this purpose we first determine the angular velocity vector $\vec{\omega} = \omega\vec{e}$, with $\vec{e}$ a vector with length 1 in the direction of the rotational axis. Thus:

$$\vec{e} = \frac{\vec{c}}{|\vec{c}|} = \frac{3}{5}\vec{e}_y + \frac{4}{5}\vec{e}_z$$

and consequently: $\vec{\omega} = 5\vec{e} = 3\vec{e}_y + 4\vec{e}_z$.
The velocity of the hammer at P can be calculated as:

$$\begin{aligned}\vec{v}_P &= \vec{\omega} \times (\vec{x}_P - \vec{x}_S)\\ &= (3\vec{e}_y + 4\vec{e}_z) \times (-2\vec{e}_x + \vec{e}_y - \vec{e}_z)\\ &= -7\vec{e}_x - 8\vec{e}_y + 6\vec{e}_z.\end{aligned}$$

Exercises

7.1 A drug is administered to the blood via an intravenous drug delivery system. In the figure, a two-dimensional schematic is given of the blood vessel (in a fixed xy-coordinate system). The stationary concentration field $c(x, y)$ of the drug in the blood is sketched by means of iso-concentration lines. It may be assumed that the concentration is constant in the z-direction. The coordinates x and y are defined in the unit [mm], the concentration is in [mMol litre^{-1}].

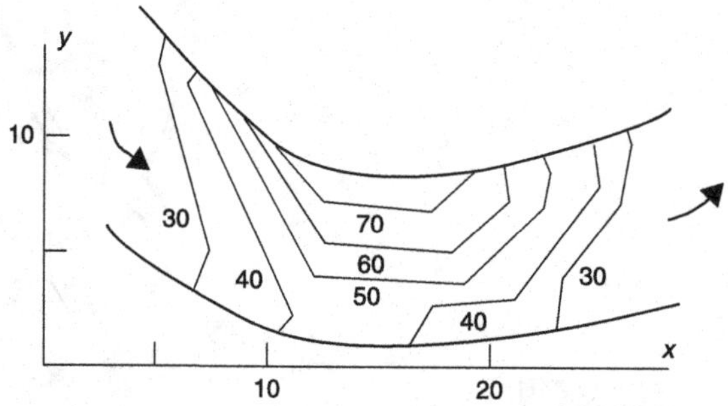

Give an estimate of the gradient $\underset{\sim}{\nabla} c$ of the concentration field in the point with coordinates $x = 20$ [mm], $y = 5$ [mm].

7.2 A cube ABCDEFGH (with edge a) rotates with angular velocity ω around the straight line ℓ, coinciding with the body diagonal CE. The points C and E, and consequently the rotation axis ℓ, have a fixed position in the xyz-coordinate system.

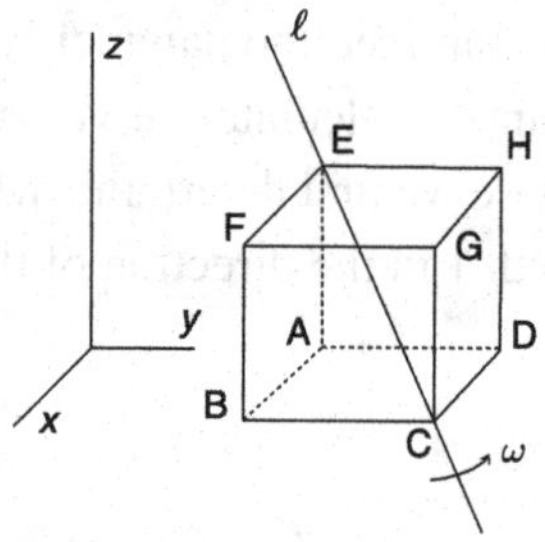

At a certain time, the edges AB, AD and AE are exactly oriented in the directions of the x-, y- and z- axes, respectively (see figure). Determine for that time instant the velocity vector $\underset{\sim}{v}_B$ of the point B.

7.3 Bone mineral density is often measured, because this quantity can be related to the strength and stiffness of bone. The measurement can also be used as a diagnostic tool for osteoporosis. Consider a rectangular piece of bone in the xy-plane of a Cartesian xyz-coordinate system, with varying density $\rho = \rho(x, y)$. For the corner points ABCD of the specimen, the density is given (expressed in [kg dm^{-3}]); see figure. Prove that it is impossible for the gradient $\vec{\nabla}\rho$ of the density field in the figure to be constant.

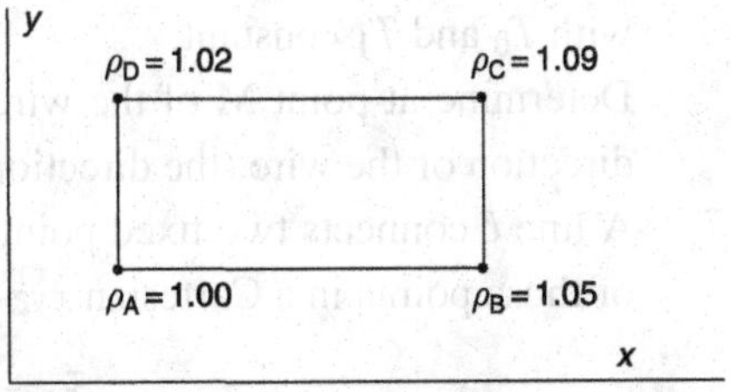

7.4 Consider a domain in the form of a cube in three-dimensional space, given by $-\ell \le x \le 3\ell, -2\ell \le y \le 2\ell$ and $-2\ell \le z \le 2\ell$ with ℓ a constant and x, y and z the Cartesian coordinates. Within this domain a temperature field exists with known gradient $\vec{\nabla}T$ which is given by

$$\vec{\nabla}T = \theta\vec{e}_y \quad \text{with } \theta \text{ a constant.}$$

In the domain, a curve is given. The points on the curve satisfy

$$(x-\ell)^2 + y^2 = \ell^2 \text{ and } z = \ell\,.$$

Calculate the coordinates of the point (or points) on the curve where the derivative of the temperature in the direction of the curve is equal to zero.

7.5 In a two-dimensional xy-coordinate system (mutually perpendicular unit vectors $\vec{e}_x$ and $\vec{e}_y$ along the axes), a two-dimensional stationary fluid flow is considered. The fluid flow is caused by a fluid source in the origin. An arbitrary point in the coordinate system is given by the vector $\vec{x} = x\vec{e}_x + y\vec{e}_y$. At some distance from the origin (with $|\vec{x}| > \lambda$), the velocity field $\vec{v}(\vec{x})$ can be written as

$$\vec{v}(\vec{x}) = \alpha\frac{\vec{x}}{|\vec{x}|^2}.$$

The location of a point P is defined by the position vector $\vec{x}_P = \ell(\vec{e}_x + \vec{e}_y)$ with $\ell > \lambda$. Determine the velocity gradient tensor $\boldsymbol{L}$ at point P.

7.6 In a Cartesian xyz-coordinate system, the geometry of a spiral-shaped wire is given by means of a parameter description according to:

$$x = R\cos(\phi); \quad y = R\sin(\phi); \quad z = 2R\phi$$

with R a positive constant and the parameter ϕ varying between -2π and 4π. For the middle point M of the wire: $\phi = \pi$. The wire is placed in a temperature field $T = T(x, y, z)$. In the close environment of the wire, the temperature field can be described by:

$$T = T_0 + T_1\frac{xy}{R^2}$$

with T_0 and T_1 constant.
Determine at point M of the wire the derivative of the temperature in the direction of the wire (the directional derivative).

7.7 A line ℓ connects two fixed points A and B in space. The position vectors of those points in a Cartesian *xyz*-coordinate system are given as:

$$\vec{x}_A = 6\vec{e}_x$$
$$\vec{x}_B = 9\vec{e}_x + 4\vec{e}_y.$$

A body rotates around this line ℓ with constant angular velocity ω. Consider a point P of this body that is currently at the origin of the coordinate system. Determine the magnitude of the velocity of point P.

7.8 In a Cartesian *xyz*-coordinate system, a (two-dimensional) stationary fluid flow is considered. The velocity field is described by:

$$\vec{v}(x, y, z) = v_x(x, y)\,\vec{e}_x + v_y(x, y)\,\vec{e}_y.$$

At three points in space (A, B and C), the velocity is measured (see table). The coordinates are in [cm], the velocities in [cm s^{-1}].

	x	y	z	v_x	v_y
A	0	2	0	10	−2
B	0	4	0	24	8
C	2	3	0	17	9

Determine the velocity gradient tensor $\boldsymbol{L}$, assuming that this tensor is constant (independent of x, y and z).

7.9 For a temperature field $T = T(x, y, z)$ in a Cartesian *xyz*-coordinate system, it is known that the gradient is constant and given by:

$$\frac{\partial T}{\partial x} = 2 \;\; [^\circ\text{C cm}^{-1}]; \qquad \frac{\partial T}{\partial y} = 0 \;\; [^\circ\text{C cm}^{-1}]; \qquad \frac{\partial T}{\partial z} = -3 \;\; [^\circ\text{C cm}^{-1}].$$

We define two points A and B in space with coordinates (in cm):

	x	y	z
A	10	10	2
B	4	2	6

The temperature in point A is given by: $T_A = 30$ [°C]. Calculate the temperature in point B.

8 Stress in Three-Dimensional Continuous Media

In this chapter, the concepts introduced in Chapter 6 for a one-dimensional continuous system are generalized to two-dimensional configurations. Extension to three-dimensional problems is briefly discussed. First, the equilibrium conditions in a two- or three-dimensional body are derived from force equilibrium of an infinitesimally small volume element. Thereafter, the concept of a stress tensor, as a sum of dyads, is introduced to compute the stress vector acting on an arbitrary surface at a material point of the body.

8.1 Stress Vector

Before examining the equilibrium conditions in a two-dimensional body, the concept of a stress vector is introduced. For this purpose we consider an infinitesimally small surface element having area ΔA: see Fig. 8.1.

On this surface an infinitesimally small force vector $\Delta\vec{F}$ is applied with components in the x-, y- and z-direction: $\Delta\vec{F} = \Delta F_x\vec{e}_x + \Delta F_y\vec{e}_y + \Delta F_z\vec{e}_z$. Following the definition of stress, Eq. (6.8), three stresses may be defined:

$$s_x = \lim_{\Delta A \to 0} \frac{\Delta F_x}{\Delta A}, \tag{8.1}$$

which acts in the x-direction, and

$$s_y = \lim_{\Delta A \to 0} \frac{\Delta F_y}{\Delta A}, \tag{8.2}$$

which acts in the y-direction, and

$$s_z = \lim_{\Delta A \to 0} \frac{\Delta F_z}{\Delta A}, \tag{8.3}$$

which acts in the z-direction. Hence, a **stress vector** may be defined:

$$\vec{s} = s_x\vec{e}_x + s_y\vec{e}_y + s_z\vec{e}_z. \tag{8.4}$$

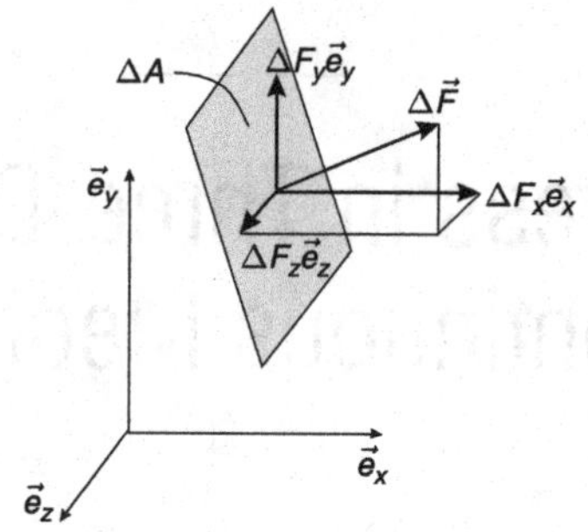

Figure 8.1

Force $\Delta\vec{F}$ on an infinitesimal surface element having area ΔA.

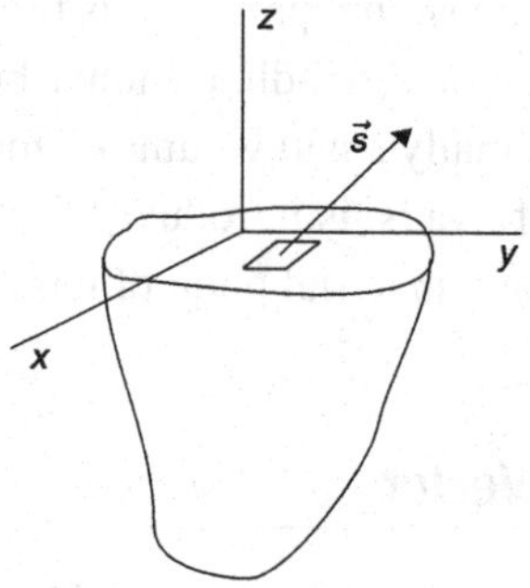

Figure 8.2

The stress vector $\vec{s}$ on a cut section of a body.

8.2 From Stress to Force

Suppose that a free body diagram is created by means of an imaginary cutting plane through a body. The cutting plane is chosen in such a way that it coincides with the xy-plane (Fig. 8.2). On the imaginary cutting plane, a stress vector $\vec{s}$ is given as a function of x and y. How can the total force vector acting on that plane be computed based on this stress vector? The complete answer to this question is somewhat beyond the scope of this course because it requires the integration of a multi-variable function. However, the simpler case in which the stress vector is a function of x (or y) only, while it acts on a rectangular plane at constant z, is easier to answer and sufficiently general to be useful in the remainder of this chapter. Therefore, suppose that the stress vector is a function of x such that it may be written as

$$\vec{s} = s_x(x)\vec{e}_x + s_y(x)\vec{e}_y + s_z(x)\vec{e}_z. \tag{8.5}$$

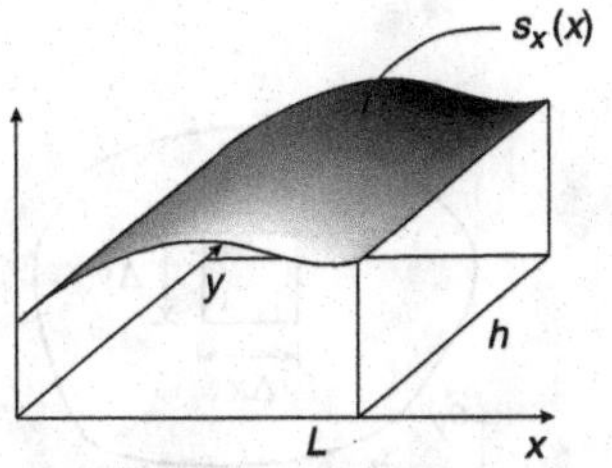

Figure 8.3

A stress (s_x) distribution where s_x is a function of x only.

Let this stress vector act on a plane $z = 0$ that spans the range $0 < x < L$ and that has a width h in the y-direction. The resulting force vector on the plane considered due to this stress vector is denoted by $\vec{F} = F_x\vec{e}_x + F_y\vec{e}_y + F_z\vec{e}_z$. If s_x, s_y and s_z are **constant**, the net force is simply computed by multiplication of the stress vector components with the surface area, hL, in this case:

$$\vec{F} = s_x hL\vec{e}_x + s_y hL\vec{e}_y + s_y hL\vec{e}_z. \tag{8.6}$$

For **non-constant** stress vector components (e.g. as visualized in Fig. 8.3), the force components in the x-, y- and z-direction due to the stress vector $\vec{s}$ are obtained via integration of these components over the domain in the x-direction and multiplication by the width h of the plane (which is allowed because the stress components are constant in the y-direction):

$$F_x = h\int_0^L s_x(x)\,dx, \tag{8.7}$$

$$F_y = h\int_0^L s_y(x)\,dx, \tag{8.8}$$

$$F_z = h\int_0^L s_z(x)\,dx, \tag{8.9}$$

respectively. In conclusion, the force vector is found by integration of the stress vector over the plane on which it acts.

8.3 Equilibrium

In Chapter 6, the equilibrium equation for the one-dimensional case was formulated by demanding the balance of forces of an isolated thin slice of a continuous body. In analogy with this, we consider a continuous, arbitrary material volume, in particular a cross section in the xy-plane as depicted in Fig. 8.4(a). In the direction perpendicular to the drawing, hence in the $\vec{e}_z$ direction, the body has a thickness

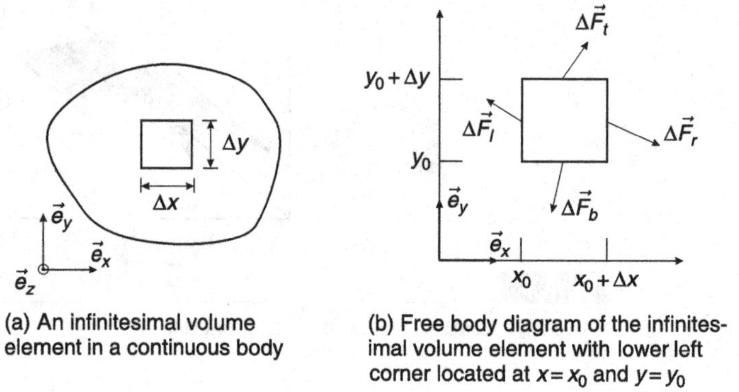

Δy
Δx
$\vec{e}_y$
$\vec{e}_x$
$\vec{e}_z$
(a) An infinitesimal volume element in a continuous body
$\Delta\vec{F}_t$
$y_0+\Delta y$
$\Delta\vec{F}_l$
$\Delta\vec{F}_r$
y_0
$\vec{e}_y$
$\Delta\vec{F}_b$
$\vec{e}_x$
x_0
$x_0+\Delta x$
(b) Free body diagram of the infinitesimal volume element with lower left corner located at $x=x_0$ and $y=y_0$

Figure 8.4

Free body diagram of an infinitesimal volume element of a continuous body.

h. It is assumed that all stress components are constant across the thickness. We examine the free body diagram of an infinitesimal volume element, which in this case is a rectangular prism having dimensions $\Delta x \times \Delta y \times h$, the cross section of which is also shown in Fig. 8.4(a). The lower left corner of the rectangle as depicted in Fig. 8.4(b) is located at $x = x_0$ and $y = y_0$. Forces have been introduced on all faces of the prism in the xz- or yz-plane: see Fig. 8.4(b). These forces are a consequence of the interaction of the prism with its surroundings.

The force vectors $\Delta\vec{F}_i$, $i = l, b, t, r$, represent forces that are acting on the left, bottom, top and right faces of the prism. These forces are infinitesimally small because they act on an infinitesimally small surface element (the faces of the prism) that experiences only a small part of the total force that is exerted on the body. Moreover, it is assumed that the infinitesimal volume element is sufficiently small that the individual force vectors can be transformed to stress vectors in the usual way: see Fig. 8.5.

Each of the stress vectors may be additively decomposed into a component acting in the $\vec{e}_x$-direction and a component acting in the $\vec{e}_y$-direction. These components are sketched in Fig. 8.6. The double subscript notation is interpreted as follows: the second subscript indicates the direction of the normal to the plane or face on which the stress component acts. The first subscript relates to the direction of the stress itself. There is also a sign convention. When both the outer normal and the stress component are oriented in positive direction relative to the coordinate axes, the stress is positive also, which is required because of Newton's law of action and reaction. When both the outer normal and the stress component are oriented in a negative direction relative to the coordinate axes, the stress is positive. When the normal points in a positive direction while the stress points in a negative direction (or vice versa), the stress is negative.

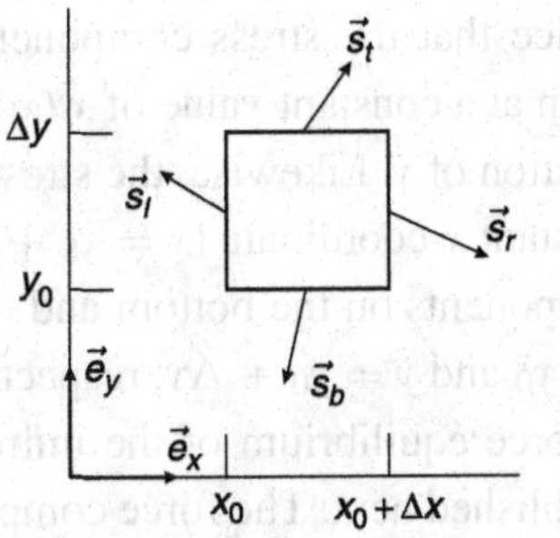

Figure 8.5
Free body diagram of the infinitesimal volume element.

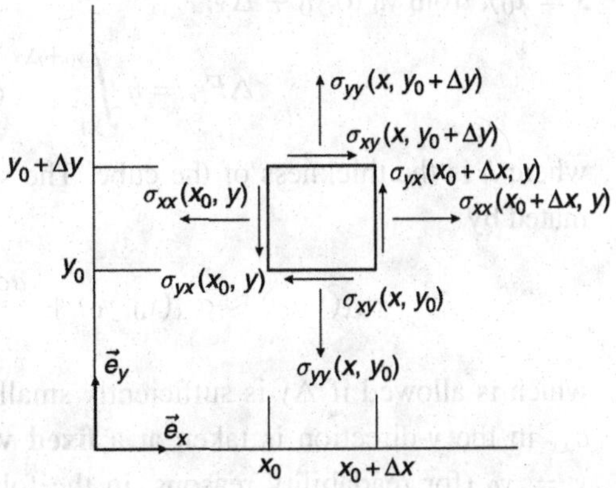

Figure 8.6
Stresses acting on the surfaces of an infinitesimal volume element.

Clearly, all stress components are a function of the position in space, hence of the x- and y-coordinate (but not of the z-coordinate because all quantities are assumed constant in the z-direction):

$$\begin{aligned}\sigma_{xx} &= \sigma_{xx}(x, y)\\ \sigma_{yy} &= \sigma_{yy}(x, y)\\ \sigma_{xy} &= \sigma_{xy}(x, y)\\ \sigma_{yx} &= \sigma_{yx}(x, y).\end{aligned} \tag{8.10}$$

Using the notation from Fig. 8.6 it holds that

$$\begin{aligned}\vec{s}_l &= -\sigma_{xx}(x_0, y)\vec{e}_x - \sigma_{yx}(x_0, y)\vec{e}_y\\ \vec{s}_b &= -\sigma_{xy}(x, y_0)\vec{e}_x - \sigma_{yy}(x, y_0)\vec{e}_y\\ \vec{s}_t &= \sigma_{xy}(x, y_0 + \Delta y)\vec{e}_x + \sigma_{yy}(x, y_0 + \Delta y)\vec{e}_y\\ \vec{s}_r &= \sigma_{xx}(x_0 + \Delta x, y)\vec{e}_x + \sigma_{yx}(x_0 + \Delta x, y)\vec{e}_y.\end{aligned} \tag{8.11}$$

Notice that the stress components of the vector $\vec{s}_l$ that act on the left face are taken at a constant value of x ($x = x_0$), but that these stress components may be a function of y. Likewise, the stress components on the right face are also taken at a constant x-coordinate ($x = x_0 + \Delta x$) and are also a function of y, while the stress components on the bottom and top faces are both a function of x but are taken at $y = y_0$ and $y = y_0 + \Delta y$, respectively.

Force equilibrium of the infinitesimal volume element depicted in Fig. 8.6 is established next. The force components acting on the faces of the prism have been identified in Fig. 8.7.

The net force in the negative x-direction acting on the left face is denoted by ΔF_{lx} and is obtained by integration of the stress σ_{xx} at constant value of x (i.e. $x = x_0$), from y_0 to $y_0 + \Delta y$:

$$\Delta F_{lx} = h \int_{y_0}^{y_0+\Delta y} \sigma_{xx}(x_0, y)\, dy, \tag{8.12}$$

where h is the thickness of the cube. The stress field $\sigma_{xx}(x_0, y)$ may be approximated by

$$\sigma_{xx}(x_0, y) \approx \sigma_{xx}(x_0, y_0) + \left.\frac{\partial \sigma_{xx}}{\partial y}\right|_{x=x_0,\, y=y_0} (y - y_0), \tag{8.13}$$

which is allowed if Δy is sufficiently small. Notice that the partial derivative of σ_{xx} in the y-direction is taken at a **fixed** value of x and y, i.e. at $x = x_0$ and $y = y_0$ (for readability reasons, in the following the addition $|_{x=x_0, y=y_0}$ to the partial derivatives is omitted from the equations). Consequently, the force ΔF_{lx} can be written as

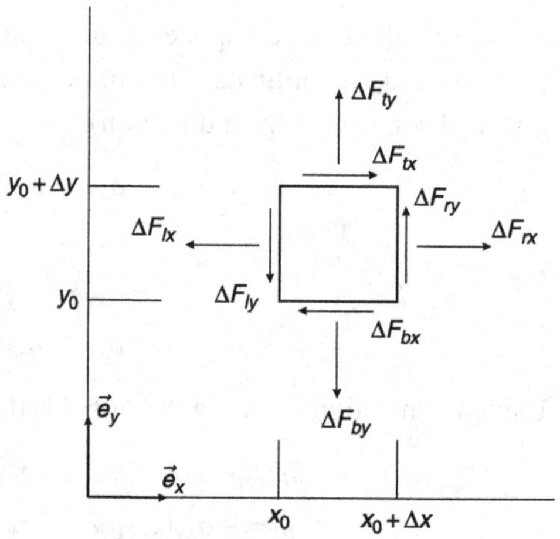

Figure 8.7

Forces acting on the surfaces of an infinitesimal volume element.

$$\Delta F_{lx} = h \int_{y_0}^{y_0+\Delta y} \left(\sigma_{xx}(x_0, y_0) + \frac{\partial \sigma_{xx}}{\partial y}(y - y_0) \right) dy. \tag{8.14}$$

To compute this integral it must be realized that $\sigma_{xx}(x_0, y_0)$, $\partial \sigma_{xx}/\partial y$ and y_0 denote quantities that are not a function of the integration parameter y. Bearing this in mind, it is straightforward to show that

$$\Delta F_{lx} = \sigma_{xx}(x_0, y_0)\, h\, \Delta y + \frac{\partial \sigma_{xx}}{\partial y}\, h\, \frac{\Delta y^2}{2}. \tag{8.15}$$

The stress on the right face of the cube, $\sigma_{xx}(x_0 + \Delta x, y)$ may be integrated to give the force in the (positive) x-direction on this face:

$$\Delta F_{rx} = h \int_{y_0}^{y_0+\Delta y} \sigma_{xx}(x_0 + \Delta x, y)\, dy. \tag{8.16}$$

Clearly, the stress component σ_{xx} on the right face, hence at constant $x_0 + \Delta x$, may be approximated by

$$\sigma_{xx}(x_0 + \Delta x, y) \approx \sigma_{xx}(x_0, y_0) + \frac{\partial \sigma_{xx}}{\partial x}\Delta x + \frac{\partial \sigma_{xx}}{\partial y}(y - y_0). \tag{8.17}$$

Therefore the force ΔF_{rx} can be computed as

$$\Delta F_{rx} = \sigma_{xx}(x_0, y_0) h \Delta y + \frac{\partial \sigma_{xx}}{\partial x} h \Delta x \Delta y + \frac{\partial \sigma_{xx}}{\partial y}\, h\, \frac{\Delta y^2}{2}. \tag{8.18}$$

A similar exercise can be performed with respect to the forces in the x-direction on the top and bottom faces, giving

$$\Delta F_{tx} = \sigma_{xy}(x_0, y_0) h \Delta x + \frac{\partial \sigma_{xy}}{\partial y} h \Delta x \Delta y + \frac{\partial \sigma_{xy}}{\partial x}\, h\, \frac{\Delta x^2}{2}, \tag{8.19}$$

while

$$\Delta F_{bx} = \sigma_{xy}(x_0, y_0) h \Delta x + \frac{\partial \sigma_{xy}}{\partial x}\, h\, \frac{\Delta x^2}{2}. \tag{8.20}$$

Force equilibrium in the x-direction now yields

$$-\Delta F_{lx} - \Delta F_{bx} + \Delta F_{rx} + \Delta F_{tx} = 0. \tag{8.21}$$

Use of the above results for the force components gives

$$\begin{aligned} &-\sigma_{xx}(x_0, y_0) h \Delta y - \frac{\partial \sigma_{xx}}{\partial y}\frac{h}{2}\Delta y^2 - \sigma_{xy}(x_0, y_0) h \Delta x \\ &- \frac{\partial \sigma_{xy}}{\partial x}\frac{h}{2}\Delta x^2 + \sigma_{xx}(x_0, y_0) h \Delta y + \frac{\partial \sigma_{xx}}{\partial x} h \Delta x \Delta y + \frac{\partial \sigma_{xx}}{\partial y}\frac{h}{2}\Delta y^2 \\ &+ \sigma_{xy}(x_0, y_0) h \Delta x + \frac{\partial \sigma_{xy}}{\partial y} h \Delta x \Delta y + \frac{\partial \sigma_{xy}}{\partial x}\frac{h}{2}\Delta x^2 = 0, \end{aligned} \tag{8.22}$$

which implies that

$$\frac{\partial \sigma_{xx}}{\partial x} + \frac{\partial \sigma_{xy}}{\partial y} = 0. \tag{8.23}$$

This should hold for any position in space of the prism, hence for all values of x and y. Performing a similar exercise in the y-direction yields the full set of partial differential equations, known as the local equilibrium equations:

$$\frac{\partial \sigma_{xx}}{\partial x} + \frac{\partial \sigma_{xy}}{\partial y} = 0 \tag{8.24}$$

$$\frac{\partial \sigma_{yx}}{\partial x} + \frac{\partial \sigma_{yy}}{\partial y} = 0. \tag{8.25}$$

Two different shear stresses are present: σ_{xy} and σ_{yx}. However, by using **equilibrium of moments** it can be proven that

$$\sigma_{xy} = \sigma_{yx}. \tag{8.26}$$

This result is revealed by considering the sum of moments with respect to the midpoint of the infinitesimal cube of Fig. 8.7. With respect to the midpoint, the moments due to the normal forces, ΔF_{lx}, ΔF_{rx}, ΔF_{ty} and ΔF_{by}, are equal to zero, while the shear forces generate a moment. Hence enforcing the sum of moments with respect to the midpoint to be equal to zero gives

$$-\frac{\Delta x}{2}(\Delta F_{ly} + \Delta F_{ry}) + \frac{\Delta y}{2}(\Delta F_{tx} + \Delta F_{bx}) = 0. \tag{8.27}$$

Using the above results for the force components it follows that

$$\begin{aligned} -\frac{\Delta x}{2}\Bigg(&\sigma_{yx}(x_0,y_0)h\Delta y + \frac{\partial \sigma_{yx}}{\partial y}\,h\,\frac{\Delta y^2}{2} + \sigma_{yx}(x_0,y_0)h\Delta y + \frac{\partial \sigma_{yx}}{\partial y}\,h\,\frac{\Delta y^2}{2} \\ &+ \frac{\partial \sigma_{yx}}{\partial x}h\Delta x\Delta y\Bigg) = \frac{\Delta y}{2}\Bigg(\sigma_{xy}(x_0,y_0)h\Delta x + \frac{\partial \sigma_{xy}}{\partial x}\,h\,\frac{\Delta x^2}{2} \\ &+ \sigma_{xy}(x_0,y_0)h\Delta x + \frac{\partial \sigma_{xy}}{\partial y}\,h\,\frac{\Delta x^2}{2} + \frac{\partial \sigma_{xy}}{\partial x}h\Delta x\Delta y\Bigg) = 0. \end{aligned} \tag{8.28}$$

Neglecting terms of order Δx^3, Δy^3 etc., reveals immediately that

$$\sigma_{yx} = \sigma_{xy}. \tag{8.29}$$

Based on this result, the equilibrium equations Eqs. (8.24) and (8.25) may be rewritten as

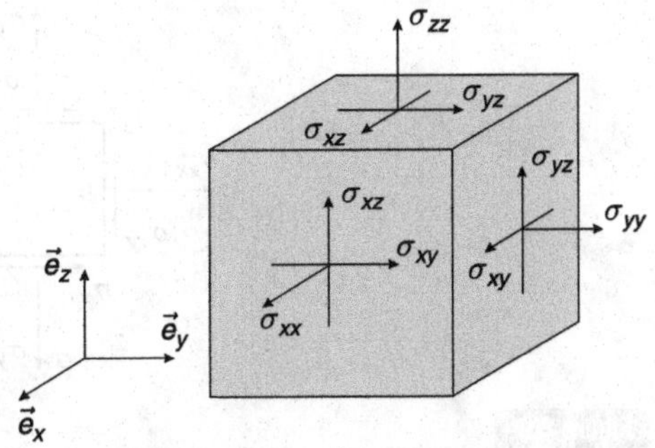

Figure 8.8

Stresses in three dimensions.

$$\frac{\partial \sigma_{xx}}{\partial x} + \frac{\partial \sigma_{xy}}{\partial y} = 0 \tag{8.30}$$

$$\frac{\partial \sigma_{xy}}{\partial x} + \frac{\partial \sigma_{yy}}{\partial y} = 0. \tag{8.31}$$

Note that, strictly speaking, the resulting forces on the faces of the prism as visualized in Fig. 8.7 are not exactly located in the midpoints of the faces. This should be accounted for in the equilibrium of moment. This would complicate the derivations considerably, but it would lead to the same conclusions.

In the three-dimensional case, a number of additional stress components are present: see Fig. 8.8. In total there are six independent stress components: σ_{xx}, σ_{xy}, σ_{xz}, σ_{yy}, σ_{yz} and σ_{zz}. Because of moment equilibrium it can be derived that

$$\sigma_{yx} = \sigma_{xy}, \qquad \sigma_{zx} = \sigma_{xz}, \qquad \sigma_{zy} = \sigma_{yz}. \tag{8.32}$$

The equilibrium equations Eqs. (8.30) and (8.31) can be generalized to three dimensions as

$$\frac{\partial \sigma_{xx}}{\partial x} + \frac{\partial \sigma_{xy}}{\partial y} + \frac{\partial \sigma_{xz}}{\partial z} = 0 \tag{8.33}$$

$$\frac{\partial \sigma_{xy}}{\partial x} + \frac{\partial \sigma_{yy}}{\partial y} + \frac{\partial \sigma_{yz}}{\partial z} = 0 \tag{8.34}$$

$$\frac{\partial \sigma_{xz}}{\partial x} + \frac{\partial \sigma_{yz}}{\partial y} + \frac{\partial \sigma_{zz}}{\partial z} = 0. \tag{8.35}$$

Some interpretation of the equilibrium equations is given by considering a number of special cases.

Example 8.1 If all the individual partial derivatives appearing in Eqs. (8.30) and (8.31) are zero, that is if

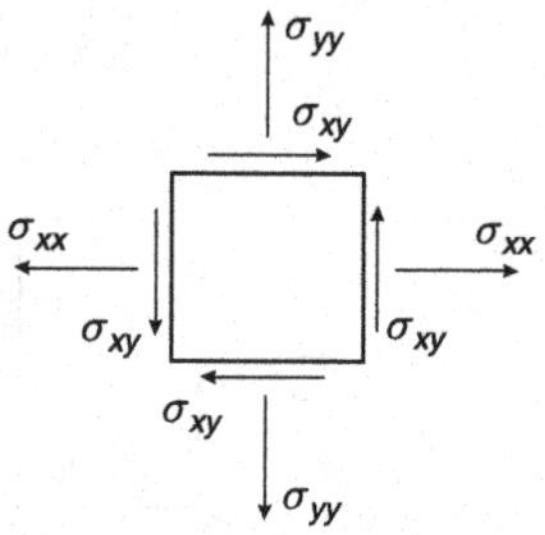

Figure 8.9
Stresses on the faces.

$$\frac{\partial \sigma_{xx}}{\partial x} = 0, \qquad \frac{\partial \sigma_{xy}}{\partial y} = 0, \qquad \frac{\partial \sigma_{yy}}{\partial y} = 0, \qquad \frac{\partial \sigma_{xy}}{\partial x} = 0,$$

the stresses on the faces of the cube are shown in Fig. 8.9. Clearly the forces on opposing faces are in equilibrium, as demanded by the equilibrium equations if the individual partial derivatives are zero.

Example 8.2 In the absence of shear stresses ($\sigma_{xy} = 0$), or if the shear stresses are constant ($\sigma_{xy} = c$) it follows that

$$\frac{\partial \sigma_{xy}}{\partial x} = \frac{\partial \sigma_{xy}}{\partial y} = 0.$$

Hence the equilibrium equations reduce to:

$$\frac{\partial \sigma_{xx}}{\partial x} = 0, \qquad \frac{\partial \sigma_{yy}}{\partial y} = 0.$$

This means that σ_{xx} may only be a function of $y : \sigma_{xx} = \sigma_{xx}(y)$. Likewise, σ_{yy} may only be a function of x.

8.4 Stress Tensor

Suppose that, in a two-dimensional configuration, all three stress components (σ_{xx}, σ_{xy} and σ_{yy}) are known. How can the resulting stress vector acting on an arbitrary cross section of a body be computed based on these stress components? To answer this question, consider an arbitrary, but infinitesimally small prism, having a triangular cross section as depicted in Fig. 8.10(a). Two faces of the prism are parallel to $\vec{e}_x$ and $\vec{e}_y$, respectively, while the third face, having length Δl, is oriented at some angle α with respect to $\vec{e}_x$. The orientation in space of this face is fully characterized by the unit outward normal vector $\vec{n}$, related to α by

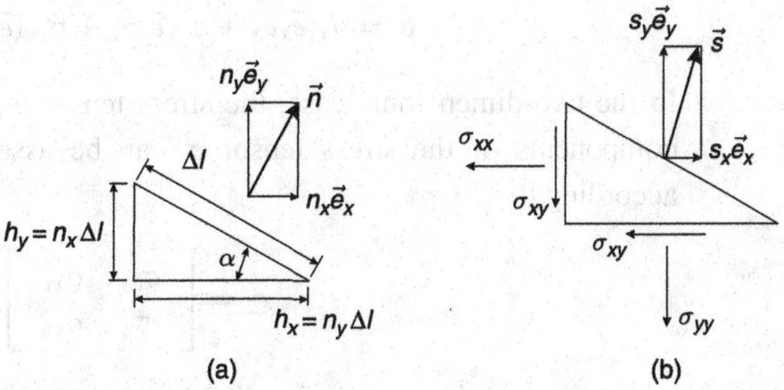

Figure 8.10

Stress vector $\vec{s}$ acting on an inclined plane with normal $\vec{n}$.

$$\vec{n} = n_x\vec{e}_x + n_y\vec{e}_y = \sin(\alpha)\,\vec{e}_x + \cos(\alpha)\,\vec{e}_y. \tag{8.36}$$

The face parallel to $\vec{e}_x$ has length $h_x = n_y\,\Delta l$, while the face parallel to $\vec{e}_y$ has length $h_y = n_x\,\Delta l$. This follows immediately from

$$n_x = \sin(\alpha) = \frac{h_y}{\Delta l} \to h_y = n_x\Delta l, \tag{8.37}$$

while

$$n_y = \cos(\alpha) = \frac{h_x}{\Delta l} \to h_x = n_y\Delta l. \tag{8.38}$$

The stresses acting on the left and bottom face of the triangular prism are depicted in Fig. 8.10(b). On the inclined face, a stress vector $\vec{s}$ is introduced. Force equilibrium in the x-direction yields

$$s_x(h\,\Delta l)\vec{e}_x = \sigma_{xx}(h\,\Delta l\,n_x)\vec{e}_x + \sigma_{xy}(h\,\Delta l\,n_y)\vec{e}_x, \tag{8.39}$$

where h denotes the thickness of the prism in the z-direction. Dividing by the area $h\Delta l$ yields

$$s_x = \sigma_{xx}n_x + \sigma_{xy}n_y. \tag{8.40}$$

A similar exercise in the y-direction gives

$$s_y(h\,\Delta l)\vec{e}_y = \sigma_{yy}(h\,\Delta l\,n_y)\vec{e}_y + \sigma_{xy}(h\,\Delta l\,n_x)\vec{e}_y, \tag{8.41}$$

hence dividing by $h\Delta l$ yields

$$s_y = \sigma_{yy}n_y + \sigma_{xy}n_x. \tag{8.42}$$

So, the stress vector $\vec{s}$ is directly related to the stress components σ_{xx}, σ_{yy} and σ_{xy} via the normal $\vec{n}$ to the infinitesimal surface element at which $\vec{s}$ acts. This can also be written in a compact form by introducing the so-called **stress tensor $\boldsymbol{\sigma}$**. Let this tensor be defined according to

$$\boldsymbol{\sigma} = \sigma_{xx}\vec{e}_x\vec{e}_x + \sigma_{yy}\vec{e}_y\vec{e}_y + \sigma_{xy}(\vec{e}_x\vec{e}_y + \vec{e}_y\vec{e}_x). \tag{8.43}$$

In the two-dimensional case, the stress tensor $\boldsymbol{\sigma}$ is the sum of four dyads. The components of the stress tensor $\boldsymbol{\sigma}$ can be assembled in the stress matrix $\underline{\sigma}$ according to

$$\underline{\sigma} = \begin{bmatrix} \sigma_{xx} & \sigma_{xy} \\ \sigma_{yx} & \sigma_{yy} \end{bmatrix}, \tag{8.44}$$

where σ_{yx} equals σ_{xy}; see Eq. (8.29). This stress tensor $\boldsymbol{\sigma}$ has been constructed such that the stress vector $\vec{s}$ (with components $\underset{\sim}{s}$) acting on an infinitesimal surface element with outward unit normal $\vec{n}$ (with components $\underset{\sim}{n}$) may be computed via

$$\vec{s} = \boldsymbol{\sigma} \cdot \vec{n}. \tag{8.45}$$

This follows immediately from

$$\begin{aligned} \boldsymbol{\sigma} \cdot \vec{n} &= (\sigma_{xx}\vec{e}_x\vec{e}_x + \sigma_{yy}\vec{e}_y\vec{e}_y + \sigma_{xy}(\vec{e}_x\vec{e}_y + \vec{e}_y\vec{e}_x)) \cdot \vec{n} \\ &= \sigma_{xx}\vec{e}_x\vec{e}_x \cdot \vec{n} + \sigma_{yy}\vec{e}_y\vec{e}_y \cdot \vec{n} + \sigma_{xy}(\vec{e}_x\vec{e}_y \cdot \vec{n} + \vec{e}_y\vec{e}_x \cdot \vec{n}) \\ &= \sigma_{xx}n_x\vec{e}_x + \sigma_{yy}n_y\vec{e}_y + \sigma_{xy}(n_y\vec{e}_x + n_x\vec{e}_y) \\ &= (\sigma_{xx}n_x + \sigma_{xy}n_y)\vec{e}_x + (\sigma_{yy}n_y + \sigma_{xy}n_x)\vec{e}_y. \end{aligned} \tag{8.46}$$

Hence, it follows immediately that, with $\vec{s} = s_x\vec{e}_x + s_y\vec{e}_y$:

$$s_x = \sigma_{xx}n_x + \sigma_{xy}n_y \tag{8.47}$$

and

$$s_y = \sigma_{xy}n_x + \sigma_{yy}n_y. \tag{8.48}$$

Equation (8.45) can also be written in column notation as

$$\underset{\sim}{s} = \underline{\sigma}\, \underset{\sim}{n}. \tag{8.49}$$

The purpose of introducing the stress tensor $\boldsymbol{\sigma}$, defined as the sum of four dyads, is to compute the stress vector that acts on an infinitesimally small area that is oriented in space as defined by the normal $\vec{n}$. For any given normal $\vec{n}$ this stress vector is computed via $\vec{s} = \boldsymbol{\sigma} \cdot \vec{n}$.

At any point in a body and for any plane in that point, this stress vector can be computed. This stress vector itself may be decomposed into a stress vector normal to the plane (normal stress) and a vector tangent to the plane (shear stress): see Fig. 8.11. Hence let

$$\vec{s} = \boldsymbol{\sigma} \cdot \vec{n}, \tag{8.50}$$

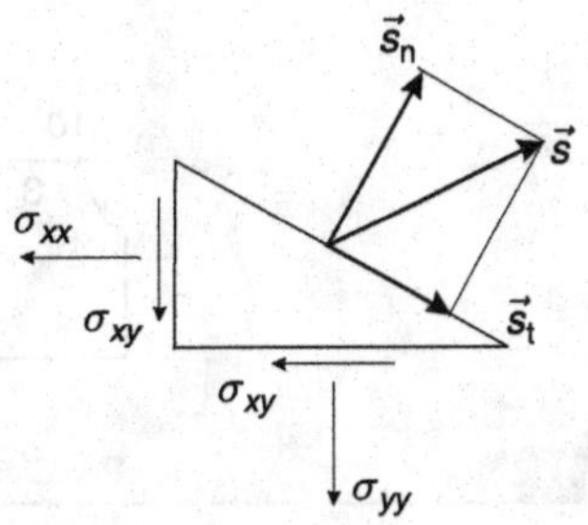

Figure 8.11

Stress vector $\vec{s}$ acting on an inclined plane with normal $\vec{n}$, decomposed into a stress vector normal to the plane and a stress vector tangent to the plane.

then the stress vector normal to the plane, $\vec{s}_n$, follows from

$$\begin{aligned}\vec{s}_n &= (\vec{s}\cdot\vec{n})\vec{n}\\ &= ((\boldsymbol{\sigma}\cdot\vec{n})\cdot\vec{n})\vec{n} = \boldsymbol{\sigma}\cdot\vec{n}\cdot(\vec{n}\vec{n}).\end{aligned} \tag{8.51}$$

The stress vector tangent to the plane is easily obtained via

$$\vec{s} = \vec{s}_n + \vec{s}_t, \tag{8.52}$$

hence

$$\begin{aligned}\vec{s}_t &= \vec{s} - \vec{s}_n\\ &= \boldsymbol{\sigma}\cdot\vec{n} - ((\boldsymbol{\sigma}\cdot\vec{n})\cdot\vec{n})\,\vec{n}\\ &= \boldsymbol{\sigma}\cdot\vec{n}\cdot(\boldsymbol{I} - \vec{n}\vec{n}).\end{aligned} \tag{8.53}$$

Example 8.3 If the stress state is specified as depicted in Fig. 8.12, then

$$\boldsymbol{\sigma} = 10\vec{e}_x\vec{e}_x + 3(\vec{e}_x\vec{e}_y + \vec{e}_y\vec{e}_x).$$

If the normal to the plane of interest equals

$$\vec{n} = \vec{e}_x,$$

then the stress vector on this plane follows from

$$\begin{aligned}\vec{s} &= \boldsymbol{\sigma}\cdot\vec{n}\\ &= (10\vec{e}_x\vec{e}_x + 3(\vec{e}_x\vec{e}_y + \vec{e}_y\vec{e}_x))\cdot\vec{e}_x\\ &= 10\vec{e}_x + 3\vec{e}_y.\end{aligned}$$

Clearly, as expected, the operation $\boldsymbol{\sigma}\cdot\vec{e}_x$ extracts the stress components acting on the right face of the rectangle shown in Fig. 8.12. The stress vector $\vec{s}$ may be decomposed into a component normal to this face and a component tangent to

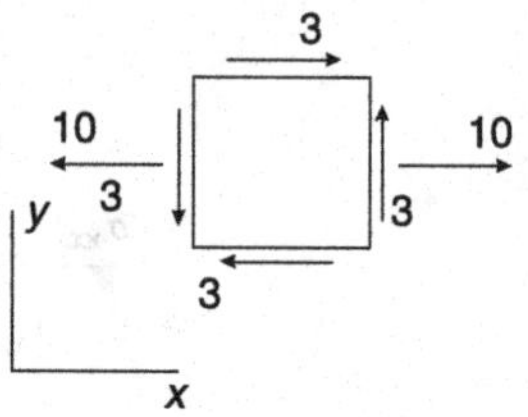

Figure 8.12

Stress components.

this face. Clearly, the normal component should be $\vec{s}_{\mathrm{n}} = 10\vec{e}_x$, while the tangent component should be $\vec{s}_{\mathrm{t}} = 3\vec{e}_y$. This also follows from

$$\begin{aligned}\vec{s}_{\mathrm{n}} &= ((\boldsymbol{\sigma} \cdot \vec{n}) \cdot \vec{n})\vec{n} \\ &= ((10\vec{e}_x + 3\vec{e}_y) \cdot \vec{e}_x)\vec{e}_x \\ &= 10\vec{e}_x.\end{aligned}$$

Generalization to three dimensions As noted before, there are six independent stress components in the three-dimensional case. These may be stored in the three-dimensional stress tensor using the sum of nine dyads:

$$\begin{aligned}\boldsymbol{\sigma} = {} & \sigma_{xx}\vec{e}_x\vec{e}_x + \sigma_{yy}\vec{e}_y\vec{e}_y + \sigma_{zz}\vec{e}_z\vec{e}_z \\ & + \sigma_{xy}(\vec{e}_x\vec{e}_y + \vec{e}_y\vec{e}_x) + \sigma_{xz}(\vec{e}_x\vec{e}_z + \vec{e}_z\vec{e}_x) + \sigma_{yz}(\vec{e}_y\vec{e}_z + \vec{e}_z\vec{e}_y),\end{aligned} \tag{8.54}$$

and also in the symmetric stress matrix:

$$\underline{\sigma} = \begin{bmatrix} \sigma_{xx} & \sigma_{xy} & \sigma_{xz} \\ \sigma_{yx} & \sigma_{yy} & \sigma_{yz} \\ \sigma_{zx} & \sigma_{zy} & \sigma_{zz} \end{bmatrix}. \tag{8.55}$$

In this case the normal to a plane has three components: $\vec{n} = n_x\vec{e}_x + n_y\vec{e}_y + n_z\vec{e}_z$, therefore

$$\begin{aligned}\vec{s} &= \boldsymbol{\sigma} \cdot \vec{n} \\ &= \big[\sigma_{xx}\vec{e}_x\vec{e}_x + \sigma_{yy}\vec{e}_y\vec{e}_y + \sigma_{zz}\vec{e}_z\vec{e}_z \\ &\quad + \sigma_{xy}(\vec{e}_x\vec{e}_y + \vec{e}_y\vec{e}_x) + \sigma_{xz}(\vec{e}_x\vec{e}_z + \vec{e}_z\vec{e}_x) \\ &\quad + \sigma_{yz}\ (\vec{e}_y\vec{e}_z + \vec{e}_z\vec{e}_y)\big] \cdot (n_x\vec{e}_x + n_y\vec{e}_y + n_z\vec{e}_z) \\ &= \sigma_{xx}n_x\vec{e}_x + \sigma_{yy}n_y\vec{e}_y + \sigma_{zz}n_z\vec{e}_z \\ &\quad + \sigma_{xy}(n_y\vec{e}_x + n_x\vec{e}_y) + \sigma_{xz}(n_z\vec{e}_x + n_x\vec{e}_z) \\ &\quad + \sigma_{yz}(n_z\vec{e}_y + n_y\vec{e}_z).\end{aligned} \tag{8.56}$$

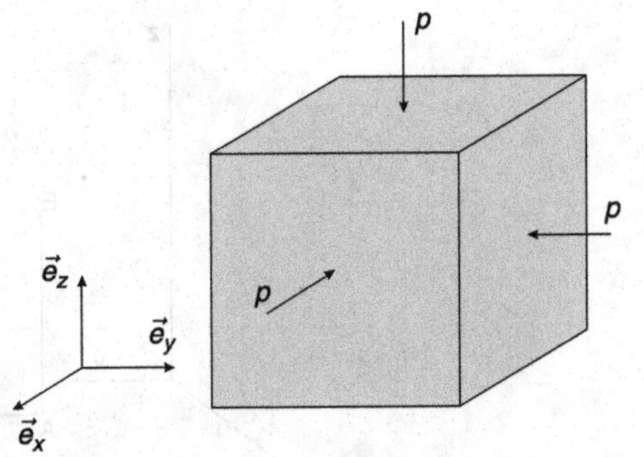

Figure 8.13

Pressure.

Hence the components of the stress vector $\vec{s} = s_x\vec{e}_x + s_y\vec{e}_y + s_z\vec{e}_z$ satisfy

$$\begin{aligned} s_x &= \sigma_{xx}n_x + \sigma_{xy}n_y + \sigma_{xz}n_z \\ s_y &= \sigma_{xy}n_x + \sigma_{yy}n_y + \sigma_{yz}n_z \\ s_z &= \sigma_{xz}n_x + \sigma_{yz}n_y + \sigma_{zz}n_z. \end{aligned} \tag{8.57}$$

Applying of Eq. (8.55) and using

$$\underset{\sim}{s} = \begin{bmatrix} s_x \\ s_y \\ s_z \end{bmatrix}, \qquad \underset{\sim}{n} = \begin{bmatrix} n_x \\ n_y \\ n_z \end{bmatrix}, \tag{8.58}$$

gives an equivalent, but much shorter, expression for Eq. (8.57):

$$\underset{\sim}{s} = \underline{\sigma}\,\underset{\sim}{n}. \tag{8.59}$$

If all the shear stress components are zero, i.e. $\sigma_{xy} = \sigma_{xz} = \sigma_{yz} = 0$, and all the normal stresses are equal, i.e. $\sigma_{xx} = \sigma_{yy} = \sigma_{zz}$, this normal stress is called the **pressure** p such that

$$p = -\sigma_{xx} = -\sigma_{yy} = -\sigma_{zz}, \tag{8.60}$$

while

$$\sigma = -p\,(\vec{e}_x\vec{e}_x + \vec{e}_y\vec{e}_y + \vec{e}_z\vec{e}_z) = -p\,\boldsymbol{I}, \tag{8.61}$$

with $\boldsymbol{I}$ the unit tensor. This is illustrated in Fig. 8.13.

Example 8.4 Consider a material cube ABCDEFGH. The edges of the cube are oriented in the direction of the axes of a Cartesian xyz-coordinate system (see Fig. 8.14). In this figure we also depict the normal and shear stresses (expressed in [MPa]) that act on the visible faces of the cube.

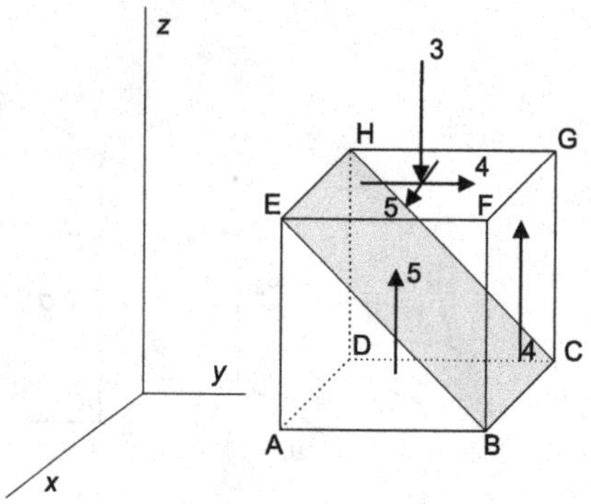

Figure 8.14

Material cube with stress components shown on the visible faces.

Consider the part ABCDEH of the cube and determine the normal stress vector and the shear stress vector on the surface BCHE. The stress matrix $\underline{\sigma}$ can be derived from the figure:

$$\underline{\sigma} = \begin{bmatrix} 0 & 0 & 5 \\ 0 & 0 & 4 \\ 5 & 4 & -3 \end{bmatrix}.$$

The outward normal $\underset{\sim}{n}$ on the surface BCHE can also be directly derived from the figure. The normal is given by:

$$\underset{\sim}{n} = \frac{1}{\sqrt{2}} \begin{bmatrix} 0 \\ 1 \\ 1 \end{bmatrix}.$$

The stress vector $\underset{\sim}{s}$ on this surface can be determined by using Eq. (8.59):

$$\underset{\sim}{s} = \begin{bmatrix} 0 & 0 & 5 \\ 0 & 0 & 4 \\ 5 & 4 & -3 \end{bmatrix} \frac{1}{\sqrt{2}} \begin{bmatrix} 0 \\ 1 \\ 1 \end{bmatrix} = \frac{1}{\sqrt{2}} \begin{bmatrix} 5 \\ 4 \\ 1 \end{bmatrix}.$$

The magnitude of the normal stress s_n follows from the inner product of $\underset{\sim}{s}$ with the normal $\underset{\sim}{n}$:

$$s_n = \underset{\sim}{s}^T \underset{\sim}{n} = \frac{1}{2}(5 \;\; 4 \;\; 1) \begin{bmatrix} 0 \\ 1 \\ 1 \end{bmatrix} = \frac{5}{2}.$$

The vector $\underset{\sim}{s}_n$ (note the tilde underneath the symbol) is found by multiplying the magnitude with the unit normal $\underset{\sim}{n}$:

$$\underset{\sim}{s}_n = \frac{5}{2\sqrt{2}} \begin{bmatrix} 0 \\ 1 \\ 1 \end{bmatrix}.$$

Finally the shear stress vector can be found by a vectorial subtraction of the stress vector $\underset{\sim}{s}$ and the normal stress vector $\underset{\sim}{s}_n$:

$$\underset{\sim}{s}_t = \underset{\sim}{s} - \underset{\sim}{s}_n = \frac{1}{2\sqrt{2}} \left(\begin{bmatrix} 10 \\ 8 \\ 2 \end{bmatrix} - \begin{bmatrix} 0 \\ 5 \\ 5 \end{bmatrix} \right) = \frac{1}{2\sqrt{2}} \begin{bmatrix} 10 \\ 3 \\ -3 \end{bmatrix}.$$

Example 8.5 In many biomechanical applications, it is important to know the stress in the fibres that constitute the material. The reasons for this are plenty. The fibres play a major role in the mechanical behaviour of the tissues as a whole, and many tissues will remodel their fibre architecture as a result of differences in loading of the cells and the fibres. Consider the material that is shown in Fig. 8.15(a). It is the collagenous fibre structure that is found in human skin and visualized with a fluorescence microscope. Figure 8.15(b) shows a detail from the image with one specific fibre bundle. We would like to determine the longitudinal stress in that particular fibre bundle.

Assume that we have done some (typically numerical) analysis and that we have found the stress matrix with respect to the two-dimensional x, y-coordinate system in point A:

$$\underset{\sim}{\sigma} = \begin{bmatrix} 4 & 1 \\ 1 & 2 \end{bmatrix} \quad \text{[MPa]}.$$

The unit vector in the direction of the fibre is given by:

$$\underset{\sim}{n} = \frac{1}{5} \begin{bmatrix} 4 \\ 3 \end{bmatrix}.$$

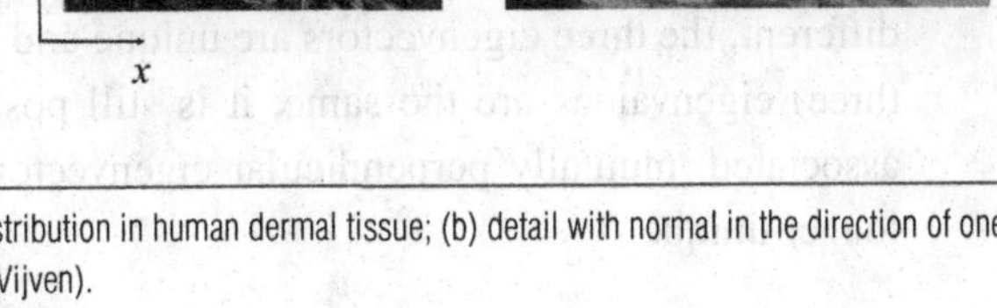

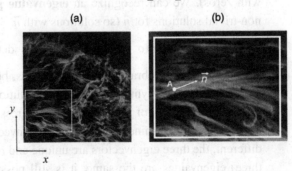

Figure 8.15

(a) Collagen fibre distribution in human dermal tissue; (b) detail with normal in the direction of one fibre bundle (courtesy: Marc van Vijven).

First, we can determine with Eq. (8.59) the stress vector on a small surface through point A for which the vector $\underset{\sim}{n}$ is the normal. It will be clear that this surface coincides with the cross section of the fibre.

$$\underset{\sim}{s} = \frac{1}{5}\begin{bmatrix} 4 & 1 \\ 1 & 2 \end{bmatrix}\begin{bmatrix} 4 \\ 3 \end{bmatrix} = \frac{1}{5}\begin{bmatrix} 19 \\ 10 \end{bmatrix}.$$

In general, the resulting stress vector will not point in the direction of the fibre. In that case we need the projection of the vector on the fibre, which can be determined from the inner product of normal $\underset{\sim}{n}$ with stress vector $\underset{\sim}{s}$:

$$s_f = \underset{\sim}{s}^T \underset{\sim}{n} = \frac{1}{25}[19 \;\; 10]\begin{bmatrix} 4 \\ 3 \end{bmatrix} = \frac{106}{25} = 4.24 \text{ [MPa]}.$$

8.5 Principal Stresses and Principal Stress Directions

Assume that at a certain point of the material volume, the stress state is known by specification of the tensor $\boldsymbol{\sigma}$. One might ask whether it is possible to choose the orientation of a surface element in such a direction, that only a normal stress acts on the surface and no shear stress. This means that we are trying to determine a vector $\vec{n}$ for which the following equation holds:

$$\boldsymbol{\sigma} \cdot \vec{n} = \lambda \vec{n}, \tag{8.62}$$

with λ to be interpreted as the normal stress. By shifting the right-hand side to the left, this equation can also be written as

$$(\boldsymbol{\sigma} - \lambda \boldsymbol{I}) \cdot \vec{n} = \vec{0} \text{ and in components as } (\underline{\sigma} - \lambda \underline{I})\underset{\sim}{n} = \underset{\sim}{0}, \tag{8.63}$$

with $\boldsymbol{I}$ the unit tensor ($\underline{I}$ the unit matrix) and $\vec{0}$ the zero vector ($\underset{\sim}{0}$ the column with zeros). We can recognize an **eigenvalue problem**. The equation only has non-trivial solutions for $\vec{n}$ (so solutions with $\vec{n} \neq \vec{0}$) if

$$\det(\boldsymbol{\sigma} - \lambda \boldsymbol{I}) = 0 \text{ and also } \det(\underline{\sigma} - \lambda \underline{I}) = 0. \tag{8.64}$$

This third-order algebraic equation for λ has, because the tensor $\boldsymbol{\sigma}$ (with matrix representation $\underline{\sigma}$) is symmetric, three real solutions (sometimes coinciding): these are the **eigenvalues** of the stress tensor (stress matrix). For each eigenvalue it is possible to determine a normalized eigenvector. If the three eigenvalues are different, the three eigenvectors are unique and mutually perpendicular. If two (or three) eigenvalues are the same, it is still possible to determine a set of three, associated, mutually perpendicular eigenvectors of unit length, but they are no longer unique.

The three eigenvalues (solutions for λ) are called the **principal stresses** and denoted as σ_1, σ_2 and σ_3. The principal stresses are arranged in such a way that $\sigma_1 \leq \sigma_2 \leq \sigma_3$.

The associated, normalized, mutually perpendicular eigenvectors are the principal stress directions and specified by $\vec{n}_1$, $\vec{n}_2$ and $\vec{n}_3$. It will be clear that

$$\boldsymbol{\sigma} \cdot \vec{n}_i = \sigma_i \vec{n}_i \quad \text{for} \quad i = 1, 2, 3 \tag{8.65}$$

and also

$$\begin{aligned} \vec{n}_i \cdot \vec{n}_j &= 1 \ \text{for} \ i = j \\ &= 0 \ \text{for} \ i \neq j. \end{aligned} \tag{8.66}$$

Based on the above, it is obvious that to every arbitrary stress cube, as depicted in Fig. 8.8, another cube can be attributed, which is differently oriented in space, upon which only normal stresses (the principal stresses) and no shear stresses are acting. Such a cube is called a principal stress cube: see Fig. 8.16. Positive principal stresses indicate extension, negative stresses indicate compression. The principal stress cube makes it easier to interpret a stress state and to identify the way a material is loaded. In the following section this will be discussed in more detail. As observed earlier, the stress state at a certain point is determined completely by the stress tensor $\boldsymbol{\sigma}$; in other words, by all stress components that act upon a cube, of which the orientation coincides with the xyz-coordinate system. Because the choice of the coordinate system is arbitrary, it can also be stated that the stress state is completely determined when the principal stresses and principal stress directions are known. How the principal stresses σ_i (with $i = 1, 2, 3$) and the

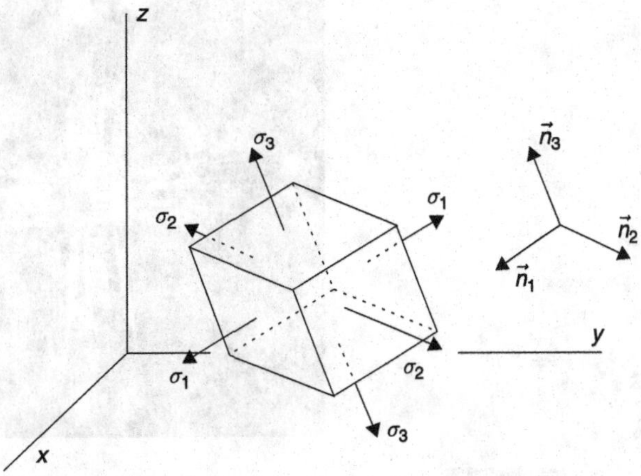

Figure 8.16

The principal stress cube.

principal stress directions $\vec{n}_i$ can be determined when the stress tensor $\boldsymbol{\sigma}$ is known is discussed above. The inverse procedure to reconstruct the stress tensor $\boldsymbol{\sigma}$, when the principal stresses and stress directions are known, will not be outlined here.

Example 8.6 Assume that, at some material point, a stress state is defined by means of the stress matrix $\underline{\sigma}$ according to:

$$\underline{\sigma} = \begin{bmatrix} 36 & 0 & 48 \\ 0 & 100 & 0 \\ 48 & 0 & 64 \end{bmatrix} \quad \text{expressed in [kPa].}$$

We want to determine the principal stresses. For this we have to solve:

$$\underline{\sigma}\,\underset{\sim}{n} = \lambda \underset{\sim}{n} \quad \rightarrow \quad (\underline{\sigma} - \lambda \underline{I})\underset{\sim}{n} = \underset{\sim}{0} \quad \text{with} \quad \underset{\sim}{n} \neq \underset{\sim}{0}.$$

This equation only has a solution when $\det(\underline{\sigma} - \lambda \underline{I}) = 0$, yielding:

$$\det \begin{bmatrix} 36-\lambda & 0 & 48 \\ 0 & 100-\lambda & 0 \\ 48 & 0 & 64-\lambda \end{bmatrix} = 0,$$

or:

$$(36-\lambda)(100-\lambda)(64-\lambda) - 48(100-\lambda)48 = 0$$

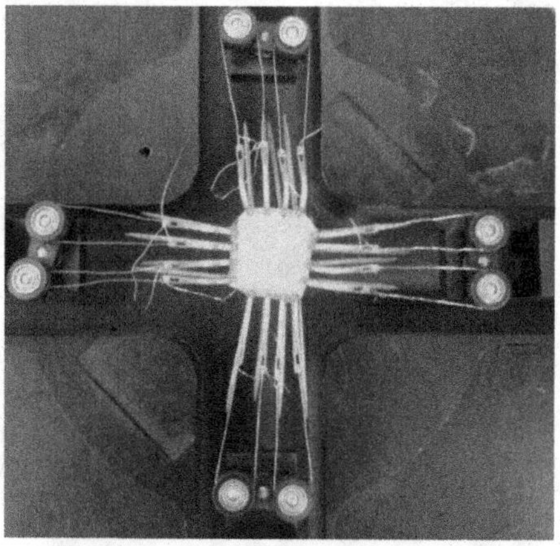

Figure 8.17

A human skin sample mounted in a biaxial testing machine. The dimensions of the sample are $15 \times 15 \times 1$ [mm^3] (courtesy: Marc van Vijven).

from which the following characteristic equation can be derived:

$$(100 - \lambda)(\lambda - 100)\lambda = 0.$$

So the eigenvalues are $\lambda_1 = \lambda_2 = 100$ [kPa] and $\lambda_3 = 0$ [kPa], and consequently $\sigma_1 = 0$[kPa] and $\sigma_2 = \sigma_3 = 100$ [kPa]. This is the kind of stress state that could be found when a thin square membrane is equibiaxially stretched in two directions (see Fig. 8.17). Because the membrane is very thin, the stress components in the direction perpendicular to the membrane are very small and usually ignored. This is called a situation of **plane stress**. The biaxial test is very common when determining material properties of soft biological tissues.

8.6 Mohr's Circles for the Stress State

When the stress tensor $\boldsymbol{\sigma}$ is given, the stress vector $\vec{s}$ on an arbitrary oriented plane, defined by the unit outward normal $\vec{n}$ can be determined by

$$\vec{s} = \boldsymbol{\sigma} \cdot \vec{n}. \tag{8.67}$$

Then the normal stress s_n and shear stress s_t can be determined. The normal stress s_n is the inner product of the stress vector $\vec{s}$ with the unit outward normal $\vec{n}$, $s_n = \vec{s} \cdot \vec{n}$, and can be either positive or negative (extension or compression). The shear stress s_t is the magnitude of the component of $\vec{s}$ tangent to that surface; see Section 8.4. In this way, it is possible to add to each $\vec{n}$ a combination (s_n, s_t) that can be regarded as a mapping in a graph with s_n along the horizontal axis and s_t along the vertical axis; see Fig. 8.18. It can be proven that all possible combinations (s_n, s_t) are located in the shaded area between

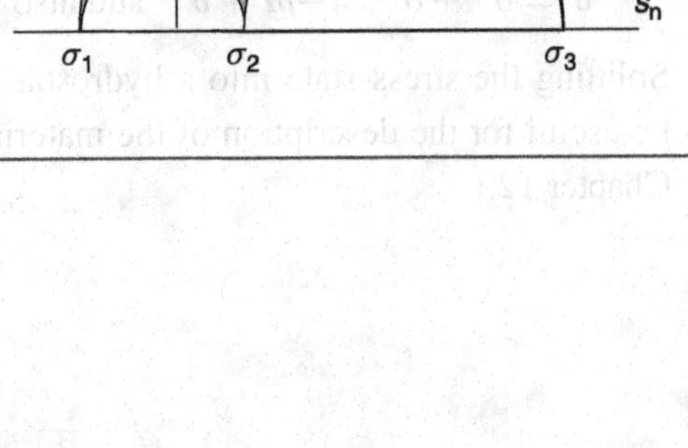

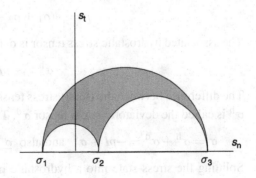

Figure 8.18

Mohr's circles for the stress.

the drawn circles (the three Mohr's circles). If the principal stresses σ_1, σ_2 and σ_3 are known, Mohr's circles (the centroid is located on the s_n-axis) can be drawn at once. For this, it is not necessary to determine the principal stress directions. In the $s_n s_t$-coordinate system the combinations $(\sigma_1, 0)$, $(\sigma_2, 0)$ and $(\sigma_3, 0)$ constitute the image points associated with the faces of the principal stress cube in Fig. 8.16. Based on Fig. 8.18, it can immediately be concluded that

$$\begin{aligned}(s_n)_{max} &= \sigma_3, \quad (s_n)_{min} = \sigma_1 \\ (s_t)_{max} &= (\sigma_3 - \sigma_1)/2.\end{aligned} \tag{8.68}$$

Example 8.7 Consider the stress state in Example 8.6. In that case, the maximum eigenvalue for the stress equals $\sigma_3 = \sigma_2 = 100$ [kPa]. The minimal eigenvalue is $\sigma_1 = 0$. So, according to criterium Eq. (8.68), the maximum shear stress in this mechanical state is:

$$(s_t)_{max} = (\sigma_3 - \sigma_1)/2 = 50 \quad \text{[kPa]}.$$

8.7 Hydrostatic Pressure and Deviatoric Stress

The hydrostatic pressure p is defined as the average of the normal stresses in Fig. 8.8. It is common practice to give the pressure the opposite sign to the average normal stresses, so:

$$p = -(\sigma_{xx} + \sigma_{yy} + \sigma_{zz})/3 = -\text{tr}(\boldsymbol{\sigma})/3 = -\text{tr}(\underline{\sigma})/3. \tag{8.69}$$

It can be shown that, if expressed in principal stresses, we find

$$p = -(\sigma_1 + \sigma_2 + \sigma_3)/3. \tag{8.70}$$

The associated hydrostatic stress tensor is denoted by $\boldsymbol{\sigma}^h$ and defined according to

$$\boldsymbol{\sigma}^h = -p\boldsymbol{I}. \tag{8.71}$$

The difference between the (total) stress tensor $\boldsymbol{\sigma}$ and the hydrostatic stress tensor $\boldsymbol{\sigma}^h$ is called the deviatoric stress tensor $\boldsymbol{\sigma}^d$. Thus

$$\boldsymbol{\sigma} = \boldsymbol{\sigma}^h + \boldsymbol{\sigma}^d = -p\boldsymbol{I} + \boldsymbol{\sigma}^d \text{ and also } \underline{\sigma} = \underline{\sigma}^h + \underline{\sigma}^d = -p\underline{I} + \underline{\sigma}^d. \tag{8.72}$$

Splitting the stress state into a hydrostatic part and a deviatoric part appears to be useful for the description of the material behaviour. This will be the theme of Chapter 12.

8.8 Equivalent Stress

In general, mechanical failure of materials is (among other things) determined by the stresses that act on the material. In principle, this means that all components of the stress tensor $\boldsymbol{\sigma}$ in one way or another may contribute to failure. It is common practice to attribute a scalar property to the stress tensor $\boldsymbol{\sigma}$ that reflects the gravity of the stress state with respect to failure. Such a scalar property is normalized, based on a consideration of a uniaxial stress state (related to the extension of a bar), and is called an equivalent stress $\overline{\sigma}$. The equivalent stress is a scalar function of the stress tensor and thus of the components of the stress matrix

$$\overline{\sigma} = \overline{\sigma}(\boldsymbol{\sigma}) \text{ and also } \overline{\sigma} = \overline{\sigma}(\underline{\sigma}). \tag{8.73}$$

The formal relationship given above has to be specified, based on physical understandings of the failure of the material. Only experimentally is it possible to assess at which stress combinations a certain material will reach the limits of its resistance to failure. It is very possible and obvious that one material may fail by means of a completely different mechanism than another material. For example, considering technical materials, it is known that for a metal the maximum shear strain $(s_t)_{max}$ is often the limiting factor, while a ceramic material might fail because of too high a maximum extensional stress $(s_n)_{max}$. Such knowledge is important for the design of hip and knee prostheses or tooth implants, where both types of materials are used. But biological materials also may have different failure mechanisms. A bone, for example, will often fail as a result of the maximum compression stress, but a tendon will usually fail because it is overstretched, i.e. owing to the maximum extensional stress or maximum shear stress. The functional relationship $\overline{\sigma}(\boldsymbol{\sigma})$ is primarily determined by the (micro)structure of the considered material. Thus, depending on the material, different specifications of $\overline{\sigma}$ may be applied. We will limit ourselves to a few examples.

According to the equivalent stress $\overline{\sigma}_T$ ascribed to Tresca (but sometimes also attributed to Coulomb, Mohr or Guest), the maximum shear stress is held responsible for failure. The definition is

$$\overline{\sigma}_T = 2(\sigma_t)_{max} = \sigma_3 - \sigma_1, \tag{8.74}$$

which is normalized in such a way that for a uniaxially loaded bar, with an extensional stress $\sigma_{ax} > 0$, we find $\overline{\sigma}_T = \sigma_{ax}$.

The equivalent stress σ_M according to von Mises (also Hüber and Hencky) is based on the deviatoric stress tensor:

$$\overline{\sigma}_M = \sqrt{\frac{3}{2}\text{tr}(\boldsymbol{\sigma}^d \cdot \boldsymbol{\sigma}^d)} = \sqrt{\frac{3}{2}\text{tr}(\underline{\sigma}^d \underline{\sigma}^d)}. \tag{8.75}$$

This can be elaborated to

$$\overline{\sigma}_{\mathrm{M}} = \sqrt{\frac{1}{2}\left[(\sigma_{xx}-\sigma_{yy})^2+(\sigma_{yy}-\sigma_{zz})^2+(\sigma_{zz}-\sigma_{xx})^2\right]+3\left[\sigma_{xy}^2+\sigma_{xz}^2+\sigma_{yz}^2\right]}. \tag{8.76}$$

Here also the specification is chosen in such a way that for a uniaxially loaded bar with axial stress σ_{ax}, the equivalent von Mises stress satisfies $\overline{\sigma}_{\mathrm{M}} = \sigma_{\mathrm{ax}}$. It can be proven that in terms of principal stresses:

$$\overline{\sigma}_{\mathrm{M}} = \sqrt{\frac{1}{2}\left[(\sigma_1-\sigma_2)^2+(\sigma_2-\sigma_3)^2+(\sigma_3-\sigma_1)^2\right]}. \tag{8.77}$$

In general (for arbitrary $\boldsymbol{\sigma}$), the differences between the equivalent stresses according to Tresca and von Mises are relatively small.

The equivalent stress $\overline{\sigma}_{\mathrm{R}}$, according to Rankine (also Galilei), expresses that, in absolute sense, the maximum principal stress determines failure. This means that

$$\begin{aligned}\overline{\sigma}_{\mathrm{R}} &= |\sigma_3| \ \text{ if } |\sigma_3| \geq |\sigma_1| \\ \overline{\sigma}_{\mathrm{R}} &= |\sigma_1| \ \text{ if } |\sigma_3| < |\sigma_1|.\end{aligned} \tag{8.78}$$

And again for a uniaxially loaded bar $\overline{\sigma}_{\mathrm{R}} = \sigma_{\mathrm{ax}}$. For an arbitrary stress state, the equivalent stress according to Rankine can be completely different from the equivalent stress according to Tresca or von Mises.

A few remarks on this subject are opportune at this point. The understanding of failure thresholds and mechanisms for biological materials involves much more than identifying a suitable equivalent stress expression. Biological materials can also fail because of a disturbance of the metabolic processes in the cells. The mechanical state in biological materials is not only determined by stresses and strains, but also by rather complicated transport processes of nutrients, oxygen and waste products and very complex biochemistry. These processes can be disturbed by mechanical deformation (for example occlusion of blood vessels, causing an ischaemic state of the tissue, resulting in lack of oxygen and nutrients and accumulation of waste products). After some time this may result in cell death and thus damaged tissues. How these processes evolve and lead to tissue remodelling and/or damage is still the subject of research.

Exercises

8.1 Consider a two-dimensional plane stress state, with stress components:

$$\begin{aligned}\sigma_{xx} &= ax^2 + by \\ \sigma_{yy} &= bx^2 + ay^2 - cx.\end{aligned}$$

Use the equilibrium equations to determine the shear stress component $\sigma_{xy}(x, y)$, satisfying $\sigma_{xy}(0, 0) = 2a$.

8.2 On an infinitesimal area segment, stress components are working as given in the figure.

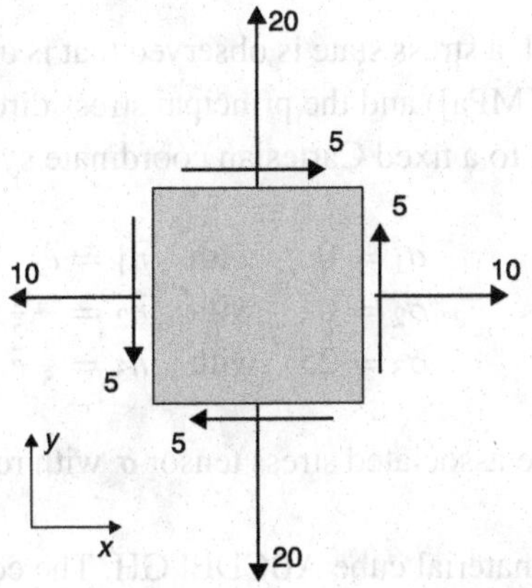

(a) Determine the stress tensor.

(b) Determine the stress vector $\vec{s}$ acting on a plane with unit normal vector $\vec{n} = \frac{1}{2}\sqrt{2}\vec{e}_x + \frac{1}{2}\sqrt{2}\vec{e}_y$.

(c) Determine the components of $\vec{s}$ perpendicular and parallel to the plane defined in item (b).

8.3 On an infinitesimal area segment, two sets of stress components are working as shown in figures (a) and (b).

(a) Determine the stress tensor for both situations.

(b) Determine the stress vector on the plane with normal $\vec{n} = \vec{e}_x \cos(\alpha) + \vec{e}_y \sin(\alpha)$. The angle α represents the angle of the normal vector with the x-axis.

(c) Determine for both situations the length of the stress vector as a function of α.

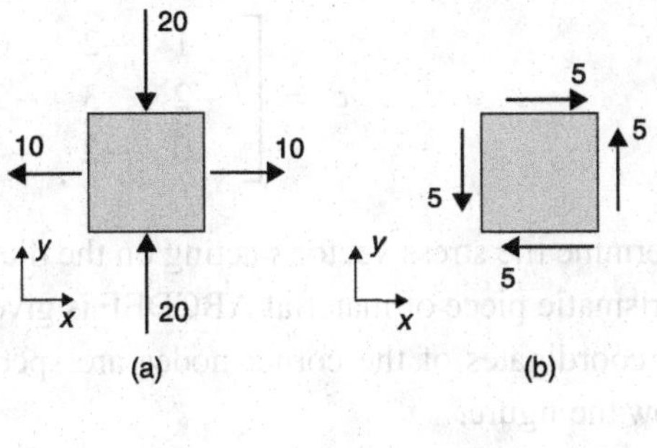

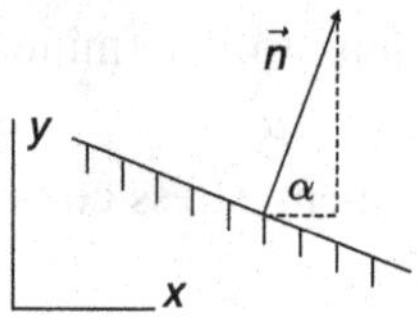

8.4 In a material, a stress state is observed that is characterized by the principal stresses (in [MPa]) and the principal stress directions (unit vectors, defined with respect to a fixed Cartesian coordinate system):

$$
\begin{aligned}
\sigma_1 &= 0 \quad \text{with} \quad \vec{n}_1 = \vec{e}_z \\
\sigma_2 &= 0 \quad \text{with} \quad \vec{n}_2 = -\tfrac{4}{5}\,\vec{e}_x + \tfrac{3}{5}\,\vec{e}_y \\
\sigma_3 &= 25 \quad \text{with} \quad \vec{n}_3 = \tfrac{3}{5}\,\vec{e}_x + \tfrac{4}{5}\,\vec{e}_y .
\end{aligned}
$$

Calculate the associated stress tensor $\boldsymbol{\sigma}$ with respect to the basis vectors $\vec{e}_x$, $\vec{e}_y$ and $\vec{e}_z$.

8.5 Consider a material cube ABCDEFGH. The edges of the cube are oriented in the direction of the axes of a Cartesian xyz-coordinate system; see the figure.

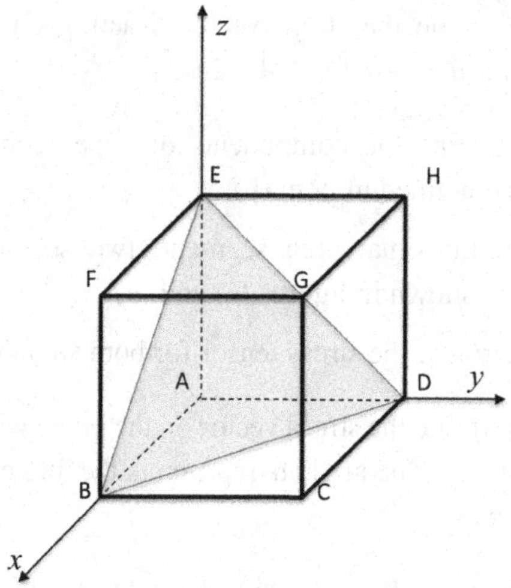

In addition, the relevant stress matrix $\underline{\sigma}$ for the particle is given:

$$
\underline{\sigma} = \begin{bmatrix} 1 & -2 & 0 \\ -2 & 3 & -2 \\ 0 & -2 & -4 \end{bmatrix} .
$$

Determine the stress vector $\vec{s}$ acting on the diagonal plane BDE.

8.6 A prismatic piece of material ABCDEF is given: see the figure.
The coordinates of the corner nodes are specified (in [mm]) in the table below the figure.

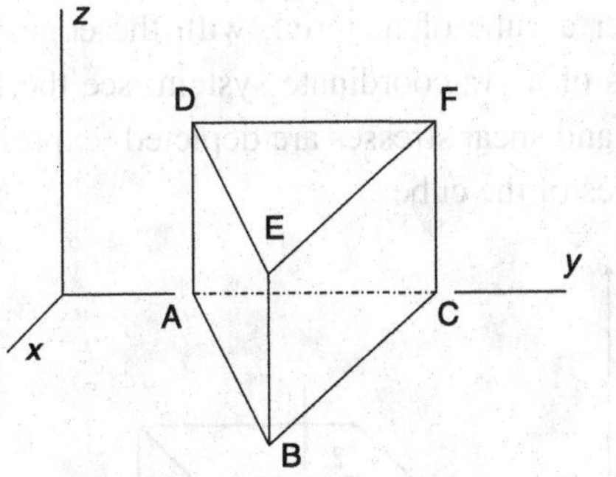

	A	B	C	D	E	F
x	0	8	0	0	8	0
y	1	7	7	1	7	7
z	0	0	0	5	5	5

For the faces of the prism that are visible in the figure, the stresses that act upon these faces (in [MPa]) are known:

$$\vec{s}_{\text{ABED}} = -5\vec{e}_x + 2\vec{e}_y + 6\vec{e}_z$$

$$\vec{s}_{\text{BCFE}} = 10\vec{e}_x + 5\vec{e}_y$$

$$\vec{s}_{\text{DEF}} = 10\vec{e}_x.$$

Calculate the associated stress tensor $\boldsymbol{\sigma}$ under the assumption that the stress state in the considered piece of material is homogeneous.

8.7 Consider a material cube ABCDEFGH. The edges of the cube are oriented in the direction of the axes of a Cartesian xyz-coordinate system: see the figure.

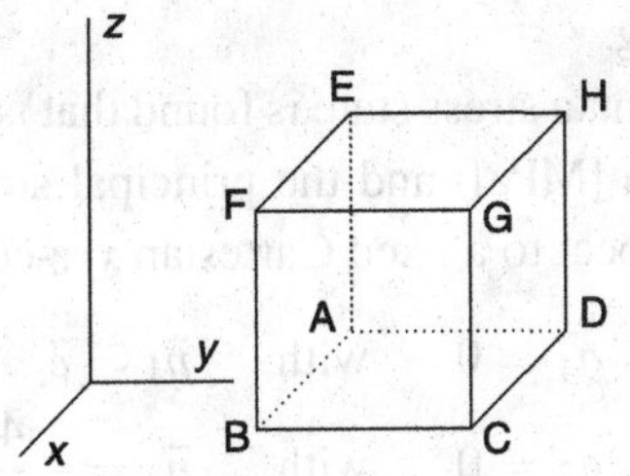

In addition, it is given that the relevant stress matrix $\underline{\sigma}$ for the particle is

$$\underline{\sigma} = \begin{bmatrix} 5 & 2 & 5 \\ 2 & 12 & 10 \\ 5 & 10 & 3 \end{bmatrix} \text{ [MPa]}.$$

Determine the magnitude s_t of the shear stress vector $\vec{s}_t$ acting on the diagonal plane BDE.

8.8 Consider a cube of material, with the edges oriented in the direction of the axes of a xyz-coordinate system: see the figure. In this figure also the normal and shear stresses are depicted (expressed in [MPa]) that act on the side faces of the cube.

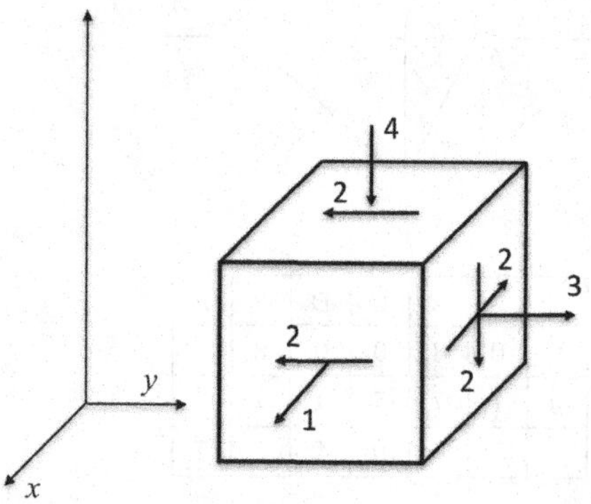

Determine the equivalent stress $\overline{\sigma}_{\mathrm{M}}$ according to von Mises.

8.9 For a material element, the stress state is defined by means of the stress matrix $\underline{\sigma}$:

$$\underline{\sigma} = \begin{bmatrix} 6 & 1 & -2 \\ 1 & 2 & 2 \\ -2 & 2 & 5 \end{bmatrix} \quad \text{[MPa]}.$$

It can be derived that the associated principal stresses are:

$$\sigma_1 = (4 - \sqrt{13}), \quad \sigma_2 = 5, \quad \sigma_3 = (4 + \sqrt{13}).$$

Determine the normal $\underset{\sim}{n}_2$ to the plane upon which the principal stress $\sigma_2 = 5$ [MPa] is acting.

8.10 In a material point, a stress state is found that is characterized with the principal stresses (in [MPa]) and the principal stress directions (unit vectors, defined with respect to a fixed Cartesian xyz-coordinate system):

$$\sigma_1 = 0 \quad \text{with} \quad \vec{n}_1 = \vec{e}_z$$
$$\sigma_2 = 0 \quad \text{with} \quad \vec{n}_2 = -\frac{4}{5}\vec{e}_x + \frac{3}{5}\vec{e}_y$$
$$\sigma_3 = 25 \quad \text{with} \quad \vec{n}_3 = \frac{3}{5}\vec{e}_x + \frac{4}{5}\vec{e}_y.$$

Calculate the associated stress tensor $\boldsymbol{\sigma}$ represented with respect to the base vectors $\vec{e}_x$, $\vec{e}_y$ and $\vec{e}_z$.

9 Motion: Time as an Extra Dimension

9.1 Introduction

Let us consider the geometrical change in time (the deformation and movement in three-dimensional space) of a coherent amount of material or material fraction, for which a continuum modelling approach is allowed. In cases where more fractions are involved, it is in principle possible to describe the behaviour of each fraction separately, as if it were isolated from the other fractions (however, it will be necessary to include interactions between fractions). The present chapter is focussed on a detailed description of motion. In addition, the consequences of configuration changes for the formulation of physical fields will be discussed. No attention will be given to the possible causes of the motion. In the present chapter, an approach will be followed that is common practice in the continuum description of solids (although it can also be applied to fluids). The specific aspects relevant to fluids will be treated at the end of the chapter.

9.2 Geometrical Description of the Material Configuration

Consider a coherent amount of material in a completely defined geometrical state (the reference configuration). For each material point P, a position vector $\vec{x}_0$ (with components stored in the column $\underset{\sim}{x}_0$) is allocated. In the following, this position vector will be used to identify the material point. The vector $\vec{x}_0$ is inextricably bound to the material point P, as if it were an attached label. With respect to a Cartesian xyz-coordinate system, $\vec{x}_0$ can be written as

$$\vec{x}_0 = x_0\vec{e}_x + y_0\vec{e}_y + z_0\vec{e}_z \quad \text{and also} \quad \underset{\sim}{x}_0 = \begin{bmatrix} x_0 \\ y_0 \\ z_0 \end{bmatrix}. \tag{9.1}$$

Because $\vec{x}_0$ is uniquely coupled to a material point, the components $\underset{\sim}{x}_0$ of $\vec{x}_0$ are called material coordinates. The set of position vectors $\vec{x}_0$ that address all the material points in the configuration comprise the reference volume V_0; see

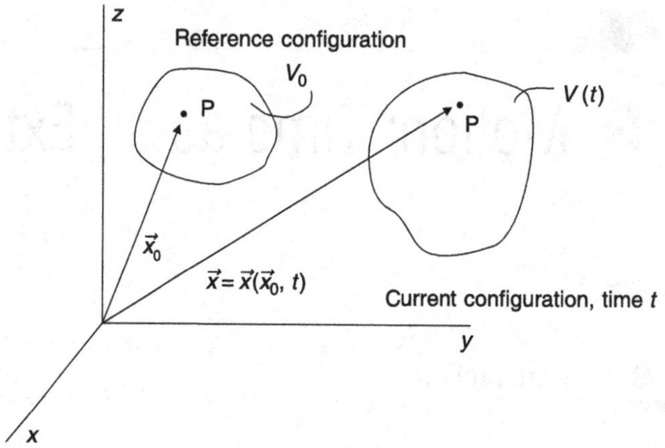

Figure 9.1

The position vector of a material point P.

Fig. 9.1. In an arbitrary current state, at time t, the position vector of the point P is specified by $\vec{x}$ and can be written as

$$\vec{x} = x\vec{e}_x + y\vec{e}_y + z\vec{e}_z \quad \text{and also} \quad \underset{\sim}{x} = \begin{bmatrix} x \\ y \\ z \end{bmatrix}. \tag{9.2}$$

With the attention focussed on a certain material point, it can be stated that

$$\vec{x} = \vec{x}(\vec{x}_0, t). \tag{9.3}$$

This functional relation expresses that the current position $\vec{x}$ of a material point is determined by the material identification $\vec{x}_0$ in V_0 of that point and by the current time t. When $\vec{x}_0$ is constant and with t passing through a certain time interval, $\vec{x} = \vec{x}(\vec{x}_0, t)$ can be considered to be a parameter description (with parameter t) of the trajectory of a material point (defined by $\vec{x}_0$) through three-dimensional space: the path of the particle.

Differentiation of the relation $\vec{x} = \vec{x}(\vec{x}_0, t)$ to the time t, with $\vec{x}_0$ taken constant (partial differentiation), results in the velocity vector $\vec{v}$ of the material point under consideration. It can be written:

$$\vec{v} = \dot{\vec{x}} \quad \text{and also} \quad \underset{\sim}{v} = \dot{\underset{\sim}{x}}, \tag{9.4}$$

with $(\dot{\ })$ the material time derivative: partial differentiation with respect to the time t with constant $\vec{x}_0$. For the acceleration vector $\vec{a}$ it follows:

$$\vec{a} = \dot{\vec{v}} = \ddot{\vec{x}} \quad \text{and also} \quad \underset{\sim}{a} = \dot{\underset{\sim}{v}} = \ddot{\underset{\sim}{x}}. \tag{9.5}$$

It will be clear that, in the context of the discussion above, the following formal relations hold for the velocity field and the acceleration field:

$$\vec{v} = \vec{v}(\vec{x}_0, t) \quad \text{and} \quad \vec{a} = \vec{a}(\vec{x}_0, t). \tag{9.6}$$

The configuration change of the material in a certain time interval can be associated with deformation. We deal with deformation when the mutual distances between material points change. The mathematical description of deformation and deformation velocity is the major theme of Chapter 10.

Example 9.1 Assume that the position of a material point as a function of time is given by:

$$\vec{x} = (x_0 + at)\vec{e}_x + bt^2\vec{e}_y$$

with a and b constants and t the time. In that case:

$$\vec{v} = \left[\frac{\partial \vec{x}}{\partial t}\right]_{\vec{x}_0 \text{ constant}} = a\vec{e}_x + 2bt\vec{e}_y$$

and:

$$\vec{a} = \left[\frac{\partial \vec{v}}{\partial t}\right]_{\vec{x}_0 \text{ constant}} = 2b\vec{e}_y.$$

9.3 Lagrangian and Eulerian Descriptions

In the previous section, the velocity and acceleration of the material are formally written as functions of the material identification $\vec{x}_0$ with material coordinates $\underset{\sim}{x}_0$ in V_0 and the time t. Obviously, this can also be done for other physical properties associated with the material, for example the temperature T. For a (time-dependent) temperature field, this can be written

$$T = T(\vec{x}_0, t). \tag{9.7}$$

The temperature field in the current configuration $V(t)$ is mapped on the reference configuration. Such a description is referred to as **Lagrangian**. Partial differentiation to time t at constant $\vec{x}_0$ results in the **material time derivative** of the temperature, $\dot{T}$:

$$\dot{T} = \lim_{\Delta t \to 0} \frac{T(\vec{x}_0, t + \Delta t) - T(\vec{x}_0, t)}{\Delta t} = \left[\frac{\partial T}{\partial t}\right]_{\vec{x}_0 \text{ constant}}. \tag{9.8}$$

This variable $\dot{T}$ has to be interpreted as the change (per unit time) of the temperature at a material point (moving through space) identified by $\vec{x}_0$.

Another approach concentrates on a fixed point in three-dimensional space. At every point in time, a different material particle may arrive at this location. For the (time-dependent) temperature field, we can write:

$$T = T(\vec{x}, t), \tag{9.9}$$

indicating the temperature of the material present at time t at the spatial point $\vec{x}$ in $V(t)$. This alternative field specification is called **Eulerian**. When the partial derivative with respect to time t of the temperature field in the Eulerian description is determined, **the spatial time derivative** $\delta T/\delta t$ is obtained:

$$\frac{\delta T}{\delta t} = \lim_{\Delta t \to 0} \frac{T(\vec{x}, t + \Delta t) - T(\vec{x}, t)}{\Delta t} = \left[\frac{\partial T}{\partial t}\right]_{\vec{x}\ \text{constant}} . \tag{9.10}$$

This result $\delta T/\delta t$ should be interpreted as the change (per unit time) of the temperature at a fixed point in space $\vec{x}$, in which, at consecutive time values t, different material points will be found.

The temperature field at time t, used as an example above, can be written in a Lagrangian description, $T = T(\vec{x}_0, t)$, and thus be mapped on the reference configuration with the domain V_0. The field can also be written in an Eulerian description, $T = T(\vec{x}, t)$, and be associated with the current configuration with domain $V(t)$: see Fig. 9.2. It should be noticed that a graphical representation of such a field in these two cases can be very different, especially in the case of large deformations and large rotations (both quite common in biological applications)

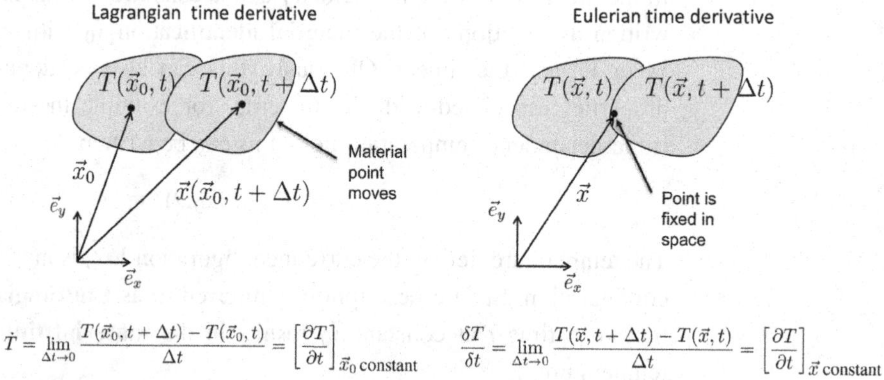

Figure 9.2

Lagrangian and Eulerian time derivatives.

In Section 9.4, we focus on the relation between the time derivatives discussed above. In Section 9.6, the relation between gradient operators applied to both descriptions will be discussed.

Example 9.2

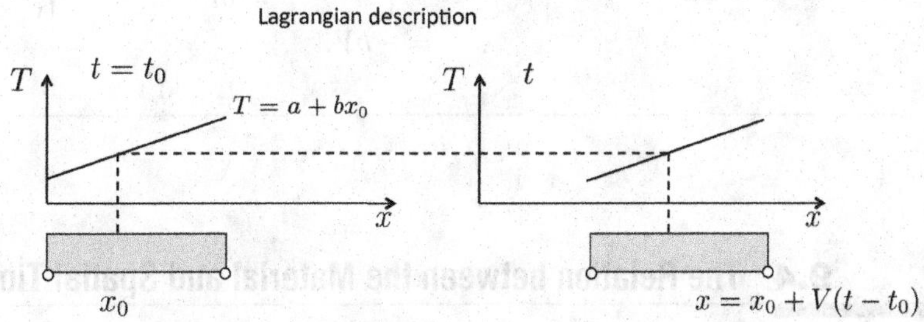

Consider an object of length ℓ that is moving in the x-direction. The object is subjected to a temperature field that can be described by $T = a + bx_0$ with a and b constant and $0 \leq x_0 \leq \ell$. Consider a material point of the object which is found at position x_0 at time $t = t_0$. At time t, the material point has reached position $x = x_0 + V(t - t_0)$ with V a constant velocity. In that case:

$$\begin{aligned}\dot{T} &= \left[\frac{\partial T}{\partial t}\right]_{x_0 \text{ constant}} \\ &= \left[\frac{\partial(a + bx_0)}{\partial t}\right]_{x_0 \text{ constant}} \\ &= 0.\end{aligned}$$

Example 9.3

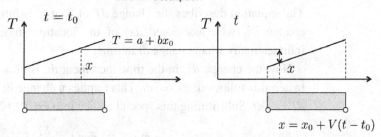

Consider the same object as in the previous example, but now we are focussing on a fixed position in space x $(x < \ell)$. In that case, the relevant time derivative is the Eulerian derivative:

$$\begin{aligned}\frac{\delta T}{\delta t} &= \left[\frac{\partial T}{\partial t}\right]_{x \text{ constant}} \\ &= \left[\frac{\partial(a + bx_0)}{\partial t}\right]_{x \text{ constant}} \\ &= \left[\frac{\partial(a + b(x - V(t - t_0)))}{\partial t}\right]_{x \text{ constant}} \\ &= -bV.\end{aligned}$$

9.4 The Relation between the Material and Spatial Time Derivatives

For the derivation of the relation between the material and spatial time derivative of, for example, the temperature (as an arbitrary physical state variable, associated with the material), we start with the Eulerian description of the temperature field $T = T(\vec{x}, t)$, in components formulated as $T = T(\underset{\sim}{x}, t) = T(x, y, z, t)$. For the total differential dT it can be written:

$$\begin{aligned}dT = &\left[\frac{\partial T}{\partial x}\right]_{y,z,t \text{ constant}} dx + \left[\frac{\partial T}{\partial y}\right]_{x,z,t \text{ constant}} dy \\ &+ \left[\frac{\partial T}{\partial z}\right]_{x,y,t \text{ constant}} dz + \left[\frac{\partial T}{\partial t}\right]_{x,y,z \text{ constant}} dt\end{aligned} \tag{9.11}$$

and in a more compact notation, using the gradient operator (see Chapter 7):

$$dT = d\vec{x} \cdot \vec{\nabla} T + \frac{\delta T}{\delta t} dt \quad \text{and also} \quad dT = d\underset{\sim}{x}^{\mathrm{T}} \underset{\sim}{\nabla} T + \frac{\delta T}{\delta t} dt. \tag{9.12}$$

This equation describes the change dT of T at an arbitrary, infinitesimally small change $d\vec{x}$ (with associated $d\underset{\sim}{x}$) of the location in space, combined with an infinitesimally small change dt in time.

Now the change $d\vec{x}$, in the time increment dt, is chosen in such a way that the material is followed: $d\vec{x} = \vec{v} dt$. This implies a change in temperature according to $dT = \dot{T} dt$. Substituting this special choice into Eq. (9.12) leads directly to

$$\dot{T} dt = \vec{v} \cdot \vec{\nabla} T dt + \frac{\delta T}{\delta t} dt, \tag{9.13}$$

and thus

$$\dot{T} = \vec{v} \cdot \vec{\nabla} T + \frac{\delta T}{\delta t} \quad \text{and also} \quad \dot{T} = \underset{\sim}{v}^{T} \underset{\sim}{\nabla} T + \frac{\delta T}{\delta t}. \tag{9.14}$$

The first term on the right-hand side, the difference between the material derivative and the spatial derivative, is called the **convective** contribution. For an arbitrary physical variable associated with the material, the following relation between the operators has to be applied:

$$(\dot{\ }) = \vec{v} \cdot \vec{\nabla}(\,) + \frac{\delta(\,)}{\delta t} \quad \text{and also} \quad (\dot{\ }) = \underset{\sim}{v}^{\mathrm{T}} \underset{\sim}{\nabla}(\,) + \frac{\delta(\,)}{\delta t}. \tag{9.15}$$

If $\vec{v} = \vec{0}$, in other words if the material is not moving in three-dimensional space, there is no difference between the material and spatial time derivative.

Applying the operator to the vector $\vec{x}$ results in an identity:

$$\dot{\vec{x}} = \vec{v} \cdot \vec{\nabla}\vec{x} + \frac{\delta \vec{x}}{\delta t} \quad \rightarrow \quad \vec{v} = \vec{v} \cdot \boldsymbol{I} + \vec{0} \quad \rightarrow \quad \vec{v} = \vec{v}. \tag{9.16}$$

Of course this also holds for application to the row $\underset{\sim}{x}^{\mathrm{T}}$ (application to the column $\underset{\sim}{x}$ is not allowed in the framework of the notation used):

$$\dot{\underset{\sim}{x}}^{\mathrm{T}} = \underset{\sim}{v}^{\mathrm{T}} \underset{\sim}{\nabla}\, \underset{\sim}{x}^{\mathrm{T}} + \frac{\delta \underset{\sim}{x}^{\mathrm{T}}}{\delta t} \quad \rightarrow \quad \underset{\sim}{v}^{\mathrm{T}} = \underset{\sim}{v}^{\mathrm{T}} \underline{I} + \underset{\sim}{0}^{\mathrm{T}} \quad \rightarrow \quad \underset{\sim}{v}^{\mathrm{T}} = \underset{\sim}{v}^{\mathrm{T}}. \tag{9.17}$$

Applying the operation to $\vec{v}$ (the velocity vector) leads to

$$\begin{aligned} \dot{\vec{v}} = \vec{v} \cdot \vec{\nabla}\vec{v} + \frac{\delta \vec{v}}{\delta t} \quad &\rightarrow \quad \vec{a} = \vec{v} \cdot \vec{\nabla}\vec{v} + \frac{\delta \vec{v}}{\delta t} \\ &\rightarrow \quad \vec{a} = \left(\vec{\nabla}\vec{v}\right)^{\mathrm{T}} \cdot \vec{v} + \frac{\delta \vec{v}}{\delta t} \end{aligned} \tag{9.18}$$

and application to the row $\underset{\sim}{v}^{\mathrm{T}}$ with velocity components results in

$$\begin{aligned} \dot{\underset{\sim}{v}}^{\mathrm{T}} = \underset{\sim}{v}^{\mathrm{T}} \underset{\sim}{\nabla} \underset{\sim}{v}^{\mathrm{T}} + \frac{\delta \underset{\sim}{v}^{\mathrm{T}}}{\delta t} \quad &\rightarrow \quad \underset{\sim}{a}^{\mathrm{T}} = \underset{\sim}{v}^{\mathrm{T}} \underset{\sim}{\nabla} \underset{\sim}{v}^{\mathrm{T}} + \frac{\delta \underset{\sim}{v}^{\mathrm{T}}}{\delta t} \\ &\rightarrow \quad \underset{\sim}{a} = \left(\underset{\sim}{\nabla} \underset{\sim}{v}^{\mathrm{T}}\right)^{\mathrm{T}} \underset{\sim}{v} + \frac{\delta \underset{\sim}{v}}{\delta t}. \end{aligned} \tag{9.19}$$

With this equation, the acceleration field can be found if the Eulerian description of the velocity field is known. In the result, the velocity gradient tensor $\boldsymbol{L}$ (with associated matrix representation $\underline{L}$), as introduced in Chapter 7, can be recognized:

$$\boldsymbol{L} = \left(\vec{\nabla}\vec{v}\right)^{\mathrm{T}} \quad \text{and also} \quad \underline{L} = \left(\underset{\sim}{\nabla} \underset{\sim}{v}^{\mathrm{T}}\right)^{\mathrm{T}}. \tag{9.20}$$

Now the equation for the acceleration vector $\vec{a}$ with the components $\underset{\sim}{a}$ becomes

$$\vec{a} = \boldsymbol{L} \cdot \vec{v} + \frac{\delta \vec{v}}{\delta t} \quad \text{and also} \quad \underset{\sim}{a} = \underline{L}\, \underset{\sim}{v} + \frac{\delta \underset{\sim}{v}}{\delta t}. \tag{9.21}$$

In the case of a stationary flow, where $\vec{v} = \vec{v}(\vec{x})$ instead of $\vec{v} = \vec{v}(\vec{x}, t)$, the equation for the acceleration reduces to

$$\vec{a} = \boldsymbol{L} \cdot \vec{v} \quad \text{and also} \quad \underset{\sim}{a} = \underline{L}\, \underset{\sim}{v}\,. \tag{9.22}$$

Example 9.4 Consider a temperature field in a fixed xy-coordinate system (Eulerian description) given by:

$$T = atx + bt^2 y$$

with a and b constants and t the time. A fluid is moving through this domain and the velocity field is given by:

$$\vec{v} = V(\vec{e}_x + \vec{e}_y)$$

with V a constant. We want to derive the material and spatial time derivative of the temperature. The spatial derivative is easy and straightforward:

$$\frac{\delta T}{\delta t} = \left[\frac{\partial T}{\partial t}\right]_{\vec{x}\ \text{constant}} = ax + 2bty$$

to be interpreted as the temperature change over time that an observer experiences at a fixed position in space. To determine the material time derivative, we need to determine the gradient of the temperature T first:

$$\begin{aligned}\vec{\nabla} T &= \vec{e}_x \frac{\partial T}{\partial x} + \vec{e}_y \frac{\partial T}{\partial y}\\ &= at\vec{e}_x + bt^2\vec{e}_y.\end{aligned}$$

Then the material time derivative yields:

$$\dot{T} = \vec{v} \cdot \vec{\nabla} T + \frac{\delta T}{\delta t} = Vat + Vbt^2 + ax + 2bty$$

which can be interpreted as the temperature change experienced by an observer moving along with the fluid.

9.5 The Displacement Vector

Consider a material point P, which is defined by the position vector $\vec{x}_0$ in the reference configuration with volume V_0. In the current configuration, the position vector of that point is denoted by $\vec{x}$; see Fig. 9.3.

The displacement vector of a point P, in the current configuration with respect to the undeformed configuration, is denoted by $\vec{u}$, satisfying

$$\vec{u} = \vec{x} - \vec{x}_0 \tag{9.23}$$

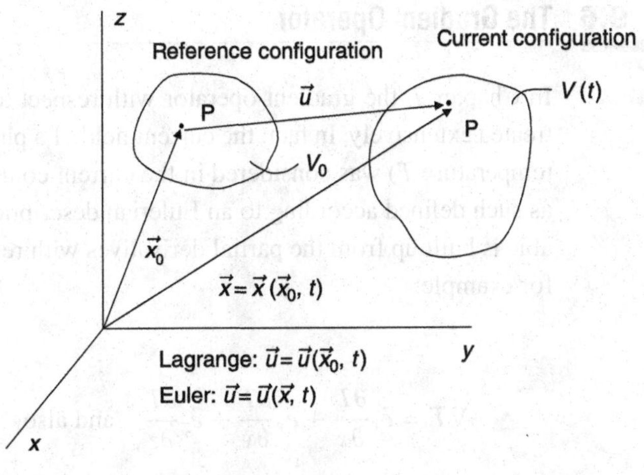

Figure 9.3

The displacement of a material point P.

and in component form:

$$\begin{bmatrix} u_x \\ u_y \\ u_z \end{bmatrix} = \begin{bmatrix} x \\ y \\ z \end{bmatrix} - \begin{bmatrix} x_0 \\ y_0 \\ z_0 \end{bmatrix}. \tag{9.24}$$

In the Lagrangian description, $\vec{u}$ is considered to be a function of $\vec{x}_0$ in V_0 and t and thus

$$\vec{u} = \vec{u}(\vec{x}_0, t) = \vec{x}(\vec{x}_0, t) - \vec{x}_0. \tag{9.25}$$

This relation makes it possible to formally calculate the displacement (in the current configuration at time t with respect to the reference configuration) of a material point, defined in the reference configuration with material identification $\vec{x}_0$. For the use of Eq. (9.25), it is assumed that $\vec{x}(\vec{x}_0, t)$ is available.

In the Eulerian description, $\vec{u}$ is considered to be a function of $\vec{x}$ in $V(t)$ and t and thus

$$\vec{u} = \vec{u}(\vec{x}, t) = \vec{x} - \vec{x}_0(\vec{x}, t). \tag{9.26}$$

With Eq. (9.26), it is possible to formally calculate the displacement (in the current configuration with respect to the reference configuration) of a material point, which is actually (at time t) at position $\vec{x}$ in the three-dimensional space. It is necessary for this that $\vec{x}_0(\vec{x}, t)$ is known, expressing which material point $\vec{x}_0$ at time t is present at the spatial point $\vec{x}$, in other words, the inverse relation of $\vec{x}(\vec{x}_0, t)$.

9.6 The Gradient Operator

In Chapter 7, the gradient operator with respect to the current configuration was treated extensively. In fact, the current field of a physical variable (for example the temperature T) was considered in the current configuration with domain $V(t)$ and as such defined according to an Eulerian description. The gradient of such a variable is built up from the partial derivatives with respect to the spatial coordinates, for example:

$$\vec{\nabla} T = \vec{e}_x \frac{\partial T}{\partial x} + \vec{e}_y \frac{\partial T}{\partial y} + \vec{e}_z \frac{\partial T}{\partial z} \quad \text{and also} \quad \underset{\sim}{\nabla} T = \begin{bmatrix} \frac{\partial T}{\partial x} \\ \frac{\partial T}{\partial y} \\ \frac{\partial T}{\partial z} \end{bmatrix} . \tag{9.27}$$

The current field (with respect to time t) can also be mapped onto the reference configuration with volume V_0 and thus formulated by means of a Lagrangian description. In this formulation, the gradient can also be defined and is built up from partial derivatives with respect to the material coordinates:

$$\vec{\nabla}_0 T = \vec{e}_x \frac{\partial T}{\partial x_0} + \vec{e}_y \frac{\partial T}{\partial y_0} + \vec{e}_z \frac{\partial T}{\partial z_0} \quad \text{and also} \quad \underset{\sim}{\nabla}_0 T = \begin{bmatrix} \frac{\partial T}{\partial x_0} \\ \frac{\partial T}{\partial y_0} \\ \frac{\partial T}{\partial z_0} \end{bmatrix} . \tag{9.28}$$

To relate the aforementioned gradient operators, the chain rule for differentiation is used. For a fixed time t we find

$$\begin{aligned} \frac{\partial T}{\partial x_0} &= \frac{\partial T}{\partial x}\frac{\partial x}{\partial x_0} + \frac{\partial T}{\partial y}\frac{\partial y}{\partial x_0} + \frac{\partial T}{\partial z}\frac{\partial z}{\partial x_0} \\ \frac{\partial T}{\partial y_0} &= \frac{\partial T}{\partial x}\frac{\partial x}{\partial y_0} + \frac{\partial T}{\partial y}\frac{\partial y}{\partial y_0} + \frac{\partial T}{\partial z}\frac{\partial z}{\partial y_0} \\ \frac{\partial T}{\partial z_0} &= \frac{\partial T}{\partial x}\frac{\partial x}{\partial z_0} + \frac{\partial T}{\partial y}\frac{\partial y}{\partial z_0} + \frac{\partial T}{\partial z}\frac{\partial z}{\partial z_0} , \end{aligned} \tag{9.29}$$

and in a more concise notation:

$$\underset{\sim}{\nabla}_0 T = \underline{F}^{\mathrm{T}} \underset{\sim}{\nabla} T \quad \text{with} \quad \underline{F}^{\mathrm{T}} = \begin{bmatrix} \frac{\partial x}{\partial x_0} & \frac{\partial y}{\partial x_0} & \frac{\partial z}{\partial x_0} \\ \frac{\partial x}{\partial y_0} & \frac{\partial y}{\partial y_0} & \frac{\partial z}{\partial y_0} \\ \frac{\partial x}{\partial z_0} & \frac{\partial y}{\partial z_0} & \frac{\partial z}{\partial z_0} \end{bmatrix} . \tag{9.30}$$

The matrix $\underline{F}$, for which the transpose is defined in Eq. (9.30), is called the **deformation matrix** or **deformation gradient matrix** for the current configuration with respect to the reference configuration. In the next chapter, this matrix will be discussed in full detail. For the deformation matrix $\underline{F}$ we can write in a more concise notation:

$$\underline{F}^{\mathrm{T}} = \underset{\sim}{\nabla}_0 \, \underset{\sim}{x}^{\mathrm{T}} \quad \text{and} \quad \underline{F} = \left(\underset{\sim}{\nabla}_0 \, \underset{\sim}{x}^{\mathrm{T}}\right)^{\mathrm{T}}. \tag{9.31}$$

By substituting $\underset{\sim}{x} = \underset{\sim}{x}_0 + \underset{\sim}{u}$ into Eq. (9.31), and using $\underset{\sim}{\nabla}_0 \underset{\sim}{x}^{\mathrm{T}} = \underline{I}$, we obtain

$$\underline{F}^{\mathrm{T}} = \underline{I} + \underset{\sim}{\nabla}_0 \, \underset{\sim}{u}^{\mathrm{T}} \quad \text{and} \quad \underline{F} = \underline{I} + \left(\underset{\sim}{\nabla}_0 \, \underset{\sim}{u}^{\mathrm{T}}\right)^{\mathrm{T}}. \tag{9.32}$$

In tensor notation, the relation between $\underset{\sim}{\nabla}_0 T$ and $\underset{\sim}{\nabla} T$ can be written as

$$\vec{\nabla}_0 T = \boldsymbol{F}^{\mathrm{T}} \cdot \vec{\nabla} T \quad \text{with} \quad \boldsymbol{F} = \left(\vec{\nabla}_0 \vec{x}\right)^{\mathrm{T}} = \boldsymbol{I} + \left(\vec{\nabla}_0 \vec{u}\right)^{\mathrm{T}}. \tag{9.33}$$

with $\boldsymbol{F}$ the **deformation tensor** (also called the **deformation gradient tensor**).

Above, a relation is derived between the gradient of a physical property (at time t) with respect to the reference configuration and the gradient of that property with respect to the current configuration. Consequently the mutual relation between the gradient operators can be written as

$$\vec{\nabla}_0(\,) = \boldsymbol{F}^{\mathrm{T}} \cdot \vec{\nabla}(\,) \quad \text{and also} \quad \underset{\sim}{\nabla}_0(\,) = \underline{F}^{\mathrm{T}} \underset{\sim}{\nabla}(\,), \tag{9.34}$$

and the inverse relation

$$\vec{\nabla}(\,) = \boldsymbol{F}^{-\mathrm{T}} \cdot \vec{\nabla}_0(\,) \quad \text{and also} \quad \underset{\sim}{\nabla}(\,) = \underline{F}^{-\mathrm{T}} \underset{\sim}{\nabla}_0(\,). \tag{9.35}$$

If the current and reference configuration are identical (in that case $\vec{u} = \vec{0}$ and $\boldsymbol{F} = \boldsymbol{I}$) the gradient operators are also identical.

Example 9.5 Consider, in a two-dimensional context, a material point P in a deforming body. The current position vector $\vec{x} = x\vec{e}_x + y\vec{e}_y$, as a function of the position vector $\vec{x}_0 = x_0\vec{e}_x + y_0\vec{e}_y$ in the reference configuration, can be written as:

$$\vec{x} = (ax_0 + by_0^2)\vec{e}_x + cy_0\vec{e}_y \qquad \text{with } a, b \text{ and } c \text{ constants.}$$

In the domain a temperature field is given by:

$$T(x, y) = px + qy \qquad \text{with } p \text{ and } q \text{ constants.}$$

The gradient of T with respect to the current configuration can be determined as:

$$\begin{aligned} \vec{\nabla} T &= \vec{e}_x \frac{\partial T}{\partial x} + \vec{e}_y \frac{\partial T}{\partial y} \\ &= p\vec{e}_x + q\vec{e}_y. \end{aligned}$$

For the gradient of T with respect to the reference configuration we need:

$$\frac{\partial T}{\partial x_0} = \frac{\partial T}{\partial x}\frac{\partial x}{\partial x_0} + \frac{\partial T}{\partial y}\frac{\partial y}{\partial x_0} = pa$$
$$\frac{\partial T}{\partial y_0} = \frac{\partial T}{\partial x}\frac{\partial x}{\partial y_0} + \frac{\partial T}{\partial y}\frac{\partial y}{\partial y_0} = 2pby_0 + qc = 2pb\frac{y}{c} + qc$$

yielding:

$$\begin{aligned}\vec{\nabla}_0 T &= \vec{e}_x\frac{\partial T}{\partial x_0} + \vec{e}_y\frac{\partial T}{\partial y_0} \\ &= pa\vec{e}_x + (2pby_0 + qc)\vec{e}_y \\ &= pa\vec{e}_x + \left(2pb\frac{y}{c} + qc\right)\vec{e}_y.\end{aligned}$$

Example 9.6 Consider a square material sample as shown by the solid line in Fig. 9.4. In a uniaxial stress test, the sample is deformed to a rectangular structure depicted by the dashed line in the figure.

From the figure it can be derived that for an arbitrary point in the sample we can write:

$$\vec{x} = \frac{\ell}{\ell_0}x_0\vec{e}_x + \frac{h}{h_0}y_0\vec{e}_y.$$

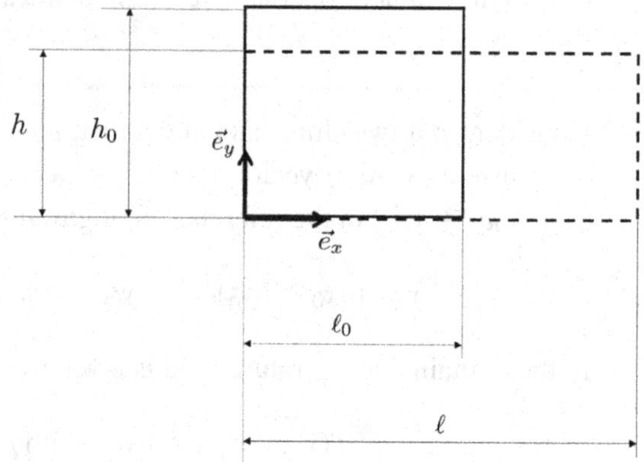

Figure 9.4

Undeformed (solid line) and deformed (dashed line) configuration of a material sample in a uniaxial stress test.

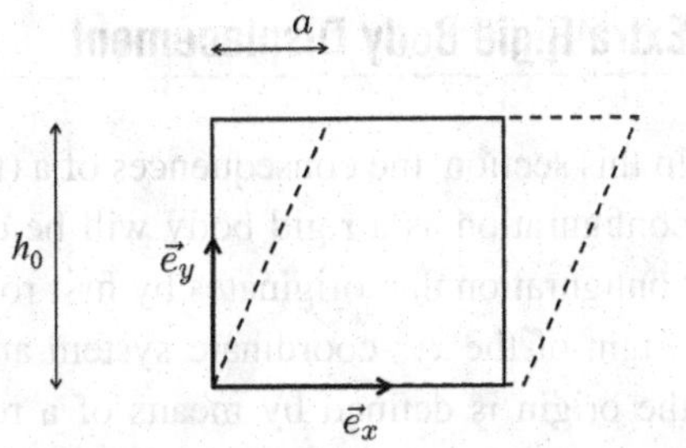

Figure 9.5

Undeformed (solid line) and deformed (dashed line) configuration of a material sample in 'simple shear'.

The deformation tensor $\boldsymbol{F}$ can be derived by using Eq. (9.31):

$$\begin{aligned}\boldsymbol{F} &= \frac{\partial x}{\partial x_0}\vec{e}_x\vec{e}_x + \frac{\partial x}{\partial y_0}\vec{e}_x\vec{e}_y + \frac{\partial y}{\partial x_0}\vec{e}_y\vec{e}_x + \frac{\partial y}{\partial y_0}\vec{e}_y\vec{e}_y \\ &= \frac{\ell}{\ell_0}\vec{e}_x\vec{e}_x + \frac{h}{h_0}\vec{e}_y\vec{e}_y\end{aligned}$$

and the deformation matrix with respect to the basis $\{\vec{e}_x, \vec{e}_y\}$ reads:

$$\underline{F} = \begin{bmatrix} \frac{\ell}{\ell_0} & 0 \\ 0 & \frac{h}{h_0} \end{bmatrix}.$$

Example 9.7 Consider the sample shown in Fig. 9.5, which is in a state of **simple shear**.

It is obvious that in this case also the deformation tensor is constant everywhere in the sample. From the figure it can be derived that for an arbitrary point in the sample we can write:

$$\vec{x} = (x_0 + \frac{a}{h_0}y_0)\vec{e}_x + y_0\vec{e}_y.$$

The deformation tensor is again determined by means of Eq. (9.31):

$$\begin{aligned}\boldsymbol{F} &= \frac{\partial x}{\partial x_0}\vec{e}_x\vec{e}_x + \frac{\partial x}{\partial y_0}\vec{e}_x\vec{e}_y + \frac{\partial y}{\partial x_0}\vec{e}_y\vec{e}_x + \frac{\partial y}{\partial y_0}\vec{e}_y\vec{e}_y \\ &= \vec{e}_x\vec{e}_x + \frac{a}{h_0}\vec{e}_x\vec{e}_y + \vec{e}_y\vec{e}_y\end{aligned}$$

and equivalently:

$$\underline{F} = \begin{bmatrix} 1 & \frac{a}{h_0} \\ 0 & 1 \end{bmatrix}.$$

Notice that this deformation state leads to a non-symmetric deformation tensor $\boldsymbol{F}$.

9.7 Extra Rigid Body Displacement

In this section, the consequences of a (fictitious) extra displacement of the current configuration as a rigid body will be discussed. Consider a hypothetical current configuration that originates by first rotating the current configuration around the origin of the *xyz*-coordinate system and then translating it. The rotation around the origin is defined by means of a rotation tensor $\boldsymbol{Q}$ (orthogonal) with matrix representation $\underline{Q}$, satisfying

$$\boldsymbol{Q}^{-1} = \boldsymbol{Q}^{\mathrm{T}} \quad \text{and} \quad \underline{Q}^{-1} = \underline{Q}^{\mathrm{T}} \quad \text{while} \quad \det(\boldsymbol{Q}) = \det(\underline{Q}) = 1, \tag{9.36}$$

and the translation by a vector $\vec{\lambda}$ with components $\underset{\sim}{\lambda}$. Figure 9.6 shows the rigid body motion. Variables associated with the extra rotated and translated configuration are indicated with the superscript *. Because of the extra rigid body displacement, the current position of a material point will change from $\vec{x}$ to $\vec{x}^*$ according to

$$\vec{x}^* = \boldsymbol{Q} \cdot \vec{x} + \vec{\lambda} = \boldsymbol{Q} \cdot (\vec{x}_0 + \vec{u}) + \vec{\lambda}, \tag{9.37}$$

while the displacement of the virtual configuration with respect to the reference configuration can be written as

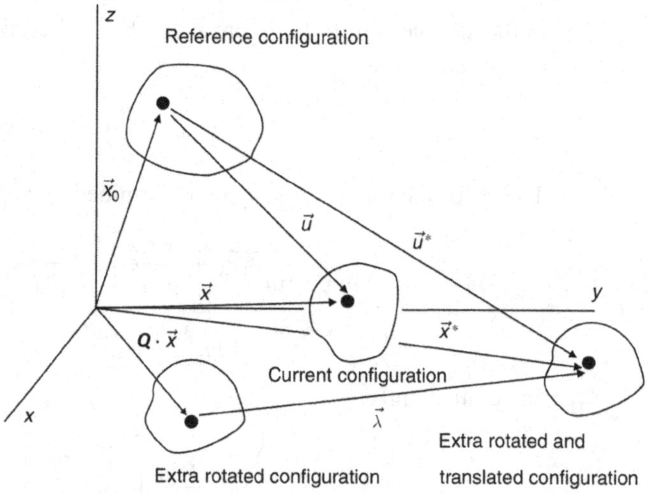

Figure 9.6

The rigid body displacements.

$$\begin{aligned}\vec{u}^* &= \vec{x}^* - \vec{x}_0 = \boldsymbol{Q}\cdot\vec{x} + \vec{\lambda} - \vec{x}_0 \\ &= \boldsymbol{Q}\cdot(\vec{x}_0 + \vec{u}) + \vec{\lambda} - \vec{x}_0 = (\boldsymbol{Q} - \boldsymbol{I})\cdot\vec{x}_0 + \boldsymbol{Q}\cdot\vec{u} + \vec{\lambda} \\ &= \boldsymbol{Q}\cdot\vec{x} + \vec{\lambda} - (\vec{x} - \vec{u}) = (\boldsymbol{Q} - \boldsymbol{I})\cdot\vec{x} + \vec{u} + \vec{\lambda}.\end{aligned} \tag{9.38}$$

In the following, the attention is focussed on a fixed material point with position vector $\vec{x}_0$ in the reference configuration, the position vector $\vec{x}$ in the current configuration at time t and the position vector $\vec{x}^*$ in the extra rotated and translated virtual configuration.

For a scalar physical variable, for example the temperature T, the value will not change as a result of an extra rigid body motion; thus, with respect to the same material point, $T^* = T$.

For the gradient operator, applied to a certain physical variable connected to the material, it follows directly that, on the basis of the relation between $\vec{x}$ and $\vec{x}^*$,

$$\left(\vec{\nabla}(\,)\right)^* = \boldsymbol{Q}\cdot\vec{\nabla}(\,) = \boldsymbol{Q}\cdot\boldsymbol{F}^{-T}\cdot\vec{\nabla}_0(\,). \tag{9.39}$$

Note that the gradient is the same operator for the real as well as for the imaginary, extra displaced, current configuration. However, the effect on (for example) the temperature field T and the field T^* is different. Equation (9.39) shows this difference.

For the deformation tensor of the virtual configuration with respect to the undeformed reference configuration, it is found that

$$\begin{aligned}\boldsymbol{F}^* &= \left(\vec{\nabla}_0\vec{x}^*\right)^{\mathrm{T}} = \left(\vec{\nabla}_0(\boldsymbol{Q}\cdot\vec{x} + \vec{\lambda})\right)^{\mathrm{T}} \\ &= \left(\vec{\nabla}_0(\vec{x}\cdot\boldsymbol{Q}^{\mathrm{T}} + \vec{\lambda})\right)^{\mathrm{T}} = (\boldsymbol{F}^{\mathrm{T}}\cdot\boldsymbol{Q}^{\mathrm{T}})^{\mathrm{T}} \\ &= \boldsymbol{Q}\cdot\boldsymbol{F}.\end{aligned} \tag{9.40}$$

Finally, the influence of the rigid body motion on the stress state will be determined. Assume that the internal interaction between the material particles will not change (except in direction) because of the motion as a rigid body. For the material point being considered, the stress tensor $\boldsymbol{\sigma}$ relates the stress vector $\vec{p}$ on a surface element to the unit normal $\vec{n}$ of that element, according to: $\vec{p} = \boldsymbol{\sigma}\cdot\vec{n}$. Because the imaginary configuration is rotated with respect to the current configuration, both vectors $\vec{n}^*$ and $\vec{p}^*$ can be written as

$$\vec{n}^* = \boldsymbol{Q}\cdot\vec{n} \text{ and } \vec{p}^* = \boldsymbol{Q}\cdot\vec{p}\,. \tag{9.41}$$

This reveals

$$\vec{p}^* = \boldsymbol{Q} \cdot \vec{p} = \boldsymbol{Q} \cdot \boldsymbol{\sigma} \cdot \vec{n} = \boldsymbol{Q} \cdot \boldsymbol{\sigma} \cdot \boldsymbol{Q}^{-1} \cdot \vec{n}^*$$
$$= \left(\boldsymbol{Q} \cdot \boldsymbol{\sigma} \cdot \boldsymbol{Q}^T\right) \cdot \vec{n}^*, \tag{9.42}$$

and so for the stress tensor $\boldsymbol{\sigma}^*$ and the associated matrix $\underline{\sigma}^*$ in the imaginary configuration it is found that

$$\boldsymbol{\sigma}^* = \boldsymbol{Q} \cdot \boldsymbol{\sigma} \cdot \boldsymbol{Q}^{\mathrm{T}} \quad \text{and also} \quad \underline{\sigma}^* = \underline{Q}\, \underline{\sigma}\, \underline{Q}^{\mathrm{T}} \tag{9.43}$$

with $\underline{Q}$ the matrix representation of the tensor $\boldsymbol{Q}$.

9.8 Fluid Flow

For fluids, it is not common practice (and in general not very useful) to define a reference state. This implies that the Lagrangian description (expressing properties as a function of $\vec{x}_0$ and t) is not commonly used for fluids. Related to this, derivatives with respect to $\vec{x}_0$ (the gradient operator $\vec{\nabla}_0$) and derivatives with respect to time under constant $\vec{x}_0$ will not appear in fluid mechanics. The deformation tensor $\boldsymbol{F}$ is not relevant for fluids. However, the material time derivative (for example to calculate the acceleration) is important nevertheless. For fluids an Eulerian description is used, meaning that all physical properties are considered in the current configuration, so as functions of $\vec{x}$ in the volume $V(t)$ and t.

The kinematic variables that generally play a role in fluid mechanics problems are the velocity $\vec{v} = \vec{v}(\vec{x}, t)$ and the acceleration $\vec{a} = \vec{a}(\vec{x}, t)$, both in an Eulerian description. Their relation is given by (see the end of Section 9.4)

$$\vec{a} = \left(\vec{\nabla}\vec{v}\right)^T \cdot \vec{v} + \frac{\delta \vec{v}}{\delta t} = \boldsymbol{L} \cdot \vec{v} + \frac{\delta \vec{v}}{\delta t}. \tag{9.44}$$

Based on the velocity field in $V(t)$, often streamlines are drawn. Streamlines are representative for the current (at time t) direction of the velocity: the direction of the velocity at a certain point $\vec{x}$ corresponds to the direction of the tangent to the streamline in point $\vec{x}$. Figure 9.7 gives an example of streamlines in a flow through a constriction.

For a stationary flow, $\vec{v} = \vec{v}(\vec{x})$ and thus $\delta\vec{v}/\delta t = 0$, the streamline pattern is the same at each time point. In that case, the material particles follow the streamlines exactly, i.e. the particle tracks coincide with the streamlines.

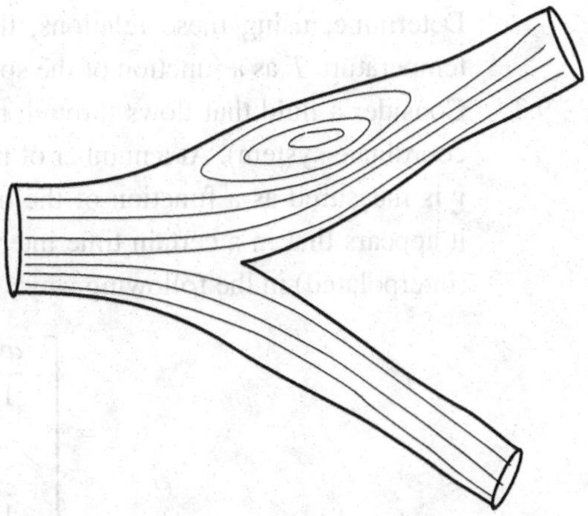

Figure 9.7
Streamlines in a model of a carotid artery bifurcation.

Exercises

9.1 The material points of a deforming continuum are identified with the position vectors $\tilde{x}_0$ of those points in the reference configuration at time $t = 0$. The deformation process is described (Lagrangian description) by the current position vectors $\tilde{x}$ as a function of $\tilde{x}_0$ and time t, according to:

$$\tilde{x}(\tilde{x}_0, t) = \begin{bmatrix} x_0 + (a + by_0)t \\ y_0 + at \\ z_0 \end{bmatrix} \quad \text{with } a \text{ and } b \text{ constant.}$$

Determine the velocity field as a function of time in an Euler description, or in other words, give an expression for $\tilde{v} = \tilde{v}(\tilde{x}, t)$.

9.2 Consider a fluid flowing through three-dimensional space (with an xyz-coordinate system). At a number of fixed points, the temperature T and the fluid velocity $\tilde{v}$ are measured as a function of time t. Based on these measurements the temperature and velocity fields can be approximated as follows:

$$T = c(x + 2y)e^{(1-t/\tau)} \quad \text{with } c \text{ and } \tau \text{ constants.}$$

$$\tilde{v} = \frac{1}{\tau} \begin{bmatrix} 0 \\ x \\ 3y \end{bmatrix}.$$

Determine, using these relations, the material time derivative $\dot{T}$ of the temperature T as a function of the spatial coordinates and time.

9.3 Consider a fluid that flows through three-dimensional space (with an xyz-coordinate system). At a number of fixed points in space, the fluid velocity $\underset{\sim}{v}$ is measured as a function of the time t. Based on these measurements, it appears that in a certain time interval the velocity can be approximated (interpolated) in the following way:

$$\underset{\sim}{v} = \begin{bmatrix} \dfrac{ay + bz}{1 + \alpha t} \\ 0 \\ \dfrac{cx}{1 + \alpha t} \end{bmatrix}.$$

Determine, based on this approximation of the velocity field as a function of time, the associated acceleration field as a function of time, thus: $\underset{\sim}{a}(\underset{\sim}{x}, t)$.

9.4 Consider a (two-dimensional) velocity field for a stationary flowing continuum:

$$v_x = \frac{x}{y^2}$$
$$v_y = \frac{1}{y}$$

with x and y spatial coordinates (expressed in [m]), while v_x and v_y are the velocity components in the x- and y-directions (expressed in [m s^{-1}]). The velocity field holds for the shaded domain in the figure given below.

Consider a material particle that at time $t = 0$ enters the domain at the position with coordinates $x = 1$ [m], $y = 1$ [m].

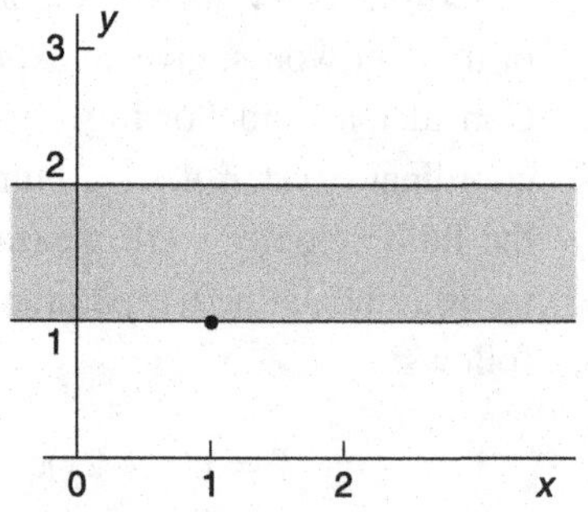

Calculate the time at which this particle leaves the shaded domain.

9.5 In a Cartesian xyz-coordinate system, a rigid body is rotating around the z-axis with constant angular velocity ω. For the velocity field, in an Eulerian description, the following expression holds:

$$\vec{v}(\vec{x}) = \mathbf{\Omega} \cdot \vec{x} \quad \text{with} \quad \mathbf{\Omega} = \omega(-\vec{e}_x\vec{e}_y + \vec{e}_y\vec{e}_x).$$

Consider the associated acceleration field $\vec{a}(\vec{x}) = \dot{\vec{v}}(\vec{x})$ and show that the result can be written as: $\vec{a}(\vec{x}) = \boldsymbol{H} \cdot \vec{x}$, with $\boldsymbol{H}$ a constant tensor.

Give an expression for $\boldsymbol{H}$ formulated as

$$\boldsymbol{H} = H_{xx}\vec{e}_x\vec{e}_x + H_{xy}\vec{e}_x\vec{e}_y + \cdots + H_{zz}\vec{e}_z\vec{e}_z\,.$$

9.6 A method that is sometimes used to study the mechanical behaviour of cells is based on the so-called cross flow experiment. An example of such an experiment is given in the right figure below (image courtesy of Patrick Anderson). In this case, a fluorescent fibroblast is positioned almost in the centre of the cross flow. There are several ways to create such a flow. The set-up shown in the left figure consists of a reservoir with four cylinders which rotate with the same angular velocity. The figure gives a top view of the set-up. A cell can be trapped in the centre of the cross flow and thus be stretched by the flow.

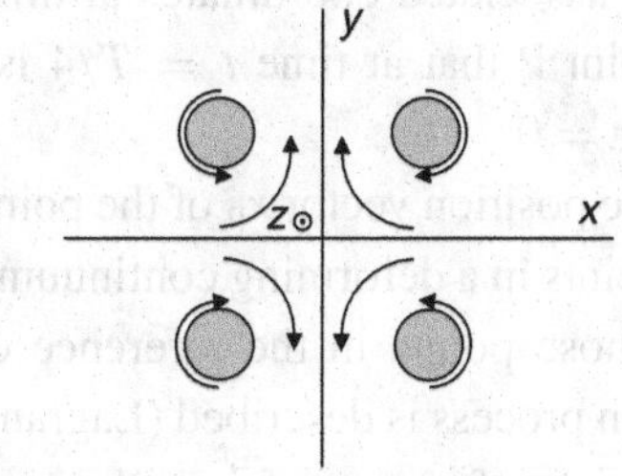

(a)

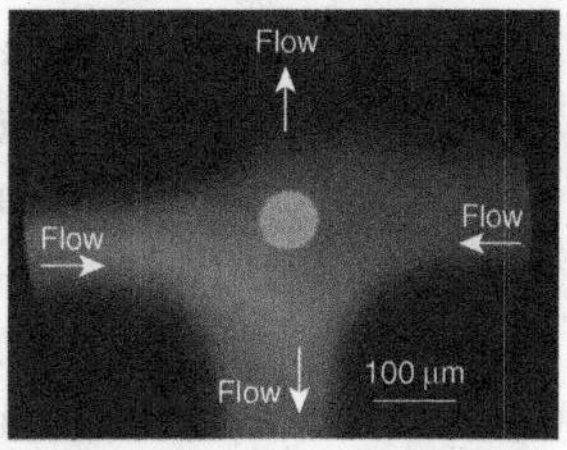

(b)

The reservoir is assumed to be filled with an incompressible fluid. Close to the origin of the xyz-coordinate system, the (stationary) two-dimensional velocity field is given by

$$\vec{v} = c(-x\vec{e}_x + y\vec{e}_y) \quad \text{with } c \text{ a constant,}$$

where $\vec{e}_x$ and $\vec{e}_y$ are unit vectors along the x- and y-axis.

Determine the associated acceleration field: $\vec{a}(\vec{x}) = \dot{\vec{v}}(\vec{x})$.

9.7 In the reference configuration (at time $t = 0$), the edges (with length ℓ) of a cubic material specimen are parallel to the axes of a Cartesian xyz-coordinate system. See the figure on the left. The specimen is loaded in shear cyclically in the time t.

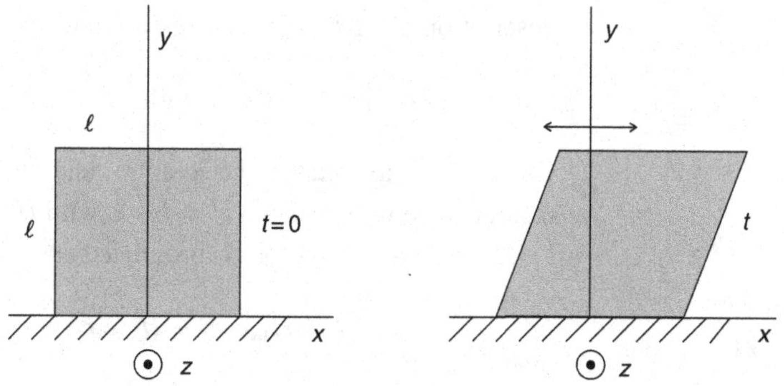

The time-dependent deformation is, in the Lagrangian approach, described by

$$x = x_0 + \frac{y_0}{2}\sin\left(2\pi\frac{t}{T}\right),$$

with T the (constant) time of one cycle,

$$y = y_0 \quad z = z_0,$$

with x_0, y_0, z_0 the coordinates of the material points at time $t = 0$ and with x, y, z the associated coordinates at time t. Attention is focussed on the material point P that at time $t = T/4$ is located at the position $x = 2\ell/3, y = \ell/3, z = 0$.

Determine the position vector $\underset{\sim}{x}_P$ of the point P as a function of the time t.

9.8 The material points in a deforming continuum are identified by the position vectors $\vec{x}_0$ of those points in the reference configuration at time $t = 0$. The deformation process is described (Lagrangian approach) by the current position vector $\vec{x}$ as a function of $\vec{x}_0$ and time t:

$$\vec{x} = (x_0 + (a + by_0)t)\,\vec{e}_x + (y_0 + at)\vec{e}_y + z_0\vec{e}_z$$

with a and b positive constants and t the time. Determine the deformation tensor $\boldsymbol{F}$.

9.9 The stress state at a point of the deformed configuration of a material sample is given by the following stress tensor:

$$\boldsymbol{\sigma} = 100\vec{e}_x\vec{e}_x + 50\vec{e}_y\vec{e}_y \quad [\text{kPa}].$$

Now the material is rotated as a rigid body described by the rotation tensor:

$$\boldsymbol{Q} = \frac{1}{2}\sqrt{2}\left(\boldsymbol{I} + \vec{e}_x\vec{e}_y - \vec{e}_y\vec{e}_x\right).$$

Calculate the stress tensor in this new rotated configuration.

9.10 On two-dimensional surfaces that have a spatial gradient in stiffness, cells are known to migrate towards areas with higher stiffness (a phenomenon called durotaxis). For a certain surface ($x \geq 0$, $y \geq 0$), the stiffness E is constant in the y-direction and varies in the x-direction according to

$$E(x, y) = a(x - 1),$$

with a a constant. A cell is originally located at (x_0, y_0) and its trajectory with time equals

$$\vec{x} = (x_0 + bt^2)\vec{e}_x + (y_0 + ct)\vec{e}_y,$$

where b and c are constants. Determine the stiffness that the cell experiences during its migration as a function of time t.

10 Deformation and Rotation, Deformation Rate and Spin

10.1 Introduction

Consider the geometrical change of a coherent amount of material or material fraction, for which modelling as a continuum is assumed to be permitted. The first part of the present chapter is focussed on the description of the local deformation (generally coupled with rotations of the material) and along with that, the introduction of a number of different strain measures.

Only after choosing a reference configuration is it possible to define deformation in a meaningful way (deformation is a relative concept). This implies that initially the theory and the accompanying application area are related to solids. When the material is a mixture of several material fractions, each fraction can, with regard to local geometrical changes, in principle be isolated from the other fractions.

The second part of this chapter discusses geometrical changes with time. Central concepts in this part are deformation rate and rotation velocity (spin). The derivations in this part are not only relevant for solids, but even more important for applications including fluids.

10.2 A Material Line Segment in the Reference and Current Configurations

Consider a coherent amount of material in a fully defined state (the reference configuration). At a material point P, with position vector $\vec{x}_0$ in the reference configuration, we focus our attention on an arbitrary infinitesimally small material line segment $d\vec{x}_0$: see Fig. 10.1.

With respect to the Cartesian xyz-coordinate system, the vector $d\vec{x}_0$ can be written as

$$d\vec{x}_0 = dx_0\vec{e}_x + dy_0\vec{e}_y + dz_0\vec{e}_z \quad \text{and also} \quad d\underset{\sim}{x}_0 = \begin{bmatrix} dx_0 \\ dy_0 \\ dz_0 \end{bmatrix}. \tag{10.1}$$

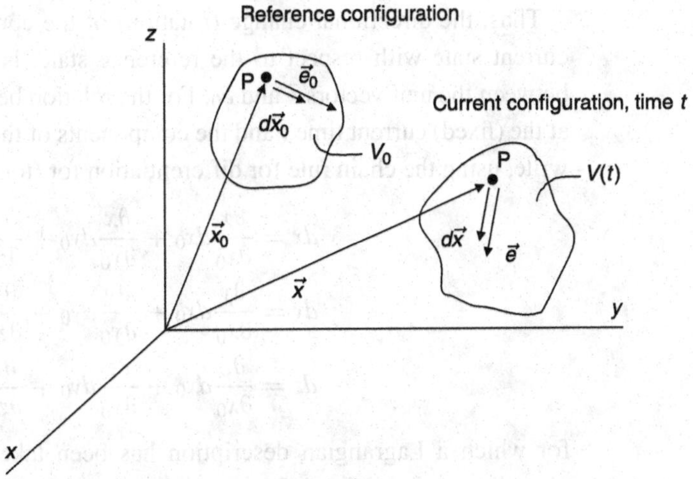

Figure 10.1

A material line segment in the reference and current configurations.

The orientation of the line segment $d\vec{x}_0$ is defined by the unit vector $\vec{e}_0$ with components in the column $\underset{\sim}{e}_0$. In that case it can be written:

$$d\vec{x}_0 = \vec{e}_0\, d\ell_0 \quad \text{and also} \quad d\underset{\sim}{x}_0 = \underset{\sim}{e}_0\, d\ell_0$$

$$\text{with} \quad d\ell_0 = \sqrt{d\vec{x}_0 \cdot d\vec{x}_0} = \sqrt{d\underset{\sim}{x}_0^T d\underset{\sim}{x}_0}, \tag{10.2}$$

where $d\ell_0$ specifies the length of the vector $d\vec{x}_0$.

The same material line segment, but now considered in the current configuration at time t, is indicated with $d\vec{x}$. It should be emphasized that the line segment $d\vec{x}$ in the current configuration is composed of the same material points as the line segment $d\vec{x}_0$ in the reference configuration.

With respect to the Cartesian xyz-coordinate system, we can write for the vector $d\vec{x}$:

$$d\vec{x} = dx\vec{e}_x + dy\vec{e}_y + dz\vec{e}_z \quad \text{and also} \quad d\underset{\sim}{x} = \begin{bmatrix} dx \\ dy \\ dz \end{bmatrix}. \tag{10.3}$$

The orientation of the line segment $d\vec{x}$ is defined by the unit vector $\vec{e}$ with components in the column $\underset{\sim}{e}$. In that case it can be written:

$$d\vec{x} = \vec{e} d\ell \quad \text{and also} \quad d\underset{\sim}{x} = \underset{\sim}{e} d\ell$$

$$\text{with} \quad d\ell = \sqrt{d\vec{x} \cdot d\vec{x}} = \sqrt{d\underset{\sim}{x}^{\mathrm{T}} d\underset{\sim}{x}}, \tag{10.4}$$

where $d\ell$ specifies the length of the vector $d\vec{x}$.

Thus, the directional change (rotation) of the considered line segment, of the current state with respect to the reference state, is described by the difference between the unit vectors $\vec{e}$ and $\vec{e}_0$. For the relation between the components of $d\vec{x}$ at the (fixed) current time t and the components of the accompanying $d\vec{x}_0$, we can write, using the chain rule for differentiation for (total) differentials:

$$\begin{aligned} dx &= \frac{\partial x}{\partial x_0}dx_0 + \frac{\partial x}{\partial y_0}dy_0 + \frac{\partial x}{\partial z_0}dz_0 \\ dy &= \frac{\partial y}{\partial x_0}dx_0 + \frac{\partial y}{\partial y_0}dy_0 + \frac{\partial y}{\partial z_0}dz_0 \\ dz &= \frac{\partial z}{\partial x_0}dx_0 + \frac{\partial z}{\partial y_0}dy_0 + \frac{\partial z}{\partial z_0}dz_0, \end{aligned} \tag{10.5}$$

for which a Lagrangian description has been taken as the point of departure according to: $\vec{x} = \vec{x}(\vec{x}_0, t)$. In a more compact form it can be formulated as

$$d\underset{\sim}{x} = \underline{F}\, d\underset{\sim}{x}_0 \quad \text{with} \quad \underline{F} = \begin{bmatrix} \frac{\partial x}{\partial x_0} & \frac{\partial x}{\partial y_0} & \frac{\partial x}{\partial z_0} \\ \frac{\partial y}{\partial x_0} & \frac{\partial y}{\partial y_0} & \frac{\partial y}{\partial z_0} \\ \frac{\partial z}{\partial x_0} & \frac{\partial z}{\partial y_0} & \frac{\partial z}{\partial z_0} \end{bmatrix} = \left(\underset{\sim}{\nabla}_0\, \underset{\sim}{x}^{\mathrm{T}}\right)^{\mathrm{T}} \tag{10.6}$$

and in tensor notation as

$$d\vec{x} = \boldsymbol{F} \cdot d\vec{x}_0 \quad \text{with} \quad \boldsymbol{F} = \left(\vec{\nabla}_0\, \vec{x}\right)^{\mathrm{T}}. \tag{10.7}$$

The tensor $\boldsymbol{F}$, the deformation tensor (or deformation gradient tensor), with matrix representation $\underline{F}$, was already introduced in Section 9.6. This tensor completely describes the (local) geometry change (deformation and rotation). After all, when $\boldsymbol{F}$ is known, it is possible, for every line segment (and therefore also for a three-dimensional element) in the reference configuration, to calculate the accompanying line segment (or three-dimensional element) in the current configuration. The tensor $\boldsymbol{F}$ describes for every material line segment the length and orientation change: $\boldsymbol{F}$ determines the transition from $d\ell_0$ to $d\ell$ and the transition from $\vec{e}_0$ to $\vec{e}$.

Figure 10.2 visualizes a uniaxially loaded bar. It is assumed that the deformation is homogeneous: for every material point of the bar the same deformation tensor $\boldsymbol{F}$ is applicable. It can easily be verified that the depicted transition from the reference configuration to the current configuration is defined by

$$\boldsymbol{F} = \lambda_x \vec{e}_x \vec{e}_x + \lambda_y \vec{e}_y \vec{e}_y + \lambda_z \vec{e}_z \vec{e}_z, \tag{10.8}$$

with the stretch ratios:

$$\lambda_x = \frac{\ell}{\ell_0} \quad \text{and} \quad \lambda_y = \lambda_z = \sqrt{\frac{A}{A_0}}. \tag{10.9}$$

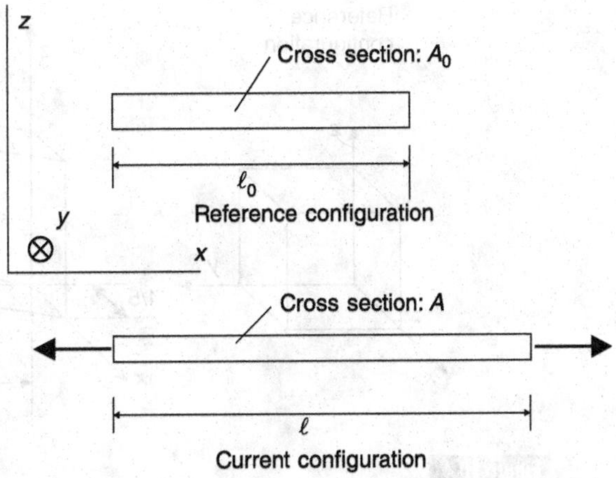

Figure 10.2

A uniaxially loaded bar.

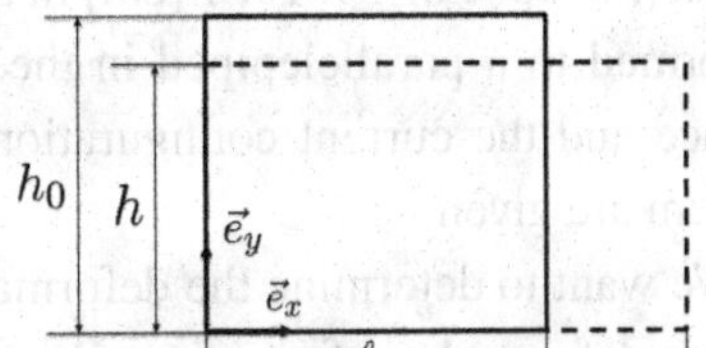

Figure 10.3

A square piece of material that is uniaxially stretched.

Example 10.1 Consider a square piece of material with dimensions $\ell_0 \times h_0$ and constant thickness that is being stretched. In the deformed state the material is a rectangle with dimensions $\ell \times h$. When the thickness is not changed, this is called a state of plane strain and the deformation can be described in two dimensions. For this case the deformation is homogeneous, meaning that the deformation tensor is not a function of $\vec{x}_0$. From Fig. 10.3 it can be derived that:

$$\vec{x} = \frac{\ell}{\ell_0} x_0 \vec{e}_x + \frac{h}{h_0} y_0 \vec{e}_y.$$

For this case, the two-dimensional deformation tensor $\boldsymbol{F}$ can be written as:

$$\begin{aligned} \boldsymbol{F} &= \frac{\partial x}{\partial x_0}\vec{e}_x\vec{e}_x + \frac{\partial x}{\partial y_0}\vec{e}_x\vec{e}_y + \frac{\partial y}{\partial x_0}\vec{e}_y\vec{e}_x + \frac{\partial y}{\partial y_0}\vec{e}_y\vec{e}_y \\ &= \frac{\ell}{\ell_0}\vec{e}_x\vec{e}_x + \frac{h}{h_0}\vec{e}_y\vec{e}_y. \end{aligned}$$

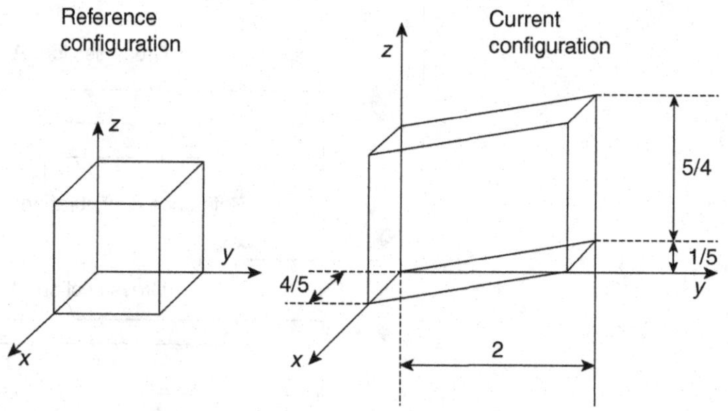

Figure 10.4

A cube in the undeformed and deformed configuration.

Example 10.2 A material cube of $1 \times 1 \times 1$ [cm] in the reference configuration is homogeneously deformed to a parallelepiped in the current configuration. In Fig. 10.4, the reference and the current configuration with respect to a Cartesian *xyz*-coordinate system are given.

We want to determine the deformation tensor $\boldsymbol{F}$ mapping the undeformed state to this deformed configuration. Because the deformation is homogeneous, we do not have to work with infinitesimally small line segments, but we can use the deformations of the basis vectors that define the cube. In matrix notation we can write, with respect to base vectors $\{\vec{e}_x, \vec{e}_y, \vec{e}_z\}$ for basis vector $\vec{e}_x$ (see figure):

$$\begin{bmatrix} F_{11} & F_{12} & F_{13} \\ F_{21} & F_{22} & F_{23} \\ F_{31} & F_{32} & F_{33} \end{bmatrix} \begin{bmatrix} 1 \\ 0 \\ 0 \end{bmatrix} = \begin{bmatrix} \frac{4}{5} \\ 0 \\ 0 \end{bmatrix}$$

showing that the current state vector associated with the original vector $\vec{e}_x$ represents the first column of matrix $\underline{F}$. We can follow a similar procedure using the vector $\vec{e}_y$:

$$\begin{bmatrix} \frac{4}{5} & F_{12} & F_{13} \\ 0 & F_{22} & F_{23} \\ 0 & F_{32} & F_{33} \end{bmatrix} \begin{bmatrix} 0 \\ 1 \\ 0 \end{bmatrix} = \begin{bmatrix} 0 \\ 2 \\ \frac{1}{5} \end{bmatrix},$$

leading to:

$$\underline{F} = \begin{bmatrix} \frac{4}{5} & 0 & F_{13} \\ 0 & 2 & F_{23} \\ 0 & \frac{1}{5} & F_{33} \end{bmatrix}.$$

Finally, applying the procedure to $\vec{e}_z$ leads to:

$$\underline{F} = \begin{bmatrix} \frac{4}{5} & 0 & 0 \\ 0 & 2 & 0 \\ 0 & \frac{1}{5} & \frac{5}{4} \end{bmatrix}$$

or

$$\boldsymbol{F} = \frac{4}{5}\vec{e}_x\vec{e}_x + 2\vec{e}_y\vec{e}_y + \frac{5}{4}\vec{e}_z\vec{e}_z + \frac{1}{5}\vec{e}_z\vec{e}_y.$$

Example 10.3 A small material volume is subjected to a homogeneous deformation, described by the following deformation tensor $\boldsymbol{F}$:

$$\boldsymbol{F} = \boldsymbol{I} + 4\vec{e}_x\vec{e}_x + 2\vec{e}_x\vec{e}_y + 2\vec{e}_y\vec{e}_x,$$

with $\boldsymbol{I}$ the unit tensor. Consider two material points P and Q in the deformed state:

$$\vec{x}_{\mathrm{P}} = 2\vec{e}_x + 2\vec{e}_y + 2\vec{e}_z \quad , \quad \vec{x}_{\mathrm{Q}} = 11\vec{e}_x + 6\vec{e}_y + 3\vec{e}_z.$$

In the undeformed state, point P is located at:

$$\vec{x}_{0\mathrm{P}} = \vec{e}_x + \vec{e}_y + \vec{e}_z.$$

Our aim is to calculate the position $\vec{x}_{0\mathrm{Q}}$ of point Q in the undeformed configuration. Because the deformation is homogeneous, meaning that $\boldsymbol{F}$ is constant, we are allowed to replace the infinitesimal line segments in Eq. (10.7) by finite line segments and write:

$$(\vec{x}_{\mathrm{P}} - \vec{x}_{\mathrm{Q}}) = \boldsymbol{F} \cdot (\vec{x}_{0\mathrm{P}} - \vec{x}_{0\mathrm{Q}}).$$

In matrix notation, with unknown $\vec{x}_{0\mathrm{Q}}$, we can write:

$$\begin{bmatrix} 2 \\ 2 \\ 2 \end{bmatrix} - \begin{bmatrix} 11 \\ 6 \\ 3 \end{bmatrix} = \begin{bmatrix} 5 & 2 & 0 \\ 2 & 1 & 0 \\ 0 & 0 & 1 \end{bmatrix} \left(\begin{bmatrix} 1 \\ 1 \\ 1 \end{bmatrix} - \begin{bmatrix} x_{0\mathrm{Q}} \\ y_{0\mathrm{Q}} \\ z_{0\mathrm{Q}} \end{bmatrix} \right)$$

This leads to a set of three equations to solve x, y and z, leading to:

$$\underset{\sim}{x}_{0\mathrm{Q}} = \begin{bmatrix} 2 \\ 3 \\ 2 \end{bmatrix}$$

or in tensor notation:

$$\vec{x}_{0\mathrm{Q}} = 2\vec{e}_x + 3\vec{e}_y + 2\vec{e}_z.$$

10.3 The Stretch Ratio and Rotation

Consider an infinitesimally small line segment $d\vec{x}_0$ in the reference configuration, directed along the unit vector $\vec{e}_0$ and with length $d\ell_0$, so $d\vec{x}_0 = \vec{e}_0 d\ell_0$. To this line segment belongs, in the current configuration, the line segment $d\vec{x} = \vec{e} d\ell$ directed along the unit vector $\vec{e}$ and with length $d\ell$. The mutual relation satisfies

$$d\vec{x} = \boldsymbol{F} \cdot d\vec{x}_0, \tag{10.10}$$

so

$$\vec{e}\, d\ell = \boldsymbol{F} \cdot \vec{e}_0\, d\ell_0. \tag{10.11}$$

The stretch ratio λ is defined as the ratio between $d\ell$ and $d\ell_0$ (and therefore it will always hold that $\lambda > 0$). Using Eq. (10.11), we can write

$$\vec{e} \cdot \vec{e}\, d\ell^2 = \vec{e}_0 \cdot \boldsymbol{F}^{\mathrm{T}} \cdot \boldsymbol{F} \cdot \vec{e}_0\, d\ell_0^2, \tag{10.12}$$

and consequently

$$\lambda = \lambda(\vec{e}_0) = \sqrt{\vec{e}_0 \cdot \boldsymbol{F}^{\mathrm{T}} \cdot \boldsymbol{F} \cdot \vec{e}_0}. \tag{10.13}$$

This equation can be used to determine the stretch ratio λ for a material line segment with direction $\vec{e}_0$ in the reference configuration (Lagrangian approach). So, for that purpose, the tensor (or tensor product) $\boldsymbol{F}^{\mathrm{T}} \cdot \boldsymbol{F}$ has to be known. The tensor $\boldsymbol{C}$ is defined as

$$\boldsymbol{C} = \boldsymbol{F}^{\mathrm{T}} \cdot \boldsymbol{F}, \tag{10.14}$$

and using this:

$$\lambda = \lambda(\vec{e}_0) = \sqrt{\vec{e}_0 \cdot \boldsymbol{C} \cdot \vec{e}_0}. \tag{10.15}$$

The tensor $\boldsymbol{C}$ is called the **right Cauchy–Green deformation tensor**. In component form, Eq. (10.15) can be written as

$$\lambda = \lambda(\underset{\sim}{e}_0) = \sqrt{\underset{\sim}{e}_0^{\mathrm{T}}\, \underline{C}\, \underset{\sim}{e}_0}, \tag{10.16}$$

with

$$\underline{C} = \underline{F}^{\mathrm{T}} \underline{F}. \tag{10.17}$$

The direction change (rotation) of a material line segment can, for the transition from the reference state to the current state, formally be stated as

$$\vec{e} = \boldsymbol{F} \cdot \vec{e}_0 \frac{d\ell_0}{d\ell} = \boldsymbol{F} \cdot \vec{e}_0 \frac{1}{\lambda} = \frac{\boldsymbol{F} \cdot \vec{e}_0}{\sqrt{\vec{e}_0 \cdot \boldsymbol{C} \cdot \vec{e}_0}}. \tag{10.18}$$

In component form this equation can be written as

$$\underset{\sim}{e} = \underline{F}\, \underset{\sim}{e}_0 \frac{d\ell_0}{d\ell} = \underline{F}\, \underset{\sim}{e}_0 \frac{1}{\lambda} = \frac{\underline{F}\, \underset{\sim}{e}_0}{\sqrt{\underset{\sim}{e}_0^{\mathrm{T}}\, \underline{C}\, \underset{\sim}{e}_0}}. \tag{10.19}$$

Above, the current state is considered as a 'function' of the reference state: for a direction $\vec{e}_0$ in the reference configuration, the associated direction $\vec{e}$ and the stretch ratio λ were determined. In the following, the 'inverse' procedure is shown. Based on

$$\boldsymbol{F}^{-1} \cdot d\vec{x} = d\vec{x}_0, \tag{10.20}$$

so

$$\boldsymbol{F}^{-1} \cdot \vec{e}\, d\ell = \vec{e}_0\, d\ell_0, \tag{10.21}$$

and subsequently

$$\vec{e} \cdot \boldsymbol{F}^{-\mathrm{T}} \cdot \boldsymbol{F}^{-1} \cdot \vec{e}\, d\ell^2 = \vec{e}_0 \cdot \vec{e}_0\, d\ell_0^2, \tag{10.22}$$

it follows for the stretch ratio $\lambda = d\ell/d\ell_0$ that

$$\lambda = \lambda(\vec{e}) = \frac{1}{\sqrt{\vec{e} \cdot \boldsymbol{F}^{-\mathrm{T}} \cdot \boldsymbol{F}^{-1} \cdot \vec{e}}}. \tag{10.23}$$

This equation can be used to determine the stretch ratio λ for a material line element with direction $\vec{e}$ in the current configuration (Eulerian description). For this, the tensor (tensor product) $\boldsymbol{F}^{-\mathrm{T}} \cdot \boldsymbol{F}^{-1}$ has to be known. The tensor $\boldsymbol{B}$ is defined according to

$$\boldsymbol{B} = \boldsymbol{F} \cdot \boldsymbol{F}^{\mathrm{T}}, \tag{10.24}$$

and using this:

$$\lambda = \lambda(\vec{e}) = \frac{1}{\sqrt{\vec{e} \cdot \boldsymbol{B}^{-1} \cdot \vec{e}}}. \tag{10.25}$$

The tensor $\boldsymbol{B}$ is called the **left Cauchy–Green deformation tensor**. In component form, Eq. (10.25) can be written as:

$$\lambda = \lambda(\underset{\sim}{e}) = \frac{1}{\sqrt{\underset{\sim}{e}^{\mathrm{T}} \underline{B}^{-1}\, \underset{\sim}{e}}}, \tag{10.26}$$

with

$$\underline{B} = \underline{F}\, \underline{F}^{\mathrm{T}}. \tag{10.27}$$

The direction change (rotation) of a material line segment with direction $\vec{e}$ in the current configuration with respect to the reference configuration can formally be calculated with

$$\vec{e}_0 = \boldsymbol{F}^{-1} \cdot \vec{e} \, \frac{d\ell}{d\ell_0} = \boldsymbol{F}^{-1} \cdot \vec{e} \, \lambda = \frac{\boldsymbol{F}^{-1} \cdot \vec{e}}{\sqrt{\vec{e} \cdot \boldsymbol{B}^{-1} \cdot \vec{e}}}, \tag{10.28}$$

and alternatively, using components:

$$\underset{\sim}{e}_0 = \underline{F}^{-1} \, \underset{\sim}{e} \, \frac{d\ell}{d\ell_0} = \underline{F}^{-1} \, \underset{\sim}{e} \, \lambda = \frac{\underline{F}^{-1} \, \underset{\sim}{e}}{\sqrt{\underset{\sim}{e}^{\mathrm{T}} \, \underline{B}^{-1} \, \underset{\sim}{e}}}. \tag{10.29}$$

Example 10.4 Nearly all biological materials contain fibres, which can either be passive (for example collagen and elastin fibres) or active (muscle fibres). Next to this, cells reorient either in the fibre direction or in a direction that is determined by the mechanical state of the tissue. This implies that mechanical properties become anisotropic, which means that the local properties of the material vary with different directions. This fibre/cell direction can be a function of the position as is illustrated in Fig. 10.5.

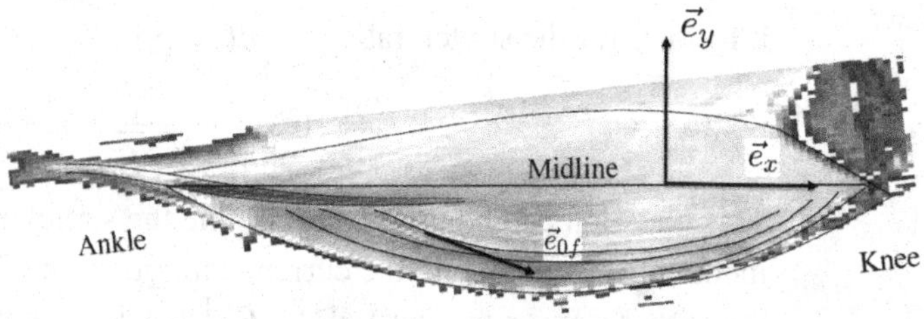

Figure 10.5

Image of a skeletal muscle in the lower leg of a rat. The solid black lines are drawn over the local muscle fibres, showing that the direction varies with the position (courtesy of Mascha Maenhout).

In a mechanical analysis, usually the mechanical state of the system is determined in a fixed, global coordinate system, defined by a basis $\{\vec{e}_x, \vec{e}_y, \vec{e}_z\}$, and the output of such an analysis is the deformation tensor $\boldsymbol{F}(\vec{x}_0)$ as a function of the position $\vec{x}_0$. However, to understand and predict how tissue responds to the mechanical load, either by adapting to the load or because it is damaged, it is often necessary to know the stretch λ along or perpendicular to the fibre direction, or how the orientation of a fibre changes during deformation. Sometimes the focus will be on a fibre or cell that is identified in the deformed situation, but it is also possible that attention is focussed on the undeformed situation. The deformation tensor $\boldsymbol{F}$ contains all the information needed, and, depending on the focus (deformed or undeformed configuration), the associated tensors $\boldsymbol{C}$ or $\boldsymbol{B}$ will be used. The next examples will illustrate some of these choices.

Example 10.5 Assume that a piece of material is subjected to 'simple shear', as illustrated in Fig. 10.6.

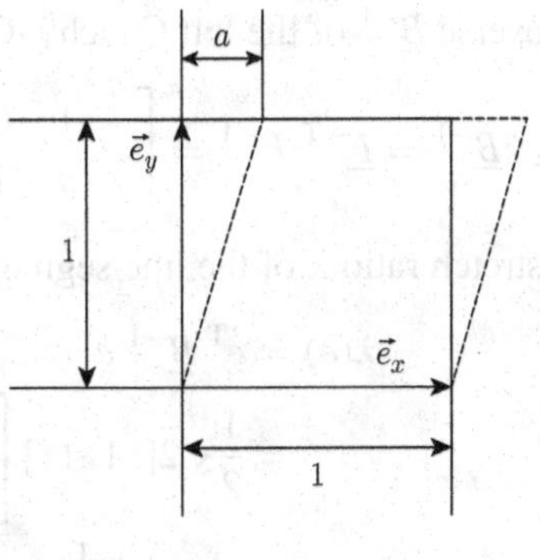

Figure 10.6

A square piece of material subjected to 'simple shear'.

For all material points in the sample, the position vector $\vec{x}(\vec{x}_0)$ can be written as:

$$\vec{x} = (x_0 + a y_0)\vec{e}_x + y_0\vec{e}_y,$$

with a a constant. The two-dimensional deformation matrix $\underline{F}$, with respect to the basis $\{\vec{e}_x, \vec{e}_y\}$ can be written as:

$$\underline{F} = \begin{bmatrix} 1 & a \\ 0 & 1 \end{bmatrix}.$$

and the right Cauchy–Green matrix $\underline{C}$ yields:

$$\underline{C} = \underline{F}^{\mathrm{T}}\,\underline{F} = \begin{bmatrix} 1 & 0 \\ a & 1 \end{bmatrix}\begin{bmatrix} 1 & a \\ 0 & 1 \end{bmatrix} = \begin{bmatrix} 1 & a \\ a & 1+a^2 \end{bmatrix}$$

Now, it is easy to determine the extension ratio λ of line segments in different directions. For a line segment that is oriented in the $\vec{e}_x$-direction in the undeformed configuration we can write:

$$\lambda(\vec{e}_x) = \sqrt{\underset{\sim}{e}_x^{\mathrm{T}}\,\underline{C}\,\underset{\sim}{e}_x} = 1,$$

so the length does not change. For a line segment in the $\vec{e}_y$-direction we find:

$$\lambda(\vec{e}_y) = \sqrt{\underset{\sim}{e}_y^{\mathrm{T}}\,\underline{C}\,\underset{\sim}{e}_y} = \sqrt{[\,0\;1\,]\begin{bmatrix} 1 & a \\ a & 1+a^2 \end{bmatrix}\begin{bmatrix} 0 \\ 1 \end{bmatrix}} = \sqrt{1+a^2}.$$

Example 10.6 Consider the same deformation case of simple shear as in the previous example. Now we would like to know the stretch ratio of a line segment that in the current deformed situation is directed according to $\vec{e} = \frac{1}{2}\sqrt{2}(\vec{e}_x + \vec{e}_y)$. With:

$$\underline{F}^{-1} = \begin{bmatrix} 1 & -a \\ 0 & 1 \end{bmatrix}$$

the inverse $\underline{B}^{-1}$ of the left Cauchy–Green strain matrix yields:

$$\underline{B}^{-1} = \underline{F}^{-\mathrm{T}}\,\underline{F}^{-1} = \begin{bmatrix} 1 & 0 \\ -a & 1 \end{bmatrix}\begin{bmatrix} 1 & -a \\ 0 & 1 \end{bmatrix} = \begin{bmatrix} 1 & -a \\ -a & 1+a^2 \end{bmatrix}$$

The stretch ratio λ of the line segment in direction $\vec{e}$ is:

$$\begin{aligned}\lambda(\vec{e}) &= \underset{\sim}{e}^{\mathrm{T}}\,\underline{B}^{-1}\,\underset{\sim}{e} \\ &= \frac{1}{2}\sqrt{2}[\,1\;\;1\,]\begin{bmatrix} 1 & -a \\ -a & 1+a^2 \end{bmatrix}\frac{1}{2}\sqrt{2}\begin{bmatrix} 1 \\ 1 \end{bmatrix} \\ &= 1 - a + \frac{1}{2}a^2.\end{aligned}$$

10.4 Strain Measures and Strain Tensors and Matrices

In the preceding sections, the stretch ratio λ is considered to be a measure for the relative length change of a material line segment in the transition from the reference configuration to the current configuration. If there is no deformation, $\lambda = 1$. Often it is more convenient to introduce a variable equal to zero when there is no deformation: the strain. In the present section, several different, generally accepted strain measures are treated.

In the previous section it was found, using the Lagrangian description:

$$\lambda^2 = \lambda^2(\vec{e}_0) = \vec{e}_0 \cdot \boldsymbol{F}^{\mathrm{T}} \cdot \boldsymbol{F} \cdot \vec{e}_0 = \vec{e}_0 \cdot \boldsymbol{C} \cdot \vec{e}_0. \tag{10.30}$$

Coupled to this relation, the Green–Lagrange strain $\varepsilon_{\mathrm{GL}}$ is defined by:

$$\begin{aligned}\varepsilon_{\mathrm{GL}} = \varepsilon_{\mathrm{GL}}(\vec{e}_0) &= \frac{\lambda^2 - 1}{2} = \frac{1}{2}\,\vec{e}_0 \cdot \left(\boldsymbol{F}^{\mathrm{T}} \cdot \boldsymbol{F} - \boldsymbol{I}\right) \cdot \vec{e}_0 \\ &= \frac{1}{2}\,\vec{e}_0 \cdot (\boldsymbol{C} - \boldsymbol{I}) \cdot \vec{e}_0\,.\end{aligned} \tag{10.31}$$

This result invites us to introduce the symmetric **Green–Lagrange strain tensor** $\boldsymbol{E}$ according to:

$$\boldsymbol{E} = \frac{1}{2}\left(\boldsymbol{F}^{\mathrm{T}} \cdot \boldsymbol{F} - \boldsymbol{I}\right) = \frac{1}{2}\,(\boldsymbol{C} - \boldsymbol{I}), \tag{10.32}$$

which implies

$$\varepsilon_{\mathrm{GL}} = \varepsilon_{\mathrm{GL}}(\vec{e}_0) = \vec{e}_0 \cdot \boldsymbol{E} \cdot \vec{e}_0\,. \tag{10.33}$$

When using matrix notation, Eq. (10.32) can be formulated as

$$\underline{E} = \frac{1}{2}\left(\underline{F}^{\mathrm{T}}\,\underline{F} - \underline{I}\right) = \frac{1}{2}\left(\underline{C} - \underline{I}\right), \tag{10.34}$$

implying

$$\varepsilon_{\rm GL} = \varepsilon_{\rm GL}(\underset{\sim}{e}_0) = \underset{\sim}{e}_0^{\rm T}\, \underline{E}\, \underset{\sim}{e}_0. \tag{10.35}$$

From a mathematical perspective, these are easily manageable relations. The Green–Lagrange strain tensor $\boldsymbol{E}$ (with matrix representation $\underline{E}$) is invariant for extra rigid body motions of the current state. The components of the (symmetric, 3 × 3) Green–Lagrange strain matrix $\underline{E}$ can be interpreted as follows:

- The terms on the diagonal are the Green–Lagrange strains of material line segments of the reference configuration in the x-, y- and z-directions respectively (the component on the first row in the first column is the Green–Lagrange strain of a line segment that is oriented in the x-direction in the reference configuration).
- The off-diagonal terms determine the shear of the material (the component on the first row of the second column is a measure for the change of the angle enclosed by material line segments that are oriented in the x- and y-direction in the undeformed configuration).

For the deformation tensor $\boldsymbol{F}$ and the displacement vector $\vec{u}$, both applying to the current configuration and related to the reference configuration, the following relation was derived in Section 9.6:

$$\boldsymbol{F} = \boldsymbol{I} + \left(\vec{\nabla}_0 \vec{u}\right)^{\rm T}. \tag{10.36}$$

Substitution into Eq. (10.32) yields

$$\boldsymbol{E} = \frac{1}{2}\left(\left(\vec{\nabla}_0\vec{u}\right) + \left(\vec{\nabla}_0\vec{u}\right)^{\rm T} + \left(\vec{\nabla}_0\vec{u}\right)\cdot\left(\vec{\nabla}_0\vec{u}\right)^{\rm T}\right). \tag{10.37}$$

It can be observed that the first two terms on the right-hand side of this equation are linear in the displacements, while the third term is non-linear (quadratic).

The linear strain $\varepsilon_{\rm lin}$ is defined according to:

$$\varepsilon_{\rm lin} = \varepsilon_{\rm lin}(\vec{e}_0) = \lambda - 1 = \sqrt{\vec{e}_0 \cdot \boldsymbol{F}^{\rm T} \cdot \boldsymbol{F} \cdot \vec{e}_0} - 1. \tag{10.38}$$

This expression is not easily manageable for mathematical elaborations.

At small deformations and small rotations, for which $\boldsymbol{F} \approx \boldsymbol{I}$ (and therefore the components of the tensor $\left(\vec{\nabla}_0\vec{u}\right)^{\rm T}$ are much smaller than 1), it can be written:

$$\begin{aligned} \varepsilon_{\rm lin} &= \varepsilon_{\rm lin}(\vec{e}_0) \\ &= \sqrt{1 + \vec{e}_0 \cdot (\boldsymbol{F}^{\rm T}\cdot\boldsymbol{F} - \boldsymbol{I})\cdot\vec{e}_0} - 1 \\ &\approx \frac{1}{2}\,\vec{e}_0 \cdot (\boldsymbol{F}^{\rm T}\cdot\boldsymbol{F} - \boldsymbol{I})\cdot\vec{e}_0 \\ &\approx \frac{1}{2}\,\vec{e}_0 \cdot (\boldsymbol{F}^{\rm T} + \boldsymbol{F} - 2\boldsymbol{I})\cdot\vec{e}_0. \end{aligned} \tag{10.39}$$

The last of these approximations for the linear strain is denoted by the symbol ε. This strain definition is used on a broad scale. Therefore, the assumption $\boldsymbol{F} \approx \boldsymbol{I}$ leads to the following, mathematically well-manageable relation:

$$\varepsilon = \varepsilon(\vec{e}_0) = \vec{e}_0 \cdot \boldsymbol{\varepsilon} \cdot \vec{e}_0 \quad \text{with} \quad \boldsymbol{\varepsilon} = \frac{1}{2}\left(\boldsymbol{F}^{\mathrm{T}} + \boldsymbol{F} - 2\boldsymbol{I}\right), \tag{10.40}$$

where the symmetric tensor $\boldsymbol{\varepsilon}$ is called the **linear strain tensor**. In displacements this tensor can also be expressed as

$$\boldsymbol{\varepsilon} = \frac{1}{2}\left(\left(\vec{\nabla}_0 \vec{u}\right) + \left(\vec{\nabla}_0 \vec{u}\right)^{\mathrm{T}}\right). \tag{10.41}$$

The strain tensor $\boldsymbol{\varepsilon}$ is linear in the displacements and can be considered (with respect to the displacements) as a linearized form of the Green–Lagrange strain tensor $\boldsymbol{E}$. In component form this results in the well-known and often-used formulation:

$$\underline{\varepsilon} = \begin{bmatrix} \frac{\partial u_x}{\partial x_0} & \frac{1}{2}\left(\frac{\partial u_x}{\partial y_0} + \frac{\partial u_y}{\partial x_0}\right) & \frac{1}{2}\left(\frac{\partial u_x}{\partial z_0} + \frac{\partial u_z}{\partial x_0}\right) \\ \frac{1}{2}\left(\frac{\partial u_y}{\partial x_0} + \frac{\partial u_x}{\partial y_0}\right) & \frac{\partial u_y}{\partial y_0} & \frac{1}{2}\left(\frac{\partial u_y}{\partial z_0} + \frac{\partial u_z}{\partial y_0}\right) \\ \frac{1}{2}\left(\frac{\partial u_z}{\partial x_0} + \frac{\partial u_x}{\partial z_0}\right) & \frac{1}{2}\left(\frac{\partial u_z}{\partial y_0} + \frac{\partial u_y}{\partial z_0}\right) & \frac{\partial u_z}{\partial z_0} \end{bmatrix}. \tag{10.42}$$

The components of the (symmetric, 3×3) linear strain matrix $\underline{\varepsilon}$ can be interpreted as follows:

- The terms on the diagonal are the linear strains of material line segments of the reference configuration in the x-, y- and z-directions respectively (the component on the first row in the first column is the linear strain of a line segment that is oriented in the x-direction in the reference configuration).
- The off-diagonal terms determine the shear of the material (the component on the first row of the second column is a measure for the change of the angle enclosed by material line segments that are oriented in the x- and y-direction in the undeformed configuration). See Fig. 10.7.

In the previous section, the Eulerian description for the stretch ratio was derived:

$$\lambda^{-2} = \lambda^{-2}(\vec{e}) = \vec{e} \cdot \boldsymbol{F}^{-\mathrm{T}} \cdot \boldsymbol{F}^{-1} \cdot \vec{e} = \vec{e} \cdot \boldsymbol{B}^{-1} \cdot \vec{e}. \tag{10.43}$$

Coupled to this relation, the Almansi–Euler strain $\varepsilon_{\mathrm{AE}}$ is defined by

$$\begin{aligned} \varepsilon_{\mathrm{AE}} = \varepsilon_{\mathrm{AE}}(\vec{e}) &= \frac{1 - \lambda^{-2}}{2} = \frac{1}{2}\vec{e} \cdot \left(\boldsymbol{I} - \boldsymbol{F}^{-\mathrm{T}} \cdot \boldsymbol{F}^{-1}\right) \cdot \vec{e} \\ &= \frac{1}{2}\vec{e} \cdot \left(\boldsymbol{I} - \boldsymbol{B}^{-1}\right) \cdot \vec{e}. \end{aligned} \tag{10.44}$$

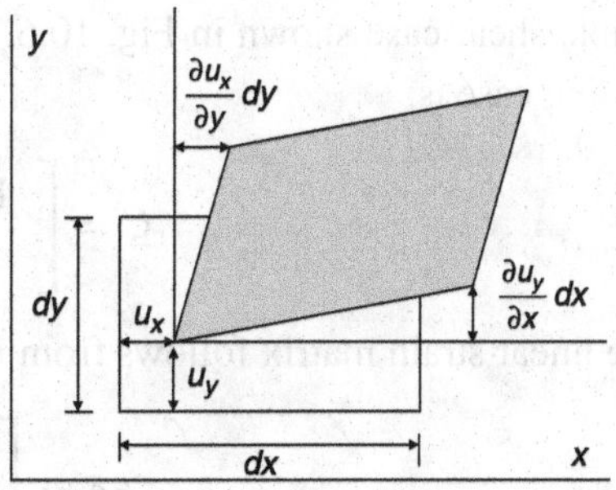

Figure 10.7

Interpretation of linear strain components.

This relation gives rise to introduce the symmetric **Almansi–Euler strain tensor** $\boldsymbol{A}$ according to

$$\boldsymbol{A} = \frac{1}{2}\left(\boldsymbol{I} - \boldsymbol{F}^{-\mathrm{T}} \cdot \boldsymbol{F}^{-1}\right) = \frac{1}{2}\left(\boldsymbol{I} - \boldsymbol{B}^{-1}\right), \tag{10.45}$$

which implies

$$\varepsilon_{\mathrm{AE}} = \varepsilon_{\mathrm{AE}}(\vec{e}) = \vec{e} \cdot \boldsymbol{A} \cdot \vec{e}\,. \tag{10.46}$$

When using matrix notation, this may be formulated as

$$\underline{A} = \frac{1}{2}\left(\underline{I} - \underline{F}^{-\mathrm{T}}\,\underline{F}^{-1}\right) = \frac{1}{2}\left(\underline{I} - \underline{B}^{-1}\right), \tag{10.47}$$

implying

$$\varepsilon_{AE} = \varepsilon_{AE}(\underset{\sim}{e}) = \underset{\sim}{e}^{\mathrm{T}}\,\underline{A}\,\underset{\sim}{e}. \tag{10.48}$$

Again, from a mathematical perspective this represents well-manageable relations. The Almansi–Euler strain is used only sporadically, in contrast to the related symmetric strain tensor $\boldsymbol{\varepsilon}_{\mathrm{F}}$ (Finger) defined by

$$\boldsymbol{\varepsilon}_{\mathrm{F}} = \frac{1}{2}\left(\boldsymbol{F} \cdot \boldsymbol{F}^{\mathrm{T}} - \boldsymbol{I}\right) = \frac{1}{2}(\boldsymbol{B} - \boldsymbol{I}) = \boldsymbol{B} \cdot \boldsymbol{A} = \boldsymbol{A} \cdot \boldsymbol{B}. \tag{10.49}$$

In this section, a number of different strain tensors have been reviewed. It is an interesting exercise to compare these different tensors for a few elementary homogeneous deformations, for example for the case of uniaxial stress (see Fig. 10.2) and for pure shear. It should be noted that for (infinitesimally) small deformations and (infinitesimally) small rotations (so the limiting case that $\boldsymbol{F} \to \boldsymbol{I}$) the difference between all treated strain tensors vanishes.

Example 10.7 It is interesting to study the implication of using the linear strain $\varepsilon_{\mathrm{lin}}$ compared to the Green–Lagrange strain for different loading cases. Let us first consider the simple shear case shown in Fig. 10.6. The right Cauchy–Green matrix $\underline{C}$ for this loading case is:

$$\underline{C} = \begin{bmatrix} 1 & a \\ a & 1+a^2 \end{bmatrix}.$$

The linear strain matrix follows from Eq. (10.38):

$$\underset{\sim}{\varepsilon} = \begin{bmatrix} 0 & \frac{a}{2} \\ \frac{a}{2} & 0 \end{bmatrix}.$$

Using the geometrically non-linear right Cauchy–Green matrix to calculate the stretch ratio for a line segment in the $\vec{e}_y$-direction leads to:

$$\lambda(\vec{e}_y) = \sqrt{\underset{\sim}{e}_y^{\mathrm{T}}\, \underline{C}\, \underset{\sim}{e}_y} = \sqrt{1+a^2} \approx 1 + \frac{a^2}{2}.$$

Using Eq. (10.38) and (10.40), the linear theory yields:

$$\lambda(\vec{e}_y) = 1 + [0 \;\; 1]\; \underset{\sim}{\varepsilon} \begin{bmatrix} 0 \\ 1 \end{bmatrix} = 1.$$

Thus, the linear theory ignores the length change of the line segment in $\vec{e}_y$-direction, and this is allowed when $\frac{a^2}{2} << 1$.

Example 10.8 Let us perform a similar evaluation of the linear theory for a rigid body rotation over an angle α. In that case the deformation matrix $\underline{F}$ and the right Cauchy–Green matrix are given by:

$$\underline{F} = \begin{bmatrix} \cos(\alpha) & -\sin(\alpha) \\ \sin(\alpha) & \cos(\alpha) \end{bmatrix} \rightarrow \underline{C} = \begin{bmatrix} 1 & 0 \\ 0 & 1 \end{bmatrix}.$$

The linear strain matrix $\underset{\sim}{\varepsilon}$ can be written as:

$$\underset{\sim}{\varepsilon} = \frac{1}{2}\left((\underset{\sim}{\nabla}\underset{\sim}{u}^{\mathrm{T}})^{\mathrm{T}} + \underset{\sim}{\nabla}\underset{\sim}{u}^{\mathrm{T}}\right) = \begin{bmatrix} \cos(\alpha)-1 & 0 \\ 0 & \cos(\alpha)-1 \end{bmatrix}.$$

Using the non-linear theory, the stretch ratio of a line segment originally in the $\vec{e}_x$ direction equals:

$$\lambda(\vec{e}_x) = \sqrt{\underset{\sim}{e}_x^{\mathrm{T}} \underline{C} \underset{\sim}{e}_x} = \sqrt{[1 \;\; 0] \begin{bmatrix} 1 & 0 \\ 0 & 1 \end{bmatrix} \begin{bmatrix} 1 \\ 0 \end{bmatrix}} = 1.$$

Indeed this results in no length change, which is correct because of the rigid body rotation. Using the linear theory we find:

$$\lambda(\vec{e}_x) = 1 + \underset{\sim}{e}_x^{\mathrm{T}} \underset{=}{\varepsilon} \underset{\sim}{e}_x = 1 + [1 \;\; 0] \begin{bmatrix} \cos(\alpha) - 1 & 0 \\ 0 & \cos(\alpha) - 1 \end{bmatrix} \begin{bmatrix} 1 \\ 0 \end{bmatrix} = \cos(\alpha).$$

This shows that for angles α very close to zero, $\cos(\alpha) \approx 1$, and thus $\lambda(\vec{e}_x)$ approaches 1, but when α becomes larger the linear theory leads to strains of line segments that are not valid.

10.5 The Volume Change Factor

Consider an arbitrary, infinitesimally small material element (parallelepiped), in the reference configuration spanned by three linearly independent vectors $d\vec{x}_0^a$, $d\vec{x}_0^b$ and $d\vec{x}_0^c$. The volume of the element is specified by dV_0 and can in principle be calculated when the vectors $d\vec{x}_0^i$ (with $i = a, b, c$) are known. The vectors $d\vec{x}^i$ in the current configuration, associated with the vectors $d\vec{x}_0^i$ in the reference configuration, can be determined using the deformation tensor $\boldsymbol{F}$ via

$$d\vec{x}^i = \boldsymbol{F} \cdot d\vec{x}_0^i \quad \text{for } i = a, b, c. \tag{10.50}$$

Based on the vectors $d\vec{x}^i$ the volume dV of the material element in the current configuration can be calculated. For the volumetric change ratio J it can be shown:

$$J = \frac{dV}{dV_0} = \det(\boldsymbol{F}). \tag{10.51}$$

The result is independent of the originally chosen shape and orientation of the element dV_0.

For **isochoric** deformation (this is a deformation such that locally the volume of the material does not change), the volume change ratio satisfies

$$J = \det(\boldsymbol{F}) = 1. \tag{10.52}$$

For materials that are incompressible, a property which is often attributed to biological materials (related to the high water content of the materials), the deformation is always isochoric.

10.6 Deformation Rate and Rotation Velocity

In the preceding sections, the current configuration or current state was considered at a (fixed) time t and compared with the reference configuration. Based on that, concepts such as deformation and rotation were defined. In this section, the attention is focussed on (infinitesimally) small changes of the current state, as seen in the time domain.

A material line segment $d\vec{x}$ (solid or fluid) in the current state at time t converts in the line segment $d\vec{x} + d\dot{\vec{x}}\, dt$ at time $t + dt$; see Fig. 10.8.

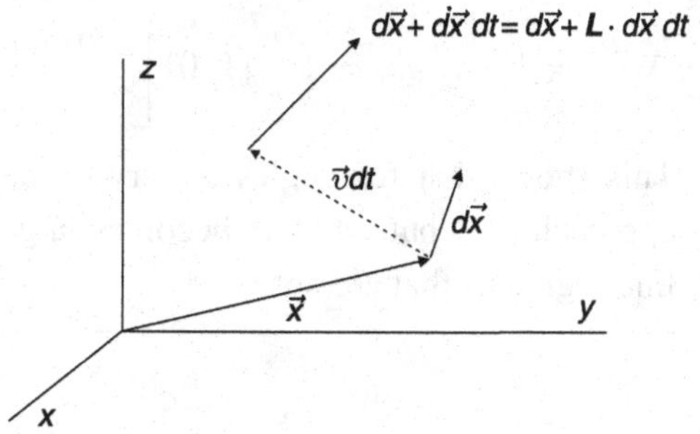

Figure 10.8

Change of a material line segment $d\vec{x}$ after a time increment dt.

It can be written (also see Chapter 7):

$$\dot{d\vec{x}} = \boldsymbol{L} \cdot d\vec{x} \quad \text{with} \quad \boldsymbol{L} = \left(\vec{\nabla}\vec{v}\right)^{\mathrm{T}}, \tag{10.53}$$

where $\vec{v} = \dot{\vec{x}}$ specifies the velocity of the material and $\boldsymbol{L}$ is the velocity gradient tensor. The tensor $\boldsymbol{L}$ is purely a current variable, not in any way related to the reference configuration.

Using the deformation tensor, we can write:

$$d\vec{x} = \boldsymbol{F} \cdot d\vec{x}_0 \tag{10.54}$$

and therefore

$$\dot{d\vec{x}} = \dot{\boldsymbol{F}} \cdot d\vec{x}_0 = \dot{\boldsymbol{F}} \cdot \boldsymbol{F}^{-1} \cdot d\vec{x}, \tag{10.55}$$

resulting in the relation between the tensors $\boldsymbol{L}$ and $\boldsymbol{F}$:

$$\boldsymbol{L} = \dot{\boldsymbol{F}} \cdot \boldsymbol{F}^{-1}. \tag{10.56}$$

It is common practice to decompose the velocity gradient tensor $\boldsymbol{L}$ into a symmetric part $\boldsymbol{D}$ and a skew symmetric part $\boldsymbol{\Omega}$. The tensor $\boldsymbol{D}$ is called the **rate of deformation tensor** and the tensor $\boldsymbol{\Omega}$ the **rotation velocity tensor** or **spin tensor**. The definitions are:

$$\begin{aligned} \boldsymbol{D} &= \frac{1}{2}\left(\boldsymbol{L} + \boldsymbol{L}^{\mathrm{T}}\right) = \frac{1}{2}\left(\left(\vec{\nabla}\vec{v}\right)^{\mathrm{T}} + \left(\vec{\nabla}\vec{v}\right)\right) = \frac{1}{2}\left(\dot{\boldsymbol{F}} \cdot \boldsymbol{F}^{-1} + \boldsymbol{F}^{-\mathrm{T}} \cdot \dot{\boldsymbol{F}}^{\mathrm{T}}\right) \\ \boldsymbol{\Omega} &= \frac{1}{2}\left(\boldsymbol{L} - \boldsymbol{L}^{\mathrm{T}}\right) = \frac{1}{2}\left(\left(\vec{\nabla}\vec{v}\right)^{\mathrm{T}} - \left(\vec{\nabla}\vec{v}\right)\right) = \frac{1}{2}\left(\dot{\boldsymbol{F}} \cdot \boldsymbol{F}^{-1} - \boldsymbol{F}^{-\mathrm{T}} \cdot \dot{\boldsymbol{F}}^{\mathrm{T}}\right) \end{aligned} \tag{10.57}$$

and so

$$\dot{d\vec{x}} = \left(\boldsymbol{D} + \boldsymbol{\Omega}\right) \cdot d\vec{x}, \tag{10.58}$$

with

$$\boldsymbol{D}^{\mathrm{T}} = \boldsymbol{D} \tag{10.59}$$

and

$$\boldsymbol{\Omega}^{\mathrm{T}} = -\boldsymbol{\Omega}. \tag{10.60}$$

For an interpretation of the symmetric tensor $\boldsymbol{D}$, we depart from the relations that have been derived in Section 10.3:

$$\boldsymbol{F} \cdot \vec{e}_0 = \lambda\, \vec{e} \tag{10.61}$$

and

$$\lambda^2 = \vec{e}_0 \cdot \boldsymbol{F}^{\mathrm{T}} \cdot \boldsymbol{F} \cdot \vec{e}_0. \tag{10.62}$$

The material time derivative of the equation for λ^2 can be elaborated as follows:

$$\begin{aligned}
2\lambda\dot{\lambda} &= \vec{e}_0 \cdot \left(\dot{\boldsymbol{F}}^{\mathrm{T}} \cdot \boldsymbol{F} + \boldsymbol{F}^{\mathrm{T}} \cdot \dot{\boldsymbol{F}}\right) \cdot \vec{e}_0 \\
&= \vec{e}_0 \cdot \left(\boldsymbol{I} \cdot \dot{\boldsymbol{F}}^{\mathrm{T}} \cdot \boldsymbol{F} + \boldsymbol{F}^{\mathrm{T}} \cdot \dot{\boldsymbol{F}} \cdot \boldsymbol{I}\right) \cdot \vec{e}_0 \\
&= \vec{e}_0 \cdot \left(\boldsymbol{F}^{\mathrm{T}} \cdot \boldsymbol{F}^{-\mathrm{T}} \cdot \dot{\boldsymbol{F}}^{\mathrm{T}} \cdot \boldsymbol{F} + \boldsymbol{F}^{\mathrm{T}} \cdot \dot{\boldsymbol{F}} \cdot \boldsymbol{F}^{-1} \cdot \boldsymbol{F}\right) \cdot \vec{e}_0 \\
&= \vec{e}_0 \cdot \boldsymbol{F}^{\mathrm{T}} \cdot \left(\boldsymbol{F}^{-\mathrm{T}} \cdot \dot{\boldsymbol{F}}^{\mathrm{T}} + \dot{\boldsymbol{F}} \cdot \boldsymbol{F}^{-1}\right) \cdot \boldsymbol{F} \cdot \vec{e}_0 \\
&= \lambda^2 \vec{e} \cdot (2\boldsymbol{D}) \cdot \vec{e},
\end{aligned} \tag{10.63}$$

eventually resulting in the simple relation:

$$\frac{\dot{\lambda}}{\lambda} = \vec{e} \cdot \boldsymbol{D} \cdot \vec{e} \quad \left(= \dot{\overline{\ln(\lambda)}}\right). \tag{10.64}$$

This equation shows that the deformation velocity tensor $\boldsymbol{D}$ completely determines the current rate of (logarithmic) strain for an arbitrary line segment in the current state with a direction specified by $\vec{e}$. The analogous equation in component form is written

$$\frac{\dot{\lambda}}{\lambda} = \underset{\sim}{e}^{\mathrm{T}}\, \underline{D}\, \underset{\sim}{e}\,. \tag{10.65}$$

The terms on the diagonal of the matrix $\underline{D}$ represent the rate of strain in the directions of the x-, y- and z-coordinates. The off-diagonal terms represent the rate of shear.

For the interpretation of the skew-symmetric spin tensor, Eq. (10.58) can be used directly. After all, it is clear that the contribution $\boldsymbol{\Omega} \cdot d\vec{x}$ to $\dot{\overline{d\vec{x}}}$ is always perpendicular to $d\vec{x}$, because for all $d\vec{x}$:

$$d\vec{x} \cdot \boldsymbol{\Omega} \cdot d\vec{x} = 0 \quad \text{because} \quad \boldsymbol{\Omega}^{\mathrm{T}} = -\boldsymbol{\Omega}, \tag{10.66}$$

meaning that the contribution $\boldsymbol{\Omega} \cdot d\vec{x}$ has to be considered as the effect of a rotation.

For the material time derivative of the volume change factor $J = \det(\boldsymbol{F})$, it can be derived (without proof):

$$\dot{J} = J\,\mathrm{tr}(\dot{\boldsymbol{F}} \cdot \boldsymbol{F}^{-1}) = J\,\mathrm{tr}(\boldsymbol{L}) = J\,\vec{\nabla} \cdot \vec{v} = J\,\mathrm{tr}(\boldsymbol{D}). \tag{10.67}$$

Example 10.9 In the neighbourhood of the origin of a *xyz*-coordinate system, the velocity of a stationary fluid flow is given by:

$$\underset{\sim}{v} = \begin{bmatrix} v_x \\ v_y \\ v_z \end{bmatrix} \quad \text{with:} \quad v_x = \frac{\alpha x}{(y-\beta)^2}\,, \quad v_y = \frac{\alpha}{y-\beta}\,, \quad v_z = 0\,,$$

with α and β positive constants.

We want to determine the associated deformation rate tensor $\boldsymbol{D}$ and the spin tensor $\boldsymbol{\Omega}$ in the point $\mathrm{P} = (\beta, 0, 4\beta)$.

The matrix representation of $(\vec{\nabla}\vec{v})^{\mathrm{T}}$ can be derived as in Eq. (7.17). After differentiating and substituting the coordinates of point P, this leads to:

$$\underline{L} = \left(\underset{\sim}{\nabla}\underset{\sim}{v}^{\mathrm{T}}\right)^{\mathrm{T}} = \begin{bmatrix} \frac{\partial v_x}{\partial x} & \frac{\partial v_x}{\partial y} & \frac{\partial v_x}{\partial z} \\ \frac{\partial v_y}{\partial x} & \frac{\partial v_y}{\partial y} & \frac{\partial v_y}{\partial z} \\ \frac{\partial v_z}{\partial x} & \frac{\partial v_z}{\partial y} & \frac{\partial v_z}{\partial z} \end{bmatrix} = \begin{bmatrix} \frac{\alpha}{\beta^2} & \frac{2\alpha}{\beta^2} & 0 \\ 0 & -\frac{\alpha}{\beta^2} & 0 \\ 0 & 0 & 0 \end{bmatrix}$$

Thus:

$$\underline{D} = \frac{1}{2}[(\underset{\sim}{\nabla}\underset{\sim}{v}^{\mathrm{T}})^{\mathrm{T}} + \underset{\sim}{\nabla}\underset{\sim}{v}^{\mathrm{T}}] = \begin{bmatrix} \frac{\alpha}{\beta^2} & \frac{\alpha}{\beta^2} & 0 \\ \frac{\alpha}{\beta^2} & -\frac{\alpha}{\beta^2} & 0 \\ 0 & 0 & 0 \end{bmatrix}$$

end:

$$\underline{\Omega} = \frac{1}{2}[(\underset{\sim}{\nabla}\underset{\sim}{v}^{T})^{T} - \underset{\sim}{\nabla}\underset{\sim}{v}^{T}] = \begin{bmatrix} 0 & \frac{\alpha}{\beta^2} & 0 \\ -\frac{\alpha}{\beta^2} & 0 & 0 \\ 0 & 0 & 0 \end{bmatrix}$$

Exercises

10.1 In a subvolume of a material continuum, the deformation of the current state, with respect to the reference state, is homogeneous. In a Cartesian *xyz*-coordinate system, the associated deformation tensor is given as

$$\boldsymbol{F} = \boldsymbol{I} + 3\vec{e}_y\vec{e}_y - 7\vec{e}_y\vec{e}_z - \vec{e}_z\vec{e}_y + \vec{e}_z\vec{e}_z,$$

with $\boldsymbol{I}$ the unit tensor.

For a material point P within the subvolume the position vectors in the reference state as well as the current state are given, respectively, as:

$$\vec{x}_{0P} = \vec{e}_x + \vec{e}_y + \vec{e}_z, \quad \vec{x}_P = 2\vec{e}_x + 3\vec{e}_y - 2\vec{e}_z.$$

Another point Q within the subvolume appears to be in the origin in the current configuration: $\vec{x}_Q = \vec{0}$.

Calculate the position vector $\vec{x}_{0Q}$ of the point Q in the reference state.

10.2 Within a subvolume of a material continuum the deformation tensor in the deformed current state, with respect to the reference state, is constant. Consider a vector of material points, denoted with $\vec{a}_0$ in the reference state and with $\vec{a}$ in the current state. The angle between $\vec{a}_0$ and $\vec{a}$ is given by ϕ.

Prove that we can write for ϕ:

$$\cos(\phi) = \frac{\vec{a}_0 \cdot \boldsymbol{F} \cdot \vec{a}_0}{\sqrt{(\vec{a}_0 \cdot \vec{a}_0)(\vec{a}_0 \cdot \boldsymbol{F}^{\mathrm{T}} \cdot \boldsymbol{F} \cdot \vec{a}_0)}}.$$

10.3 Within a subvolume of a material continuum, the deformation tensor in the deformed current state, with respect to the reference state, is constant. In a Cartesian xyz-coordinate system, the following deformation tensor is given:

$$\boldsymbol{F} = \boldsymbol{I} + 4\,\vec{e}_x\vec{e}_x + 2\,\vec{e}_x\vec{e}_y + 2\,\vec{e}_y\vec{e}_x,$$

with $\boldsymbol{I}$ the unit tensor. Within the subvolume, two material points P and Q are considered. The position vectors in the reference state are given as:

$$\vec{x}_{0P} = \vec{e}_x + \vec{e}_y + \vec{e}_z, \quad \vec{x}_{0Q} = 2\vec{e}_x + 3\vec{e}_y + 2\vec{e}_z.$$

In addition, the position vector for the point P in the current state is given as:

$$\vec{x}_P = 2\vec{e}_x + 2\vec{e}_y + 2\vec{e}_z.$$

Calculate the position vector $\vec{x}_Q$ of the point Q in the current state.

10.4 The deformation of a material particle, in the current state with respect to the reference state, is fully described by the Green–Lagrange strain matrix $\underline{E}$, with respect to a Cartesian xyz-coordinate system, with

$$\underline{E} = \frac{1}{2}\begin{bmatrix} 4 & 3 & 0 \\ 3 & 1 & 0 \\ 0 & 0 & 3 \end{bmatrix}.$$

Calculate the volume change factor J for this particle.

10.5 A tendon is stretched in a uniaxial stress test. The tendon behaves like an incompressible material. The length axis of the tendon coincides with the x-axis. For this test, the following (time-independent) deformation rate matrix $\underline{D}$ is applicable (expressed in [s^{-1}]):

$$\underline{D} = \begin{bmatrix} 0.02 & 0 & 0 \\ 0 & -0.01 & 0 \\ 0 & 0 & -0.01 \end{bmatrix}.$$

At time $t = 0$ [s], the tendon has a length ℓ_0 equal to 3 [cm]. From this time on, the above given matrix $\underline{D}$ can be applied. Calculate the length ℓ of the tendon as a function of the time t.

10.6 At some material point, the local deformation process is described by means of the deformation tensor as a function of time: $\boldsymbol{F}(t)$. Based on this deformation tensor, the left Cauchy–Green tensor $\boldsymbol{B} = \boldsymbol{F}\cdot\boldsymbol{F}^{\mathrm{T}}$ can be derived and subsequently the Finger tensor $\boldsymbol{\varepsilon}_{\mathrm{F}} = \frac{1}{2}(\boldsymbol{B} - \boldsymbol{I})$.

Prove that $\dot{\boldsymbol{\varepsilon}}_{\mathrm{F}} = \frac{1}{2}(\boldsymbol{L}\cdot\boldsymbol{B} + \boldsymbol{B}\cdot\boldsymbol{L}^{\mathrm{T}})$, with $\boldsymbol{L} = \dot{\boldsymbol{F}}\cdot\boldsymbol{F}^{-1}$ the velocity gradient tensor.

10.7 Consider a homogeneously deforming continuum. In the figure, the initial undeformed configuration (at $t = 0$ [s]) and the deformed current configuration (at $t = 1$ [s]) are given in a Cartesian xyz-coordinate system. During deformation, the edges of the cube do not change direction.

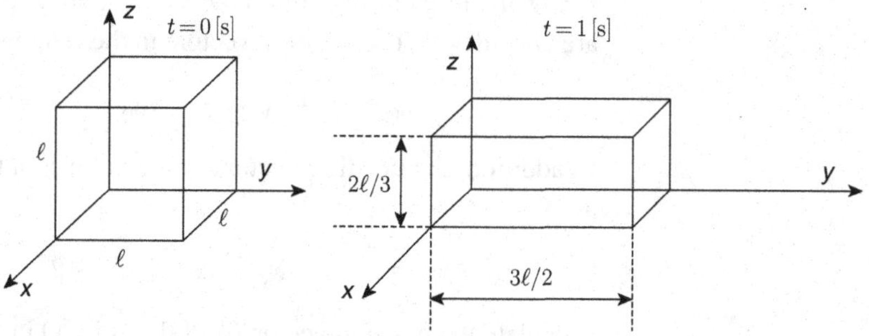

The deformation proces for $0 \leq t \leq 1$, with t expressed in seconds, is described by the deformation matrix $\underline{F}(t)$. The components of the matrix $\underline{F}(t)$ change linearly with time.

Determine the volume change factor $J = \det(\underline{F})$ as a function of time t.

10.8 In rheological tests, often a Couette flow is used. This shear flow can be modelled as a stationary flow between two parallel plates. The lower plate is fixed in space and the top plate is moving with a velocity V in the x-direction. The distance between the plates is H.

Give expressions for the fluid velocity $\vec{v}$, the rate of deformation tensor $\boldsymbol{D}$ and the rotation velocity tensor $\boldsymbol{\Omega}$ as a function of the parameters V and H and the coordinates x, y, z.

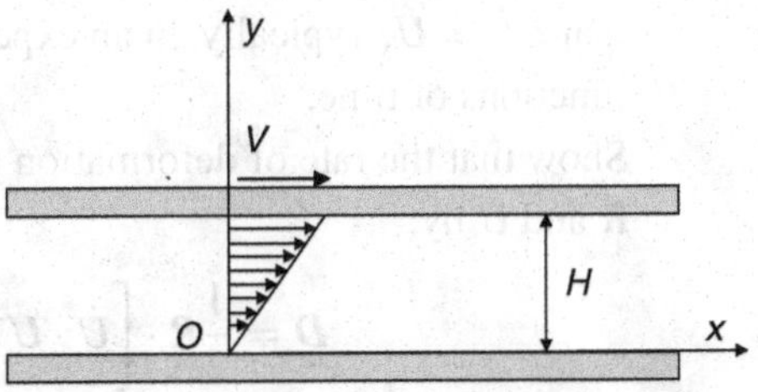

10.9 A popular test in soft tissue biomechanics is the indentation test (see figure below). A spherical indentor is pushed into a tissue sample that is positioned on a smooth glass plate. From the combination of the indentation force, the vertical displacement of the indentor and horizontal displacements, viewed from below with a microscope, it is possible to examine quite complex material behaviour. Because the deformation is heterogeneous (***F*** depends on the location), the analysis of this experiment has to be done by means of a computer. We focus on the point P and its environment in the form of a small rectangle around the point. It is possible to decompose the deformation tensor ***F*** at point P in a rigid body rotation ***R*** and a dilatation ***U***:

$$\boldsymbol{F} = \boldsymbol{R} \cdot \boldsymbol{U}$$

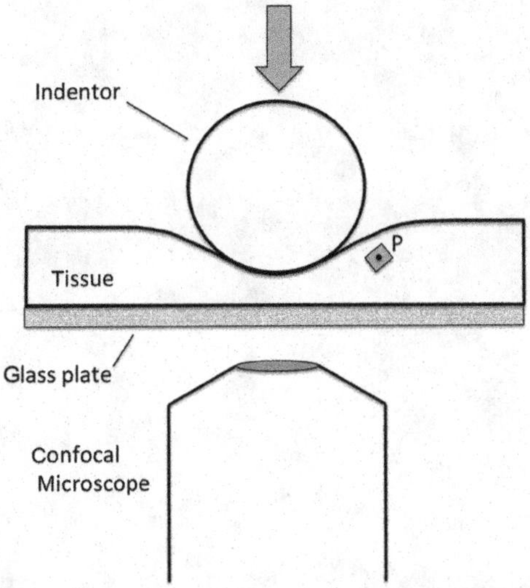

This separation into a rotation and dilatation tensor is called **polar decomposition**. Because ***R*** describes a rotation, the tensor is orthogonal, which means that $\boldsymbol{R}^{\mathrm{T}} = \boldsymbol{R}^{-1}$. The dilation tensor ***U*** is symmetric, which means that $U^T = U$. Typically, in an experiment both $R = R(t)$ and $U = U(t)$ are functions of time.

Show that the rate of deformation tensor D can be expressed in the tensors R and U by:

$$D = \frac{1}{2} R \cdot \left[\dot{U} \cdot U^{-1} + \left(\dot{U} \cdot U^{-1} \right)^{\mathrm{T}} \right] \cdot R^{\mathrm{T}}$$

11 Local Balance of Mass, Momentum and Energy

11.1 Introduction

A coherent amount of material (a material body or possibly a distinguishable material fraction) is considered to be a continuum with current volume V in three-dimensional space. In Chapter 8, attention was focussed on the local stress state (the internal interaction between neighbouring volume elements), whereas Chapters 9 and 10 treated the local kinematics (shape and volume changes of material particles). To determine the stresses and kinematic variables as a function of the position in the three-dimensional space, we need a description of the material behaviour, which will be the subject of subsequent chapters, and we need local balance laws. In the present chapter, the balance of mass (leading to the continuity equation) and the balance of momentum (leading to the equations of motion) for a continuum will be formulated. In addition, the balance of mechanical power will be derived based on the balance of momentum.

11.2 The Local Balance of Mass

Let us focus our attention on an infinitesimally small rectangular material element $dV = dxdydz$ in the current state: see Fig. 11.1.

The mass in the current volume element dV equals the mass in the reference configuration of the associated volume element dV_0, while the volumes are related by

$$dV = JdV_0 \text{ with } J = \det(\boldsymbol{F}), \tag{11.1}$$

where $\boldsymbol{F}$ is the deformation tensor. It is assumed that during the transformation from the reference state to the current deformed state no material of the considered type is created or lost. So there is no mass exchange with certain other fractions, for example in the form of a chemical reaction. Based on mass conservation,

$$\rho_0 dV_0 = \rho dV = \rho J dV_0, \tag{11.2}$$

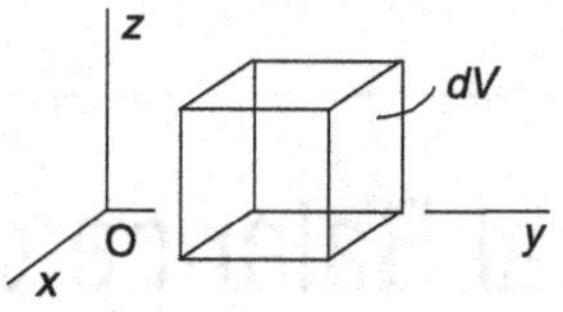

Figure 11.1

Infinitesimally small element $dV = dxdydz$ in the current state.

with ρ_0 and ρ the (mass) densities in the reference and current configurations, respectively. So, balance of mass leads to the statement that the product ρJ is time-independent and that the material time derivative of the product equals zero:

$$\dot{\rho} J + \rho \dot{J} = 0. \tag{11.3}$$

Using Eq. (10.67), with $\boldsymbol{D}$ the deformation rate tensor, we obtain:

$$\dot{\rho} = -\rho \, \mathrm{tr}(\boldsymbol{D}). \tag{11.4}$$

In this final result, the reference configuration is no longer represented: all variables in this equation are related to the current configuration.

11.3 The Local Balance of Momentum

In Chapter 8, the equilibrium equations were derived by considering the force equilibrium of a material cube. The equilibrium equations are a special form of the balance of momentum equations, which are the subject of the present section. This means that some overlap between the two derivations can be recognized.

Again, the material volume element dV from Fig. 11.1 is considered. Using the balance of mass allows us to write for the current momentum of the element:

$$\vec{v}\rho dV = \vec{v}\rho J dV_0, \tag{11.5}$$

with $\vec{v}$ the velocity of the material. Using that $\rho J dV_0 = \rho_0 dV_0$ is constant, the change per unit time of the momentum of the volume element equals

$$\dot{\vec{v}}\rho J dV_0 = \dot{\vec{v}}\rho dV = \vec{a}\rho dV, \tag{11.6}$$

with $\vec{a}$ the acceleration vector (see Section 9.2). In the following, the momentum change given above will be postulated, according to Newton's second law, to be equal to the total 'external' force acting on the volume element.

For the element in Fig. 11.1, the resulting force in the x-direction is considered. This force is the sum of the forces working in the x-direction on the outer surfaces of the element and the distributed load (force per unit mass) acting on the element in the x-direction, so successively (also see Chapter 8):

$$
\begin{array}{ll}
-\sigma_{xx}dydz & \text{back plane} \\
\left(\sigma_{xx} + \frac{\partial \sigma_{xx}}{\partial x}dx\right)dydz & \text{frontal plane} \\
-\sigma_{xy}dxdz & \text{left plane} \\
\left(\sigma_{xy} + \frac{\partial \sigma_{xy}}{\partial y}dy\right)dxdz & \text{right plane} \\
-\sigma_{xz}dxdy & \text{bottom plane} \\
\left(\sigma_{xz} + \frac{\partial \sigma_{xz}}{\partial z}dz\right)dxdy & \text{top plane} \\
q_x\rho dxdydz & \text{volume,}
\end{array}
$$

with resultant (in x-direction):

$$
\begin{aligned}
&\left(\frac{\partial \sigma_{xx}}{\partial x} + \frac{\partial \sigma_{xy}}{\partial y} + \frac{\partial \sigma_{xz}}{\partial z} + q_x\rho\right)dxdydz \\
&= \left(\frac{\partial \sigma_{xx}}{\partial x} + \frac{\partial \sigma_{xy}}{\partial y} + \frac{\partial \sigma_{xz}}{\partial z} + q_x\rho\right)dV.
\end{aligned} \tag{11.7}
$$

Similarly, the resulting forces in the y- and z-directions can be determined. All external forces applied to the volume element are stored in a column. Using the gradient operator $\underset{\sim}{\nabla}$ introduced earlier and the symmetrical Cauchy stress matrix $\underline{\sigma}$, the column with external forces can be written as:

$$
\begin{bmatrix}
\frac{\partial \sigma_{xx}}{\partial x} + \frac{\partial \sigma_{xy}}{\partial y} + \frac{\partial \sigma_{xz}}{\partial z} + q_x\rho \\
\frac{\partial \sigma_{xy}}{\partial x} + \frac{\partial \sigma_{yy}}{\partial y} + \frac{\partial \sigma_{yz}}{\partial z} + q_y\rho \\
\frac{\partial \sigma_{xz}}{\partial x} + \frac{\partial \sigma_{yz}}{\partial y} + \frac{\partial \sigma_{zz}}{\partial z} + q_z\rho
\end{bmatrix} dV = \left(\left(\underset{\sim}{\nabla}^T \underline{\sigma}\right)^T + \rho \underset{\sim}{q}\right) dV \tag{11.8}
$$

with the column $\underset{\sim}{q}$ for the distributed load defined according to

$$
\underset{\sim}{q} = \begin{bmatrix} q_x \\ q_y \\ q_z \end{bmatrix}.
$$

In vector/tensor notation, the following expression can be given for the resulting force on the element dV:

$$
\left(\vec{\nabla} \cdot \boldsymbol{\sigma} + \rho \vec{q}\right) dV.
$$

For the element under consideration, the time derivative of the momentum equals the external load, leading to

$$
\vec{\nabla} \cdot \boldsymbol{\sigma} + \rho \vec{q} = \rho \vec{a}, \tag{11.9}
$$

and also

$$(\underset{\sim}{\nabla}^{\mathrm{T}}\underline{\sigma})^{\mathrm{T}} + \rho\underset{\sim}{q} = \rho\underset{\sim}{a}. \qquad (11.10)$$

This equation is called the local equation of motion. Again, all variables in this equation refer to the current configuration; the defined reference configuration (as usual for a solid) is not relevant for this.

11.4 The Local Balance of Mechanical Power

It has to be stated explicitly that no new balance law is introduced here; use will be made only of relations that were already introduced previously. Again the element is considered that was introduced in Fig. 11.1 (volume in reference state dV_0, current volume $dV = JdV_0$). The dot product of the local equation of motion Eq. (11.9) with the velocity vector $\vec{v}$ is taken. The result is multiplied by the current volume of the element, yielding

$$\vec{v} \cdot \left(\vec{\nabla} \cdot \boldsymbol{\sigma}\right) dV + \vec{v} \cdot \vec{q}\rho dV = \vec{v} \cdot \vec{a}\rho dV. \qquad (11.11)$$

The terms in this equation will be interpreted separately.

- For the term on the right-hand side, we can write:

$$\vec{v} \cdot \vec{a}\rho dV = d\dot{U}_{\mathrm{kin}}, \qquad (11.12)$$

  with

$$dU_{\mathrm{kin}} = \frac{1}{2}\vec{v} \cdot \vec{v}\rho dV = \frac{1}{2}\vec{v} \cdot \vec{v}\rho J dV_0 = \frac{1}{2}\vec{v} \cdot \vec{v}\rho_0 dV_0, \qquad (11.13)$$

  where dU_{kin} is the current kinetic energy of the volume element being considered. With respect to the material time derivative that is used in Eq. (11.12), it should be realized that $\rho J dV_0 = \rho_0 dV_0$ is constant.
  The term on the right-hand side of Eq. (11.11) can be interpreted as the change of the kinetic energy per unit time.
- The second term on the left-hand side of Eq. (11.11) can be directly interpreted as the mechanical power, externally applied to the volume element by the distributed load $\vec{q}$. It can be noted that

$$dP_{\mathrm{ext},\,\vec{q}} = \vec{v} \cdot \vec{q}\rho dV. \qquad (11.14)$$

- For the first term on the left-hand side of Eq. (11.11), some careful mathematical elaboration, using the symmetry of the stress tensor $\boldsymbol{\sigma}$, leads to

$$\vec{v}\cdot\left(\vec{\nabla} \cdot \boldsymbol{\sigma}\right) dV = \vec{\nabla}\cdot(\boldsymbol{\sigma} \cdot \vec{v})\, dV - \mathrm{tr}\left(\boldsymbol{\sigma} \cdot (\vec{\nabla}\vec{v})\right) dV = \vec{\nabla}\cdot(\boldsymbol{\sigma} \cdot \vec{v})\, dV - \mathrm{tr}(\boldsymbol{\sigma} \cdot \boldsymbol{D})\, dV. \qquad (11.15)$$

  For this, the definition of the deformation rate tensor $\boldsymbol{D}$ of Section 10.6 is used.

The first term on the right-hand side of Eq. (11.15) represents the resulting, externally applied power to the volume element by the forces (originating from neighbouring elements) acting on the outer surfaces of the element:

$$dP_{\text{ext},\sigma} = \vec{\nabla} \cdot (\vec{\sigma} \cdot \vec{v})\, dV. \tag{11.16}$$

This can be proven with a similar strategy to that used in the previous section, to derive the resultant of the forces acting on the outer surfaces of the volume element. Interpretation of the second term on the right-hand side,

$$-\text{tr}(\boldsymbol{\sigma} \cdot \boldsymbol{D})\, dV,$$

will be given below.

Summarizing, after reconsidering Eq. (11.11), and using the results of the above, it can be stated:

$$dP_{ext,\boldsymbol{\sigma}} - \text{tr}(\boldsymbol{\sigma} \cdot \boldsymbol{D})dV + dP_{\text{ext},\vec{q}} = d\dot{U}_{\text{kin}}, \tag{11.17}$$

and after some re-ordering:

$$\text{tr}(\boldsymbol{\sigma} \cdot \boldsymbol{D})dV + d\dot{U}_{\text{kin}} = dP_{\text{ext}}, \tag{11.18}$$

with

$$dP_{\text{ext}} = dP_{\text{ext},\sigma} + dP_{\text{ext},\vec{q}}. \tag{11.19}$$

So, it can be observed that the total externally applied mechanical power is partly used for a change in the kinetic energy. The remaining part is stored internally, so:

$$dP_{\text{int}} = \text{tr}(\boldsymbol{\sigma} \cdot \boldsymbol{D})dV. \tag{11.20}$$

This internally stored mechanical power (increase of the internal mechanical energy per unit of time) can be completely reversible (for elastic behaviour), partly reversible and partly irreversible (for visco-elastic behaviour), or fully irreversible (for viscous behaviour). In the latter case, all externally applied mechanical energy to the material is dissipated and converted into other forms of energy (in general, a large part is converted into heat).

11.5 Lagrangian and Eulerian Descriptions of the Balance Equations

In summary, the balance equations of mass and momentum as derived in the Sections 11.2 and 11.3 can be written as:

$$\text{balance of mass} \quad \dot{\rho} = -\rho \text{tr}(\boldsymbol{D}) = -\rho \vec{\nabla} \cdot \vec{v},$$
$$\text{balance of momentum} \quad \vec{\nabla} \cdot \boldsymbol{\sigma} + \rho \vec{q} = \rho \vec{a} = \rho \dot{\vec{v}},$$

and in column/matrix notation:

$$\text{balance of mass} \quad \dot{\rho} = -\rho \operatorname{tr}(\underset{\sim}{D}) = -\rho \underset{\sim}{\nabla}^{\mathrm{T}} \underset{\sim}{v},$$
$$\text{balance of momentum} \quad \left(\underset{\sim}{\nabla}^{\mathrm{T}} \underset{\sim}{\sigma}\right)^{\mathrm{T}} + \rho \underset{\sim}{q} = \rho \underset{\sim}{a} = \rho \dot{\underset{\sim}{v}}.$$

In a typical Lagrangian description, the field variables are considered to be a function of the material coordinates $\vec{x}_0$, defined in the reference configuration, and time t. It can be stated that the balance laws have to be satisfied for all $\vec{x}_0$ within the domain V_0 and for all points in time.

Because of the physical relevance (for solids) of the deformation tensor $\boldsymbol{F}$, the balance of mass will usually not be used in the differential form as given above, but rather as:

$$\rho = \frac{\rho_0}{\det(\boldsymbol{F})} \quad \text{and also} \quad \rho = \frac{\rho_0}{\det(\underline{F})}. \tag{11.21}$$

The gradient operator $\vec{\nabla}$ (and also $\underset{\sim}{\nabla}$) in the balance of momentum equation, built up from derivatives with respect to the spatial coordinates, can be transformed into the gradient operator $\vec{\nabla}_0$ (and $\underset{\sim}{\nabla}_0$) with respect to the material coordinates, see Section 9.6. The balance of momentum can then be formulated according to:

$$\left(\boldsymbol{F}^{-\mathrm{T}} \cdot \vec{\nabla}_0\right) \cdot \boldsymbol{\sigma} + \rho \vec{q} = \rho \vec{a} = \rho \dot{\vec{v}}, \tag{11.22}$$

and also

$$\left(\left(\underline{F}^{-\mathrm{T}} \underset{\sim}{\nabla}_0\right)^{\mathrm{T}} \underline{\sigma}\right)^{\mathrm{T}} + \rho \underset{\sim}{q} = \rho \underset{\sim}{a} = \rho \dot{\underset{\sim}{v}}. \tag{11.23}$$

In a typical Eulerian description, the field properties are considered to be a function of the spatial coordinates $\vec{x}$, indicating locations in the current configuration, and time t. It can be stated that the balance laws have to be satisfied for all $\vec{x}$ within the domain V and for all points in time.

To reformulate the balance laws, the material time derivative is 'replaced' by the spatial time derivative: see Section 9.4. For the mass balance, this yields:

$$\frac{\delta \rho}{\delta t} + \vec{v} \cdot \vec{\nabla} \rho = -\rho \operatorname{tr}(\boldsymbol{D}) = -\rho \vec{\nabla} \cdot \vec{v}, \tag{11.24}$$

and so

$$\frac{\delta \rho}{\delta t} + \vec{\nabla} \cdot (\rho \vec{v}) = 0. \tag{11.25}$$

In column/matrix notation:

$$\frac{\delta \rho}{\delta t} + \underset{\sim}{v}^{T} \underset{\sim}{\nabla} \rho = -\rho \operatorname{tr}(\underline{D}) = -\rho \underset{\sim}{\nabla}^{\mathrm{T}} \underset{\sim}{v}, \tag{11.26}$$

and so

$$\frac{\delta \rho}{\delta t} + \underset{\sim}{\nabla}^{\mathrm{T}} (\rho \underset{\sim}{v}) = 0. \tag{11.27}$$

For the balance of momentum, this yields:

$$\vec{\nabla} \cdot \boldsymbol{\sigma} + \rho \vec{q} = \rho \left(\left(\vec{\nabla} \vec{v} \right)^{\mathrm{T}} \cdot \vec{v} + \frac{\delta \vec{v}}{\delta t} \right), \tag{11.28}$$

and in column/matrix notation:

$$\left(\underset{\sim}{\nabla}^{\mathrm{T}} \underset{\sim}{\sigma} \right)^{\mathrm{T}} + \rho \underset{\sim}{q} = \rho \left(\left(\underset{\sim}{\nabla} \underset{\sim}{v}^{\mathrm{T}} \right)^{\mathrm{T}} \underset{\sim}{v} + \frac{\delta \underset{\sim}{v}}{\delta t} \right). \tag{11.29}$$

For the special case of a stationary flow of a material, the following balance equation results for the mass balance:

$$\vec{\nabla} \cdot (\rho \vec{v}) = 0, \tag{11.30}$$

and also

$$\underset{\sim}{\nabla}^{\mathrm{T}} (\rho \underset{\sim}{v}) = 0. \tag{11.31}$$

The momentum equation reduces to:

$$\vec{\nabla} \cdot \boldsymbol{\sigma} + \rho \vec{q} = \rho \left(\vec{\nabla} \vec{v} \right)^{\mathrm{T}} \cdot \vec{v}, \tag{11.32}$$

and also

$$\left(\underset{\sim}{\nabla}^{\mathrm{T}} \underset{\sim}{\sigma} \right)^{\mathrm{T}} + \rho \underset{\sim}{q} = \rho \left(\underset{\sim}{\nabla} \underset{\sim}{v}^{\mathrm{T}} \right)^{\mathrm{T}} \underset{\sim}{v}. \tag{11.33}$$

Exercises

11.1 Compressible air is flowing through the bronchi. A bronchus is modelled as a straight cylindrical tube. We consider a stationary flow. For each cross section of the tube the velocity of the air V and the density ρ is constant over the cross section. In the direction of the flow, the velocity of the air and the density vary, because of temperature differences. Two cross sections A and B at a separation L are considered. See the figure below.

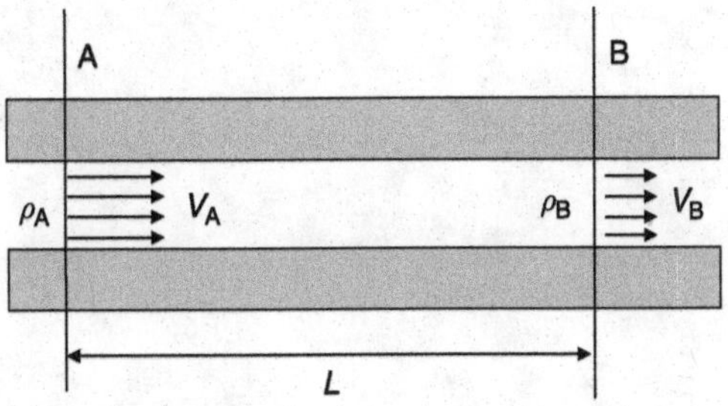

What relation can be derived for the variables indicated in the figure?

11.2 For a solid element, the deformation matrix $\underline{F}$ is given as a function of time t, with reference to the undeformed configuration at time $t = 0$:

$$\underline{F} = \begin{bmatrix} 1+\alpha t & 0 & \alpha t \\ 0 & 1 & 0 \\ \alpha t & 0 & 1+\alpha t \end{bmatrix} \quad \text{with} \quad \alpha = 0.01 \text{ [s}^{-1}\text{]}.$$

In the reference configuration, the density ρ_0 equals 1500 [kg m^{-3}]. As a result of the deformation process, the density ρ of the material will change. Determine the density as a function of the time.

12 Constitutive Modelling of Solids and Fluids

12.1 Introduction

In the first section of this chapter, the (biological) material under consideration can be regarded as a solid. This implies that it is possible to define a reference configuration and local deformations can be related to this reference configuration (see Chapter 10). It is assumed that the deformations (or more precisely, the path along which the actual deformations are reached: the deformation history) fully determine the current stress state (see Chapter 8), except for the special case when we deal with incompressible material behaviour. A number of non-mechanical phenomena are not included, such as the influence of temperature variations.

The **constitutive equations** discussed in the present chapter address a relation that can be formally written as

$$\boldsymbol{\sigma}(t) = \mathcal{F}\{\boldsymbol{F}(\tau);\ \tau \leq t\}, \tag{12.1}$$

with $\boldsymbol{\sigma}(t)$ the current Cauchy stress tensor at time t and with $\boldsymbol{F}(\tau)$ the deformation tensor at the relevant (previous) times τ up to the time t, assuming compressible material behaviour. A specification of this relation (the material behaviour) can only be obtained by means of experimental studies. In the current chapter we restrict our attention to **elastic behaviour**. In that case, the deformation history is not relevant, and we can formally write

$$\boldsymbol{\sigma} = \boldsymbol{\sigma}(\boldsymbol{F}). \tag{12.2}$$

Further elaboration of this relation will initially be done for the case of small deformations and rotations, so under the condition $\boldsymbol{F} \approx \boldsymbol{I}$. After that, the consequences of including large deformations (and large rotations) will be discussed. In that case, it is important to account for the fact that large additional rigid body motions are not allowed to induce extra stresses in the material.

In the second part of this chapter, we focus on fluid behaviour. Since for a fluid the reference configuration is usually not defined, an Eulerian approach will be adopted, and the velocity gradient tensor $\boldsymbol{L}$ and the deformation rate tensor $\boldsymbol{D}$ play a central role.

12.2 Elastic Behaviour at Small Deformations and Rotations

To describe the current deformation state of the material, a reference configuration has to be specified. Although, in principle, the choice for the reference configuration is free, in this case a fixed, stress-free state is chosen. This is not as trivial as it may seem, because for many biological materials a zero-stress state does not exist in vivo; however, this discussion is not within the scope of the current book. Interested readers are referred to e.g. [8]. So

$$\boldsymbol{\sigma}(\boldsymbol{F} = \boldsymbol{I}) = \boldsymbol{0}. \tag{12.3}$$

It should be noted that this statement only applies to purely elastic material behaviour. If the deformation history is important for the current stress state, the relation given above is generally certainly not valid.

The current deformation tensor $\boldsymbol{F}$ fully describes the local deformations with respect to the reference configuration. This also holds for the linear strain tensor $\boldsymbol{\epsilon}$ that was introduced in Section 10.4 under the strict condition that $\boldsymbol{F} \approx \boldsymbol{I}$. Because rotations and large deformations can be neglected under these conditions, the Cauchy stress tensor $\boldsymbol{\sigma}$ can be coupled directly to the linear strain tensor $\boldsymbol{\epsilon}$ via an expression according to the format:

$$\boldsymbol{\sigma} = \boldsymbol{\sigma}(\boldsymbol{\epsilon}) \quad \text{with} \quad \boldsymbol{\sigma}(\boldsymbol{\epsilon} = \boldsymbol{0}) = \boldsymbol{0}. \tag{12.4}$$

The exact specification of Eq. (12.4) has to be derived from experimental work and is a major issue in biomechanics. In the current section, we will explore the commonly used **Hooke's law**, which is often adopted as the first approximation to describe the material behaviour. Hooke's law supplies a linear relation between the components of the stress tensor $\boldsymbol{\sigma}$, stored in the (3×3) matrix $\underline{\sigma}$, and the components of the linear strain tensor $\boldsymbol{\epsilon}$, stored in the (3×3) matrix $\underline{\epsilon}$. In addition, it is assumed that the material behaviour is **isotropic** (the behaviour is identical in all directions).

Hooke's law will be specified, using the decomposition of the stress tensor in a hydrostatic and a deviatoric part, as described in Section 8.7:

$$\boldsymbol{\sigma} = \boldsymbol{\sigma}^{\mathrm{h}} + \boldsymbol{\sigma}^{\mathrm{d}} = -p\boldsymbol{I} + \boldsymbol{\sigma}^{\mathrm{d}}, \tag{12.5}$$

with p the hydrostatic pressure, defined as

$$p = -\frac{\mathrm{tr}(\boldsymbol{\sigma})}{3}. \tag{12.6}$$

The linear strain tensor $\boldsymbol{\epsilon}$ can be split in a similar way:

$$\boldsymbol{\epsilon} = \boldsymbol{\epsilon}^{\mathrm{h}} + \boldsymbol{\epsilon}^{\mathrm{d}} = \frac{\varepsilon^{\mathrm{v}}}{3}\boldsymbol{I} + \boldsymbol{\epsilon}^{\mathrm{d}}, \tag{12.7}$$

with ε^{v} the (relative) volume change:

$$\varepsilon^{\mathrm{v}} = \mathrm{tr}(\boldsymbol{\epsilon}). \tag{12.8}$$

This requires some extra elucidation. In general, the relative volume change is defined as (see Sections 10.4 and 10.5)

$$\varepsilon^{\mathrm{v}} = \frac{dV - dV_0}{dV_0} = J - 1 = \det(\boldsymbol{F}) - 1 = \det\left(\boldsymbol{I} + \left(\vec{\nabla}_0 \vec{u}\right)^{\mathrm{T}}\right) - 1. \tag{12.9}$$

However, because the components of $\vec{\nabla}_0 \vec{u}$ can be neglected with respect to 1, we can write (after linearization):

$$\begin{aligned} \varepsilon^{\mathrm{v}} &= 1 + \mathrm{tr}\left(\left(\vec{\nabla}_0 \vec{u}\right)^{\mathrm{T}}\right) - 1 \\ &= \mathrm{tr}\left(\frac{1}{2}\left(\left(\vec{\nabla}_0 \vec{u}\right) + \left(\vec{\nabla}_0 \vec{u}\right)^{\mathrm{T}}\right)\right) = \mathrm{tr}(\boldsymbol{\epsilon}). \end{aligned} \tag{12.10}$$

Hooke's law for linearly elastic isotropic behaviour can now be written as

$$p = -K\varepsilon^{\mathrm{v}}, \quad \boldsymbol{\sigma}^{\mathrm{d}} = 2G\,\boldsymbol{\epsilon}^{\mathrm{d}}, \tag{12.11}$$

and in matrix notation:

$$p = -K\varepsilon^{\mathrm{v}}, \quad \underline{\sigma}^{\mathrm{d}} = 2G\,\underline{\epsilon}^{\mathrm{d}}, \tag{12.12}$$

with K the **compression modulus** or **bulk modulus** of the material and G the **shear modulus** (for Hooke's law the relevant material parameters; both positive). In the present section, it is assumed that the material is compressible, meaning that the compression modulus K has a finite value. Based on Hooke's law, the above equations can be written as

$$\boldsymbol{\sigma} = \left(K - \frac{2G}{3}\right)\mathrm{tr}(\boldsymbol{\epsilon})\boldsymbol{I} + 2G\boldsymbol{\epsilon}, \tag{12.13}$$

and also

$$\underline{\sigma} = \left(K - \frac{2G}{3}\right)\mathrm{tr}(\underline{\epsilon})\underline{I} + 2G\underline{\epsilon}, \tag{12.14}$$

thus fully establishing the desired tensor relation $\boldsymbol{\sigma} = \boldsymbol{\sigma}(\boldsymbol{\epsilon})$, and $\underline{\sigma} = \underline{\sigma}(\underline{\epsilon})$ in matrix notation. For the inverse relationship $\boldsymbol{\epsilon} = \boldsymbol{\epsilon}(\boldsymbol{\sigma})$, and $\underline{\epsilon} = \underline{\epsilon}(\underline{\sigma})$, it can easily be derived that:

$$\boldsymbol{\epsilon} = \left(\frac{1}{9K} - \frac{1}{6G}\right)\mathrm{tr}(\boldsymbol{\sigma})\boldsymbol{I} + \frac{1}{2G}\boldsymbol{\sigma}, \tag{12.15}$$

and also

$$\underline{\epsilon} = \left(\frac{1}{9K} - \frac{1}{6G}\right)\mathrm{tr}(\underline{\sigma})\underline{I} + \frac{1}{2G}\underline{\sigma}. \tag{12.16}$$

In the following, an interpretation of Hooke's law will be given. For this purpose we focus on the matrix formulations $\underline{\sigma} = \underline{\sigma}(\underline{\epsilon})$ and $\underline{\epsilon} = \underline{\epsilon}(\underline{\sigma})$. The symmetrical matrices $\underline{\sigma}$ and $\underline{\epsilon}$ are composed according to:

$$\underline{\sigma} = \begin{bmatrix} \sigma_{xx} & \sigma_{xy} & \sigma_{xz} \\ \sigma_{yx} & \sigma_{yy} & \sigma_{yz} \\ \sigma_{zx} & \sigma_{zy} & \sigma_{zz} \end{bmatrix}, \ \underline{\epsilon} = \begin{bmatrix} \varepsilon_{xx} & \varepsilon_{xy} & \varepsilon_{xz} \\ \varepsilon_{yx} & \varepsilon_{yy} & \varepsilon_{yz} \\ \varepsilon_{zx} & \varepsilon_{zy} & \varepsilon_{zz} \end{bmatrix}. \tag{12.17}$$

Using $\underline{\epsilon} = \underline{\epsilon}(\underline{\sigma})$ it follows from Eq. (12.16) for the diagonal components (the strains in the x-, y- and z-directions):

$$\begin{aligned} \varepsilon_{xx} &= \left(\frac{1}{9K} - \frac{1}{6G}\right)(\sigma_{xx} + \sigma_{yy} + \sigma_{zz}) + \frac{1}{2G}\sigma_{xx} \\ \varepsilon_{yy} &= \left(\frac{1}{9K} - \frac{1}{6G}\right)(\sigma_{xx} + \sigma_{yy} + \sigma_{zz}) + \frac{1}{2G}\sigma_{yy} \\ \varepsilon_{zz} &= \left(\frac{1}{9K} - \frac{1}{6G}\right)(\sigma_{xx} + \sigma_{yy} + \sigma_{zz}) + \frac{1}{2G}\sigma_{zz}. \end{aligned} \tag{12.18}$$

Equation (12.18) shows that the strains in the x-, y- and z-directions are determined solely by the normal stresses in the x-, y- and z-directions, which is a consequence of assuming isotropy. It is common practice to use an alternative set of material parameters, namely the Young's modulus E and the Poisson's ratio ν. The Young's modulus follows from

$$\frac{1}{E} = \left(\frac{1}{9K} - \frac{1}{6G}\right) + \frac{1}{2G} = \frac{3K + G}{9KG}, \tag{12.19}$$

and therefore

$$E = \frac{9KG}{3K + G}. \tag{12.20}$$

The Poisson's ratio is defined by

$$\frac{\nu}{E} = -\left(\frac{1}{9K} - \frac{1}{6G}\right) = \frac{3K - 2G}{18KG}, \tag{12.21}$$

so

$$\nu = \frac{3K - 2G}{6K + 2G}. \tag{12.22}$$

From $K > 0$ and $G > 0$, it can easily be derived that $E > 0$ and $-1 < \nu < 0.5$. Using E and ν leads to the commonly used formulation for Hooke's equations:

$$\begin{aligned}
\varepsilon_{xx} &= \frac{1}{E}(\quad \sigma_{xx} - \nu\,\sigma_{yy} - \nu\,\sigma_{zz}) \\
\varepsilon_{yy} &= \frac{1}{E}(-\nu\,\sigma_{xx} + \quad \sigma_{yy} - \nu\,\sigma_{zz}) \\
\varepsilon_{zz} &= \frac{1}{E}(-\nu\,\sigma_{xx} - \nu\,\sigma_{yy} + \quad \sigma_{zz}).
\end{aligned} \tag{12.23}$$

The strain in a certain direction is directly coupled to the stress in that direction via the Young's modulus. The stresses in the other directions cause a transverse strain. The reverse equations can also be derived:

$$\begin{aligned}
\sigma_{xx} &= \frac{E}{(1+\nu)(1-2\nu)}((1-\nu)\varepsilon_{xx} + \nu\,\varepsilon_{yy} + \nu\,\varepsilon_{zz}) \\
\sigma_{yy} &= \frac{E}{(1+\nu)(1-2\nu)}(\nu\,\varepsilon_{xx} + (1-\nu)\varepsilon_{yy} + \nu\,\varepsilon_{zz}) \\
\sigma_{zz} &= \frac{E}{(1+\nu)(1-2\nu)}(\nu\,\varepsilon_{xx} + \nu\,\varepsilon_{yy} + (1-\nu)\varepsilon_{zz}).
\end{aligned} \tag{12.24}$$

For the shear strains, the off-diagonal components of the matrix $\underset{\sim}{\epsilon}$, it follows from Eq. (12.16):

$$\begin{aligned}
\varepsilon_{xy} &= \varepsilon_{yx} = \frac{1}{2G}\sigma_{xy} = \frac{1}{2G}\sigma_{yx} \\
\varepsilon_{xz} &= \varepsilon_{zx} = \frac{1}{2G}\sigma_{xz} = \frac{1}{2G}\sigma_{zx} \\
\varepsilon_{zy} &= \varepsilon_{yz} = \frac{1}{2G}\sigma_{yz} = \frac{1}{2G}\sigma_{zy}.
\end{aligned} \tag{12.25}$$

It is clear that shear strains are coupled directly to shear stresses (owing to the assumption of isotropy). The inverse relations are trivial. If required, the shear modulus can be written as a function of the Young's modulus E and the Poisson's ratio ν by means of the following equation (which follows from Eqs. (12.20) and (12.21) by eliminating K):

$$G = \frac{E}{2(1+\nu)}. \tag{12.26}$$

Example 12.1 A frequently applied test to determine stiffness properties of biological materials is the 'confined compression test'. A schematic of such a test is given in Fig. 12.1. A cylindrical specimen of the material is placed in a tight-fitting die, which can be regarded as a rigid mould with a smooth inner wall (no friction at the wall).

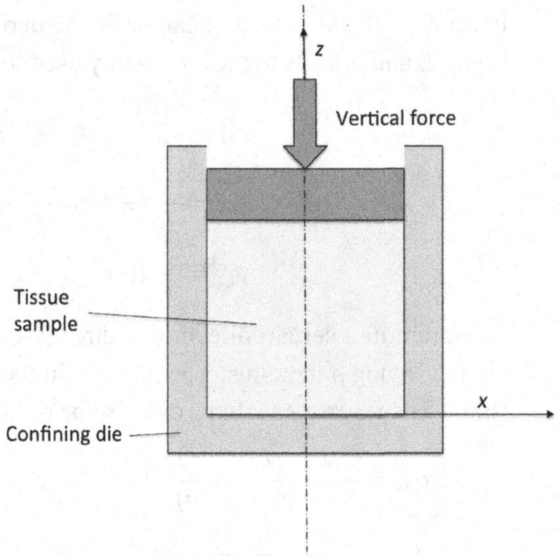

Figure 12.1

Schematic of a confined compression test.

The sample is loaded with a vertical force leading to an average stress $\sigma_{zz} = -48$ [kPa]. We consider the material as an isotropic elastic material with a Young's modulus $E = 8$ [MPa] and a Poisson's ratio $\nu = 1/4$ [-]. The deformations of the sample are small.

We want to determine the strain ε_{zz} in the z-direction. Because the displacements of the sample are suppressed in the x- and y-directions, the strains in those directions are zero, thus $\varepsilon_{xx} = \varepsilon_{yy} = 0$. In the deformed configuration, the shear strain components are also zero: $\varepsilon_{xy} = \varepsilon_{xz} = \varepsilon_{yz} = 0$. The required strain ε_{zz} can be directly determined, using the last equation of the set Eq. (12.24):

$$\sigma_{zz} = -48 \times 10^3 = \frac{8 \times 10^6}{\frac{5}{4} \times \frac{1}{2}} \left(\frac{3}{4} \varepsilon_{zz} \right),$$

yielding $\varepsilon_{zz} = -0.005$.

By substituting ε_{zz} in the first two equations of the set Eq. (12.24), we find that $\sigma_{xx} = \sigma_{yy} = -16$ [kPa].

Example 12.2 Consider a square membrane with a thickness that is much smaller than the in-plane dimensions. The membrane is parallel to the xy-plane of a Cartesian xyz-coordinate system. The material is loaded in its plane with four equal forces in the corners pointing in the diagonal directions: see Fig.12.2(a). Figure 12.2(b) depicts a deformed mesh, which is determined by means of the finite element method, a numerical method to solve partial differential equations (see Chapters 14 to 18). The greyscale plot in Fig. 12.2(c) shows the distribution of the shear strain $\varepsilon_{xy}(x, y)$

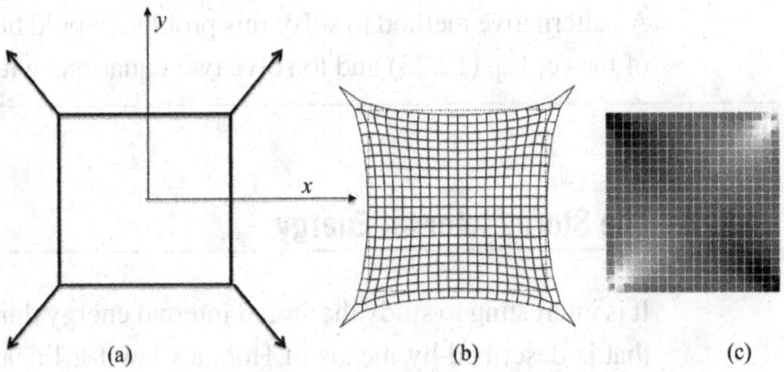

Figure 12.2

(a) Square sample with forces acting on the corners. (b) A sample deformed as a result of this load as calculated with a finite element program. (c) Greyscale plot of the shear strain distribution $\varepsilon_{x,y}(x,y)$. White is high positive strain, black is high negative strain. The grey colour around the midpoint represents a shear strain that is almost zero.

as a function of x and y. The grey shades represent strains within a certain interval. It is clear that, in the corners, the absolute strains are much larger than the absolute strains in the centre. This means that we are dealing with a heterogeneous strain field. Far away from the corners, in the centre of the sample, however, we are allowed to assume a homogeneous, biaxial strain field with shear strain $\varepsilon_{xy} = 0$ and $\varepsilon_{xx} = \varepsilon_{yy}$. Assume that in the centre the strain in the x-direction is measured: $\varepsilon_{xx} = 0.0005$. In addition, assume that the material behaviour of the plate can be described by Hooke's law, with Young's modulus $E = 8$ [MPa] and Poisson's ratio $\nu = 1/3$ [-]. Based on this information, we would like to determine the full stress matrix $\underline{\sigma}$ at the centre point. The deformation field of the sample is symmetric, implying that $\varepsilon_{yy} = \varepsilon_{xx} = 0.0005$. It is clear that the deformations are small and linear theory can be used.

The membrane can freely expand or shrink in the z-direction, meaning that σ_{zz} is zero. In the centre point, also, $\sigma_{xy} = \sigma_{xz} = \sigma_{yz} = 0$. First, the unknown strain ε_{zz} can be directly determined, using the last equation of the set Eq. (12.24):

$$\frac{1}{3} \times 0.0005 + \frac{1}{3} \times 0.0005 + \frac{2}{3}\varepsilon_{zz} = 0 \quad \rightarrow \quad \varepsilon_{zz} = -0.0005.$$

Substituting the results in the first two equations leads to $\sigma_{xx} = \sigma_{yy} = 6$ [kPa]. Thus the stress matrix $\underline{\sigma}$ at the centre point yields:

$$\underline{\sigma} = \begin{bmatrix} 6 & 0 & 0 \\ 0 & 6 & 0 \\ 0 & 0 & 0 \end{bmatrix} \quad \text{[kPa]}.$$

An alternative method to solve this problem would be to use the first two equations of the set Eq. (12.23) and to solve two equations with two unknowns.

12.3 The Stored Internal Energy

It is interesting to study the stored internal energy during deformation of a material that is described by means of Hooke's law for linearly elastic behaviour. In Section 11.4, the balance of power was derived for an infinitesimally small element with reference volume dV_0 and current volume $dV = JdV_0$. By integrating the internally stored power dP_{int} over the relevant time domain $0 \leq \tau \leq t$, the process time between the reference state and the current state, we find that the total internal energy dE_{int} that is stored by the element is

$$dE_{\text{int}}(t) = \int_{\tau=0}^{t} \text{tr}(\boldsymbol{\sigma} \cdot \boldsymbol{D}) dV d\tau. \tag{12.27}$$

For this relation, it is assumed that initially (so for $\tau = 0$) the internal mechanical energy was zero. For further elaboration, we have to account for dV being a function of time. That is why dV is expressed in terms of dV_0, resulting in

$$dE_{\text{int}}(t) = \int_{\tau=0}^{t} \text{tr}(\boldsymbol{\sigma} \cdot \boldsymbol{D}) J dV_0 d\tau = \int_{\tau=0}^{t} \text{tr}(\boldsymbol{\sigma} \cdot \boldsymbol{D}) J d\tau dV_0. \tag{12.28}$$

Subsequently, Φ_{int_0} is introduced as the internal mechanical energy per unit of reference volume, also called the internal energy density (the subscript 0 refers to the fact that the density is defined with respect to the volume in the reference state):

$$\Phi_{\text{int}_0}(t) = \int_{\tau=0}^{t} \text{tr}(\boldsymbol{\sigma} \cdot \boldsymbol{D}) J d\tau. \tag{12.29}$$

For a deformation history, given by means of specifying the deformation tensor as a function of time: $\boldsymbol{F}(\tau)$; $0 \leq \tau \leq t$ (with $\boldsymbol{F}(\tau = 0) = \boldsymbol{I}$) and for known material behaviour by specifying the constitutive equations, $\Phi_{\text{int}_0}(t)$ can be calculated. In the following, this will be performed using Hooke's law.

Assuming small deformations ($\boldsymbol{F} \approx \boldsymbol{I}$), the rate of deformation tensor $\boldsymbol{D}$ can be written as

$$\boldsymbol{D} = \frac{1}{2}\left(\dot{\boldsymbol{F}} \cdot \boldsymbol{F}^{-1} + \boldsymbol{F}^{-\text{T}} \cdot \dot{\boldsymbol{F}}^{\text{T}}\right) \approx \frac{1}{2}\left(\dot{\boldsymbol{F}} + \dot{\boldsymbol{F}}^{\text{T}}\right) = \dot{\boldsymbol{\epsilon}}. \tag{12.30}$$

The volume change factor J can be approximated according to

$$J \approx 1 + \text{tr}(\boldsymbol{\epsilon}) \approx 1. \tag{12.31}$$

Using the definition of the deviatoric form of the strain tensor $\boldsymbol{\epsilon}$, as defined in Eq. (1.49), and Eq. (12.11), Hooke's law can be written as

$$\boldsymbol{\sigma} = K\,\mathrm{tr}(\boldsymbol{\epsilon})\boldsymbol{I} + 2G\boldsymbol{\epsilon}^{\mathrm{d}}. \tag{12.32}$$

Substituting these results for $\boldsymbol{\sigma}$, $\boldsymbol{D}$ and $\boldsymbol{J}$ into the expression for Φ_{int_0} yields

$$\begin{aligned}\Phi_{\mathrm{int}_0} &= \int_{\tau=0}^{t}\left(K\,\mathrm{tr}(\boldsymbol{\epsilon})\mathrm{tr}(\dot{\boldsymbol{\epsilon}}) + 2G\mathrm{tr}(\boldsymbol{\epsilon}^{\mathrm{d}}\cdot\dot{\boldsymbol{\epsilon}})\right)d\tau \\ &= \int_{\tau=0}^{t}\left(K\,\mathrm{tr}(\boldsymbol{\epsilon})\mathrm{tr}(\dot{\boldsymbol{\epsilon}}) + 2G\mathrm{tr}(\boldsymbol{\epsilon}^{\mathrm{d}}\cdot\dot{\boldsymbol{\epsilon}}^{\mathrm{d}})\right)d\tau \\ &= \left(\frac{1}{2}K\,\mathrm{tr}^2(\boldsymbol{\epsilon}) + G\mathrm{tr}(\boldsymbol{\epsilon}^{\mathrm{d}}\cdot\boldsymbol{\epsilon}^{\mathrm{d}})\right)\Bigg|_{\tau=0}^{t}.\end{aligned} \tag{12.33}$$

Taking into account that $\boldsymbol{\epsilon}(\tau = 0) = \mathbf{0}$, it appears that the 'internal elastic energy per unit volume' Φ_{int_0} at time t is fully determined by the components of the strain tensor at that specific time t; this means that the indication t in this case is redundant and, without any problem the energy density can be written:

$$\Phi_{\mathrm{int}_0}(\boldsymbol{\epsilon}) = \frac{1}{2}K\,\mathrm{tr}^2(\boldsymbol{\epsilon}) + G\,\mathrm{tr}(\boldsymbol{\epsilon}^{\mathrm{d}}\cdot\boldsymbol{\epsilon}^{\mathrm{d}}) = \frac{1}{2}K(\varepsilon^{v})^2 + G\,\mathrm{tr}(\boldsymbol{\epsilon}^{\mathrm{d}}\cdot\boldsymbol{\epsilon}^{\mathrm{d}}). \tag{12.34}$$

It can be established that in this energy density the hydrostatic (volumetric) part and the deviatoric part are separately identifiable; just like in Hooke's law, there is a decoupling. Both parts deliver a non-negative contribution to the energy density, for every arbitrary $\boldsymbol{\epsilon}$.

Using Hooke's law, the energy density $\Phi_{\mathrm{int}_0}(\boldsymbol{\epsilon})$, according to the above given equation, can be transformed to $\Phi_{\mathrm{int}_0}(\boldsymbol{\sigma})$ resulting in

$$\Phi_{\mathrm{int}_0}(\boldsymbol{\sigma}) = \frac{1}{2}\frac{1}{K}p^2 + \frac{1}{4G}\,\mathrm{tr}(\boldsymbol{\sigma}^{\mathrm{d}}\cdot\boldsymbol{\sigma}^{\mathrm{d}}). \tag{12.35}$$

Finally, discussion above from the it can be concluded that a cyclic process in the deformation or in the stress will always be energetically neutral. This means that the 'postulated' constitutive equation (Hooke's law) does indeed give a correct description of elastic material behaviour. Every cyclic process is reversible. No energy is dissipated or released. If we compare the second term on the right-hand side of Eq. (12.35), representing the 'distortion energy density', with Eq. (8.75), defining the von Mises stress, it is clear that they are related. In other words, if we would like to define some threshold based on the maximum amount of distortional energy that can be stored in a material before it becomes damaged, the von Mises stress can be used for this purpose.

12.4 Elastic Behaviour at Large Deformations and/or Large Rotations

Large displacements and rotations play a major role in biomechanics. Most tissues of the human body (except for bone and teeth) are fairly soft materials, repeatedly undergoing very high strains in everyday life. Large strains and rotations require special care when constitutive equations are derived. High deformations usually include large rigid body motions, and the modelling must be done in such a way that these rigid body motions do not lead to extra stresses in the material. Only real deformations – shape and volume changes – may result in stress. This is related to the concept of objectivity or material frame indifference of constitutive equations. In the current section, the formulation of constitutive equations for materials subjected to large deformations will be treated, but we will start with the concept of material frame indifference.

12.4.1 Material Frame Indifference

According to the principle of material frame indifference, constitutive equations must be invariant under changes of the coordinate system to which the stresses are referred. An equivalent statement says that the state of stress in a material does not change if it is subjected to a rigid translation or rotation. Figure 12.3

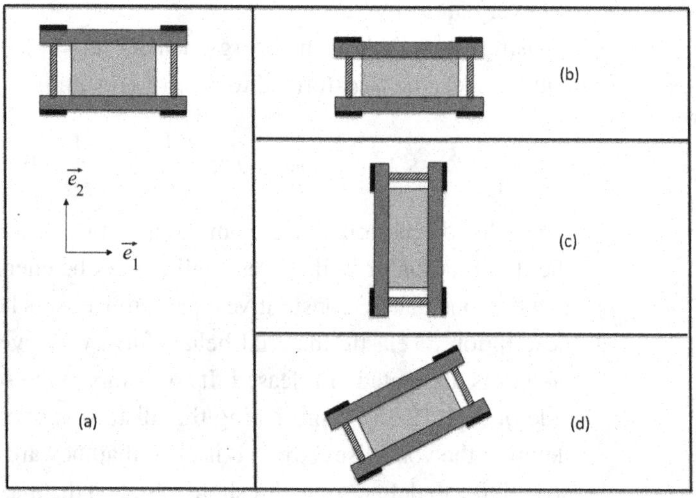

Figure 12.3

(a) Unloaded configuration; (b) loaded configuration; (c) loaded configuration after an extra rigid body rotation of 90°; (d) loaded configuration after an extra rigid body rotation through an arbitrary angle α.

illustrates the meaning of the above statements. Let us assume we have a soft tissue sample between two rigid plates with an undeformed configuration as given in Fig. 12.3(a). We define a fixed coordinate system as shown in the figure. By tightening the screws on the side, we squeeze the sample, and it becomes thinner and wider. We assume the friction between the sample and the plates to be negligible. This will lead to a uniaxial stress state in the sample that is given by:

$$\underline{\sigma} = \begin{bmatrix} 0 & 0 & 0 \\ 0 & \sigma_0 & 0 \\ 0 & 0 & 0 \end{bmatrix}, \tag{12.36}$$

with σ_0 the negative normal stress in the $\vec{e}_2$-direction. If we allow the entire loaded set-up to make a rigid body rotation at an angle of 90° around the $\vec{e}_3$-axis without changing the rest of the loading condition (Fig. 12.3(c)), the stress state can be described by:

$$\underline{\sigma} = \begin{bmatrix} \sigma_0 & 0 & 0 \\ 0 & 0 & 0 \\ 0 & 0 & 0 \end{bmatrix}. \tag{12.37}$$

So, although the stress σ_0 changes position in the stress matrix $\underline{\sigma}$, meaning that it rotates with the body, the magnitude of that stress does not change. This is essential, because the deformed state of the material also did not change. This can be generalized by looking at a small surface element inside a cross section of a deformed body, as shown in Fig. 12.4. If we know the stress state in a body, we know the Cauchy stress tensor $\boldsymbol{\sigma}$ in each point. Any infinitesimal surface passing through a point can be identified with the normal vector $\vec{n}$ perpendicular to that surface. When $\boldsymbol{\sigma}$ and $\vec{n}$ are known, we can determine the stress vector $\vec{s}$ on the surface, by means of:

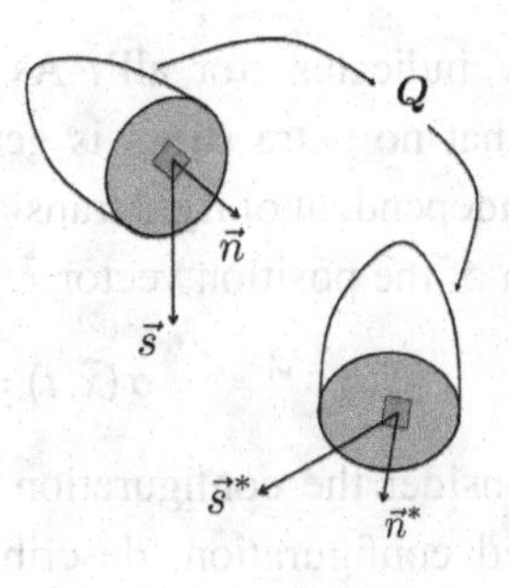

Figure 12.4

Stress vector and normal after a rigid body rotation.

$$\boldsymbol{\sigma} \cdot \vec{n} = \vec{s}. \tag{12.38}$$

After an extra rigid body motion, characterized by an orthogonal rotation tensor $\boldsymbol{Q}$ onto the motion of the entire system (including the external load), it follows that:

$$\boldsymbol{\sigma}^* \cdot \vec{n}^* = \vec{s}^*. \tag{12.39}$$

Clearly, the unit outward normal and the stress vector in the rotated configuration, $\vec{n}^*$ and $\vec{s}^*$ respectively, are related to the unrotated normal $\vec{n}$ and load vector $\vec{s}$ by:

$$\vec{n}^* = \boldsymbol{Q} \cdot \vec{n}, \qquad \vec{s}^* = \boldsymbol{Q} \cdot \vec{s}. \tag{12.40}$$

Substitution of this result into Eq. (12.39) yields:

$$\boldsymbol{\sigma}^* \cdot \boldsymbol{Q} \cdot \vec{n} = \boldsymbol{Q}.\vec{s}. \tag{12.41}$$

and with the definition of $\vec{s}$:

$$\boldsymbol{\sigma}^* \cdot \boldsymbol{Q} \cdot \vec{n} = \boldsymbol{Q} \cdot \boldsymbol{\sigma} \cdot \vec{n}. \tag{12.42}$$

This must hold for *any* $\vec{n}$, hence:

$$\boldsymbol{\sigma}^* \cdot \boldsymbol{Q} = \boldsymbol{Q} \cdot \boldsymbol{\sigma}, \tag{12.43}$$

or, equivalently, because $\boldsymbol{Q}$ is an orthogonal rotation tensor (as $\boldsymbol{Q}^{-1} = \boldsymbol{Q}^{\mathrm{T}}$)

$$\boldsymbol{\sigma}^* = \boldsymbol{Q} \cdot \boldsymbol{\sigma} \cdot \boldsymbol{Q}^{\mathrm{T}}. \tag{12.44}$$

A tensor that after a rigid body rotation transforms as $\boldsymbol{\sigma}$ in Eq. (12.44) is called an **objective** tensor.

When we try to solve a mechanical problem, we require a constitutive relation defining the stress as a function of the deformation history, formally written as:

$$\boldsymbol{\sigma}(\vec{x}, t) = \mathcal{N}\{\boldsymbol{F}(\vec{x}, \tau), \vec{x} \ |\forall\ \tau \leq t\}, \tag{12.45}$$

where $\forall$ indicates 'for all'. As mentioned above, material frame indifference means that no extra stress is generated after rigid body translation or rotation. Being independent of rigid translation means that the stress $\boldsymbol{\sigma}$ cannot be an explicit function of the position vector $\vec{x}$. So the constitutive equation reduces to:

$$\boldsymbol{\sigma}(\vec{x}, t) = \mathcal{N}\{\boldsymbol{F}(\vec{x}, \tau) \ |\forall\ \tau \leq t\}. \tag{12.46}$$

Now consider the configuration undergoing an extra rigid body rotation of the deformed configuration, described by the rotation tensor $\boldsymbol{Q}(t)$. Owing to this rigid rotation, the stress tensor $\boldsymbol{\sigma}$ and the deformation tensor $\boldsymbol{F}$ at the time t are converted to:

$$\boldsymbol{\sigma} \overset{\boldsymbol{Q}(t)}{\rightarrow} \boldsymbol{\sigma}^* \tag{12.47}$$

$$\boldsymbol{F} \overset{\boldsymbol{Q}(t)}{\rightarrow} \boldsymbol{F}^*. \tag{12.48}$$

Then the condition that the constitutive relation must be invariant for rigid rotations can be formulated as:

$$\boldsymbol{\sigma}^*(t) = \boldsymbol{N}\{\boldsymbol{F}^*(\tau) \;|\forall\, \tau \leq t\}. \tag{12.49}$$

This means that it is required that the stresses $\boldsymbol{\sigma}^*$ in the rotated body can be calculated from the deformation gradient tensor $\boldsymbol{F}^*$ in the rotated body according to the same constitutive relation $\boldsymbol{N}$. We have already shown that $\boldsymbol{\sigma}$ behaves like an **objective** tensor; see Eq. (12.44).

Let us consider how the deformation tensor $\boldsymbol{F}$ transforms after a rigid body rotation:

$$\begin{aligned} \boldsymbol{F}^* &= (\vec{\nabla}_0 \vec{x}^*)^T \\ &= (\vec{\nabla}_0 \boldsymbol{Q} \cdot \vec{x})^T \\ &= \boldsymbol{Q} \cdot (\vec{\nabla}_0 \vec{x})^T \\ &= \boldsymbol{Q} \cdot \boldsymbol{F}. \end{aligned} \tag{12.50}$$

Clearly $\boldsymbol{F}$ is not an objective tensor, which can be explained as follows. The deformation tensor $\boldsymbol{F}$ maps an infinitesimally small line segment $d\vec{x}_0$ in the undeformed configuration on a line segment $d\vec{x}$ in the deformed configuration. An **extra** rigid body rotation only affects $d\vec{x}$ and not $d\vec{x}_0$, leading to only one appearance of $\boldsymbol{Q}$ in Eq. (12.50). This means that we have to be careful when we define a relation between $\boldsymbol{\sigma}$ and $\boldsymbol{F}$ because it is not trivial that this will lead to a constitutive equation that is material frame indifferent.

Example 12.3 Suppose we were to propose a constitutive law as follows:

$$\boldsymbol{\sigma} = \alpha(\boldsymbol{F} - \boldsymbol{I}) \tag{12.51}$$

with α a constant material property. After rigid body rotation, this material law would become:

$$\boldsymbol{\sigma}^* = \alpha(\boldsymbol{F}^* - \boldsymbol{I})$$

This leads to:

$$\boldsymbol{\sigma}^* = \alpha(\boldsymbol{Q} \cdot \boldsymbol{F} - \boldsymbol{I}) \neq \alpha \boldsymbol{Q} \cdot (\boldsymbol{F} - \boldsymbol{I}) \cdot \boldsymbol{Q}^{\mathrm{T}} = \boldsymbol{Q} \cdot \boldsymbol{\sigma} \cdot \boldsymbol{Q}^{\mathrm{T}}$$

In other words, as Eq. (12.44) is violated, the material law is not material-space-frame indifferent.

Example 12.4 Let us propose a constitutive law as follows:

$$\boldsymbol{\sigma} = \alpha(\boldsymbol{F} \cdot \boldsymbol{F}^{\mathrm{T}} - \boldsymbol{I}). \tag{12.52}$$

Then:

$$\boldsymbol{\sigma}^* = \alpha(\boldsymbol{F}^* \cdot \boldsymbol{F}^{*\mathrm{T}} - \boldsymbol{I}),$$

or:

$$\boldsymbol{\sigma}^* = \alpha(\boldsymbol{Q} \cdot \boldsymbol{F} \cdot \boldsymbol{F}^{\mathrm{T}} \cdot \boldsymbol{Q}^{\mathrm{T}} - \boldsymbol{I}) = \boldsymbol{Q} \cdot \boldsymbol{\sigma} \cdot \boldsymbol{Q}^{\mathrm{T}}.$$

And this eventually will lead to:

$$\boldsymbol{\sigma} = \alpha(\boldsymbol{F} \cdot \boldsymbol{F}^{\mathrm{T}} - \boldsymbol{I}).$$

which is exactly the same as Eq. (12.52). The above shows that $\boldsymbol{B} = \boldsymbol{F} \cdot \boldsymbol{F}^{\mathrm{T}}$, which we know as the left Cauchy–Green deformation tensor, also behaves as an objective tensor.

The result from the previous example can be generalized. Because $\boldsymbol{\sigma}$ behaves as an objective tensor, any material law that uses objective tensors to describe the deformation history will satisfy the rule of material frame indifference. However, this does not guarantee that the constitutive law actually describes the correct material behaviour, because the final form of the equation and the value of the material parameters have to be determined by experiment.

It is worthwhile to investigate how some of the strain tensors defined in continuum mechanics behave after rigid body rotation. If we assume that both the deformation tensor $\boldsymbol{F} = \boldsymbol{F}(t)$ and the rotation tensor $\boldsymbol{Q} = \boldsymbol{Q}(t)$ are functions of time, we can investigate the behaviour of the velocity gradient tensor $\boldsymbol{L}$:

$$\begin{aligned} \boldsymbol{L}^* &= \dot{\boldsymbol{F}}^* \cdot \boldsymbol{F}^{*-1} \\ &= \left(\dot{\boldsymbol{Q}} \cdot \boldsymbol{F} + \boldsymbol{Q} \cdot \dot{\boldsymbol{F}}\right) \cdot \left(\boldsymbol{F}^{-1} \cdot \boldsymbol{Q}^T\right) \\ &= \dot{\boldsymbol{Q}} \cdot \boldsymbol{Q}^T + \boldsymbol{Q} \cdot \boldsymbol{L} \cdot \boldsymbol{Q}^T, \end{aligned} \tag{12.53}$$

where use is made of the relation $\boldsymbol{L} = \dot{\boldsymbol{F}} \cdot \boldsymbol{F}^{-1}$. Clearly $\boldsymbol{L}$ is not an objective tensor.

The spin tensor $\boldsymbol{\Omega}^* = \frac{1}{2}\left(\boldsymbol{L}^* - \boldsymbol{L}^{*\mathrm{T}}\right)$ can be written as:

$$\begin{aligned}\boldsymbol{\Omega}^* &= \frac{1}{2}\left(\dot{\boldsymbol{F}}^* \cdot \boldsymbol{F}^{*-1} - \boldsymbol{F}^{*-\mathrm{T}} \cdot \dot{\boldsymbol{F}}^{*\mathrm{T}}\right) \\ &= \frac{1}{2}\Big(\dot{\boldsymbol{Q}} \cdot \boldsymbol{F} \cdot \boldsymbol{F}^{-1} \cdot \boldsymbol{Q}^{\mathrm{T}} + \boldsymbol{Q} \cdot \dot{\boldsymbol{F}} \cdot \boldsymbol{F}^{-1} \cdot \boldsymbol{Q}^{\mathrm{T}} \\ &\qquad - \boldsymbol{Q} \cdot \boldsymbol{F}^{-\mathrm{T}} \cdot \dot{\boldsymbol{F}}^{\mathrm{T}} \cdot \boldsymbol{Q}^{\mathrm{T}} - \boldsymbol{Q} \cdot \boldsymbol{F}^{-\mathrm{T}} \cdot \boldsymbol{F}^{\mathrm{T}} \cdot \dot{\boldsymbol{Q}}^{\mathrm{T}}\Big) \\ &= \boldsymbol{Q} \cdot \boldsymbol{\Omega} \cdot \boldsymbol{Q}^{\mathrm{T}} + \dot{\boldsymbol{Q}} \cdot \boldsymbol{Q}^{\mathrm{T}}.\end{aligned} \tag{12.54}$$

This shows that $\boldsymbol{\Omega}$ is also not an objective tensor. For the rate of deformation tensor $\boldsymbol{D}$ we find:

$$\begin{aligned}\boldsymbol{D}^* &= \frac{1}{2}(\boldsymbol{L}^* + \boldsymbol{L}^{*\mathrm{T}}) \\ &= \underbrace{\frac{1}{2}\left[\dot{\boldsymbol{Q}} \cdot \boldsymbol{Q}^{\mathrm{T}} + \boldsymbol{Q} \cdot \dot{\boldsymbol{Q}}^{\mathrm{T}}\right]}_{=\boldsymbol{0}\ (\boldsymbol{Q}\cdot\boldsymbol{Q}^{\mathrm{T}}=\boldsymbol{I})} + \frac{1}{2}\left[\boldsymbol{Q} \cdot \boldsymbol{L} \cdot \boldsymbol{Q}^{\mathrm{T}} + \boldsymbol{Q} \cdot \boldsymbol{L}^{\mathrm{T}} \cdot \boldsymbol{Q}^{\mathrm{T}}\right] \\ &= \boldsymbol{Q} \cdot \boldsymbol{D} \cdot \boldsymbol{Q}^{\mathrm{T}}.\end{aligned} \tag{12.55}$$

Clearly, $\boldsymbol{D}$ is an objective tensor. This tensor is extensively used for constitutive equations for fluids (see Section 12.5).

Invariant tensors also exist, meaning that they do not change at all after a rigid body rotation. Examples are the right Cauchy–Green strain tensor $\boldsymbol{C}$ and the Green–Lagrange strain tensor.
For the right Cauchy–Green tensor, we find:

$$\boldsymbol{C}^* = (\boldsymbol{F}^*)^{\mathrm{T}} \cdot \boldsymbol{F}^* = \boldsymbol{F}^{\mathrm{T}} \cdot \boldsymbol{Q}^{\mathrm{T}} \cdot \boldsymbol{Q} \cdot \boldsymbol{F} = \boldsymbol{C}. \tag{12.56}$$

The Green–Lagrange strain tensor yields:

$$\boldsymbol{E}^* = \frac{1}{2}(\boldsymbol{C}^* - \boldsymbol{I}) = \frac{1}{2}(\boldsymbol{C} - \boldsymbol{I}) = \boldsymbol{E}. \tag{12.57}$$

It should be clear from the arguments above that it is also possible to define material frame indifferent constitutive laws when purely invariant stress and deformation definitions are used. In that case, we need a stress definition that is also invariant. Such a stress definition exists and is called the second Piola–Kirchhoff stress $\boldsymbol{S}$:

$$\boldsymbol{S} = \boldsymbol{F}^{-1} \cdot J\boldsymbol{\sigma} \cdot \boldsymbol{F}^{-\mathrm{T}} \tag{12.58}$$

with $J = \det(\boldsymbol{F})$. Using:

$$(\boldsymbol{F}^*)^{-1} = (\boldsymbol{Q} \cdot \boldsymbol{F})^{-1} = \boldsymbol{F}^{-1} \cdot \boldsymbol{Q}^{-1} \tag{12.59}$$

we can write:

$$S^* = F^{*-1} \cdot J\sigma^* \cdot F^{*-T} = F^{-1} \cdot Q^{-1} \cdot JQ \cdot \sigma \cdot Q^T \cdot Q^{-T} \cdot F^{-T} = S \quad (12.60)$$

which is clearly an invariant tensor.
The second Piola–Kirchhoff tensor has been extensively used in the biomechanics literature from the early seventies until the present day [9]. The advantage is that by using only invariant formulations the constitutive equations have a relatively simple form, and discretization using finite time steps does not influence the material-frame indifference of the equation. A disadvantage is that it is hard to give a physical interpretation.

12.4.2 Strain Energy Function

When a material is purely elastic, all energy that is supplied to it during loading, and which is stored as elastic energy, can be regained after unloading. So, during deformation, no energy is lost in the form of heat. It can be shown that the stress in such a material can be derived from a scalar function representing the specific elastic energy density Φ (elastic energy per unit undeformed volume) by means of:

$$\sigma = \frac{2}{J} F \cdot \frac{\partial \Phi}{\partial C} \cdot F^T. \quad (12.61)$$

Materials for which a function Φ can be proposed are called **hyperelastic** materials.

In the case of isotropic materials, the strain energy Φ depends only on the invariants I_1, I_2 and I_3 of the tensor C.

$$\Phi = \Phi(I_1, I_2, I_3). \quad (12.62)$$

These invariants of C are the coefficients of the characteristic equation (used to determine the eigenvalues of the tensor C; see Section 1.5). The starting point for the procedure is:

$$C \cdot \vec{n} = \lambda \vec{n}. \quad (12.63)$$

A non-trivial solution of this equation only exists if:

$$\det(C - \lambda I) = 0. \quad (12.64)$$

After some elaboration, this leads to the **characteristic equation**:

$$\lambda^3 - I_1\lambda^2 + I_2\lambda - I_3 = 0 \quad (12.65)$$

with:

$$I_1 = \text{tr}(\boldsymbol{C}) \tag{12.66}$$
$$I_2 = \frac{1}{2}\left[\text{tr}^2(\boldsymbol{C}) - \text{tr}(\boldsymbol{C}^2)\right]$$
$$I_3 = \det(\boldsymbol{C}) = \frac{1}{6}\left[\text{tr}^3(\boldsymbol{C}) - 3\text{tr}(\boldsymbol{C})\text{tr}(\boldsymbol{C}^2) + 2\text{tr}(\boldsymbol{C}^3)\right].$$

Note: Deriving these equations is straightforward, but quite a lot of work. The easiest way is to write the equation in component form, using the components of matrix $\underline{C}$ with respect to a Cartesian basis, and elaborate Eq. (12.64). It is clear that I_i are scalar functions of the tensor $\boldsymbol{C}$ and thus invariant for rigid body rotations.

To calculate the stresses, when the relationship for the specific elastic energy Φ is known, it is necessary to determine the derivatives of the invariants:

$$\frac{\partial I_1}{\partial \boldsymbol{C}} = \boldsymbol{I} \tag{12.67}$$
$$\frac{\partial I_2}{\partial \boldsymbol{C}} = I_1\boldsymbol{I} - \boldsymbol{C}$$
$$\frac{\partial I_3}{\partial \boldsymbol{C}} = I_2\boldsymbol{I} - I_1\boldsymbol{C} + \boldsymbol{C}^2.$$

Then the derivative of the stored energy with respect to $\boldsymbol{C}$ can be written as:

$$\frac{\partial \Phi}{\partial \boldsymbol{C}} = \frac{\partial \Phi}{\partial I_1}\frac{\partial I_1}{\partial \boldsymbol{C}} + \frac{\partial \Phi}{\partial I_2}\frac{\partial I_2}{\partial \boldsymbol{C}} + \frac{\partial \Phi}{\partial I_3}\frac{\partial I_3}{\partial \boldsymbol{C}}. \tag{12.68}$$

By substituting Eq. (12.67) into Eq. (12.68), we find:

$$\frac{\partial \Phi}{\partial \boldsymbol{C}} = \left(\frac{\partial \Phi}{\partial I_1} + I_1\frac{\partial \Phi}{\partial I_2} + I_2\frac{\partial \Phi}{\partial I_3}\right)\boldsymbol{I} - \left(\frac{\partial \Phi}{\partial I_2} + I_1\frac{\partial \Phi}{\partial I_3}\right)\boldsymbol{C} + \left(\frac{\partial \Phi}{\partial I_3}\right)\boldsymbol{C}^2. \tag{12.69}$$

As is clear from the previous section, it is useful to couple the Cauchy stress tensor $\boldsymbol{\sigma}$ to the left Cauchy–Green strain tensor $\boldsymbol{B} = \boldsymbol{F} \cdot \boldsymbol{F}^{\mathrm{T}}$ instead of $\boldsymbol{C}$, because $\boldsymbol{B}$ transforms in the same objective way after rotation as $\boldsymbol{\sigma}$ does. By substituting Eq. (12.69) into Eq. (12.61) and rearranging, we find,

$$\boldsymbol{\sigma} = \frac{2}{\sqrt{I_3}}\left[\left(\frac{\partial \Phi}{\partial I_1} + I_1\frac{\partial \Phi}{\partial I_2} + I_2\frac{\partial \Phi}{\partial I_3}\right)\boldsymbol{B} - \left(\frac{\partial \Phi}{\partial I_2} + I_1\frac{\partial \Phi}{\partial I_3}\right)\boldsymbol{B}^2 + \left(\frac{\partial \Phi}{\partial I_3}\right)\boldsymbol{B}^3\right] \tag{12.70}$$

with $\sqrt{I_3} = \sqrt{\det(\boldsymbol{F} \cdot \boldsymbol{F}^T)} = \sqrt{\det(\boldsymbol{F})\det(\boldsymbol{F}^{\mathrm{T}})} = \det(\boldsymbol{F}) = J$.

This expression can be used to derive the stress–strain relationship when the function for the specific energy is known. In the next subsections, some of the commonly used isotropic models for large deformations will be discussed.

12.4.3 The Incompressible Neo-Hookean Model

The Neo-Hookean model provides a reasonable description of the behaviour of rubber-like materials at finite strains and is also quite often used for biological materials. Rubbers behave nearly incompressibly, which is also valid for many biological materials containing 70–80% water. This means that the volume cannot change and there can be a stress in the material without deformation. To introduce the model, we split the Cauchy stress tensor $\boldsymbol{\sigma}$ into an extra stress part $\boldsymbol{\tau}$ and a hydrostatic stress tensor $-p\boldsymbol{I}$:

$$\boldsymbol{\sigma} = -p\boldsymbol{I} + \boldsymbol{\tau}. \tag{12.71}$$

This pressure p does not follow directly from the elastic energy and has to be determined via an analysis of the force equilibrium on an object. The extra stress $\boldsymbol{\tau}$ can be derived from a specific elastic energy function, given by:

$$\Phi = \frac{1}{2}G(I_1 - 3), \tag{12.72}$$

which is a considerable simplification of Eq. (12.62), leading, with Eq. (12.70), to:

$$\boldsymbol{\tau} = G\boldsymbol{B}. \tag{12.73}$$

The shear modulus G can be derived from the kinetic theory of rubber elasticity:

$$G = \frac{2\Gamma N k \theta}{\rho_0} \tag{12.74}$$

where Γ is a number of order one, k is Boltzmann's constant, θ is the absolute temperature, N is the density of cross-links per unit volume [22], and ρ_0 is the mass per unit volume in the undeformed configuration. Because the pressure p is defined up to an arbitrary constant, there is no objection to using the following equation for the extra stress:

$$\boldsymbol{\tau} = G(\boldsymbol{B} - \boldsymbol{I}). \tag{12.75}$$

The strain measure $\boldsymbol{B} - \boldsymbol{I}$ has been chosen such that, if no deformation is applied ($\boldsymbol{F} = \boldsymbol{I}$), the strain $\boldsymbol{B} - \boldsymbol{I} = \boldsymbol{0}$.

Note that this material law represents a linear relation between the strain tensor $\boldsymbol{B}$ and the Cauchy stress $\boldsymbol{\sigma}$. That is why the model is often referred to as a **physically** linear model. However, the model does account for large deformations, and by using $\boldsymbol{B} = \boldsymbol{F} \cdot \boldsymbol{F}^{\mathrm{T}}$ the stresses will be a non-linear function of the deformations. That makes the model a **geometrically** non-linear model.

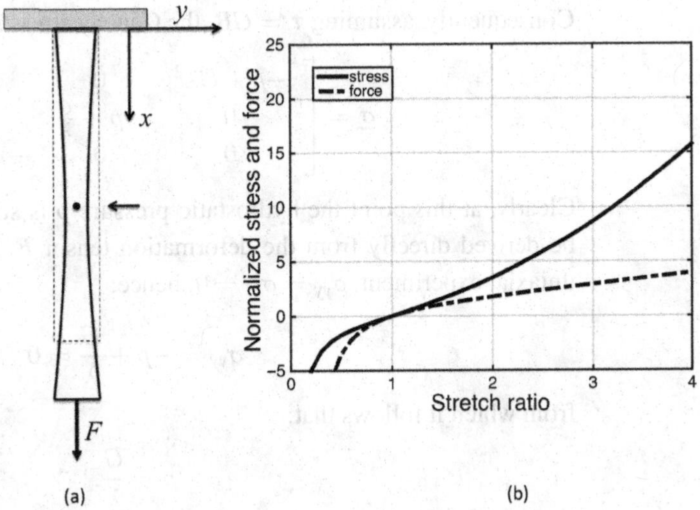

Figure 12.5

(a) A uniaxial stretch test. The dotted line represents the undeformed configuration. (b) The non-dimensional Cauchy stress $\sigma = \sigma_{xx}/G$ and non-dimensional force $F = \sigma_{xx}\lambda$ as functions of the stretch λ for the neo-Hookean model during uniaxial elongation

Example 12.5 When a long slender structure is stretched as shown in Fig.12.5(a), we can assume that in the middle of the structure (arrow), far away from the clamps, a uniaxial stress state exists. In this case, the normal stress σ_{xx} is larger than zero, and all other stress components vanish. The material is stretched in the x-direction and will shrink equally in the y- and z- directions, because of symmetry. When the material is incompressible the volume is not allowed to change, implying $\det(\underline{F}) = 1$. For this reason, a uniaxial elongation of an incompressible solid can be described by the following deformation matrix:

$$\underline{F} = \begin{bmatrix} \lambda & 0 & 0 \\ 0 & \frac{1}{\sqrt{\lambda}} & 0 \\ 0 & 0 & \frac{1}{\sqrt{\lambda}} \end{bmatrix},$$

with λ the axial stretch ratio.

The left Cauchy–Green matrix $\underline{B}$ satisfies:

$$\underline{B} = \begin{bmatrix} \lambda^2 & 0 & 0 \\ 0 & \frac{1}{\lambda} & 0 \\ 0 & 0 & \frac{1}{\lambda} \end{bmatrix}.$$

Consequently, assuming $\boldsymbol{\tau} = G\boldsymbol{B}$, the Cauchy stress matrix is given by:

$$\underline{\sigma} = \begin{bmatrix} -p + G\lambda^2 & 0 & 0 \\ 0 & -p + \frac{G}{\lambda} & 0 \\ 0 & 0 & -p + \frac{G}{\lambda} \end{bmatrix}.$$

Clearly, at this point the hydrostatic pressure p is still undetermined and cannot be derived directly from the deformation tensor $\boldsymbol{F}$. However, by definition in a uniaxial experiment, $\sigma_{yy} = \sigma_{zz} = 0$, hence:

$$\sigma_{yy} = -p + \frac{G}{\lambda} = 0$$

from which it follows that:

$$p = \frac{G}{\lambda}.$$

Therefore the Cauchy stress in the principal stretch direction is given by:

$$\sigma_{xx} = G\left(\lambda^2 - \frac{1}{\lambda}\right).$$

Figure 12.5(b) shows σ_{xx}/G as a function of the stretch λ. It is clear that the force against the stretch (basically the physical property that is measured during a uniaxial stress test) shows an almost linear behaviour in tension. However, the Cauchy stress as a function of the stretch is clearly non-linear.

Example 12.6 Assume we have a Cartesian basis $\{\vec{e}_x, \vec{e}_y, \vec{e}_z\}$. Consider a case of simple shear: see Fig. 12.6. Then the deformation tensor $\boldsymbol{F}$ can be represented by the following matrix $\underline{F}$:

$$\underline{F} = \begin{bmatrix} 1 & \gamma & 0 \\ 0 & 1 & 0 \\ 0 & 0 & 1 \end{bmatrix}$$

where γ is the applied shear.

For this case, the left Cauchy–Green matrix $\underline{B}$ equals:

$$\underline{B} = \begin{bmatrix} 1 + \gamma^2 & \gamma & 0 \\ \gamma & 1 & 0 \\ 0 & 0 & 1 \end{bmatrix}.$$

Hence, the Cauchy stresses satisfy:

$$\underline{\sigma} = \begin{bmatrix} -p + G\gamma^2 & G\gamma & 0 \\ G\gamma & -p & 0 \\ 0 & 0 & -p \end{bmatrix}.$$

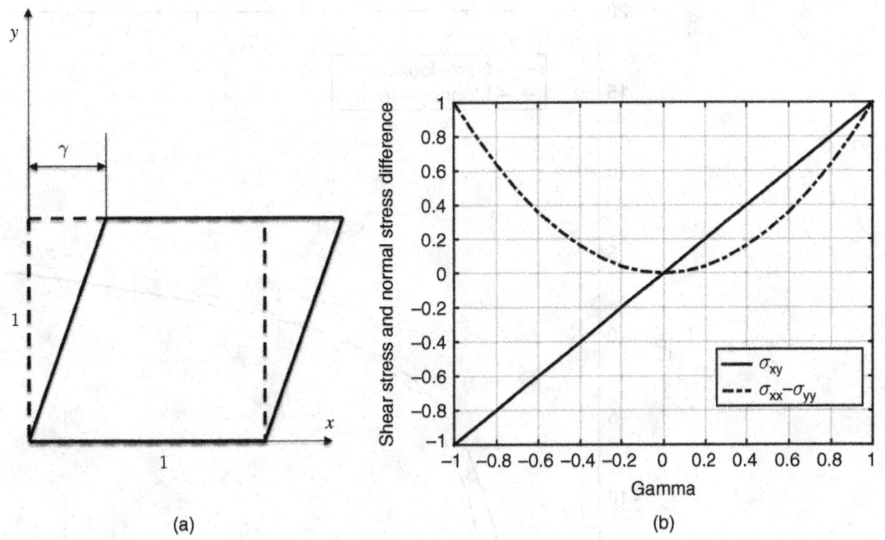

Figure 12.6

(a) Schematic of a simple shear test. (b) Normalized shear stress σ_{xy}/G and first normal stress difference $(\sigma_{xx} - \sigma_{yy})/G$ as a function of the shear strain γ.

It can be observed that the shear stress appears to be linear with respect to the applied shear. The first normal stress difference, $N_1 = \sigma_{xx} - \sigma_{yy}$ on the other hand, depends quadratically on the applied shear. This is depicted graphically in Fig. 12.6.

12.4.4 The Incompressible Mooney–Rivlin Model

The neo-Hookean model gives a reasonable description of the behaviour of natural rubbers up to strains of approximately 10%. For larger deformations the Mooney–Rivlin model is often applied, usually formulated as:

$$\Phi = C_{10}(I_1 - 3) + C_{01}(I_2 - 3). \tag{12.76}$$

Typically, biological materials show a very strong stiffening effect at high strains. The model in Eq. (12.76) describes only a minor stiffening effect. See for example Fig. 12.7. A uniaxial elongation test is simulated with a neo-Hookean and a Mooney–Rivlin model. The figure gives the dimensionless force that is necessary to extend the material as a function of the extension ratio $\lambda = l/l_0$.

That is why for biological materials often higher-order models are used, which are called extended Mooney models, for example:

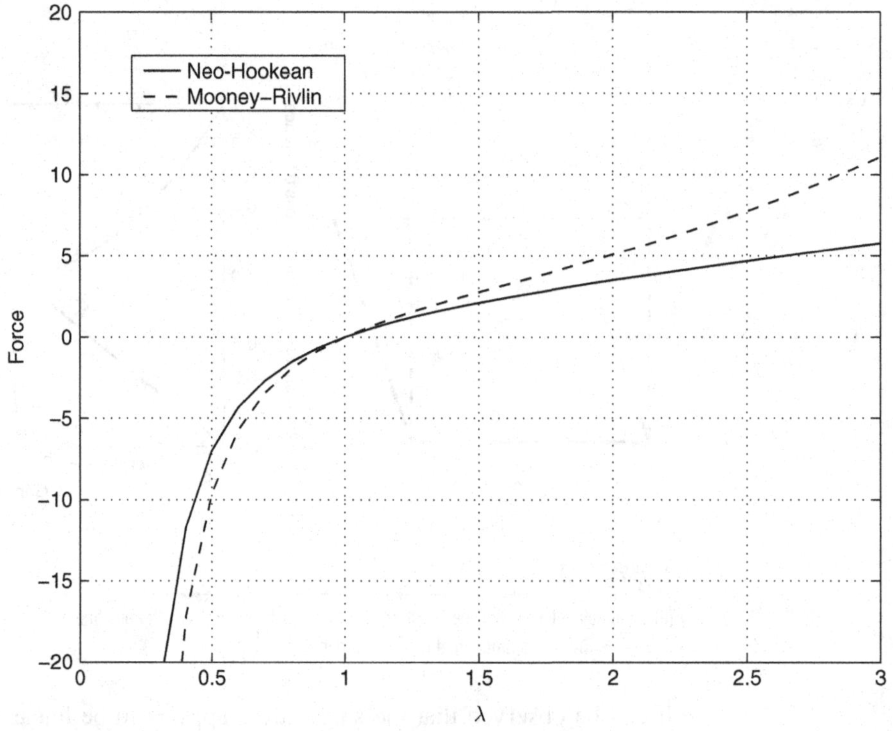

Figure 12.7

Force against extension ratio in a uniaxial test.

$$\Phi = C_{10}(I_1 - 3) + C_{01}(I_2 - 3) + C_{11}(I_1 - 3)(I_2 - 3) + C_{20}(I_1 - 3)^2 + C_{30}(I_1 - 3)^3 \quad (12.77)$$

but also models based on polynomial or exponential functions, like the model from Delfino *et al.* [6]:

$$\Phi = \frac{a}{b}\left[\exp\left(\frac{b}{2}(I_1 - 3)\right) - 1\right]. \quad (12.78)$$

Which model is best for the description of a specific material has to follow from experiments.

12.4.5 Compressible Neo-Hookean Elastic Solid

Most biological materials contain a large amount of water and behave nearly incompressibly. That is why incompressible models are often used in biomechanics. However, it is worthwhile also to consider compressible models, because many biological materials are bi-phasic mixtures of a solid and a fluid (for example cartilage, skin, intervertebral disk tissue). In this case the porous solid material (including the pores) is compressible. In the current section a few compressible models will be discussed.

Model 1 The specific elastic energy for model 1 can be described as:

$$\Phi = \frac{1}{4}\kappa\left[I_3 - 1 - \ln(I_3)\right] + \frac{1}{2}G\left[I_1 I_3^{-1/3} - 3\right] \tag{12.79}$$

with κ the compression modulus and G the shear modulus. It can easily be verified that in the undeformed state the elastic energy vanishes.
By definition $J = \det(\boldsymbol{F})$ equals the volume ratio. Hence a volume-preserving deformation tensor $\tilde{\boldsymbol{F}}$ may be defined as **isochoric deformation tensor**:

$$\tilde{\boldsymbol{F}} = J^{-1/3}\boldsymbol{F}. \tag{12.80}$$

This means that:

$$\det(\tilde{\boldsymbol{F}}) = \det(J^{-1/3}\boldsymbol{I})\det(\boldsymbol{F}) = 1. \tag{12.81}$$

And with the use of $I_3^{1/2} = J = \det(\boldsymbol{F})$, and $\tilde{\boldsymbol{B}} = \tilde{\boldsymbol{F}}\cdot\tilde{\boldsymbol{F}}^{\mathrm{T}}$, Eq. (12.79) can be written in the simple format:

$$\boldsymbol{\sigma} = \kappa\frac{1}{2}\left(J - \frac{1}{J}\right)\boldsymbol{I} + \frac{G}{J}\tilde{\boldsymbol{B}}^{\mathrm{d}}. \tag{12.82}$$

The result satisfies the requirement $\boldsymbol{\sigma} = \boldsymbol{0}$ for $\boldsymbol{B} = \boldsymbol{I}$. Furthermore, the volumetric part and the deviatoric part of the stress tensor are clearly separated.

Model 2 The specific energy for model 2 can be given as:

$$\Phi = \frac{1}{2}\kappa\left[(I_3^{1/2} - 1)^2\right] + \frac{1}{2}G\left[I_1 - 3I_3^{1/3}\right]. \tag{12.83}$$

This leads to the following stress tensor:

$$\boldsymbol{\sigma} = \kappa(J-1)\boldsymbol{I} + \frac{G}{J}\left(\boldsymbol{B} - J^{2/3}\boldsymbol{I}\right). \tag{12.84}$$

This expression also satisfies $\boldsymbol{\sigma} = \boldsymbol{0}$ for $\boldsymbol{B} = \boldsymbol{I}$. In the case of a pure triaxial volumetric deformation (with $\boldsymbol{B} = J^{2/3}\boldsymbol{I}$), the distortional part vanishes.

Although these models are elegant and thermodynamically admissible, it does not mean that they are suitable for every material under consideration. This can be made clear if we investigate the model from Eq. (12.84) for a uniaxial test. Let us assume that we apply a compressive load in the $\vec{e}_1$-direction, leading to a compression ratio of $\lambda_1 < 1$. The material will also deform in the $\vec{e}_2$-direction and

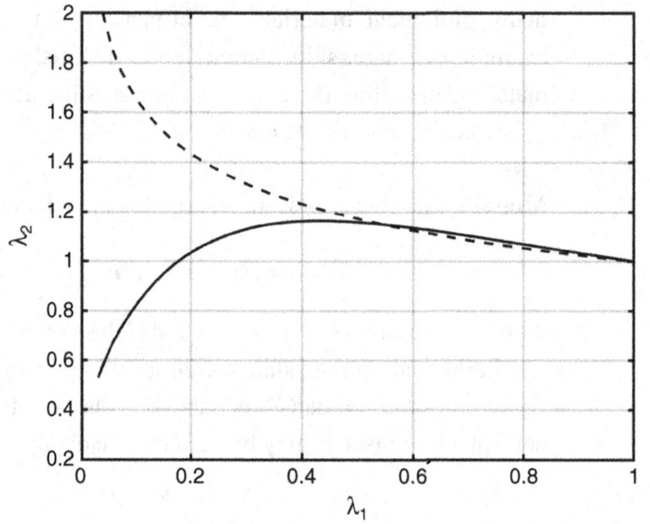

Figure 12.8

The extension ratio λ_2 as a function of λ_1 for two different models. Model 1 (solid line) uses $\kappa(J-1)$ for the volume change. Model 2 (dashed line) uses $\kappa\ \ln(J)/J$. Calculations were done with $\kappa = 5/3$ and $G = 1/2.8$, which corresponds to Young's modulus $E = 1$ and Poisson's ratio $\nu = 0.4$.

the extension ratio will be λ_2. We assume a plane strain situation, so the extension ratio in $\vec{e}_3$-direction is equal to one. Because the $\vec{e}_2$-direction is stress-free ($\sigma_{22} = 0$), it is possible to determine the ratio between λ_1 and λ_2. This relationship is given in Fig. 12.8, solid line. When λ_1 decreases, at first λ_2 will increase as can be expected. However, when λ_1 becomes smaller than 0.5, λ_2 starts to decrease and also becomes smaller than one. You could say that the material implodes. In numerical simulations, this will lead to convergence problems and instabilities. This problem can be solved by changing the volumetric part of the constitutive equation and using:

$$\boldsymbol{\sigma} = \kappa \frac{\ln(J)}{J} \boldsymbol{I} + \frac{G}{J} \left(\boldsymbol{B} - J^{2/3} \boldsymbol{I} \right). \tag{12.85}$$

Now λ_2 will increase further, leading to stable solutions for large deformations. Up to an extension ratio of 0.8, both models perform equally well.

From the example above, it will be clear that being thermodynamically correct does not mean that a constitutive equation is suitable to describe the mechanical behaviour of materials. It is a necessary requirement, but not sufficient. Whether or not a constitutive equation is suitable to describe the mechanical material behaviour of a specific material, typically has to follow from experiments.

So far, the constitutive equations, apart from the last one, have been derived from a strain energy function, but often it is tempting and easier to derive constitutive equations by just proposing a relation between the Cauchy stress $\boldsymbol{\sigma}$ and some objective strain tensor. This, however, implies the risk of defining a constitutive relation that is thermodynamically inadmissible. An example of this procedure is dealt with below.

Just as above, the contribution originating from the volume change to the stress tensor is considered separately from the distortion (shape change) by decomposing tensor $\boldsymbol{F}$ according to:

$$\boldsymbol{F} = J^{1/3}\tilde{\boldsymbol{F}} \quad \text{with} \quad J = \det(\boldsymbol{F}). \tag{12.86}$$

The associated objective, isochoric left Cauchy–Green tensor $\tilde{\boldsymbol{B}}$ is defined as:

$$\tilde{\boldsymbol{B}} = \tilde{\boldsymbol{F}} \cdot \tilde{\boldsymbol{F}}^{\mathrm{T}} = J^{-2/3}\boldsymbol{F} \cdot \boldsymbol{F}^{\mathrm{T}} = J^{-2/3}\boldsymbol{B}, \tag{12.87}$$

and subsequently the isochoric strain tensor $\tilde{\boldsymbol{\epsilon}}_{\mathrm{F}}$ is defined according to:

$$\tilde{\boldsymbol{\epsilon}}_{\mathrm{F}} = \frac{1}{2}\left(\tilde{\boldsymbol{B}} - \boldsymbol{I}\right) = \frac{1}{2}\left(J^{-2/3}\boldsymbol{B} - \boldsymbol{I}\right). \tag{12.88}$$

Analogous to the formulation in Section 12.2, linear relations for the hydrostatic and deviatoric part of the stress tensor can be postulated:

$$\boldsymbol{\sigma}^{\mathrm{h}} = -p\boldsymbol{I} = \kappa(J-1)\boldsymbol{I}, \tag{12.89}$$

$$\boldsymbol{\sigma}^{\mathrm{d}} = 2G\tilde{\boldsymbol{\epsilon}}_{\mathrm{F}}^{\mathrm{d}} = G\tilde{\boldsymbol{B}}^{\mathrm{d}}, \tag{12.90}$$

with κ the compression modulus and G the shear modulus of the material. Summation leads to:

$$\boldsymbol{\sigma} = \kappa(J-1)\boldsymbol{I} + G\tilde{\boldsymbol{B}}^{\mathrm{d}}. \tag{12.91}$$

This coupling between the stress and deformation state is often described as 'compressible neo-Hookean' material behaviour, thus referring to the linearity.

It can be shown that Eq. (12.91) does not exactly satisfy the requirement that in a cyclic process no energy is dissipated, showing that merely extrapolating from a linear model may lead to problems. However, it can be proven that a small (but non-trivial) modification, according to:

$$\boldsymbol{\sigma} = K(J-1)\boldsymbol{I} + \frac{G}{J}\tilde{\boldsymbol{B}}^{\mathrm{d}} \tag{12.92}$$

does satisfy this requirement.

Substituting Eq. (12.92) into the definition equation for $\Phi_{\mathrm{int}_0}(t)$, Eq. (12.29), yields

$$\Phi_{\mathrm{int}_0}(t) = \int_{\tau=0}^{t} \left(K(J-1)J\,\mathrm{tr}(\boldsymbol{D}) + G\,\mathrm{tr}(\tilde{\boldsymbol{B}}^{\mathrm{d}} \cdot \boldsymbol{D})\right) d\tau. \tag{12.93}$$

Based on the relations given in Chapter 10, the following expressions can be derived for the first and second terms in the integrand:

$$J\,\mathrm{tr}(\boldsymbol{D}) = \dot{J}, \tag{12.94}$$

and

$$\begin{aligned}
\mathrm{tr}(\tilde{\boldsymbol{B}}^{\mathrm{d}} \cdot \boldsymbol{D}) &= \mathrm{tr}\left(\left(\tilde{\boldsymbol{B}} - \frac{1}{3}\,\mathrm{tr}(\tilde{\boldsymbol{B}})\boldsymbol{I}\right) \cdot \boldsymbol{D}\right) \\
&= J^{-2/3}\,\mathrm{tr}(\boldsymbol{B} \cdot \boldsymbol{D}) - \frac{1}{3}J^{-2/3}\,\mathrm{tr}(\boldsymbol{B})\mathrm{tr}(\boldsymbol{D}) \\
&= \frac{1}{2}J^{-2/3}\,\mathrm{tr}\left(\boldsymbol{F} \cdot \boldsymbol{F}^{\mathrm{T}} \cdot \dot{\boldsymbol{F}} \cdot \boldsymbol{F}^{-1} + \boldsymbol{F} \cdot \boldsymbol{F}^{\mathrm{T}} \cdot \boldsymbol{F}^{-\mathrm{T}} \cdot \dot{\boldsymbol{F}}^{\mathrm{T}}\right) \\
&\quad - \frac{1}{3}J^{-5/3}\dot{J}\,\mathrm{tr}(\boldsymbol{B}) \\
&= \frac{1}{2}J^{-2/3}\,\mathrm{tr}\left(\dot{\boldsymbol{F}} \cdot \boldsymbol{F}^{\mathrm{T}} + \boldsymbol{F} \cdot \dot{\boldsymbol{F}}^{\mathrm{T}}\right) + \frac{1}{2}(\dot{J^{-2/3}})\,\mathrm{tr}(\boldsymbol{B}) \\
&= \frac{1}{2}J^{-2/3}\,\mathrm{tr}\left(\dot{\boldsymbol{B}}\right) + \frac{1}{2}(\dot{J^{-2/3}})\,\mathrm{tr}(\boldsymbol{B}).
\end{aligned} \tag{12.95}$$

With Eq. (12.94) and Eq. (12.95), the integral expression for $\Phi_{\mathrm{int}_0}(t)$ can be elaborated further:

$$\Phi_{\mathrm{int}_0} = \left(\frac{1}{2}K(J-1)^2 + \frac{1}{2}GJ^{-2/3}\,\mathrm{tr}(\boldsymbol{B})\right)\Bigg|_{\tau=0}^{t}. \tag{12.96}$$

Using $J = 1$ and $\boldsymbol{B} = \boldsymbol{I}$ for $\tau = 0$ results in the current energy density Φ_{int_0}, which only depends on the current left Cauchy–Green tensor $\boldsymbol{B}(t)$, so it can be noted:

$$\Phi_{\mathrm{int}_0}(\boldsymbol{B}) = \frac{1}{2}K(J-1)^2 + \frac{1}{2}G\left(J^{-2/3}\,\mathrm{tr}(\boldsymbol{B}) - 3\right), \tag{12.97}$$

with

$$J = (\det(\boldsymbol{B}))^{1/2}. \tag{12.98}$$

Again, it can be established that in Φ_{int_0} the volumetric and the deviatoric part are clearly distinguishable, and that both parts deliver an always positive contribution to the energy density, for every arbitrary deformation process. Finally, it is clear that a cyclic process in the deformation or in the stress will always be energetically neutral.

More general expressions for (non-linear) elastic behaviour can formally be written as

$$p = p(J), \quad \boldsymbol{\sigma}^{\mathrm{d}} = \boldsymbol{\sigma}^{\mathrm{d}}(J, \tilde{\boldsymbol{B}}). \tag{12.99}$$

For a detailed specification, many possibilities exist and have been published in the scientific literature. An extensive treatment for biological materials is beyond the scope of the present discussion. For this the reader is referred to more advanced textbooks on biomechanics.

12.5 Constitutive Modelling of Viscous Fluids

For viscous fluids, as considered in this book, in contrast to solids a reference state is not important. Therefore an Eulerian approach is pursued. Accordingly, the velocity field is written as

$$\vec{v} = \vec{v}(\vec{x}, t) \quad \text{and also} \quad \underset{\sim}{v} = \underset{\sim}{v}(\underset{\sim}{x}, t). \tag{12.100}$$

So the velocity is a function of the coordinates x, y and z associated with a fixed coordinate system in space, and the time t. The current local velocity does not include any information with respect to deformation changes in the fluid, contrary to the velocity gradient tensor $\boldsymbol{L}$ with matrix representation $\underline{L}$, defined in Sections 7.5 and 10.6 according to:

$$\boldsymbol{L} = \left(\vec{\nabla}\vec{v}\right)^{\mathrm{T}} \quad \text{and also} \quad \underline{L} = \left(\underset{\sim}{\nabla}\underset{\sim}{v}^{\mathrm{T}}\right)^{\mathrm{T}}. \tag{12.101}$$

In Section 10.6, the velocity gradient tensor $\boldsymbol{L}$ is split into the symmetrical rate of deformation tensor $\boldsymbol{D}$ and the skew-symmetric spin tensor $\boldsymbol{\Omega}$. The tensor $\boldsymbol{D}$ is a measure for the deformation changes. After all, the tensor $\boldsymbol{D}$ determines the current length changes of all material line segments, and if $\boldsymbol{D} = \boldsymbol{0}$, all those line segments have a (temporarily) constant length, independent of $\boldsymbol{\Omega}$. In components, the matrix $\underline{D}$ associated with tensor $\boldsymbol{D}$ can be written as

$$\underline{D} = \begin{bmatrix} \frac{\partial v_x}{\partial x} & \frac{1}{2}\left(\frac{\partial v_x}{\partial y} + \frac{\partial v_y}{\partial x}\right) & \frac{1}{2}\left(\frac{\partial v_x}{\partial z} + \frac{\partial v_z}{\partial x}\right) \\ \frac{1}{2}\left(\frac{\partial v_y}{\partial x} + \frac{\partial v_x}{\partial y}\right) & \frac{\partial v_y}{\partial y} & \frac{1}{2}\left(\frac{\partial v_y}{\partial z} + \frac{\partial v_z}{\partial y}\right) \\ \frac{1}{2}\left(\frac{\partial v_z}{\partial x} + \frac{\partial v_x}{\partial z}\right) & \frac{1}{2}\left(\frac{\partial v_z}{\partial y} + \frac{\partial v_y}{\partial z}\right) & \frac{\partial v_z}{\partial z} \end{bmatrix}. \tag{12.102}$$

When comparing this elaborated expression for matrix $\underline{D}$ with the linear strain matrix $\underline{\epsilon}$ (see Section 10.4), it is not surprising that $\underline{D}$ is called the rate of deformation matrix. Should the current state be chosen to be the reference state (in that special case in the current state: $\boldsymbol{F} = \boldsymbol{I}$), the following relation holds: $\underline{D} = \dot{\underline{\epsilon}}$ (and in tensor notation $\boldsymbol{D} = \dot{\boldsymbol{\epsilon}}$).

As a constitutive equation for the behaviour of incompressible viscous fluids, based on the above considerations, the following (general) expression for the Cauchy stress tensor will be used:

$$\boldsymbol{\sigma} = -p\boldsymbol{I} + \boldsymbol{\sigma}^{\mathrm{d}}(\boldsymbol{D}). \tag{12.103}$$

Because of the assumption of incompressibility, the pressure p is undetermined, while the deformation rate tensor $\boldsymbol{D}$ has to satisfy the constraint $\mathrm{tr}(\boldsymbol{D}) = 0$. It should be noted that both the Cauchy stress tensor and the deformation rate tensor are objective tensors. In Sections 12.6 and 12.7, two types of constitutive behaviour for fluids will be discussed by means of a specification of $\boldsymbol{\sigma}^{\mathrm{d}}(\boldsymbol{D})$.

12.6 Newtonian Fluids

For a Newtonian fluid, the relation between the deviatoric stress tensor and the deformation rate tensor is linear, yielding:

$$\boldsymbol{\sigma} = -p\boldsymbol{I} + 2\eta\boldsymbol{D} \quad \text{and also} \quad \underline{\sigma} = -p\underline{I} + 2\eta\underline{D}, \tag{12.104}$$

with η the **viscosity** (a material parameter that is assumed to be constant) of the fluid. The typical behaviour of a Newtonian fluid can be demonstrated by applying Eq. (12.104) to two elementary examples of fluid flow: pure shear and uniaxial extensional flow.

A pure shear flow 'in the xy-plane' can be created with the following velocity field (in column notation):

$$\underset{\sim}{v} = \begin{bmatrix} v_x \\ v_y \\ v_z \end{bmatrix} = \dot{\gamma} \begin{bmatrix} y \\ 0 \\ 0 \end{bmatrix}, \tag{12.105}$$

with $\dot{\gamma}$ (the shear velocity) constant. For the associated deformation rate matrix $\underline{D}$, it can easily be derived that

$$\underline{D} = \frac{1}{2} \begin{bmatrix} 0 & \dot{\gamma} & 0 \\ \dot{\gamma} & 0 & 0 \\ 0 & 0 & 0 \end{bmatrix} \tag{12.106}$$

and verified that the constraint $\mathrm{tr}(\boldsymbol{D}) = \mathrm{tr}(\underline{D}) = 0$ is satisfied. Applying the constitutive equation, it follows for the relevant shear stress $\sigma_{xy} = \sigma_{yx}$ in the fluid:

$$\sigma_{xy} = \eta\dot{\gamma}, \tag{12.107}$$

which is constant. Thus, the viscosity can be interpreted as the 'resistance' of the fluid against 'shear' (shear rate, actually).

To a uniaxial extensional flow (incompressible) in the x-direction, the following deformation rate matrix is applicable:

$$\underline{D} = \begin{bmatrix} \dot{\epsilon} & 0 & 0 \\ 0 & -\dot{\epsilon}/2 & 0 \\ 0 & 0 & -\dot{\epsilon}/2 \end{bmatrix}, \tag{12.108}$$

with $\dot{\epsilon}$ (the rate of extension) constant. For the stress matrix, it is immediately found that

$$\underline{\sigma} = -p\,\underline{I} + 2\eta \begin{bmatrix} \dot{\epsilon} & 0 & 0 \\ 0 & -\dot{\epsilon}/2 & 0 \\ 0 & 0 & -\dot{\epsilon}/2 \end{bmatrix}. \tag{12.109}$$

Assuming that in the y- and z-directions the flow can develop freely, $\sigma_{yy} = \sigma_{zz} = 0$, and with that it follows for the hydrostatic pressure:

$$p = -\eta\dot{\epsilon}. \tag{12.110}$$

This leads to the required uniaxial stress for the extensional flow:

$$\sigma_{xx} = 3\eta\dot{\epsilon}. \tag{12.111}$$

The fact that the effective uniaxial extensional viscosity (3η) is three times as high as the shear viscosity (η) is known as **Trouton's law.**

12.7 Non-Newtonian Fluids

For a non-Newtonian incompressible viscous fluid, the constitutive equation has the same form as the equation for a Newtonian fluid:

$$\boldsymbol{\sigma} = -p\boldsymbol{I} + 2\eta\boldsymbol{D} \qquad \text{and also} \qquad \underline{\sigma} = -p\underline{I} + 2\eta\underline{D}; \tag{12.112}$$

however, the viscosity η is now a function of the deformation rate tensor: $\eta = \eta(\boldsymbol{D})$. Specification of the relation for the viscosity has to be based on experimental research. Here we limit ourselves to an example, the three-parameter 'power law' model (with the temperature influence T according to Arrhenius):

$$\eta = m e^{(A/T)} \left(\sqrt{2\,\mathrm{tr}(\boldsymbol{D}\cdot\boldsymbol{D})}\right)^{(n-1)}, \tag{12.113}$$

with m, A and n material constants (for $n = 1$ the viscosity is independent of $\boldsymbol{D}$ and the behaviour is 'Newtonian' again). Substitution of the deformation rate tensor for pure shear (see previous section) leads to

$$\eta = m e^{(A/T)} \, |\dot{\gamma}|^{(n-1)}, \tag{12.114}$$

making the mathematical format of the equation more transparent.

12.8 Diffusion and Filtration

Although somewhat beyond the scope of the present chapter, in this last section material flow due to diffusion or filtration is considered, i.e. transport of material through a stationary porous medium (no convection). The flowing material can indeed be considered to be a continuum, but the constitutive equations are completely different from those treated before. Here, the constitutive equations describe the transport of material depending on the driving mechanisms, whereas above, the constitutive equations related the characteristics of the flow to the internal stresses.

Diffusion of a certain material through a porous medium is generated by concentration differences of the material in the medium. The material will in general strive for a homogeneous density distribution (provided that the porous medium is homogeneous), implying that material will flow from regions with a high concentration to regions with a low concentration. The mathematical form for this phenomenon is given by **Fick's law**:

$$\rho\vec{v} = -D\vec{\nabla}\rho, \tag{12.115}$$

with, on the left-hand side, the mass flux vector (see also Section 7.5), and on the right-hand side, the driving mechanism for transport, $\vec{\nabla}\rho$, multiplied by a certain factor D. This factor D is called the diffusion coefficient, and can be considered to be a constitutive parameter that is determined by the combination of materials (and the temperature).

In the case of filtration of material through a stationary porous medium, the driving force is formed by pressure differences in that material. The material strives for an equal pressure, so in the presence of pressure differences, a flow will occur from areas with a high pressure to areas with a lower pressure. This is expressed by **Darcy's law**:

$$\rho\vec{v} = -\kappa\vec{\nabla}p \tag{12.116}$$

with the permeability κ as a constitutive parameter.

Exercises

12.1 Consider a cubic material element of which the edges are oriented in the directions of a Cartesian xyz-coordinate system: see figure. The figure also depicts the normal and shear stresses (expressed in [kPa]) acting on the (visible) faces of the element. The mechanical behaviour of the element is described by the linear Hooke's law, with Young's modulus $E = 8$ [MPa] and Poisson's ratio $\nu = 1/3$.

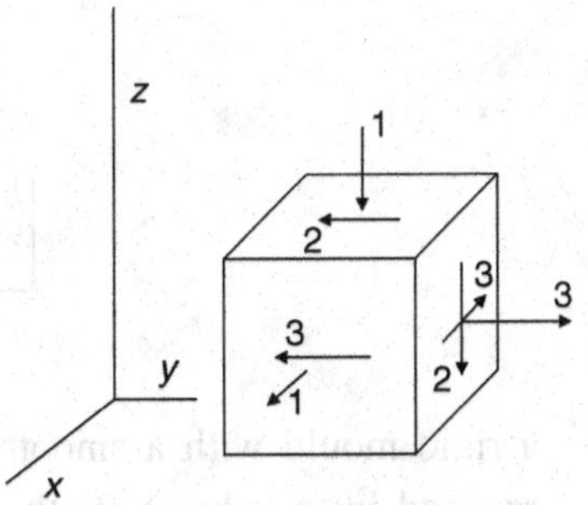

Determine the volume change V/V_0, with V_0 the volume of the element in the unloaded reference configuration and V the volume in the current loaded configuration, assuming small deformations.

12.2 An element of an incompressible material (in the reference state, a cube $\ell \times \ell \times \ell$) is placed in a Cartesian *xyz*-coordinate system as given in the figure below. Because of a load in the *z*-direction, the height of the element is reduced to $2\ell/3$. In the *x*-direction, the displacement is suppressed. The element can expand freely in the *y*-direction. The deformation is assumed to be homogeneous.

The material behaviour is described by a neo-Hookean relation, according to:

$$\underline{\sigma} = -p\underline{I} + G\underline{B}^{\mathrm{d}},$$

with $\underline{\sigma}$ the stress matrix, p the hydrostatic pressure (to be determined), $\underline{I}$ the unit matrix, G the shear modulus and $\underline{B}$ the left Cauchy–Green deformation matrix.

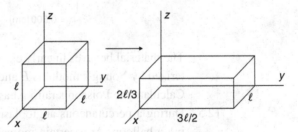

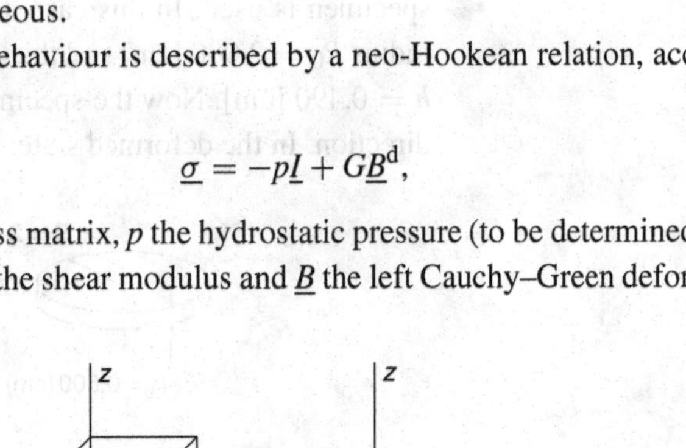

Determine the compressive force F_y in the *z*-direction that is necessary to realize this deformation.

12.3 A frequently applied test to determine the stiffness properties of biological materials is the 'confined compression test'. A schematic of such a test is given in the figure. A cylindrical specimen, Young's modulus E (cylindrically shaped with a circular cross section of radius R, height h), of the material is placed in a tight-fitting die, which can be regarded as

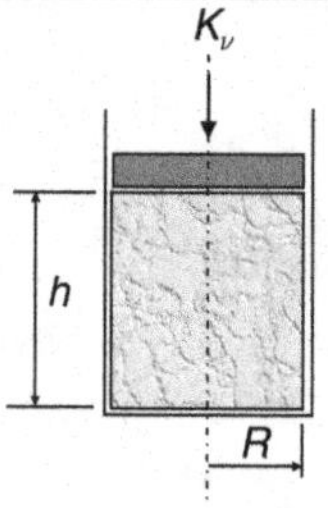

a rigid mould with a smooth inner wall. The height of the specimen is reduced by a value δ (with $\delta \ll h$) by means of an indenter. A vertical force K_ν acts on the indenter in the direction of the arrow. It can be assumed that the deformation of the specimen is homogeneous. As the strains are small, the material behaviour is described by the linear Hooke's law. The Poisson's ratio is known (measured in another experiment): $\nu = 1/4$.

Express the Young's modulus E of the material in terms the parameters R, h, δ and K_ν (i.e. in the parameters that can easily be measured).

12.4 Because confined compression tests have several disadvantages, unconfined compression tests are also frequently performed. Again, a cylindrical specimen is used. In this case, the specimen has, in the reference state, a radius $R_0 = 2.000$ [cm] and thickness $h_0 = 0.500$ [cm]. It is compressed to $h = 0.490$ [cm]. Now the specimen is allowed to expand freely in a radial direction. In the deformed state: $R = 2.008$ [cm].

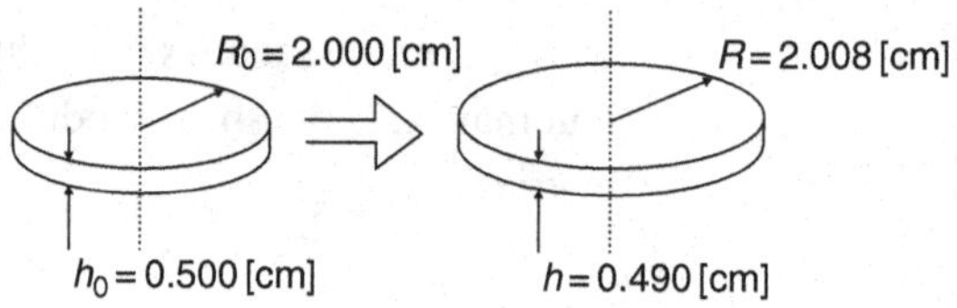

The material behaves linearly elastically according to Hooke's law, characterized by Young's modulus E and Poisson's ratio ν.

Calculate the Poisson's ratio ν based on the above-described experiment.

12.5 During a percutaneous angioplasty, the vessel wall is expanded by inflating a balloon. At a certain moment during inflation, the internal pressure $p_{\mathrm{i}} = 0.2$ [MPa], and the internal radius R_{i} of the vessel is increased by 10%. Assume that the length of the balloon and the length of the vessel wall in contact with the balloon do not change. Further, it is known that the wall behaviour can be modelled according to Hooke's law, with Young's modulus of the wall $E = 8$ [MPa] and the Poisson's ratio of the wall $\nu = 1/3$.

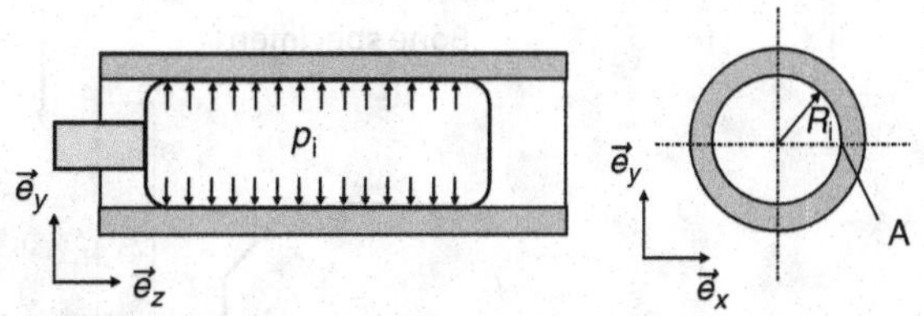

(a) Calculate the strain in the circumferential direction at the inner side of the wall.

(b) Calculate the strain in the $\vec{e}_x$-direction at point A at the inner side of the vessel wall.

(c) Calculate the stress in circumferential direction at point A.

12.6 Consider a bar (length $2L$) with a circular cross section (radius R). The axis of the bar coincides with the z-axis of a xyz-coordinate system. With respect to the mid-plane ($z = 0$), the top and bottom plane of the bar are rotated around the z-axis by an angle α (with $\alpha \ll 1$), thus loading the bar with torsion. See the figure.

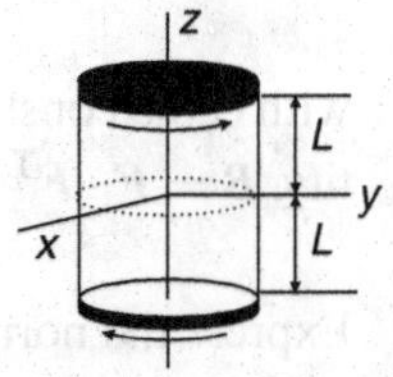

The position vector $\vec{x}$ of a material point in the deformed configuration is

$$\vec{x} = \vec{x}_0 - \frac{\alpha(\vec{x}_0 \cdot \vec{e}_y)(\vec{x}_0 \cdot \vec{e}_z)}{L}\vec{e}_x + \frac{\alpha(\vec{x}_0 \cdot \vec{e}_x)(\vec{x}_0 \cdot \vec{e}_z)}{L}\vec{e}_y,$$

with $\vec{x}_0$ the position vector of that point in the undeformed state. The material behaviour is linearly elastic according to Hooke's law, with Young's modulus E and Poisson's ratio ν.

Determine, for the material point defined by $\vec{x}_0 = R\vec{e}_y$, the linear strain tensor $\boldsymbol{\epsilon}$, the stress tensor $\boldsymbol{\sigma}$ and the equivalent stress, according to the von Mises $\bar{\sigma}_M$.

12.7 Consider an experimental set-up in which a rectangular piece of bone exactly fits between two spatially fixed parallel glass plates.

The original dimensions of the bone are $\ell_0 \times h_0 \times d_0$. The piece of bone is mechanically loaded with a compressive force in the x-direction. After loading, the stretch ratio in the x-direction is given by $\lambda = \ell/\ell_0$. The bone can freely extend in the z-direction. Friction between the glass plates and the bone can be ignored. The material behaviour can be described

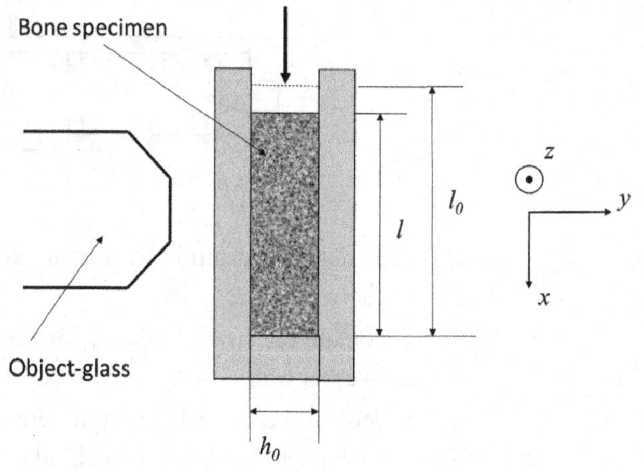

by an incompressible neo-Hookean model, according to the following constitutive law:

$$\boldsymbol{\sigma} = -p\boldsymbol{I} + G(\boldsymbol{B} - \boldsymbol{I})$$

with G the constant and positive shear modulus, p a hydrostatic stress quantity, $\boldsymbol{B} = \boldsymbol{F} \cdot \boldsymbol{F}^{\mathrm{T}}$ the left Cauchy–Green tensor, and $\boldsymbol{I}$ the unit tensor.

Express the normal stress component σ_{yy} in the y-direction in λ and G.

12.8 A popular device in tissue engineering is a cell stretching device, using membranes (see the figure). Cells are seeded on a circular membrane that rests on a cylindrically shaped post. On applying a vacuum pressure, the membrane is sucked into the gap between the post and the outer cylinder, and in this way the area with the monolayer of cells is equally stretched in the plane of the membrane (in this case the xy-plane). The layer can be considered to be in a state of **plane stress** ($\sigma_{zz} = \sigma_{xz} = \sigma_{yz} = 0$).

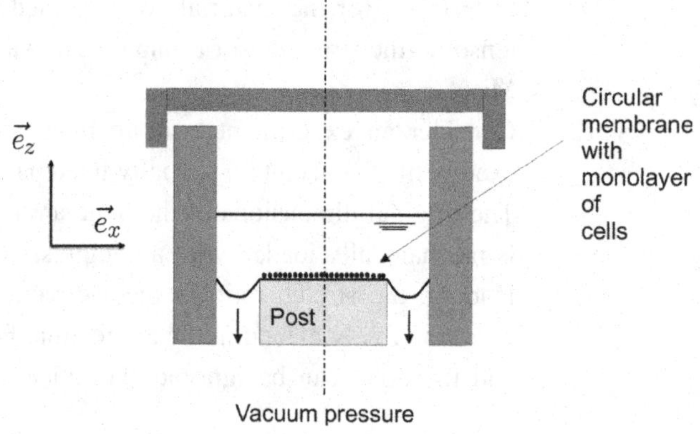

The deformation tensor for this state of strain can be given as:

$$\boldsymbol{F} = \lambda(\vec{e}_x\vec{e}_x + \vec{e}_y\vec{e}_y) + \mu\vec{e}_z\vec{e}_z$$

with λ the stretch ratio in the x- and y-directions and μ the stretch ratio in the z-direction. The material behaviour of the membrane and the monolayer is modelled by:

$$\boldsymbol{\sigma} = \kappa(J-1) + G\boldsymbol{B}^{\mathrm{d}}$$

with κ the bulk modulus, G the shear modulus and $J = \det(\boldsymbol{F})$ and $\boldsymbol{B}^{\mathrm{d}} = \boldsymbol{B} - \frac{1}{3}\mathrm{tr}(\boldsymbol{B})\boldsymbol{I}$.

Derive an expression for μ as a function of λ, κ and μ.

12.9 A soft tissue sample, with dimensions 10 × 10 × 1 [mm^3], is clamped in a biaxial testing machine (see figure). An xyz-coordinate system is defined with the origin in the centre of the sample and the axes as defined in the figure. After stretching, the sample is still square with a new length and width 15 × 15 [mm^2]. It can be assumed that a constant stress state occurs in the sample, described by the stress matrix $\underline{\sigma}$, with respect to the basis vectors $\{\vec{e}_x, \vec{e}_y, \vec{e}_z\}$. The sample can be considered to be in a plane stress situation, meaning that the stress components satisfy:

$$\sigma_{zz} = \sigma_{xz} = \sigma_{yz} = 0.$$

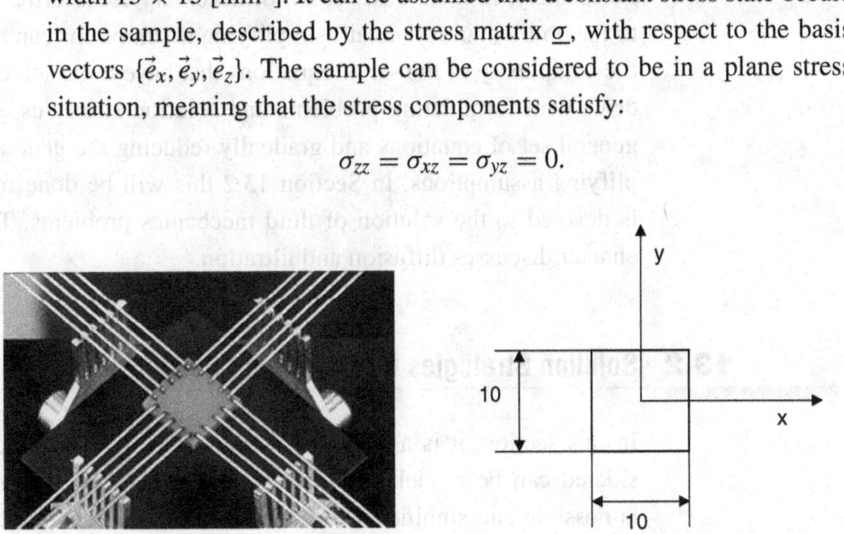

The material can be considered to be incompressible, described by means of a neo-Hookean material law:

$$\boldsymbol{\sigma} = -p\boldsymbol{I} + G(\boldsymbol{B} - \boldsymbol{I})$$

with $\boldsymbol{B} = \boldsymbol{F} \cdot \boldsymbol{F}^{\mathrm{T}}$, $\boldsymbol{F}$ the deformation tensor, $\boldsymbol{I}$ the unit tensor, the shear modulus $G = 1$ [MPa] and p a hydrostatic pressure quantity.

Determine p, σ_{xx} and σ_{yy}.

13 Solution Strategies for Solid and Fluid Mechanics Problems

13.1 Introduction

The goal of the present chapter is to describe a procedure to formally determine solutions for solid mechanics problems, fluid mechanics problems and problems with filtration and diffusion. Mechanical problems in biomechanics can be very diverse, and most problems are so complex that it is impossible to derive analytical solutions and often very complicated to determine numerical solutions. Fortunately, in most cases it is not necessary to describe all phenomena related to the problem in full detail; simplifying assumptions can be made, thus reducing the complexity of the set of equations that have to be solved. The present chapter deals with formulating problems and solution strategies, starting from the most general set of equations and gradually reducing the generality by imposing simplifying assumptions. In Section 13.2 this will be done for solids. Section 13.3 is devoted to the solution of fluid mechanics problems. The last section of this chapter discusses diffusion and filtration.

13.2 Solution Strategies for Deforming Solids

In this section, it is assumed that the material (or material fraction) to be considered can be modelled as a deforming solid continuum. This implies that it is possible and significant to define a reference configuration or reference state. With respect to the reference state, the displacement field as a function of time supplies a full description of the deformation process to which the continuum is subjected (at least under the restrictions given in previous chapters, such as a constant temperature). After all, for a displacement field that is known as a function of time, it is possible to directly calculate the local deformation history (applying the kinematics: see Chapter 10) and subsequently, the stress state as a function of time (using the constitutive equations: see Chapter 12). The relevant fields, from a mechanical point of view, that are obtained in this way for the continuum have to satisfy the balance equations (see Chapter 11). In addition, the initial conditions have to be fulfilled: at the beginning of the process, the positions of the material points in space have to be prescribed, usually in accordance with the reference configuration, and also the initial velocities of the material points have to be in agreement with the specification of the initial state. In addition, during the entire process the boundary conditions have to be satisfied: along the outer surface of the continuum, the displacement field and stress field have to be consistent with the process specification.

The goal of the present section is to outline a procedure to formally determine (as a function of time) the displacement field, such that all requirements are satisfied. For (almost) no realistic problem can exact analytical solutions be found, not even when the mathematical description is very much simplified via assumptions. A global description will be given of strategies to derive approximate solutions.

In Section 13.2.1 the general (complete) formulation of the problem will be given. Subsequently, in the sections that follow, the generality will gradually be reduced. Initially, in Section 13.2.2, the general description will be restricted with respect to the magnitude of the displacements, deformations and rotations (geometrically linear behaviour). In Section 13.2.3, the restriction to linearly elastic behaviour follows (physical linearity), leading to the set of equations establishing the so-called 'linear elasticity theory'. In Section 13.2.4, processes are considered for which inertia effects can be neglected (quasi-static processes). Time (and thus also process rate) is then no longer relevant. In Section 13.2.5, the attention is concentrated on configurations that, because of geometry and external loading, can be regarded as two-dimensional (plane stress theory). Finally, Section 13.2.6 is focussed on formulating boundary conditions for continuum problems. Sometimes, extra constitutive equations are necessary to describe the interaction of the considered continuum with the environment.

13.2.1 General Formulation for Solid Mechanics Problems

For general solid mechanics problems addressing the deformation process in a Lagrange description, the following fields have to be determined:

- the displacement field: $\vec{u}(\vec{x}_0, t)$ for all $\vec{x}_0$ in V_0 and for all t, and
- the stress field: $\boldsymbol{\sigma}(\vec{x}_0, t)$ for all $\vec{x}_0$ in V_0 and for all t.

These fields have to be connected for all $\vec{x}_0$ in V_0 in accordance with the local constitutive equation (see Chapter 12):

$$\boldsymbol{\sigma}(\vec{x}_0, t) = \mathcal{F}\{\boldsymbol{F}(\vec{x}_0, \tau);\ \tau \leq t\}, \tag{13.1}$$

with

$$\boldsymbol{F} = \boldsymbol{I} + \left(\vec{\nabla}_0 \vec{u}\right)^{\mathrm{T}}, \tag{13.2}$$

reflecting history-dependent material behaviour. In addition, the local balance of momentum has to be satisfied (see Chapter 11), as well as the local mass balance:

$$\left(\boldsymbol{F}^{-\mathrm{T}} \cdot \vec{\nabla}_0\right) \cdot \boldsymbol{\sigma} + \rho \vec{q} = \rho \ddot{\vec{u}}, \tag{13.3}$$

with

$$\rho = \frac{\rho_0}{\det(\boldsymbol{F})}. \tag{13.4}$$

The equations given above form a set of non-linear, coupled partial differential equations (derivatives with respect to the three material coordinates in $\vec{x}_0$ and the time t are dealt with). Consequently, for a unique solution of the displacement field $\vec{u}(\vec{x}_0, t)$ and the stress field $\boldsymbol{\sigma}(\vec{x}_0, t)$, boundary conditions and initial conditions are required. With respect to the boundary conditions, for all t at every point of the outer surface of V_0 three (scalar) relations have to be specified: either completely formulated in terms of stresses (dynamic or natural boundary conditions), completely formulated in terms of displacements (kinematic or essential boundary conditions) or in mixed formulations. With respect to the initial conditions, at the initial time point ($t = 0$), for all the points in V_0, the displacement and velocity have to be specified. If the initial state is used as the reference configuration, $\vec{u}(\vec{x}_0, t = 0) = \vec{0}$ for all $\vec{x}_0$ in V_0.

13.2.2 Geometrical Linearity

Provided that displacements, strains and rotations are small (so $\boldsymbol{F} \approx \boldsymbol{I}$ and consequently $\det(\boldsymbol{F}) \approx 1$), the general set for solid continuum problems as presented in the previous section can be written as

$$\boldsymbol{\sigma}(\vec{x}_0, t) = \mathcal{G}\{\boldsymbol{\varepsilon}(\vec{x}_0, \tau)\ ;\ \tau \leq t\}, \tag{13.5}$$

where $\mathcal{F}\{\}$, as used in Eq. (13.1), has been replaced by $\mathcal{G}\{\}$ owing to adaptations in the argument, with

$$\boldsymbol{\varepsilon} = \frac{1}{2}\left(\left(\vec{\nabla}_0 \vec{u}\right) + \left(\vec{\nabla}_0 \vec{u}\right)^{\mathrm{T}}\right), \tag{13.6}$$

and

$$\vec{\nabla}_0 \cdot \boldsymbol{\sigma} + \rho_0 \vec{q} = \rho_0 \ddot{\vec{u}}. \tag{13.7}$$

Again, these equations have to be satisfied for all $\vec{x}_0$ in V_0 and for all times t. Equation (13.5) (a formal functional expression) indicates that the current local Cauchy stress tensor $\boldsymbol{\sigma}$ is fully determined by the (history of the) local linear strain tensor $\boldsymbol{\varepsilon}$. Equation (13.7) actually implies that the balance of momentum (equation of motion) is now related to the reference configuration instead of the current configuration. The balance of mass is now redundant through replacing ρ by ρ_0. With respect to the boundary conditions and initial conditions, the same statements apply as in the previous section.

13.2.3 Linear Elasticity Theory, Dynamic

In Section 13.2.2, geometrical linearity has been introduced to simplify the general set of equations given in Section 13.2.1. If we add physical linearity to this (meaning that the relation between stress and strain is described by Hooke's law as formulated in Chapter 12), linear elasticity theory results. The associated (linear) equations, which have to be satisfied for all $\vec{x}_0$ in V_0 and for all times t, read:

$$\boldsymbol{\sigma} = \left(K - \frac{2G}{3}\right) \mathrm{tr}(\boldsymbol{\varepsilon})\boldsymbol{I} + 2G\boldsymbol{\varepsilon}, \tag{13.8}$$

with

$$\boldsymbol{\varepsilon} = \frac{1}{2}\left(\left(\vec{\nabla}_0\vec{u}\right) + \left(\vec{\nabla}_0\vec{u}\right)^{\mathrm{T}}\right),$$

and

$$\vec{\nabla}_0 \cdot \boldsymbol{\sigma} + \rho_0\vec{q} = \rho_0\ddot{\vec{u}}. \tag{13.9}$$

With respect to boundary conditions and initial conditions, the same procedure must be followed as in the previous section.

13.2.4 Linear Elasticity Theory, Static

For slowly evolving (quasi-static) processes, inertia effects can be neglected. In that case it is sufficient to consider only the relevant current state, and time is no longer of interest. The current displacement field $\vec{u}(\vec{x}_0)$ and the current stress field $\boldsymbol{\sigma}(\vec{x}_0)$ have to be determined on the basis of the equations:

$$\boldsymbol{\sigma} = \left(K - \frac{2G}{3}\right) \mathrm{tr}(\boldsymbol{\varepsilon})\boldsymbol{I} + 2G\boldsymbol{\varepsilon}, \tag{13.10}$$

with

$$\boldsymbol{\varepsilon} = \frac{1}{2}\left(\left(\vec{\nabla}_0\vec{u}\right) + \left(\vec{\nabla}_0\vec{u}\right)^{\mathrm{T}}\right), \tag{13.11}$$

and

$$\vec{\nabla}_0 \cdot \boldsymbol{\sigma} + \rho_0\,\vec{q} = \vec{0}. \tag{13.12}$$

The equation of motion is reduced to the equilibrium equation. Only boundary conditions are necessary to solve this set. Initial conditions are no longer applicable. At every point of the outer surface of V_0, three (scalar) relations have to be specified. In this context, to find a unique solution for the displacement field, it is necessary to suppress movement of the configuration as a rigid body. This will be further elucidated in the following example.

Example 13.1 Consider a homogeneous body with reference volume V_0 under a given hydrostatic pressure p. For the stress field in V_0 satisfying the (dynamic) boundary conditions and the equilibrium equations, we can write:

$$\boldsymbol{\sigma}(\vec{x}_0) = -p\boldsymbol{I} \text{ for all } \vec{x}_0 \text{ in } V_0.$$

For the strain field, according to Hooke's law it is found:

$$\boldsymbol{\varepsilon}(\vec{x}_0) = -\frac{p}{3K}\boldsymbol{I} \text{ for all } \vec{x}_0 \text{ in } V_0.$$

A matching displacement field is, for example,

$$\vec{u}(\vec{x}_0) = -\frac{p}{3K}\vec{x}_0.$$

It is easy to verify that the solution given above satisfies all equations. However, because in the given problem description the displacement as a rigid body is not prescribed, the solution is not unique. It can be proven that the general solution for the components of the displacement vector reads:

$$\begin{aligned} u_x &= -\frac{p}{3K}x_0 + c_1 - c_6 y_0 + c_5 z_0 \\ u_y &= -\frac{p}{3K}y_0 + c_2 + c_6 x_0 - c_4 z_0 \\ u_z &= -\frac{p}{3K}z_0 + c_3 - c_5 x_0 + c_4 y_0, \end{aligned}$$

with c_i $(i = 1, 2, \ldots, 6)$ as-yet undetermined constants. The constants c_1, c_2 and c_3 represent translations in the coordinate directions and the constants c_4, c_5 and c_6 (small) rotations around the coordinate axes.

13.2.5 Linear Plane Stress Theory, Static

Consider a flat membrane with, in the reference configuration, constant thickness h. The 'midplane' of the membrane coincides with the x_0y_0-plane, while in the direction perpendicular to that plane the domain for z_0 is given by: $-h/2 \leq z_0 \leq$

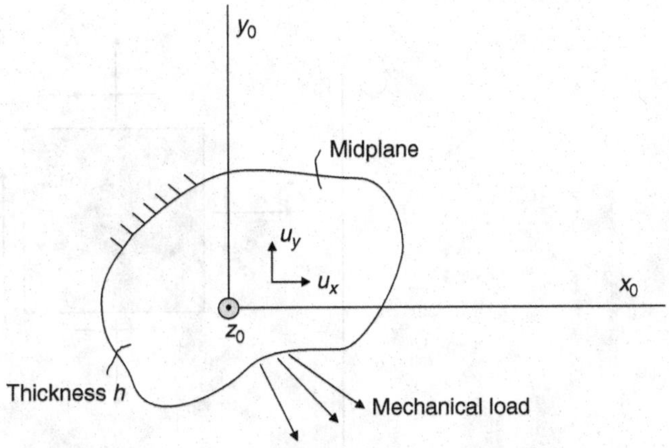

Figure 13.1
Membrane in plane stress state.

$h/2$. The thickness h is assumed to be small with respect to the dimensions 'in the plane'. The loading is parallel to the midplane of the membrane; see Fig. 13.1.

The midplane will continue to be a symmetry plane after deformation. It is assumed that straight material line segments, initially perpendicular to the midplane, will remain straight (and perpendicular to the midplane) after deformation. For the displacements, this means

$$u_x = u_x(x_0, y_0), \qquad u_y = u_y(x_0, y_0). \tag{13.13}$$

The relevant strain components of the linear strain matrix $\underline{\varepsilon}$ for the membrane are

$$\begin{aligned} \varepsilon_{xx} &= \frac{\partial u_x}{\partial x_0} \\ \varepsilon_{yy} &= \frac{\partial u_y}{\partial y_0} \\ \varepsilon_{xy} = \varepsilon_{yx} &= \frac{1}{2}\left(\frac{\partial u_x}{\partial y_0} + \frac{\partial u_y}{\partial x_0}\right). \end{aligned} \tag{13.14}$$

In general, out of the other components, ε_{zz} will certainly not be zero, while $\varepsilon_{xz} = \varepsilon_{zx}$ and $\varepsilon_{yz} = \varepsilon_{zy}$ will vanish. Actually, these other strain components are not important to set up the theory.

With respect to the stress, it is assumed that only components 'acting in the x_0y_0-plane' play a role, and that for those components (see Fig. 13.2) we can write:

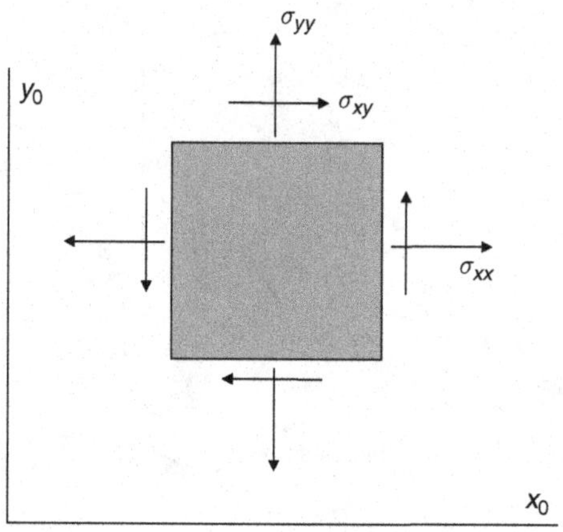

Figure 13.2

Stress components in a plane stress state.

$$\begin{aligned}\sigma_{xx} &= \sigma_{xx}(x_0, y_0)\\ \sigma_{yy} &= \sigma_{yy}(x_0, y_0)\\ \sigma_{xy} &= \sigma_{yx} = \sigma_{xy}(x_0, y_0).\end{aligned} \tag{13.15}$$

The stress components σ_{zz}, $\sigma_{xy} = \sigma_{yx}$ and $\sigma_{xz} = \sigma_{zx}$ are assumed to be zero (negligible).

Based on Hooke's law (see Section 12.2 for the fully elaborated expression in components), using $\sigma_{zz} = 0$ it is found:

$$\varepsilon_{zz} = -\frac{\nu}{1-\nu}(\varepsilon_{xx} + \varepsilon_{yy}), \tag{13.16}$$

and by exploiting this equation, the description of the (linearly elastic) material behaviour for plane stress becomes

$$\sigma_{xx} = \frac{E}{1-\nu^2}(\varepsilon_{xx} + \nu\varepsilon_{yy}) \tag{13.17}$$

$$\sigma_{yy} = \frac{E}{1-\nu^2}(\nu\varepsilon_{xx} + \varepsilon_{yy}) \tag{13.18}$$

$$\sigma_{xy} = \frac{E}{(1+\nu)}\varepsilon_{xy}, \tag{13.19}$$

thus coupling the relevant stresses and strains. It should be noted that in this case, for the material parameters, the Young's modulus E and the Poisson's ratio ν have been used, instead of the compression modulus K and the shear modulus G used in previous sections.

The stresses have to satisfy the equilibrium equations. For plane stress, this means that only equilibrium 'in the plane' (and therefore, in the case considered, in the x_0- and y_0-directions) results in non-trivial equations:

$$\frac{\partial \sigma_{xx}}{\partial x_0} + \frac{\partial \sigma_{xy}}{\partial y_0} + \rho_0\, q_x = 0 \tag{13.20}$$

$$\frac{\partial \sigma_{xy}}{\partial x_0} + \frac{\partial \sigma_{yy}}{\partial y_0} + \rho_0\, q_y = 0. \tag{13.21}$$

Summarizing, it can be stated that, in the case of plane stress, eight scalar functions of the coordinates in the midplane have to be calculated: the displacements u_x and u_y, the strains ε_{xx}, ε_{yy} and ε_{xy}, and the stresses σ_{xx}, σ_{yy} and σ_{xy}. For this objective we have eight equations at our disposal: the strain–displacement relations (three), the constitutive equations (three) and the equilibrium equations (two). In addition, for each boundary point of the midplane, two scalar boundary conditions have to be specified, and for uniqueness of the displacement field, rigid body motion has to be suppressed.

Example 13.2 In Fig. 13.3, a simple plane stress problem is defined for a rectangular membrane (length 2ℓ, width $2b$ and thickness h) with linearly elastic material behaviour (Young's modulus E and Poisson's ratio ν). The mathematical form for the boundary conditions reads:

$$\text{for } x_0 = \pm\,\ell \quad \text{and} \quad -b \le y_0 \le b \ \text{ it holds that: } \ \sigma_{xx} = \alpha + \beta\frac{y_0}{b}$$
$$\sigma_{xy} = 0$$
$$\text{for } y_0 = \pm\,b \quad \text{and} \quad -\ell \le x_0 \le \ell \ \text{ it holds that: } \ \sigma_{yy} = 0$$
$$\sigma_{xy} = 0.$$

Herewith, the external load is specified using the constants α and β; the constant α is representative for the 'normal' force in the x_0 direction and the constant β for the 'bending'. With the above-given boundary conditions, the displacement of the membrane as a rigid body is not suppressed; uniqueness of the displacement solution is obtained if it is additionally required that the material point coinciding with origin O is fixed in space and if in the deformed state the symmetry with respect to the y_0-axis is maintained. For this problem, an exact analytical solution can be calculated. It is easy to verify that the solution has the following form:

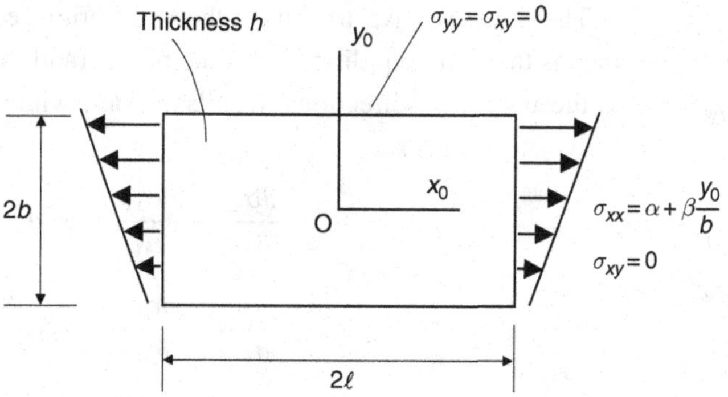

Figure 13.3

A simple plane stress problem.

$$u_x(x_0,y_0) = \frac{1}{E}\left(\alpha + \beta\frac{y_0}{b}\right)x_0$$
$$u_y(x_0,y_0) = -\frac{1}{E}\left(\nu\alpha y_0 + \nu\beta\frac{y_0^2}{2b} + \beta\frac{x_0^2}{2b}\right)$$

$$\varepsilon_{xx}(x_0,y_0) = \frac{1}{E}\left(\alpha + \beta\frac{y_0}{b}\right)$$
$$\varepsilon_{yy}(x_0,y_0) = -\frac{\nu}{E}\left(\alpha + \beta\frac{y_0}{b}\right)$$
$$\varepsilon_{xy}(x_0,y_0) = 0$$

$$\sigma_{xx}(x_0,y_0) = \alpha + \beta\frac{y_0}{b}$$
$$\sigma_{yy}(x_0,y_0) = 0$$
$$\sigma_{xy}(x_0,y_0) = 0.$$

It has to be considered as an exception when an analytical solution exists for a specified plane stress problem. In general, only approximate solutions can be determined. A technique to do this is the finite element method, which is the subject of Chapters 14 to 18. Chapter 18, especially, is devoted to the solution of linear elasticity problems as described in the present chapter.

13.2.6 Boundary Conditions

In the previous sections, only simply formulated boundary conditions have been considered. In the case of dynamic boundary conditions, the components of the

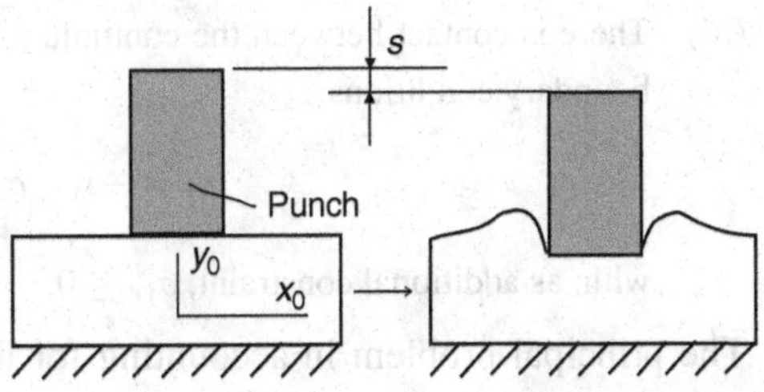

Figure 13.4
Rigid indenter impressing a deforming continuum.

stress vector $\vec{s}$ (see Fig. 8.2) are prescribed at a point along the boundary of the volume of the continuum (in the case of plane stress along the boundary of the configuration surface). In the case of kinematic boundary conditions, the displacement vector $\vec{u}$ is prescribed. Sometimes the boundary conditions are less explicitly defined, however, for example when the considered continuum interacts with its environment. Below, an example of such a situation will be outlined.

Consider a plane stress continuum of which the midplane (coinciding with the x_0y_0-plane) has a rectangular shape in the reference configuration. The current state arises because the top edge is indented by means of a rigid punch. The punch displacement is specified by s (Fig. 13.4 shows the deformation process; the displacements are magnified). At the location of the contact between indenter and continuum, the interaction is described with a friction model according to Coulomb (which can be considered to be a constitutive description for the contact interaction). For a material point at the top contour of the continuum, the following three distinguishable situations may arise:

(i) There is no contact between the point of the continuum and the indenter. In this case the boundary conditions are

$$\sigma_{yy} = 0, \quad \sigma_{xy} = 0, \tag{13.22}$$

with the extra constraint that in the current state the vector $\vec{x}_0 + \vec{u}$ does not cross the edge of the indenter.

(ii) There is contact between the point of the continuum and the indenter, which occurs with 'stick' boundary conditions (no relative tangential displacement between continuum and indenter):

$$u_x = 0, \quad u_y = -s, \tag{13.23}$$

with, as extra constraints, $\sigma_{yy} \leq 0$ and $|\sigma_{xy}| \leq -\mu\sigma_{yy}$ with μ the friction coefficient.

(iii) There is contact between the continuum and the indenter, which occurs with 'slip' boundary conditions:

$$u_y = -s, \quad \sigma_{xy} = \mu \frac{u_x}{|u_x|}\sigma_{yy}, \tag{13.24}$$

with, as additional constraint, $\sigma_{yy} \leq 0$.

The principal problem in accounting for the interaction between the continuum and the indenter is that it is not a priori known to which of these three categories the points of the top layer of the continuum belong. In general, an estimation is made for this that is updated if the constraints are violated. In this way, an iterative solution can eventually be determined.

13.3 Solution Strategies for Viscous Fluids

Consider a fixed volume V in three-dimensional space, through which (or within which) a certain amount of material flows, while this material can be considered as an incompressible viscous fluid (see Section 12.5). Because for such a fluid a (possibly defined) reference state is of no interest, an Eulerian description is used for relevant fields within the volume V.

The velocity field must fulfil the incompressibility constraint at each point in time. The current velocity field fully determines the deviatoric part of the stress state (via the constitutive modelling). The hydrostatic part of the stress field cannot be determined on the basis of the velocity field. The stress field (the combination of the hydrostatic and deviatoric part) has to satisfy the momentum balance equation (see Section 11.3). The problem definition is completed by means of initial conditions and boundary conditions. The description of the initial velocity field with the incompressibility constraint must be consistent and along the boundary of V, for every point in time velocities and/or stresses have to be in agreement with reality. It must be emphasized that considering a fixed volume in space implies a serious limitation for the prospect of applying the theory.

The goal of the present section is to outline a routine to formally determine the velocity field and the (hydrostatic) stress field, both as a function of time, such that all the above-mentioned equations are satisfied. However, for (almost all) realistic problems it is not possible to derive an exact analytical solution, not even via assumptions that severely simplify the mathematical description. A global description will be given of strategies to derive approximate solutions.

In Section 13.3.1, the general (complete) formulation of the problem will be given, including the relevant equations. Thereupon, in the following sections the complexity of the formulation will be gradually reduced. For this, first in

Section 13.3.2 the material will be modelled as a Newtonian fluid (see Section 12.6). This leads to the so-called Navier–Stokes equation (an equation with the pressure field and the velocity field as unknowns) that has to be solved in combination with the continuity equation (mass balance). In Section 13.3.3 the limitation for a stationary flow is dealt with (the time dependency, including the need for initial conditions, is no longer relevant). After a section on boundary conditions, a few elementary analytical solutions of the equations for a stationary viscous flow are presented in Section 13.3.5.

13.3.1 General Equations for Viscous Flow

Consider the flow of an incompressible fluid through (or within) a spatially fixed, time-independent volume V. For general viscous flow problems, it can be stated that, in an Eulerian description, the following physical fields have to be determined:

- the velocity field: $\vec{v}(\vec{x}, t)$ for all $\vec{x}$ in V and all t; and
- the stress field: $\boldsymbol{\sigma}(\vec{x}, t)$ for all $\vec{x}$ in V and all t.

Starting from the incompressibility condition, the velocity field has to satisfy the continuity equation (mass balance) for all $\vec{x}$ in V for all t:

$$\mathrm{tr}(\boldsymbol{D}) = 0 \quad \text{with} \quad \boldsymbol{D} = \frac{1}{2}\left(\left(\vec{\nabla}\vec{v}\right)^{\mathrm{T}} + \vec{\nabla}\vec{v}\right), \tag{13.25}$$

while that, as well as the velocity field and the stress field, should be related for all $\vec{x}$ in V and for all times t according to the local constitutive equation (see Section 12.5):

$$\boldsymbol{\sigma} = -p\boldsymbol{I} + \boldsymbol{\sigma}^{\mathrm{d}}(\boldsymbol{D}). \tag{13.26}$$

The local balance of momentum (see Section 11.5 for the Eulerian description) must also be satisfied; so, for all $\vec{x}$ in V and for all t:

$$\vec{\nabla} \cdot \boldsymbol{\sigma} + \rho\vec{q} = \rho\left(\left(\vec{\nabla}\vec{v}\right)^{\mathrm{T}} \cdot \vec{v} + \frac{\delta\vec{v}}{\delta t}\right), \tag{13.27}$$

with the (mass) density ρ constant.

The equations above form a set of coupled partial differential equations. Consequently, for a unique solution of the velocity field $\vec{v}(\vec{x}, t)$ and the stress field $\boldsymbol{\sigma}(\vec{x}, t)$, boundary conditions and initial conditions are indispensable.

With respect to boundary conditions, it can be stated that for each t at every point on the outer surface of V, three (scalar) relations have to be specified: either completely formulated in terms of stresses (dynamic boundary conditions), or completely expressed in terms of velocities (kinematic boundary conditions) or in a mixed format.

A detailed description of the interpretation of the initial conditions is not considered.

13.3.2 The Equations for a Newtonian Fluid

For a Newtonian fluid (see Section 12.5), the stress tensor can be written as

$$\boldsymbol{\sigma} = -p\boldsymbol{I} + 2\eta\boldsymbol{D}, \tag{13.28}$$

where the viscosity η is constant. Substitution of this constitutive equation into the local momentum balance leads to the equation:

$$-\vec{\nabla}p + \eta\vec{\nabla}\cdot\left(\left(\vec{\nabla}\vec{v}\right)^{\mathrm{T}} + \vec{\nabla}\vec{v}\right) + \rho\vec{q} = \rho\left(\left(\vec{\nabla}\vec{v}\right)^{\mathrm{T}}\cdot\vec{v} + \frac{\delta\vec{v}}{\delta t}\right). \tag{13.29}$$

The left-hand side of this equation can be simplified by using the following identities (to be derived by elaboration in components):

$$\vec{\nabla}\cdot\left(\vec{\nabla}\vec{v}\right)^{\mathrm{T}} = \vec{\nabla}\left(\vec{\nabla}\cdot\vec{v}\right) \tag{13.30}$$

$$\vec{\nabla}\cdot\left(\vec{\nabla}\vec{v}\right) = \left(\vec{\nabla}\cdot\vec{\nabla}\right)\vec{v}. \tag{13.31}$$

This leads to:

$$-\vec{\nabla}p + \eta\left(\vec{\nabla}\left(\vec{\nabla}\cdot\vec{v}\right) + \left(\vec{\nabla}\cdot\vec{\nabla}\right)\vec{v}\right) + \rho\vec{q} = \rho\left(\left(\vec{\nabla}\vec{v}\right)^{\mathrm{T}}\cdot\vec{v} + \frac{\delta\vec{v}}{\delta t}\right). \tag{13.32}$$

Using the expression for incompressibility of the fluid,

$$\vec{\nabla}\cdot\vec{v} = 0, \tag{13.33}$$

results in the so-called Navier–Stokes equation:

$$-\vec{\nabla}p + \eta\left(\vec{\nabla}\cdot\vec{\nabla}\right)\vec{v} + \rho\vec{q} = \rho\left(\left(\vec{\nabla}\vec{v}\right)^{\mathrm{T}}\cdot\vec{v} + \frac{\delta\vec{v}}{\delta t}\right). \tag{13.34}$$

The last two relations, the incompressibility condition (continuity equation) and the Navier–Stokes equation, together form a set that allows the determination of the velocity field $\vec{v}(\vec{x}, t)$ and the pressure field $p(\vec{x}, t)$. For the solution, boundary conditions and initial conditions have to be supplied to the equations.

13.3.3 Stationary Flow of an Incompressible Newtonian Fluid

For a stationary flow, the relevant field variables are only a function of the position vector $\vec{x}$ within the volume V and no longer a function of time. To determine the velocity field $\vec{v}(\vec{x})$ and the pressure field $p(\vec{x})$, the set of equations that has to be solved is reduced to:

$$\vec{\nabla} \cdot \vec{v} = 0 \tag{13.35}$$

$$-\vec{\nabla}p + \eta\left(\vec{\nabla} \cdot \vec{\nabla}\right)\vec{v} + \rho\vec{q} = \rho\left(\vec{\nabla}\vec{v}\right)^{\mathrm{T}} \cdot \vec{v}. \tag{13.36}$$

In addition, it is necessary to specify a full set of boundary conditions. Initial conditions do not apply for stationary problems. Note that the equation is non-linear as a consequence of the term on the right-hand side of the last equation; this has a seriously complicating effect on the solution process. Exact analytical solutions can be found only for very simple problems.

13.3.4 Boundary Conditions

In the present section, the attention is focussed on the formulations of simple boundary conditions with respect to an arbitrary point on the outer surface of the considered (fixed) volume V, with local outward unit normal $\vec{n}$. A number of different possibilities will separately be reviewed.

- Locally prescribed velocity $\vec{v}$ along the boundary, i.e. the component $\vec{v} \cdot \vec{n}$ in the normal direction as well as the component $\vec{v} - (\vec{v} \cdot \vec{n})\vec{n}$ in the tangential direction. A well-known example of this is the set of boundary conditions for 'no slip' contact of a fluid with a fixed wall: $\vec{v} = \vec{0}$. In fact, the impermeability of the wall (there is no flux through the outer surface; also see Fig. 7.11) is expressed by $\vec{v} \cdot \vec{n} = 0$, while suppression of slip is expressed by $\vec{v} - (\vec{v} \cdot \vec{n})\vec{n} = \vec{0}$.
- Locally prescribed stress vector $\boldsymbol{\sigma} \cdot \vec{n}$ along the boundary, i.e. the component $\vec{n} \cdot \boldsymbol{\sigma} \cdot \vec{n}$ in the normal direction as well as the components of $\boldsymbol{\sigma} \cdot \vec{n} - (\vec{n} \cdot \boldsymbol{\sigma} \cdot \vec{n})\vec{n}$ in the tangential direction. Boundary conditions of this type can be transformed into boundary conditions expressed in $\vec{v}$ and p by means of the constitutive equations. A known example of this is the set of boundary conditions at a free surface: the normal component of the stress vector is related to the atmospheric pressure (equal with opposite sign), and the tangential components are equal to zero.
- For a frictionless flow along a fixed wall, it should be required that $\vec{v} \cdot \vec{n} = 0$ combined with $\boldsymbol{\sigma} \cdot \vec{n} - (\vec{n} \cdot \boldsymbol{\sigma} \cdot \vec{n})\vec{n} = \vec{0}$. Again the last condition can, by using the constitutive equation, be expressed in terms of $\vec{v}$ and p.

13.3.5 Elementary Analytical Solutions

Figure 13.5 visualizes a stationary flow of a fluid between two 'infinitely extended' stationary parallel flat plates (mutual distance h). The flow in the

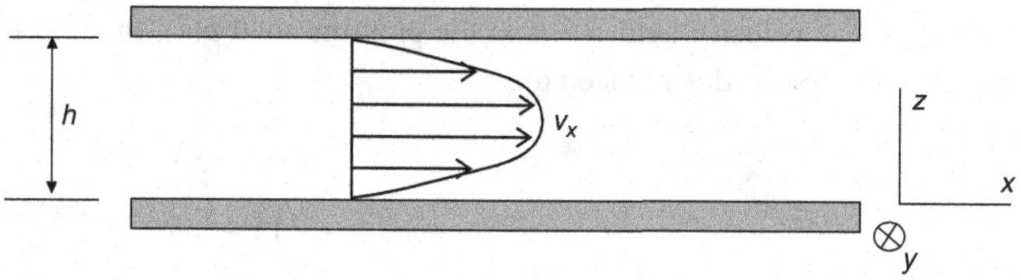

Figure 13.5

Flow between two stationary parallel plates.

positive x-direction is activated by means of an externally applied pressure gradient. We consider that part between the plates (the specific domain with x- and z-coordinates) where the flow is fully developed. This means that no influence is noticeable any longer from the detailed conditions near the inflow or outflow of the fluid domain: for all relevant values of x, the velocity profile is the same. For $z = 0$ and $z = h$, the fluid adheres to the bottom and the top plate respectively ('no slip' boundary conditions; see above). It is simple to verify that the solution for (the components of) the velocity field $\vec{v}(\vec{x})$ and the pressure field $p(\vec{x})$ according to:

$$v_x = -\frac{1}{2\eta}\frac{\partial p}{\partial x} z(h - z), \qquad v_y = 0, \qquad v_z = 0 \tag{13.37}$$

$$\frac{\partial p}{\partial x} \text{ constant}, \qquad \frac{\partial p}{\partial y} = 0, \qquad \frac{\partial p}{\partial z} = 0, \tag{13.38}$$

satisfies exactly the continuity equation, the Navier–Stokes equation (with $\vec{q} = \vec{0}$) and the prescribed boundary conditions. It can be observed that the inertia term (the right-hand side of the Navier–Stokes equation) for the defined problem vanishes. This is quite obvious as material particles of the fluid move with constant velocity (no acceleration) in the x-direction.

The situation described above is called the 'plane Poiseuille flow'. The flow is characterized by a parabolic (in the z-coordinate) profile for the velocity in the x-direction, coupled to a constant pressure gradient in the x-direction. Note that the pressure itself cannot be determined. In order to do that, the pressure should be prescribed for a certain value of x, in combination with the pressure gradient in the x-direction or in combination with, for example, the total mass flux (per unit time) in the x-direction.

In Fig. 13.6, the stationary flow is depicted of a fluid between two parallel flat plates (distance h) in the case where the bottom plate ($z = 0$) is spatially fixed and the top plate ($z = h$) translates in the x-direction with a constant velocity V. There is no pressure gradient. In this example, a fully developed flow is the starting point

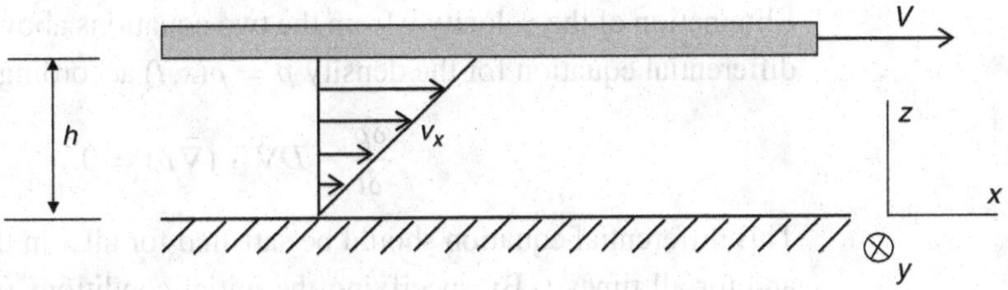

Figure 13.6

Flow between two mutually translating parallel plates.

(plates are assumed to be 'infinitely wide', and peripheral phenomena at the inflow and outflow are left out of consideration). As before, in this case for $z = 0$ and $z = h$, a perfect adhesion between the fluid and plates occurs. Again, it is simple to verify that for the velocity components the following solution holds:

$$v_x = \frac{z}{h}V, \qquad v_y = 0, \qquad v_z = 0. \tag{13.39}$$

Thus, for the velocity in x-direction, a linear profile with respect to the z-coordinate is found. The flow in this example is known as 'Couette flow'.

In this section, two very special cases (yet both of practical interest) of fluid flows have been treated that permit an analytical solution. This is possible for very few practical problems. The finite element method as discussed in Chapters 14 to 18 enables us to construct approximate solutions for very complex flows, but the specific algorithms to do that for viscous flows are beyond the scope of the contents of this book.

13.4 Diffusion and Filtration

In Section 12.8, the constitutive equations for diffusion and filtration to describe transport of material ('fluid') through a stationary porous medium have been discussed. By adding the relevant balance laws (see Chapter 11), the problem description can be further elaborated. This will be the subject of the present section.

For diffusion, Fick's law can be applied:

$$\rho\vec{v} = -D\vec{\nabla}\rho, \tag{13.40}$$

with D the diffusion coefficient.

The mass balance (see Section 11.2) can be written as

$$\frac{\delta\rho}{\delta t} + \vec{\nabla} \cdot (\rho\vec{v}) = 0. \tag{13.41}$$

Elimination of the velocity $\vec{v}$ from the two equations above leads to a linear partial differential equation for the density $\rho = \rho(\vec{x}, t)$ according to:

$$\frac{\delta\rho}{\delta t} - D\vec{\nabla} \cdot (\vec{\nabla}\rho) = 0. \tag{13.42}$$

This differential equation should be satisfied for all $\vec{x}$ in the considered domain V and for all times t. By specifying the initial conditions (prescribed ρ for all $\vec{x}$ in V at $t = 0$) and with one single boundary condition for each boundary point of V (either the density ρ or the outward mass flux $\rho\vec{v} \cdot \vec{n} = -D(\vec{\nabla}\rho) \cdot \vec{n}$ should be prescribed), in principle a solution for $\rho(\vec{x}, t)$ can be calculated.

To illustrate some of the problems that arise, we confine ourselves to an attempt to solve a simple one-dimensional problem. A domain is given by $0 \leq x \leq L$. Diffusion of a certain material (diffusion coefficient D) in the x-direction can take place. For the density $\rho = \rho(x, t)$, the following partial differential equation holds:

$$\frac{\partial\rho}{\partial t} - D\frac{\partial^2\rho}{\partial x^2} = 0, \tag{13.43}$$

emphasizing that the spatial derivative $\delta\rho/\delta t$ is written here as the partial derivative of ρ with respect to the time t (with constant x). Misunderstandings because of this will not be introduced, because an exclusively Eulerian description will be used. The differential equation in this example is completed with:

- the initial condition:

$$\rho = 0 \text{ for } 0 \leq x \leq L \text{ and } t = 0,$$

- the boundary conditions:

$$\rho = \rho_0 \text{ (with } \rho_0 \text{ a constant) for } x = 0 \text{ and } t > 0$$
$$\frac{\partial\rho}{\partial x} = 0 \text{ (no outflow of material) for } x = L \text{ and } t > 0.$$

Even for this very simple situation, an exact solution is very difficult to determine. A numerical approach (for example by means of the finite element method) can lead to a solution in a simple way. This is the topic of Chapter 14. Here, it can be stated that the solution for $\rho(x, t)$ at $t \to \infty$ has to satisfy $\rho(x, t) = \rho_0$ for all x. A large number of closed-form solutions can be found in [4].

For filtration problems, Darcy's law can be applied:

$$\rho\vec{v} = -\kappa\vec{\nabla}p, \tag{13.44}$$

with κ the permeability. The mass balance (see Section 11.2) can be written as follows:

$$\frac{\delta\rho}{\delta t} + \vec{\nabla} \cdot (\rho\vec{v}) = 0. \tag{13.45}$$

Further elaboration is limited to stationary filtration (time t does not play a role). In that case

$$\frac{\delta \rho}{\delta t} = 0 \text{ and so } \vec{\nabla} \cdot (\rho \vec{v}) = 0.$$

Combination of this equation with Darcy's constitutive law leads to:

$$\vec{\nabla} \cdot \left(\vec{\nabla} p \right) = 0. \tag{13.46}$$

This equation for the pressure p can formally be solved when for every boundary point of the volume V one single condition is specified. This can either be formulated in terms of the pressure p, or the outward mass flux $\rho \vec{v} \cdot \vec{n} = -\kappa(\vec{\nabla} p) \cdot \vec{n}$. When the solution for p is determined, it is easy to calculate directly the mass flux $\rho \vec{v}$ with Darcy's law.

In the one-dimensional case (with x as the only relevant independent variable), the differential equation for p reduces to

$$\frac{d^2 p}{dx^2} = 0. \tag{13.47}$$

In this case p will be a linear function of x.

Exercises

13.1 Consider a material element in the shape of a cube (edge length ℓ). The cube is placed in a Cartesian xyz-coordinate system (see figure).

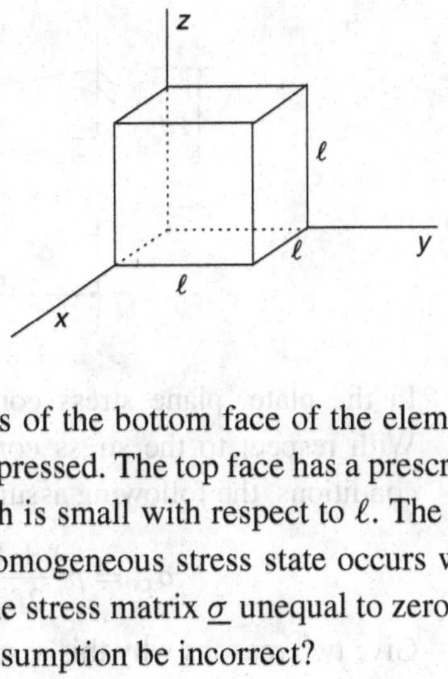

All displacements of the bottom face of the element (coinciding with the xy-plane) are suppressed. The top face has a prescribed displacement in the y-direction, which is small with respect to ℓ. The side faces are unloaded. Assume that a homogeneous stress state occurs with $\sigma_{yz} = \sigma_{zy}$, the only components of the stress matrix $\underline{\sigma}$ unequal to zero.
Why must this assumption be incorrect?

13.2 Consider a thin rectangular piece of material (constant thickness h). The midplane of the material coincides with the xy-plane of a Cartesian xyz-coordinate system. The material behaviour is described by means of Hooke's law (Young's modulus E and Poisson's ratio ν). The plate is statically loaded with pure shear (plane stress state). The accompanying boundary conditions are shown in the figure below.
For an arbitrary point of the midplane, with the coordinates (x_0, y_0) in the unloaded reference configuration, the displacements in the x- and y-directions, as a result of the stress τ, are indicated by $u_x(x_0, y_0)$, $u_y(x_0, y_0)$, respectively.
Why does no unique solution exist for the fields $u_x(x_0, y_0)$, $u_y(x_0, y_0)$, based on the information that is given above?

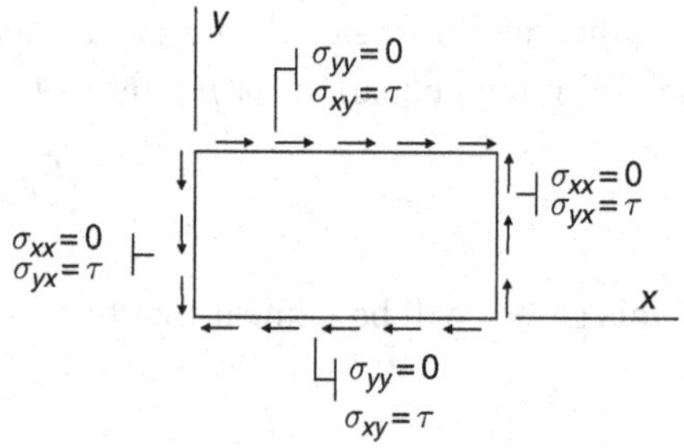

13.3 A thin trapezium-shaped plate is clamped on the left side and statically loaded on the right side with an extensional load (local extensional stress p), as given in the figure below.

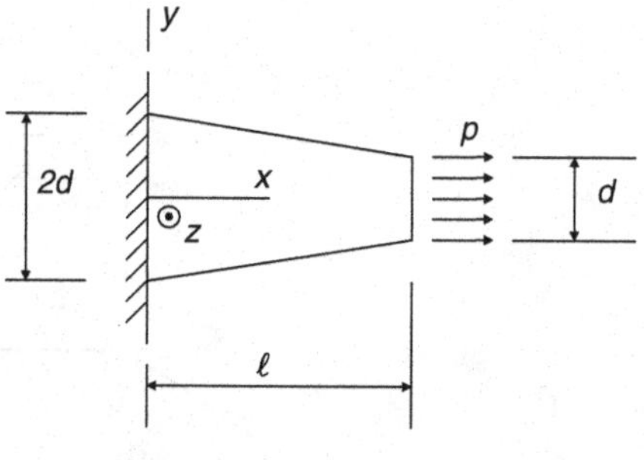

In the plate, plane stress conditions occur ($\sigma_{xz} = \sigma_{yz} = \sigma_{zz} = 0$). With respect to the stress components still relevant for the plane stress conditions, the following assumptions are made:

$$\sigma_{xx} = p\frac{x+\ell}{2\ell}, \qquad \sigma_{yy} = 0, \qquad \sigma_{xy} = 0.$$

Give two reasons why this assumed stress field cannot be correct.

13.4 Consider a rectangular plate of some material ($\ell \times b$). The thickness h is small with respect to ℓ and b: see figure.

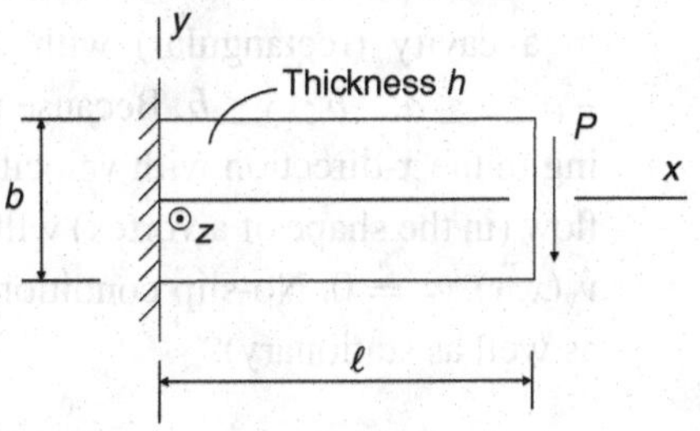

The left side ($x = 0$) is clamped. The right side ($x = \ell$) is loaded with a distributed tangential load (the resultant force P is known) in the negative y-direction. The top and bottom surface of the plate are unloaded. A plane stress condition is supposed. With respect to the stress field, the following assumption is proposed:

$$\sigma_{xx} = c_1(\ell - x)y$$

$$\sigma_{yy} = 0$$

$$\sigma_{xy} = c_2\left(\frac{b^2}{4} - y^2\right),$$

with c_1 and c_2 constants.

Determine c_2 based on the relationship between σ_{xy} and P and subsequently, determine c_1 by means of the local equilibrium equations.

13.5 In the environment of the origin of a Cartesian xyz-coordinate system it is given that a (two-dimensional) stationary velocity field in an incompressible Newtonian fluid (with density ρ) can be described as

$$\vec{v} = \alpha(-y\vec{e}_x + x\vec{e}_y) \quad \text{with } \alpha \text{ a constant.}$$

Based on this velocity field, the deformation rate tensor $\boldsymbol{D}$ can be calculated: $\boldsymbol{D} = \mathbf{0}$ with $\mathbf{0}$ the zero tensor. With substitution into the constitutive equation:

$$\boldsymbol{\sigma} = -p\boldsymbol{I} + 2\eta\boldsymbol{D},$$

with η the viscosity and $\boldsymbol{\sigma}$ the stress tensor (and with $p = 0$ originated by the applied boundary conditions), it follows that $\boldsymbol{\sigma} = \mathbf{0}$.

Then, by means of the Navier–Stokes equation, it is found that the distributed load $\vec{q}$ (force per unit mass) necessary to realize the described flow field is given by

$$\vec{q} = -\alpha^2\vec{r} \quad \text{with} \quad \vec{r} = x\vec{e}_x + y\vec{e}_y.$$

Interpret the stated results: why is $\boldsymbol{D} = \boldsymbol{0}$ obtained for the given flow field and what is the physical meaning of $\vec{q} = -\alpha^2\vec{r}$?

13.6 An incompressible Newtonian fluid (density ρ, viscosity η) is located in a cavity (rectangular) with a cross section as given in the figure: $-a \le x \le a, -b \le y \le b$. Because the top wall closing the cavity is translating in the x-direction with velocity V, a stationary (two-dimensional) fluid flow (in the shape of a vortex) will develop in the cavity: $v_x = v_x(x, y), v_y = v_y(x, y), v_z = 0$. No-slip conditions hold at all walls of the cavity (moving as well as stationary).

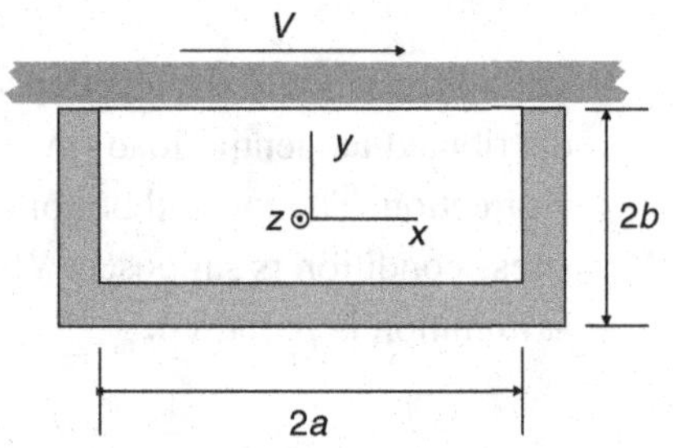

Which inconsistency ('something leading to contradiction') can be observed in the problem definition?

13.7 A compressible medium (for example a gas) is forced in the x-direction through a contraction (see figure). With respect to the components of the velocity field $\underset{\sim}{v}$ of the medium, defined in the Cartesian xyz-coordinate system, it is assumed that v_x only depends on x, thus $v_x = v_x(x)$, completed with $v_y = v_z = 0$.

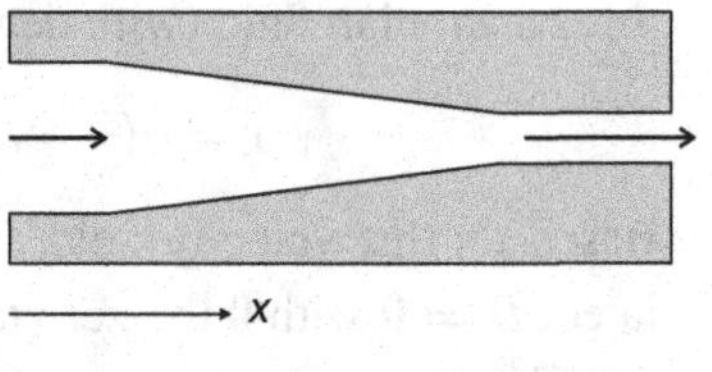

Why is it impossible for this assumption to be correct?

13.8 In the neighbourhood of the origin of a Cartesian xyz-coordinate system, the stationary (two-dimensional) flow field in an incompressible Newtonian fluid (with density ρ) can be described as

$$\vec{v} = \alpha(-y\vec{e}_x + x\vec{e}_y) \quad \text{with } \alpha \text{ a constant.}$$

With respect to the constitutive equation

$$\boldsymbol{\sigma} = -p\boldsymbol{I} + 2\eta\boldsymbol{D},$$

with η the viscosity, $\boldsymbol{\sigma}$ the stress tensor and $\boldsymbol{D}$ the deformation rate tensor, it can be stated (based on the prescribed boundary conditions) that p satisfies $p = 0$.

Determine, by means of the Navier–Stokes equation, the distributed load $\vec{q}$ (per unit mass) necessary to realize the given flow pattern.

14 Solution of the One-Dimensional Diffusion Equation by Means of the Finite Element Method

In the present and following chapters, extensive use will be made of a simple finite element code `mlfem_nac`. This code, including a manual, can be freely downloaded from the website: www.tue.nl/biomechanicsbook.

The code is written in the program environment MATLAB. To be able to use this environment, a licence for MATLAB has to be obtained. For information about MATLAB see: www.mathworks.com.

14.1 Introduction

It will be clear from the previous chapters that many problems in biomechanics are described by (sets of) partial differential equations, and for most problems it is difficult or impossible to derive closed-form (analytical) solutions. However, by means of computers, approximate solutions can be determined for a very large range of complex problems, which is one of the reasons that biomechanics as a discipline has grown so fast in the past three decades. These computer-aided solutions are called numerical solutions, as opposed to analytical or closed-form solutions of equations. The present and following chapters are devoted to the numerical solution of partial differential equations, for which several methods exist. The most important ones are the finite difference method and the finite element method. The latter is especially suitable for partial differential equations on domains with complex geometries, material properties and boundary conditions (which is nearly always the case in biomechanics). For this reason the next chapters focus on the finite element method. The basic concepts of the method are explained in the present chapter. The one-dimensional diffusion equation will be used as an example to illustrate the key features of the finite element method. Clearly, the one-dimensional diffusion problem can be solved analytically for a wide variety of parameter choices, but the structure of the differential equation is representative of a much larger class of problems to be discussed later. In addition, the diffusion equation and the more extended, non-stationary (convection) diffusion equation play an important role in many processes in biomechanics.

14.2 The Diffusion Equation

The differential equation that describes the one-dimensional diffusion problem is given by

$$\frac{d}{dx}\left(c\frac{du}{dx}\right) + f = 0, \tag{14.1}$$

where $u(x)$ is the unknown function, $c(x) > 0$ a given material characteristic function and $f(x)$ a given source term. This differential equation is defined on a one-dimensional domain Ω that spans the x-axis between $x = a$ and $x = b$, while the boundary, which is formally denoted by Γ, is located at $x = a$ and $x = b$.

Equation (14.1) is an adapted form of the diffusion equation Eq. (13.43), introduced in the previous chapter. Different symbols for the unknown (u instead of ρ) and coefficient (c instead of D) are used to emphasize the general character of the equation, applicable to different kinds of problems (see below). Furthermore, the coefficient c can be a function of x, and a source term $f(x)$ is introduced.

Two types of boundary conditions can be discerned. First, there is the essential boundary condition, which must be specified in terms of u. For the derivations that follow, the boundary at $x = a$ is chosen to specify this type of boundary condition:

$$u = U \text{ at } \Gamma_u, \tag{14.2}$$

where Γ_u denotes the boundary of the domain Ω at $x = a$. Second, a natural boundary condition may be specified. Here the boundary at $x = b$ is chosen to specify the flux $c\, du/dx$:

$$c\frac{du}{dx} = P \text{ at } \Gamma_p, \tag{14.3}$$

where Γ_p denotes the boundary at $x = b$. For diffusion problems, an essential boundary condition must be specified to have a well-posed boundary value problem. This is not necessarily the case for natural boundary conditions; they may be absent.

Example 14.1 The diffusion equation describes a large range of problems in biomechanics and is applicable in many different areas. In the way it was introduced in Section 13.4, the unknown u represents the concentration of some matter, for example oxygen in blood; proteins or other molecules in an extracellular matrix or inside a cell; medication in blood or tissue. In that case, the term f can be either a source term (where matter is produced) or a sink term (where matter is consumed). In the human body, diffusion is very important, but this is also the case for in

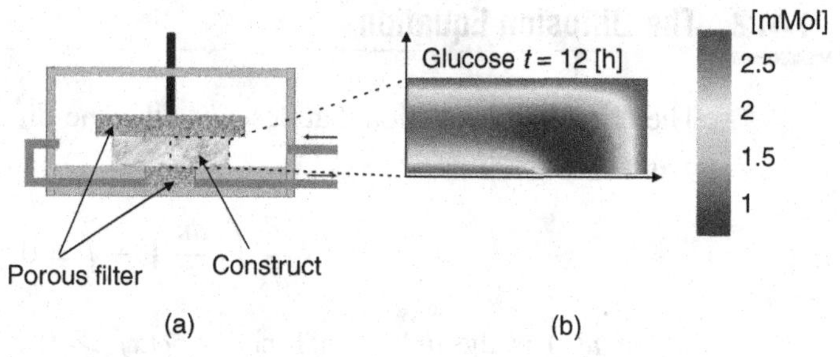

Figure 14.1

(a) Schematic view of a bioreactor system designed for growing articular cartilage tissue. (b) Result of a numerical calculation of the concentration of glucose in the tissue engineered construct. In the analysis, it is assumed that the glucose concentration in the medium surrounding the construct is constant (essential boundary condition) and that glucose is consumed by the cells in the construct (sink term). Because of its symmetry, only the right half of the construct is modelled. Adapted from [20].

vitro experimental set-ups in the laboratory, for example in tissue engineering applications (see Fig. 14.1).

Example 14.2 A completely different process which is also relevant in biomechanics is steady-state one-dimensional heat conduction with a source term:

$$\frac{d}{dx}\left(\lambda\frac{dT}{dx}\right) + f = 0,$$

where λ is the Fourier coefficient of heat conduction and f a heat source term. Further, $T(x)$ represents the temperature. The boundary conditions might be formulated as: prescribed temperature at Γ_u and prescribed heat flux at Γ_p. It should be noticed that the mathematical structure of this heat conduction problem is fully equivalent to the structure of the diffusion problem.

Example 14.3 A third example of a completely different diffusion type problem is the uniaxial tension or compression of a bar as introduced in Chapter 6, governed by the equation:

$$\frac{d}{dx}\left(EA\frac{du}{dx}\right) + f = 0,$$

where E is the Young's modulus, A the cross section of the bar and f a distributed force per unit length. The unknown $u(x)$ is the displacement field of the bar (also see Eq. (6.20)). Boundary conditions may be imposed as a prescribed displacement on Γ_u and a prescribed force at Γ_p.

The approximate solution of Eq. (14.1) is found by transforming the differential equation into a discrete set of ordinary equations:

$$\frac{d}{dx}\left(c\frac{du}{dx}\right) + f = 0 \longrightarrow \underset{\sim}{\underline{K}}\,\underset{\sim}{u} = \underset{\sim}{f}. \tag{14.4}$$

The array $\underset{\sim}{u}$ contains approximations of the continuous solution u of the differential equation at a finite number of locations on the x-axis. Increasing the number of points defining $\underset{\sim}{u}$ should lead to an increased accuracy of the approximation of u. A particularly attractive feature of the finite element method is that the spatial distribution of these points does not need to be equidistant and can be chosen such that accurate solutions can be obtained with a limited number of points, even for complex geometries (in the multi-dimensional case) or problems with large gradients in the solution.

The finite element method proceeds along three well-defined steps:

(i) Transformation of the original differential equation into an integral equation by means of the principle of **weighted residuals**.

(ii) **Discretization** of the solution u by interpolation. If an approximation of the solution u is known at a finite number of points (nodes), an approximation field may be constructed by interpolation between these point (nodal) values.

(iii) Using the discretization, the integral equation is transformed into a linear set of equations from which the nodal values $\underset{\sim}{u}$ can be solved.

14.3 Method of Weighted Residuals and Weak Form

First of all, the differential equation is transformed into an integral equation by means of the weighted residuals method. Suppose that a given function $g(x) = 0$ on a certain domain $a \leq x \leq b$; then this formulation is equivalent to requiring

$$\int_a^b w(x)g(x)\,dx = 0 \quad \text{for all } w, \tag{14.5}$$

and to emphasize this important equivalence:

$$g(x) = 0 \quad \text{on} \quad a \leq x \leq b \Leftrightarrow \int_a^b w(x)g(x)\,dx = 0 \quad \text{for all } w. \tag{14.6}$$

The function $w(x)$ is called the weighting function and is assumed to be a continuous function on the integration domain. The equivalence originates from the requirement that the integral equation must hold for all possible weighting functions w. It therefore should also hold for $w = g(x)$. For this particular choice of w, the integral expression yields

$$\int_a^b g^2(x)\,dx = 0 \;\Rightarrow\; g(x) = 0 \quad \text{on } a \leq x \leq b. \tag{14.7}$$

This follows immediately from the observation that the square of a function $g(x)$ is always greater than or equal to zero for any value of x, i.e. $g^2(x) \geq 0$, such that the integral of $g^2(x)$ over the domain $a \leq x \leq b$ must be greater than or equal to zero, i.e.

$$\int_a^b g^2(x)\, dx \geq 0, \tag{14.8}$$

and can only be equal to zero if $g(x) = 0$ for all $a \leq x \leq b$.

Effectively, the method of weighted residuals transforms the requirement that a function, say $g(x)$, must be equal to zero on a given domain at an infinite number of points into a single evaluation of the integral that must be equal to zero.

Using the method of weighted residuals, the differential equation, Eq. (14.1), is transformed into an integral equation:

$$\int_a^b w \left[\frac{d}{dx}\left(c\frac{du}{dx} \right) + f \right] dx = 0, \tag{14.9}$$

which should hold for all weighting functions $w(x)$. The term between the square brackets contains second-order derivatives d^2u/dx^2 of the function u. As has been outlined in the introduction, approximate solutions of u are sought by defining an interpolation of u on the domain of interest and transforming the integral equation into a discrete set of linear equations. Defining interpolation functions that are both second-order differentiable and still integrable is far from straightforward, in particular in the multi-dimensional case on arbitrarily shaped domains. Fortunately, the second-order derivatives can be removed by means of an integration by parts:

$$w\left(c\frac{du}{dx} \right)\Big|_a^b - \int_a^b \frac{dw}{dx} c \frac{du}{dx}\, dx + \int_a^b wf\, dx = 0. \tag{14.10}$$

This introduces the boundary terms:

$$w\left(c\frac{du}{dx} \right)\Big|_a^b = -w(a)\left(c\frac{du}{dx} \right)\Big|_{x=a} + w(b)\left(c\frac{du}{dx} \right)\Big|_{x=b}. \tag{14.11}$$

At the boundary, either u is prescribed (i.e. the essential boundary condition at $x = a$) or the derivative $c\, du/dx$ is prescribed (i.e. the natural boundary condition at $x = b$). Along the boundary where u is prescribed the corresponding flux, say p_u, with

$$p_u = c\frac{du}{dx}\Big|_{x=a}, \tag{14.12}$$

is unknown, while along the other boundary the flux $c\frac{du}{dx}\Big|_{x=b} = P$ is known. For the time being, the combination of boundary terms in Eq. (14.11) is written as

$$B = w\left(c\frac{du}{dx}\right)\Big|_a^b = -w(a)p_u + w(b)P, \tag{14.13}$$

realizing that p_u is considered as yet unknown. The term B is introduced as an abbreviation for the boundary contribution.

Consequently, the following integral equation, known as the **weak form**, results:

$$\int_a^b \frac{dw}{dx} c \frac{du}{dx}\, dx = \int_a^b wf\, dx + B. \tag{14.14}$$

Equation (14.14) is the starting point for the finite element method, and is called the weak form because the differentiability requirements imposed on u have been reduced: in the original differential equation, Eq. (14.1), the second-order derivative d^2u/dx^2 appears, while in the weak form, Eq. (14.14), only the first-order derivative du/dx has to be dealt with.

14.4 Polynomial Interpolation

Suppose that at a finite number of points x_i in the domain Ω, the function values $u_i = u(x_i)$ are known: then a polynomial approximation, denoted by u_h, of $u(x)$ can be constructed. The polynomial approximation u_h of degree $n-1$ can be constructed by

$$u_h(x) = a_0 + a_1x + a_2x^2 + \cdots + a_{n-1}x^{n-1}, \tag{14.15}$$

if u is known at n points. The coefficients a_i can be identified uniquely and expressed in terms of u_i, by solving the set of equations:

$$\begin{bmatrix} 1 & x_1 & x_1^2 & \cdots & x_1^{n-1} \\ 1 & x_2 & x_2^2 & \cdots & x_2^{n-1} \\ 1 & x_3 & x_3^2 & \cdots & x_3^{n-1} \\ \vdots & \vdots & \vdots & \vdots & \vdots \\ 1 & x_n & x_n^2 & \cdots & x_n^{n-1} \end{bmatrix} \begin{bmatrix} a_0 \\ a_1 \\ a_2 \\ \vdots \\ a_{n-1} \end{bmatrix} = \begin{bmatrix} u_1 \\ u_2 \\ u_3 \\ \vdots \\ u_n \end{bmatrix}. \tag{14.16}$$

An example is given in Fig. 14.2, where, in the domain $x_{i-1} \le x \le x_{i+2}$, a third-order polynomial (dashed curve) is used to approximate a given function (solid curve) based on the function values u_{i-1}, u_i, u_{i+1} and u_{i+2}.

Clearly, the coefficients a_i are linearly dependent on the values u_i, and therefore the polynomial may be rewritten in terms of u_i by

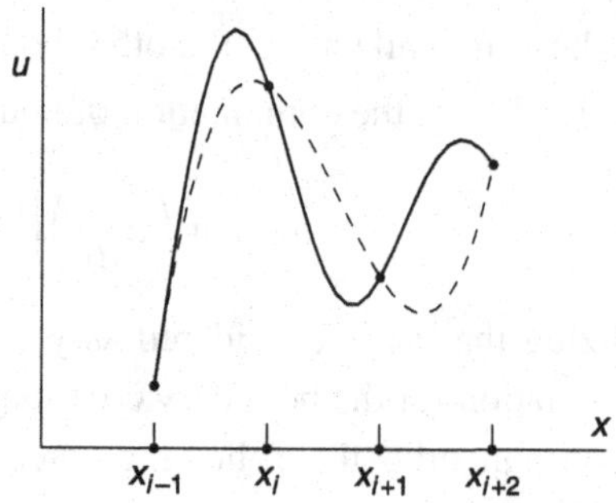

Figure 14.2

Solid line, $u(x)$; dashed line, polynomial approximation of $u(x)$.

$$u_{\text{h}}(x) = \sum_{i=1}^{n} N_i(x) u_i, \tag{14.17}$$

where the functions $N_i(x)$ are polynomial expressions of order $n-1$ in terms of the coordinate x. These functions $N_i(x)$ are called **shape functions** because they define the shape of the interpolation of u_h, for instance linear, quadratic etc. To illustrate this, consider a first-order polynomial on the domain $[x_1, x_2]$, with $u(x)$ known at x_1 and x_2. In that case

$$u_{\text{h}}(x) = \left(1 - \frac{x - x_1}{x_2 - x_1}\right) u_1 + \frac{x - x_1}{x_2 - x_1} u_2, \tag{14.18}$$

implying that

$$N_1 = \left(1 - \frac{x - x_1}{x_2 - x_1}\right), \quad N_2 = \frac{x - x_1}{x_2 - x_1}. \tag{14.19}$$

Rather than approximating $u(x)$ with a single polynomial of a certain degree over the entire domain Ω, the domain Ω may also be divided into a number of non-overlapping subdomains, say Ω_e. Within each subdomain Ω_e a local polynomial approximation of u may be constructed. A typical example is a piecewise linear approximation within each subdomain Ω_e: see Fig. 14.3. Consider one of the subdomains $\Omega_e = [x_i, x_{i+1}]$. Then within Ω_e the function $u(x)$ is approximated by

$$u_{\text{h}}(x)|_{\Omega_e} = N_1(x) u_i + N_2(x) u_{i+1}, \tag{14.20}$$

with, in conformity with Eq. (14.19):

$$N_1 = \left(1 - \frac{x - x_i}{x_{i+1} - x_i}\right) \quad N_2 = \frac{x - x_i}{x_{i+1} - x_i}. \tag{14.21}$$

More generally, if within a subdomain Ω_e an n-th order polynomial approximation of u is applied, the subdomain should have $n+1$ points at which the function u is

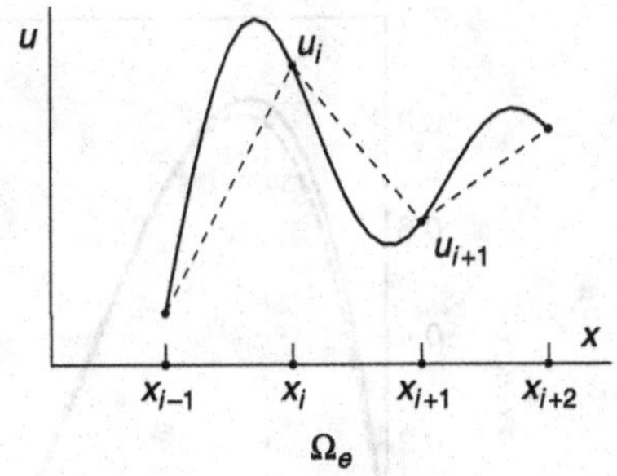

Figure 14.3

Solid line, $u(x)$; dashed line, piecewise linear approximation of $u(x)$.

known. For instance, to use a second-order (quadratic) polynomial, the subdomain should cover at least three consecutive points, for example $\Omega_e = [x_i, x_{i+2}]$, such that

$$u_h(x)|_{\Omega_e} = N_1(x)u_i + N_2(x)u_{i+1} + N_3(x)u_{i+2}. \tag{14.22}$$

The subdomain Ω_e within which a certain polynomial approximation is used is referred to as an **element**. The points at which the values of u are defined are called **nodes**.

The shape functions N_i cannot be chosen arbitrarily. The most stringent requirement is that u_h must be interpolated continuously over the total domain Ω (the first-order derivative of $u_h(x)$ should exist). Suppose that the nodes x_j within an element are numbered $j = 1, \ldots, n$ and that the associated shape functions N_i are numbered $i = 1, \ldots, n$. In that case, for consistency, the shape functions must be chosen such that

$$N_i(x_j) = \delta_{ij}, \tag{14.23}$$

with:

$$\begin{aligned} \delta_{ij} &= 0 \quad \text{if} \quad i \neq j \\ &= 1 \quad \text{if} \quad i = j. \end{aligned} \tag{14.24}$$

Example 14.4 Consider the function:

$$f(x) = \frac{2\sin(x)}{x+1}$$

on the interval $0 \leq x \leq 8$. We want to approximate the function by means of a polynomial using five nodes at equal distances. The positions of these nodes are given by $x = 0, 2, 4, 6, 8$, respectively. If we use Eqs. (14.23) and (14.24), it is straightforward to derive the following shape functions:

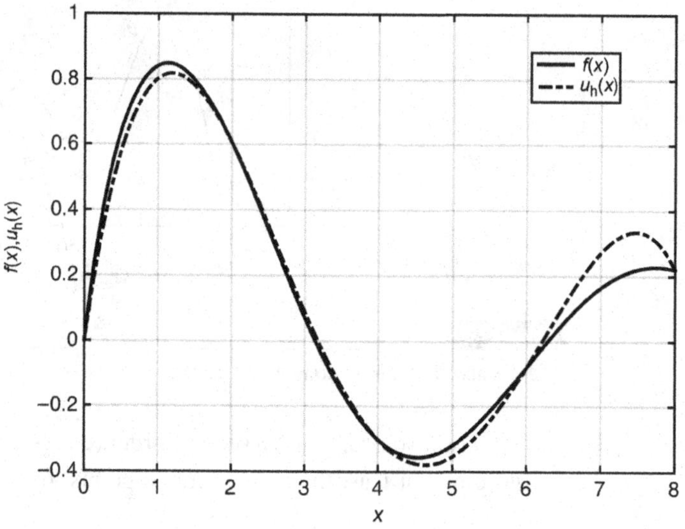

Figure 14.4

Non-linear function $f(x)$ and its polynomial approximation $u_{\mathrm{h}}(x)$.

$$\begin{aligned}
N_1(x) &= \quad (x-2)(x-4)(x-6)(x-8)/384 \\
N_2(x) &= -x(x-4)(x-6)(x-8)/96 \\
N_3(x) &= \quad x(x-2)(x-6)(x-8)/64 \\
N_4(x) &= -x(x-2)(x-4)(x-8)/96 \\
N_5(x) &= \quad x(x-2)(x-4)(x-6)/384.
\end{aligned}$$

Each shape function N_i equals 1 in the node to which the function belongs and zero in all the other nodes. The nodal values that we use in the approximation are $u_i = f(x_i)$ with x_i the coordinates of the nodes. The polynomial approximation $u_{\mathrm{h}}(x)$ can now be written as:

$$u_{\mathrm{h}}(x) = \underset{\sim}{N}^{\mathrm{T}} \underset{\sim}{u} = N_1(x)u_1 + N_2(x)u_2 + N_3(x)u_3 + N_4(x)u_4 + N_5(x)u_5.$$

The figure shows the exact function $f(x)$ and the polynomial approximation $u_{\mathrm{h}}(x)$.

14.5 Galerkin Approximation

To transform the weak form into a linear set of equations in order to derive an approximate solution, the following steps are taken.

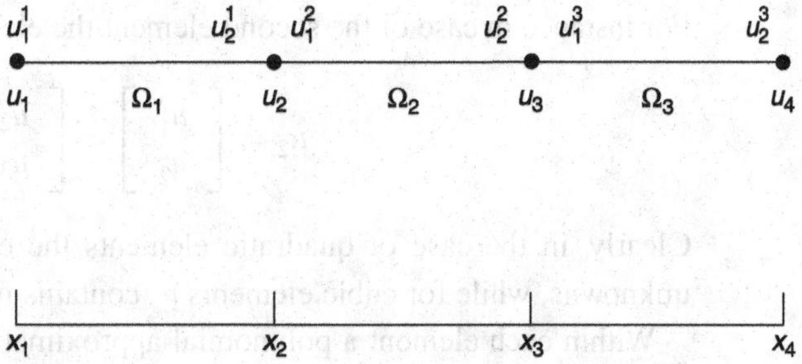

Figure 14.5

Element distribution and unknowns at local and global levels.

Step 1. Element division As shown in the previous section, the domain Ω may be split into a number of subdomains Ω_e, elements, and within each element a polynomial interpolation can be made of the function u. An example of such an element division is given in Fig. 14.5. This distribution of elements is called a **mesh**. Then, the integration over the domain Ω can be performed by summing up the integrals over each element. Consequently, Eq. (14.14) yields:

$$\sum_{e=1}^{N_{\text{el}}} \int_{\Omega_e} \frac{dw}{dx} c \frac{du}{dx}\, dx = \sum_{e=1}^{N_{\text{el}}} \int_{\Omega_e} wf\, dx + B, \tag{14.25}$$

where N_{el} denotes the number of elements.

Step 2. Interpolation Suppose that the domain Ω has been divided into three linear elements, as depicted in Fig. 14.5. Then the nodal values u_i may be collected in an array $\underset{\sim}{u}$:

$$\underset{\sim}{u} = \begin{bmatrix} u_1 \\ u_2 \\ u_3 \\ u_4 \end{bmatrix}. \tag{14.26}$$

The unknowns associated with each of the elements Ω_e are collected in the arrays $\underset{\sim}{u}_e$, such that

$$\underset{\sim}{u}_1 = \begin{bmatrix} u_1 \\ u_2 \end{bmatrix}, \quad \underset{\sim}{u}_2 = \begin{bmatrix} u_2 \\ u_3 \end{bmatrix}, \quad \underset{\sim}{u}_3 = \begin{bmatrix} u_3 \\ u_4 \end{bmatrix}. \tag{14.27}$$

So, it is important to realize that each particular element array $\underset{\sim}{u}_e$ contains a subset of the total, or global, array $\underset{\sim}{u}$. Within each element array $\underset{\sim}{u}_e$ a local numbering may be used, such that for this particular example with linear elements:

$$\underset{\sim}{u}_e = \begin{bmatrix} u_1^e \\ u_2^e \end{bmatrix}. \tag{14.28}$$

For instance in case of the second element the element array $\underset{\sim}{u}_e$ contains

$$\underset{\sim}{u}_2 = \begin{bmatrix} u_1^2 \\ u_2^2 \end{bmatrix} = \begin{bmatrix} u_2 \\ u_3 \end{bmatrix}. \tag{14.29}$$

Clearly, in the case of quadratic elements the element array $\underset{\sim}{u}_e$ contains three unknowns, while for cubic elements $\underset{\sim}{u}_e$ contains four unknowns, and so on.

Within each element a polynomial approximation for both the unknown function $u(x)$ and the weighting function $w(x)$ is introduced. Use of the shape function approach as outlined in the previous section yields

$$u_{\mathrm{h}}|_{\Omega_e} = \sum_{i=1}^{n} N_i(x) u_i^e = \underset{\sim}{N}^{\mathrm{T}} \underset{\sim}{u}_e, \tag{14.30}$$

$$w_{\mathrm{h}}|_{\Omega_e} = \sum_{i=1}^{n} N_i(x) w_i^e = \underset{\sim}{N}^{\mathrm{T}} \underset{\sim}{w}_e, \tag{14.31}$$

where N_i are the **shape** functions

$$\underset{\sim}{N}^{\mathrm{T}} = [N_1 \; N_2 \cdots N_n], \tag{14.32}$$

and $\underset{\sim}{u}_e$ and $\underset{\sim}{w}_e$ contain the unknowns u_i^e and the weighting values w_i^e, respectively, associated with element Ω_e. The fact that the same shape functions (and hence polynomial interpolation order) are chosen for both the unknown function u_{h} and the weighting function w_{h} means that the so-called **(Bubnov) Galerkin** method is used.

As a consequence of the interpolation of u_{h} governed by the shape functions N_i, the differentiation of u_{h} is straightforward since the nodal values u_i are independent of the coordinate x, while the shape functions N_i are simple, known, functions of x:

$$\begin{aligned} \left.\frac{du_{\mathrm{h}}}{dx}\right|_{\Omega_e} &= \sum_{i=1}^{n} \frac{d}{dx}\left(N_i(x) u_i^e\right) \\ &= \sum_{i=1}^{n} \frac{dN_i(x)}{dx} u_i^e \\ &= \frac{d\underset{\sim}{N}^{\mathrm{T}}}{dx} \underset{\sim}{u}_e. \end{aligned} \tag{14.33}$$

Clearly a similar expression holds for the weighting function:

$$\left.\frac{dw_{\mathrm{h}}}{dx}\right|_{\Omega_e} = \frac{d\underset{\sim}{N}^{\mathrm{T}}}{dx} \underset{\sim}{w}_e. \tag{14.34}$$

Substitution of this result into the left-hand side of Eq. (14.25) and considering one element only yields:

$$\begin{aligned}\int_{\Omega_e} \frac{dw_h}{dx} c \frac{du_h}{dx}\, dx &= \int_{\Omega_e} \frac{d\underset{\sim}{N}^{\mathrm{T}}}{dx} \underset{\sim}{w}_e\, c\, \frac{d\underset{\sim}{N}^{\mathrm{T}}}{dx} \underset{\sim}{u}_e\, dx \\ &= \int_{\Omega_e} \underset{\sim}{w}_e^{\mathrm{T}} \frac{d\underset{\sim}{N}}{dx}\, c\, \frac{d\underset{\sim}{N}^{\mathrm{T}}}{dx} \underset{\sim}{u}_e\, dx \\ &= \underset{\sim}{w}_e^{\mathrm{T}} \int_{\Omega_e} \frac{d\underset{\sim}{N}}{dx}\, c\, \frac{d\underset{\sim}{N}^{\mathrm{T}}}{dx}\, dx\, \underset{\sim}{u}_e. \end{aligned} \tag{14.35}$$

In the last step, use has been made of the fact that $\underset{\sim}{w}_e$ and $\underset{\sim}{u}_e$ are both independent of the coordinate x. Likewise, the integral expression on the right-hand side of Eq. (14.25) yields for a single element:

$$\int_{\Omega_e} w_h f\, dx = \int_{\Omega_e} \underset{\sim}{N}^{\mathrm{T}} \underset{\sim}{w}_e f\, dx = \underset{\sim}{w}_e^{\mathrm{T}} \int_{\Omega_e} \underset{\sim}{N} f\, dx. \tag{14.36}$$

Notice that the integral appearing on the right-hand side of Eq. (14.35) is in fact a matrix, called the element coefficient or (in mechanical terms) stiffness matrix:

$$\underline{K}_e = \int_{\Omega_e} \frac{d\underset{\sim}{N}}{dx}\, c\, \frac{d\underset{\sim}{N}^{\mathrm{T}}}{dx}\, dx. \tag{14.37}$$

Similarly, the integral on the right-hand side of Eq. (14.36) is the element array corresponding to the internal source or distributed load:

$$\underset{\sim}{f}_e = \int_{\Omega_e} \underset{\sim}{N} f\, dx. \tag{14.38}$$

Substitution of the expression for the element coefficient matrix and element column in Eq. (14.25) yields

$$\sum_{e=1}^{n_{\mathrm{el}}} \underset{\sim}{w}_e^T \underline{K}_e \underset{\sim}{u}_e = \sum_{e=1}^{n_{\mathrm{el}}} \underset{\sim}{w}_e^{\mathrm{T}} \underset{\sim}{f}_e + B. \tag{14.39}$$

Step 3. Assembling the global set of equations The individual element contributions

$$\underset{\sim}{w}_e^{\mathrm{T}} \underline{K}_e \underset{\sim}{u}_e, \tag{14.40}$$

using the local unknowns and weighting values ($\underset{\sim}{u}_e$ and $\underset{\sim}{w}_e$, respectively) only, may also be rewritten in terms of the global unknowns $\underset{\sim}{u}$ and the associated weighting function values $\underset{\sim}{w}$. This can be done by introducing an auxiliary matrix $\hat{\underline{K}}_e$ on element level such that

$$\underset{\sim}{w}_e^{\mathrm{T}} \underline{K}_e \underset{\sim}{u}_e = \underset{\sim}{w}^T \hat{\underline{K}}_e \underset{\sim}{u}. \tag{14.41}$$

To illustrate this, consider once more the element distribution of Fig. 14.5. For the second element, it holds that

$$\underset{\sim}{w}_2^{\mathrm{T}}\underline{K}_2\underset{\sim}{u}_2 = \begin{bmatrix} w_1^2 & w_2^2 \end{bmatrix} \begin{bmatrix} K_{11}^2 & K_{12}^2 \\ K_{21}^2 & K_{22}^2 \end{bmatrix} \begin{bmatrix} u_1^2 \\ u_2^2 \end{bmatrix} \tag{14.42}$$

$$= \begin{bmatrix} w_2 & w_3 \end{bmatrix} \begin{bmatrix} K_{11}^2 & K_{12}^2 \\ K_{21}^2 & K_{22}^2 \end{bmatrix} \begin{bmatrix} u_2 \\ u_3 \end{bmatrix}. \tag{14.43}$$

Notice that in Eq. (14.42) the local nodal values have been used, while in Eq. (14.43) the global values have been used. Equation (14.43) may also be rewritten as

$$\underset{\sim}{w}_2^{\mathrm{T}}\underline{K}_2\underset{\sim}{u}_2 = \begin{bmatrix} w_2 & w_3 \end{bmatrix} \begin{bmatrix} K_{11}^2 & K_{12}^2 \\ K_{21}^2 & K_{22}^2 \end{bmatrix} \begin{bmatrix} u_2 \\ u_3 \end{bmatrix}$$

$$= \begin{bmatrix} w_1 & w_2 & w_3 & w_4 \end{bmatrix} \underbrace{\begin{bmatrix} 0 & 0 & 0 & 0 \\ 0 & K_{11}^2 & K_{12}^2 & 0 \\ 0 & K_{21}^2 & K_{22}^2 & 0 \\ 0 & 0 & 0 & 0 \end{bmatrix}}_{\hat{\underline{K}}_2} \begin{bmatrix} u_1 \\ u_2 \\ u_3 \\ u_4 \end{bmatrix}$$

$$= \underset{\sim}{w}^{\mathrm{T}}\hat{\underline{K}}_2\,\underset{\sim}{u}. \tag{14.44}$$

Consequently,

$$\sum_{e=1}^{N_{\mathrm{el}}} \underset{\sim}{w}_e^{\mathrm{T}}\underline{K}_e\,\underset{\sim}{u}_e = \sum_{e=1}^{N_{\mathrm{el}}} \underset{\sim}{w}^{T}\hat{\underline{K}}_e\,\underset{\sim}{u}, \tag{14.45}$$

and, by summing the individual $\hat{\underline{K}}_e$ matrices, the result

$$\sum_{e=1}^{N_{\mathrm{el}}} \underset{\sim}{w}_e^{\mathrm{T}}\underline{K}_e\underset{\sim}{u}_e = \underset{\sim}{w}^{\mathrm{T}}\underline{K}\,\underset{\sim}{u}, \tag{14.46}$$

is obtained, with

$$\underline{K} = \sum_{e=1}^{N_{\mathrm{el}}} \hat{\underline{K}}_e. \tag{14.47}$$

In the case of the element distribution of Fig. 14.5, it holds that

$$\hat{\underline{K}}_1 = \begin{bmatrix} K_{11}^1 & K_{12}^1 & 0 & 0 \\ K_{21}^1 & K_{22}^1 & 0 & 0 \\ 0 & 0 & 0 & 0 \\ 0 & 0 & 0 & 0 \end{bmatrix} \tag{14.48}$$

$$\underline{\hat{K}}_2 = \begin{bmatrix} 0 & 0 & 0 & 0 \\ 0 & K_{11}^2 & K_{12}^2 & 0 \\ 0 & K_{21}^2 & K_{22}^2 & 0 \\ 0 & 0 & 0 & 0 \end{bmatrix} \tag{14.49}$$

$$\underline{\hat{K}}_3 = \begin{bmatrix} 0 & 0 & 0 & 0 \\ 0 & 0 & 0 & 0 \\ 0 & 0 & K_{11}^3 & K_{12}^3 \\ 0 & 0 & K_{21}^3 & K_{22}^3 \end{bmatrix}, \tag{14.50}$$

such that

$$\underline{K} = \underline{\hat{K}}_1 + \underline{\hat{K}}_2 + \underline{\hat{K}}_3, \tag{14.51}$$

leading to

$$\underline{K} = \begin{bmatrix} K_{11}^1 & K_{12}^1 & 0 & 0 \\ K_{21}^1 & K_{22}^1 + K_{11}^2 & K_{12}^2 & 0 \\ 0 & K_{21}^2 & K_{22}^2 + K_{11}^3 & K_{12}^3 \\ 0 & 0 & K_{21}^3 & K_{22}^3 \end{bmatrix}. \tag{14.52}$$

In computer codes, however, the individual matrices $\underline{\hat{K}}_i$ are never formed explicitly. The non-trivial components of $\underline{\hat{K}}_i$ are supplied directly to the appropriate position within $\underline{K}$. This process of assembling the global matrix $\underline{K}$ based on the contributions of the individual element matrices $\underline{K}_e$ is called the **assembly** process.

For the right-hand side of Eq. (14.39), the same procedure is followed. For element 2 it can be written analogously:

$$\begin{aligned} \underset{\sim}{w}_2^{\mathrm{T}} \underset{\sim}{f}_2 &= \begin{bmatrix} w_1^2 & w_2^2 \end{bmatrix} \begin{bmatrix} f_1^2 \\ f_2^2 \end{bmatrix} = \begin{bmatrix} w_2 & w_3 \end{bmatrix} \begin{bmatrix} f_1^2 \\ f_2^2 \end{bmatrix} \\ &= \begin{bmatrix} w_1 & w_2 & w_3 & w_4 \end{bmatrix} \underbrace{\begin{bmatrix} 0 \\ f_1^2 \\ f_2^2 \\ 0 \end{bmatrix}}_{\underset{\sim}{\hat{f}}_2} = \underset{\sim}{w}^{\mathrm{T}} \underset{\sim}{\hat{f}}_2. \end{aligned} \tag{14.53}$$

Following the same procedure for the other elements and adding up the contribution for the internal source term for all (three) elements gives

$$\underset{\sim}{f} = \underset{\sim}{\hat{f}}_1 + \underset{\sim}{\hat{f}}_2 + \underset{\sim}{\hat{f}}_3. \tag{14.54}$$

This leads to

$$\tilde{w}^{\mathrm{T}}\tilde{f}_{\mathrm{int}} = \begin{bmatrix} w_1 & w_2 & w_3 & w_4 \end{bmatrix} \begin{bmatrix} f_1^1 \\ f_2^1 + f_1^2 \\ f_2^2 + f_1^3 \\ f_2^3 \end{bmatrix}. \tag{14.55}$$

What remains is the term B in Eq. (14.25). The effect of this boundary term B may be included via

$$B = -w(a)p_a + w(b)p_b = \tilde{w}^T \tilde{f}_{\mathrm{ext}}, \tag{14.56}$$

where $\tilde{f}_{\mathrm{ext}}$ contains p_a and p_b at the appropriate positions, according to

$$B = \begin{bmatrix} w_1 & w_2 & w_3 & w_4 \end{bmatrix} \begin{bmatrix} -p_a \\ 0 \\ 0 \\ p_b \end{bmatrix}. \tag{14.57}$$

This finally leads to an equation of the form:

$$\tilde{w}^{\mathrm{T}}\underline{K}\tilde{u} = \tilde{w}^{\mathrm{T}}(\tilde{f}_{\mathrm{int}} + \tilde{f}_{\mathrm{ext}}). \tag{14.58}$$

Using the fact that Eq. (14.58) must hold for 'all' $\tilde{w}$, this results in the so-called **discrete set of equations**:

$$\begin{bmatrix} K_{11}^1 & K_{12}^1 & 0 & 0 \\ K_{21}^1 & K_{22}^1 + K_{11}^2 & K_{12}^2 & 0 \\ 0 & K_{21}^2 & K_{22}^2 + K_{11}^3 & K_{12}^3 \\ 0 & 0 & K_{21}^3 & K_{22}^3 \end{bmatrix} \begin{bmatrix} u_1 \\ u_2 \\ u_3 \\ u_4 \end{bmatrix} = \begin{bmatrix} f_1^1 - p_a \\ f_2^1 + f_1^2 \\ f_2^2 + f_1^3 \\ f_2^3 + p_b \end{bmatrix}. \tag{14.59}$$

Assume that in the example of Fig. 14.5 at $x = x_1$, an essential boundary condition is prescribed: $u(x_1) = U$. This means that $p_a = p_u$ is unknown beforehand. At $x = x_4$, a natural boundary condition is prescribed, so $p_b = P$ is known beforehand. Then

$$\begin{bmatrix} K_{11}^1 & K_{12}^1 & 0 & 0 \\ K_{21}^1 & K_{22}^1 + K_{11}^2 & K_{12}^2 & 0 \\ 0 & K_{21}^2 & K_{22}^2 + K_{11}^3 & K_{12}^3 \\ 0 & 0 & K_{21}^3 & K_{22}^3 \end{bmatrix} \begin{bmatrix} U \\ u_2 \\ u_3 \\ u_4 \end{bmatrix} = \begin{bmatrix} f_1^1 - p_u \\ f_2^1 + f_1^2 \\ f_2^2 + f_1^3 \\ f_2^3 + P \end{bmatrix}. \tag{14.60}$$

It is clear that in Eq. (14.60) the unknowns are u_2, u_3, u_4 on the left-hand side of the equation, and p_u on the right-hand side of the equation. So both columns can be divided into a known part and an unknown part, depending on the essential and natural boundary conditions that have been prescribed. The next section will outline how this equation is partitioned to facilitate the solution process.

14.6 Solution of the Discrete Set of Equations

Let Eq. (14.60) be written as

$$\underline{K}\,\underset{\sim}{u} = \underset{\sim}{f} \tag{14.61}$$

where $\underset{\sim}{f} = \underset{\sim}{f}_{\text{int}} + \underset{\sim}{f}_{\text{ext}}$.

The unknowns can be partitioned into two groups. First, some of the components of the column $\underset{\sim}{u}$ will be prescribed. This subset of $\underset{\sim}{u}$ is labelled $\underset{\sim}{u}_{\text{p}}$. The remaining components of $\underset{\sim}{u}$ are the actual unknowns, labelled $\underset{\sim}{u}_{\text{u}}$. In a similar manner $\underline{K}$ and $\underset{\sim}{f}$ can be partitioned. Consequently, Eq. (14.60) can be rewritten as

$$\begin{bmatrix} \underline{K}_{\text{uu}} & \underline{K}_{\text{up}} \\ \underline{K}_{\text{pu}} & \underline{K}_{\text{pp}} \end{bmatrix} \begin{bmatrix} \underset{\sim}{u}_{\text{u}} \\ \underset{\sim}{u}_{\text{p}} \end{bmatrix} = \begin{bmatrix} \underset{\sim}{f}_{\text{u}} \\ \underset{\sim}{f}_{\text{p}} \end{bmatrix} \tag{14.62}$$

It is emphasized that the right-hand side partition $\underset{\sim}{f}_{\text{u}}$ associated with the unknowns $\underset{\sim}{u}_{\text{u}}$ will be known, and conversely, $\underset{\sim}{f}_{\text{p}}$ will be unknown as $\underset{\sim}{u}_{\text{p}}$ is known. Since $\underset{\sim}{u}_{\text{p}}$ is known, the actual unknowns $\underset{\sim}{u}_{\text{u}}$ can be solved from

$$\underline{K}_{\text{uu}}\,\underset{\sim}{u}_{\text{u}} = \underset{\sim}{f}_{\text{u}} - \underline{K}_{\text{up}}\,\underset{\sim}{u}_{\text{p}}. \tag{14.63}$$

Notice that at the part of the boundary where u is prescribed, the associated external load f_{p} is unknown.

The components of $\underset{\sim}{f}_{\text{p}}$ can be obtained by simple multiplication, after having solved $\underset{\sim}{u}_{\text{u}}$ from Eq. (14.63):

$$\underset{\sim}{f}_{\text{p}} = \underline{K}_{\text{pu}}\,\underset{\sim}{u}_{\text{u}} + \underline{K}_{\text{pp}}\,\underset{\sim}{u}_{\text{p}}. \tag{14.64}$$

Example 14.5 Suppose we would like to solve the diffusion equation (14.1) with $c = 1$ and $f = 0$ on an interval $0 \le x \le 1$ and with an essential boundary condition $u = 0$ at $x = 0$ and natural boundary condition $du/dx = 1$ at $x = 1$. We want to solve the problem with three linear elements. For this case, Eq. (14.60) yields:

$$\begin{bmatrix} 3 & -3 & 0 & 0 \\ -3 & 6 & -3 & 0 \\ 0 & -3 & 6 & -3 \\ 0 & 0 & -3 & 3 \end{bmatrix} \begin{bmatrix} 0 \\ u_2 \\ u_3 \\ u_4 \end{bmatrix} = \begin{bmatrix} f_1 \\ 0 \\ 0 \\ 1 \end{bmatrix},$$

with f_1, u_2, u_3, u_4 unknowns.

In this case, because $u_1 = 0$, Eq. (14.63) yields:

$$\begin{bmatrix} 6 & -3 & 0 \\ -3 & 6 & -3 \\ 0 & -3 & 3 \end{bmatrix} \begin{bmatrix} u_2 \\ u_3 \\ u_4 \end{bmatrix} = \begin{bmatrix} 0 \\ 0 \\ 1 \end{bmatrix}.$$

This set of equations can be solved easily, leading to the following array of nodal solutions:

$$\underset{\sim}{u} = \begin{bmatrix} 0 \\ 1/3 \\ 2/3 \\ 1 \end{bmatrix}$$

and $f_1 = -1$.

14.7 Isoparametric Elements and Numerical Integration

In Section 14.4, the concept of shape functions has been introduced. Within each element u_h has been written as

$$u_h|_{\Omega_e} = \underset{\sim}{N}^T(x)\underset{\sim}{u}_e. \tag{14.65}$$

The shape functions are simple polynomial expressions in terms of the coordinate x. For instance, for a linear interpolation the shape functions are linear polynomials, according to

$$N_1 = 1 - \frac{x - x_1}{x_2 - x_1}, \quad N_2 = \frac{x - x_1}{x_2 - x_1}, \tag{14.66}$$

where x_1 and x_2 denote the position of the nodes of the element. In this case the shape functions are linear functions of the global coordinate x. It is appropriate in the context of a generalization to more-dimensional problems to introduce a local coordinate $-1 \le \xi \le 1$ within each element such that $\xi = -1$ and $\xi = 1$ correspond to the edges of the element. With respect to this local coordinate system, the shape functions may be written as

$$N_1 = -\frac{1}{2}(\xi - 1), \quad N_2 = \frac{1}{2}(\xi + 1). \tag{14.67}$$

This is visualized in Fig. 14.6.

Computation of components of the element coefficient matrix and the element load array requires the evaluation of integrals of the form

$$\int_{\Omega_e} f(x)\, dx\,. \tag{14.68}$$

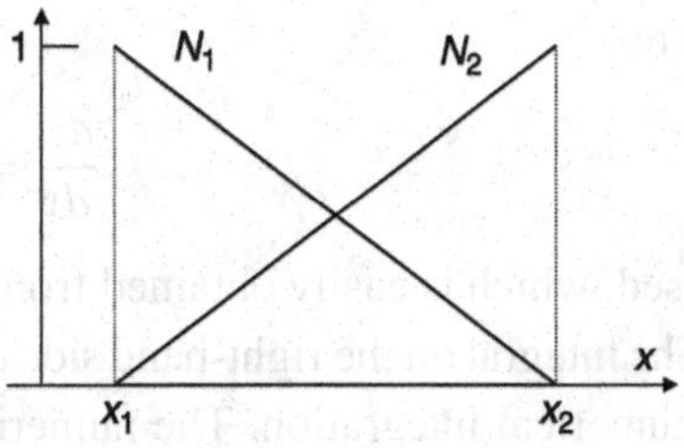

Figure 14.6

Shape functions with respect to the global x-coordinate.

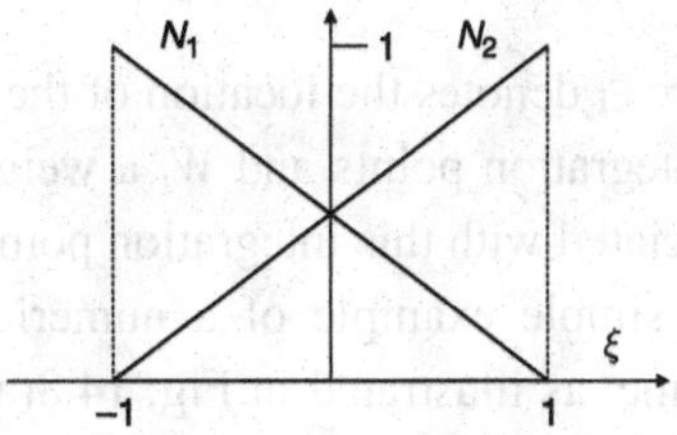

Figure 14.7

Shape functions with respect to the local ξ-coordinate.

These integrals may be transformed into integrals using the local coordinate system, according to

$$\int_{\Omega_e} f(x)\, dx = \int_{-1}^{1} f(x(\xi))\, \frac{dx}{d\xi}\, d\xi. \tag{14.69}$$

This requires the computation of the derivative $dx/d\xi$. For this purpose the concept of **isoparametric elements** is introduced. The shape functions N_i introduced for the interpolation of the unknown function u_{h} are also used for the relation between the coordinates x and the coordinates ξ within an element:

$$x(\xi) = \sum_{i=1}^{n} N_i(\xi) x_i^e = \underset{\sim}{N}^{\mathrm{T}} \underset{\sim}{x}_e, \tag{14.70}$$

where x_i^e are the coordinates of the nodes of the element (see Fig. 14.7). As a result, the derivative $dx/d\xi$ is obtained easily:

$$\frac{dx}{d\xi} = \frac{d\underset{\sim}{N}^{\mathrm{T}}}{d\xi} \underset{\sim}{x}_e. \tag{14.71}$$

The element coefficient matrix requires the derivatives of the shape functions with respect to the coordinate x. For this purpose

$$\frac{dN_i}{dx} = \frac{dN_i}{d\xi} \frac{d\xi}{dx}, \tag{14.72}$$

where

$$\frac{d\xi}{dx} = \left(\frac{dx}{d\xi}\right)^{-1}, \tag{14.73}$$

is used, which is easily obtained from Eq. (14.71).

The integral on the right-hand side of Eq. (14.69) can be approximated by means of numerical integration. The numerical integration of an arbitrary function $g(\xi)$ over the domain $-1 \leq \xi \leq 1$ is approximated by

$$\int_{-1}^{1} g(\xi)\, d\xi = \sum_{i=1}^{n_{\text{int}}} g(\xi_i)\, W_i, \tag{14.74}$$

where ξ_i denotes the location of the i-th integration point, n_{int} is the total number of integration points and W_i a weighting factor, i.e. the length of the ξ-domain, associated with this integration point.

A simple example of a numerical integration is the trapezoidal integration scheme, as illustrated in Fig. 14.8(a). Integration of $g(\xi)$ over the domain $-1 \leq \xi \leq 1$ using the trapezoidal integration rule yields:

$$\int_{-1}^{1} g(\xi)\, d\xi \approx g(\xi = -1) + g(\xi = 1), \tag{14.75}$$

which corresponds to the shaded area in Fig. 14.8(a). For trapezoidal integration, the integration point positions ξ_i are given by

$$\xi_1 = -1, \quad \xi_2 = 1, \tag{14.76}$$

while the associated weighting factors are

$$W_1 = 1, \quad W_2 = 1. \tag{14.77}$$

The trapezoidal integration rule integrates a linear function exactly. A two-point Gaussian integration rule, as depicted in Fig. 14.8(b), may yield a more accurate

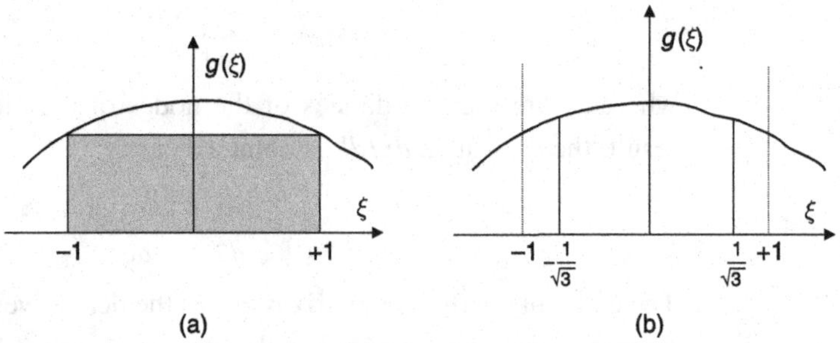

Figure 14.8

(a) Trapezoidal integration; (b) two-point Gaussian integration.

Table 14.1 Gaussian quadrature points up to $n_{\text{int}} = 3$.

n_{int}	ξ_i	W_i
1	$\xi_1 = 0$	$W_1 = 2$
2	$\xi_1 = \frac{-1}{\sqrt{3}},\ \xi_2 = \frac{1}{\sqrt{3}}$	$W_1 = W_2 = 1$
3	$\xi_1 = -\sqrt{\frac{3}{5}},\ \xi_2 = 0,\ \xi_3 = \sqrt{\frac{3}{5}}$	$W_1 = W_3 = \frac{5}{9},\ W_2 = \frac{8}{9}$

result since this integration rule integrates up to a third-order function exactly using only two integration points. In this case, the integral is approximated by

$$\int_{-1}^{1} g(\xi)\, d\xi \approx g\left(\xi = \frac{-1}{\sqrt{3}}\right) + g\left(\xi = \frac{1}{\sqrt{3}}\right). \tag{14.78}$$

The location of the Gaussian integration (quadrature) points and the associated weighting factors are summarized in Table 14.1.

Use of the local coordinate system, with isoparametric formulation and numerical integration to the element coefficient matrix, yields

$$\begin{aligned} \underset{\sim}{K_e} &= \int_{\Omega_e} \frac{d\underset{\sim}{N}}{dx} c \frac{d\underset{\sim}{N}^{\mathrm{T}}}{dx}\, dx \\ &= \int_{\xi=-1}^{1} \frac{d\underset{\sim}{N}}{d\xi}\frac{d\xi}{dx} c \frac{d\underset{\sim}{N}^{\mathrm{T}}}{d\xi}\frac{d\xi}{dx} \left(\frac{d\xi}{dx}\right)^{-1} d\xi \\ &\approx \sum_{i=1}^{n_{\text{int}}} \left(\frac{d\underset{\sim}{N}}{d\xi} c \frac{d\underset{\sim}{N}^{\mathrm{T}}}{d\xi} \frac{d\xi}{dx} \right)_{\xi=\xi_i} W_i. \end{aligned} \tag{14.79}$$

Example 14.6 Let us find a numerical approximation of the following integral using Gaussian integration:

$$I = \int_{-1}^{1} \frac{\cos(\xi)}{\sqrt{\xi + 4}}\, d\xi.$$

For two-point Gaussian integration, the integral can be approximated by:

$$I \approx f(-1/\sqrt{3})W_1 + f(1/\sqrt{3})W_2,$$

with: $f(\xi) = \cos(\xi)/\sqrt{\xi + 4}$ and $W_1 = W_2 = 1$. A MATLAB script for this problem reads:

```
ksi=[-1/sqrt(3)  1/sqrt(3)]; W=[1 1]; I=0;
for i=1:length(ksi)
    I=I+(cos(ksi(i))/sqrt(ksi(i)+4))*W(i);
end
disp(['integral = ',num2str(I)])
```

leading to: $I = 0.84456$.

A MATLAB script for three-point Gaussian integration yields:

```
ksi=[-sqrt(3/5)  0 sqrt(3/5)] ; W=[5/9 8/9 5/9] ; I=0;
for i=1:length(ksi)
    I=I+(cos(ksi(i))/sqrt(ksi(i)+4))*W(i);
end
disp(['integral = ',num2str(I)])
```

leading to: $I = 0.84724$.

14.8 Basic Structure of a Finite Element Program

The objective of a finite element program is to compute the coefficient matrix $\underline{K}$ and the right-hand side array $\underset{\sim}{f}$, and eventually to solve the resulting system of equations taking the boundary conditions into account. To illustrate the typical data structure and the layout of a finite element program, consider, as an example, the mesh depicted in Fig. 14.9.

The MATLAB programming language is used for explanation purposes. The following data input is needed:

- **Element topology** First of all the domain is divided into a number of elements and each node is given a unique global number. In this example two elements have been used, the first element Ω_1 is a quadratic element connecting nodes 3, 4 and 2 (in that order) and the second element is a linear element having nodes 1 and 3 (again, in that order). The node numbers of each element are stored in the topology array `top`, such that the i-th row of this array corresponds to the i-th element. In the current example the topology array would be:

$$\texttt{top} = \begin{bmatrix} 3 & 4 & 2 \\ 1 & 3 & 0 \end{bmatrix}.$$

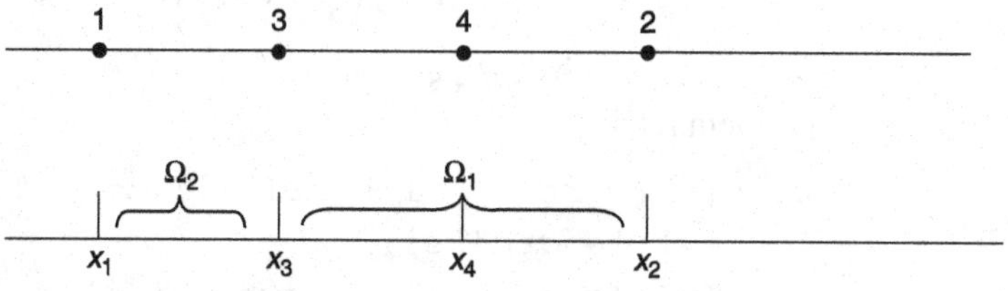

Figure 14.9

Mesh for a one-dimensional problem, consisting of a linear and a quadratic element.

Besides the node numbers of the element, a number of identifiers may be included for each element, for instance to refer to different material parameters c or different element types, e.g. linear versus quadratic elements. In fact, the MATLAB code provided to experiment with has two identifiers per element. Please consult the manual of the code: `mlfem_nac`.

- **Nodal coordinates** The nodal coordinates $\underset{\sim}{x}$ are stored in the array `coord`, hence in this example:

$$\underset{\sim}{x} = \texttt{coord} = \begin{bmatrix} x_1 \\ x_2 \\ x_3 \\ x_4 \end{bmatrix}.$$

The nodal coordinates associated with the element e can be retrieved from `coord` using `top`. The node numbers of element e can be extracted from the array `top` by:

```
ii = nonzeros(top(e,:)),
```

such that the nodal coordinates of the e-th element are obtained via

$$\underset{\sim}{x}_e = \texttt{nodcoord = coord(ii,:)}.$$

- **Solution array** The nodal unknowns, also called degrees of freedom, $\underset{\sim}{u}$ are stored in the array `sol`:

$$\underset{\sim}{u} = \texttt{sol} = \begin{bmatrix} u_1 \\ u_2 \\ u_3 \\ u_4 \end{bmatrix}.$$

It is not necessary to store the degrees of freedom in a sequential manner; in fact, any ordering may be chosen, as long as each array component corresponds to a unique degree of freedom. To extract the nodal degrees of freedom for element e, $\underset{\sim}{u}_e$ from `sol`, a separate index array is needed: `pos`. The e-th row of the array `pos` contains the location of the nodal degrees of freedom of element e in the array `sol`. For this example

$$\underset{\sim}{u}_1 = \begin{bmatrix} u_3 \\ u_4 \\ u_2 \end{bmatrix}, \quad \underset{\sim}{u}_2 = \begin{bmatrix} u_1 \\ u_3 \end{bmatrix};$$

hence the index array `pos` should contain

$$\texttt{pos} = \begin{bmatrix} 3 & 4 & 2 \\ 1 & 3 & 0 \end{bmatrix}.$$

Using this array, the nodal degrees of freedom of element e can be extracted from `sol` via

```
ii = nonzeros(pos(e,:))
```

$$\underset{\sim}{u}_e = \texttt{nodu} = \texttt{sol(ii)}$$

- **Shape functions** The shape functions and their derivatives are needed at the integration points $\underset{\sim}{N}(\xi_i)$ and $d\underset{\sim}{N}/d\xi$. The shape function values are stored in the array n such that at the i-th integration point

$$\underset{\sim}{N}^{\mathrm{T}}(\xi_i) = \texttt{n(i,:)}.$$

Using these shape functions the coordinate $x(\xi_i)$ within element e can be computed at the i-th integration point:

$$x(\xi_i) = \underset{\sim}{N}^{\mathrm{T}}(\xi_i)\underset{\sim}{x}_e = \texttt{n(i,:)*nodcoord}.$$

Similarly, the value of the solution at the i-th integration point of element e is obtained via

$$u(\xi_i) = \underset{\sim}{N}^{\mathrm{T}}(\xi_i)\underset{\sim}{u}_e = \texttt{n(i,:)*nodu}.$$

In a similar fashion the shape function derivatives with respect to the local coordinate ξ are stored in an array, called dndxi.

Structure of the finite element code Typically the structure of a finite element programme is as follows.

(i) Pre-processing: mesh generation, boundary condition specification and parameter declaration. This should provide the topology array top, the coordinate array coord and a number of auxiliary arrays containing boundary conditions and material parameters.

(ii) Based on the mesh and element types used, the index array pos can be computed.

(iii) Assembly of the coefficient matrix $\underline{K}$=q and the element array $\underset{\sim}{f}$=rhs. Let $\underline{K}_e$=qe and $\underset{\sim}{f}_e$=rhse; then the assembly process in a MATLAB environment would look like

```
% nelem: the number of elements

for ielem=1:nelem

  % compute qe and rhse

  [qe,rhse]=<elementfunction>(ielem,coord,top,.....)

  % get the location of the degrees of freedom
  % in the solution array

  ii = nonzeros(pos(ielem,:));

  % add the element coefficient matrix
  % and the element right-hand side array
  % to the total coefficient matrix q
```

```
    % and the load array rhs
    q(ii,ii)  = q(ii,ii)  + qe;
    rhs(ii)   = rhs(ii)   + rhse;

  end
```

(iv) Solution of the set of equations taking into account the boundary conditions.

(v) Post-processing based on the solution, for instance by computing associated quantities such as heat fluxes or stresses.

Example 14.7 As an example, consider the diffusion problem with the following parameter setting. We consider the domain $\Omega : 0 \leq x \leq 1$, with prescribed essential boundary conditions at $x = 0$ and $x = 1$. These conditions are: $u(0) = 0$ and $u(1) = 0$. There are no natural boundary conditions. The material constant satisfies: $c = 1$ and the source term: $f = 1$.

The domain Ω is divided into five elements of equal length. Figure 14.10 shows the solution. The left part displays the computed solution u (solid line) as well as the exact solution (dashed line). Remarkably, in this one-dimensional case with the current choice of parameters, the nodal solutions are exact. The right part of the figure shows the computed flux, say flux $p = c\, du/dx$. Again, the solid line denotes the computed flux p, which is clearly discontinuous from one element to the next, and the dashed line denotes the exact solution. The discontinuity of the computed flux field is obvious: the field u is piecewise linear, therefore the derivative du/dx is piecewise constant. The flux p does not necessarily have to be piecewise constant: if the parameter c is a function of x, the flux p will be varying within an element.

Mesh refinement leads to an improved approximate solution. For instance, using ten rather than five elements yields the results depicted in Fig. 14.11. The impact of a varying c, say $c = 1 + x$, is depicted in Fig. 14.12.

Changing the interpolation order of the shape functions N_i (see Fig. 14.13), from linear to quadratic, also has a significant impact on the results, in particular on the quality of the flux prediction. For the constant c case, the solution becomes exact. For $c = 1 + x$ a significant improvement can also be observed, as depicted in Fig. 14.14.

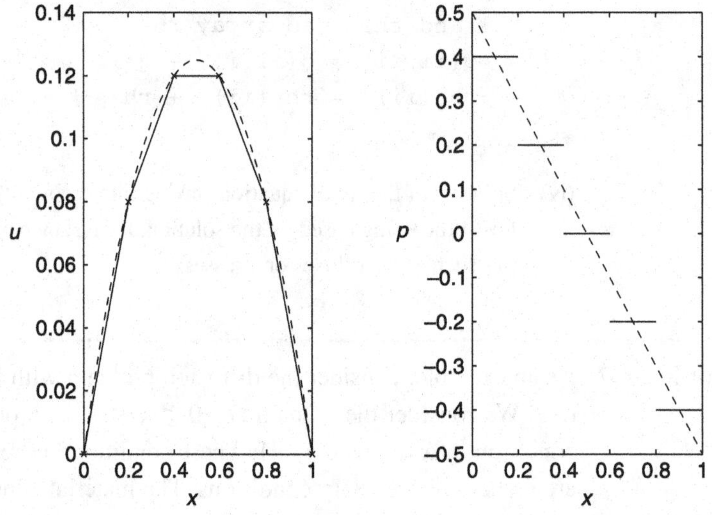

Figure 14.10

Five-element solution. Left: (solid line) approximate solution $u_h(x)$, (dashed line) exact solution $u(x)$. Right: (solid line) approximate flux $p_h(x)$, (dashed line) exact flux $p(x)$.

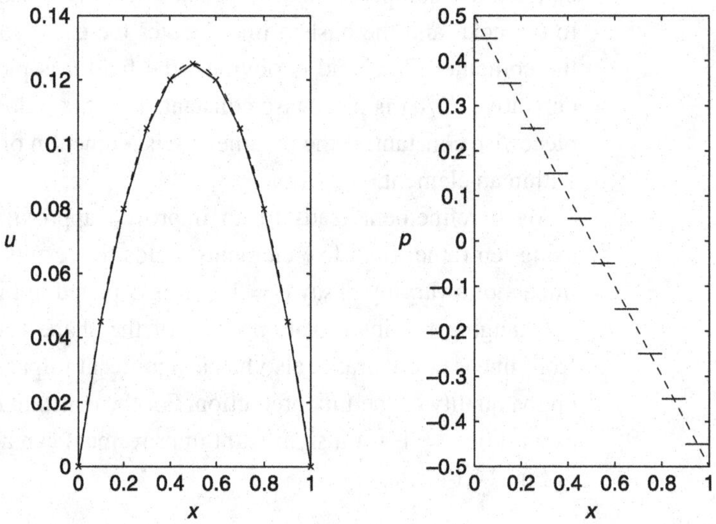

Figure 14.11

Ten-element solution. Left: (solid line) approximate solution $u_h(x)$, (dashed line) exact solution $u(x)$. Right: (solid line) approximate flux $p_h(x)$, (dashed line) exact flux $p(x)$.

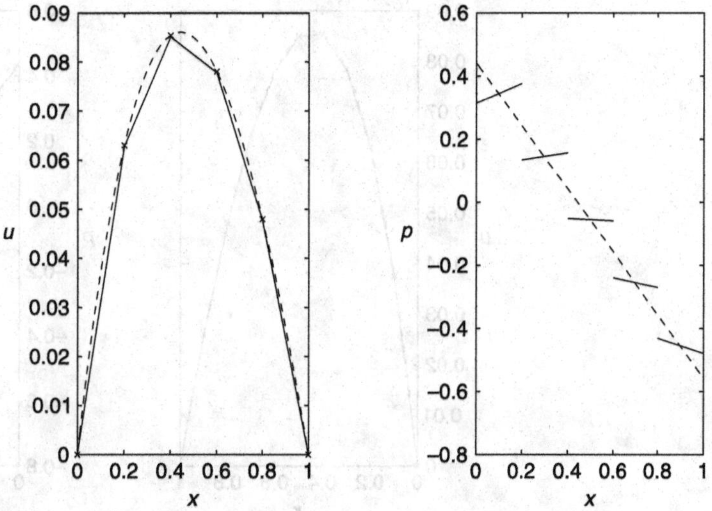

Figure 14.12

Five-element solution for $c = 1 + x$. Left: (solid line) approximate solution $u_h(x)$, (dashed line) exact solution $u(x)$. Right: (solid line) approximate flux $p_h(x)$, (dashed line) exact flux $p(x)$.

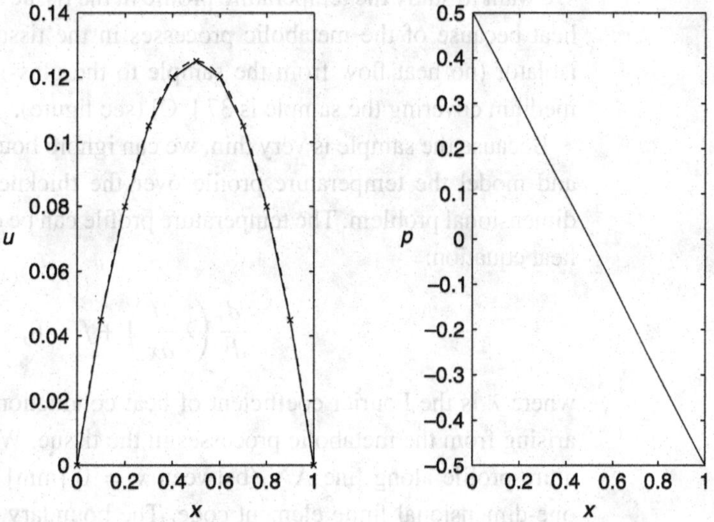

Figure 14.13

Solution for $c = 1$ using five quadratic elements. Left: (solid line) approximate solution $u_h(x)$, (dashed line) exact solution $u(x)$. Right: (solid line) approximate flux $p_h(x)$, (dashed line) exact flux $p(x)$; the lines coincide.

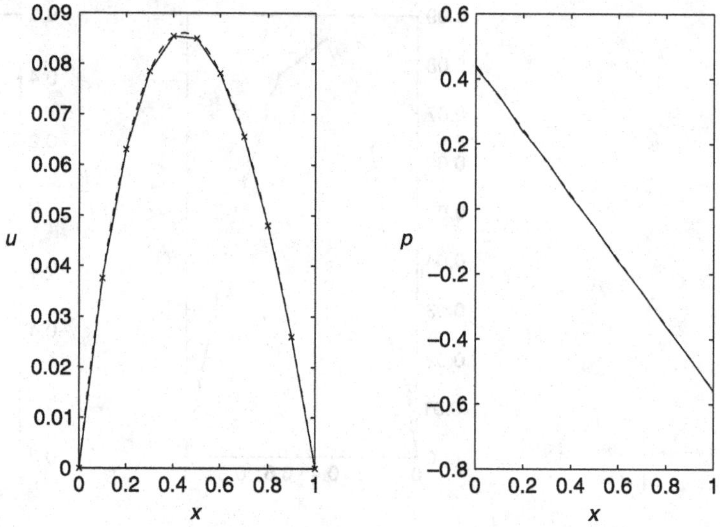

Figure 14.14

Solution for $c = 1 + x$ using five quadratic elements. Left: (solid line) approximate solution $u_h(x)$, (dashed line) exact solution $u(x)$. Right: (solid line) approximate flux $p_h(x)$, (dashed line) exact flux $p(x)$.

Example 14.8 Consider a thin piece of tissue in a glass tray that is covered with a certain medium. We want to study the temperature profile in the tissue sample, while it is producing heat because of the metabolic processes in the tissue. The glass behaves as an isolator (no heat flow from the sample to the glass) and the temperature of the medium covering the sample is 37 [°C] (see figure).

Because the sample is very thin, we can ignore boundary effects (left and right) and model the temperature profile over the thickness of the sample as a one-dimensional problem. The temperature profile can be described with the stationary heat equation:

$$\frac{d}{dx}\left(\lambda \frac{dT}{dx}\right) + f = 0,$$

where λ is the Fourier coefficient of heat conduction and f the heat source term arising from the metabolic processes in the tissue. We can determine the temperature profile along line AA′ (between $x = 0$ [mm] and $x = 10$ [mm]) using a one-dimensional finite element code. The boundary condition at $x = 0$ [mm] is $T = 37$ [°C], and the boundary condition at $x = 10$ [mm] is $dT/dx = 0$ [°C mm^{-1}]. With $\lambda = 2 \times 10^{-3}$ [J mm^{-3} s^{-1}] and the heat source term $f = 1 \times 10^{-5}$ [J °C^{-1} mm $^{-1}$ s^{-1}] and a finite element mesh of ten quadratic elements, the temperature profile as given in the figure is found with a maximum temperature at the bottom side of the sample of 37.25 [°C].

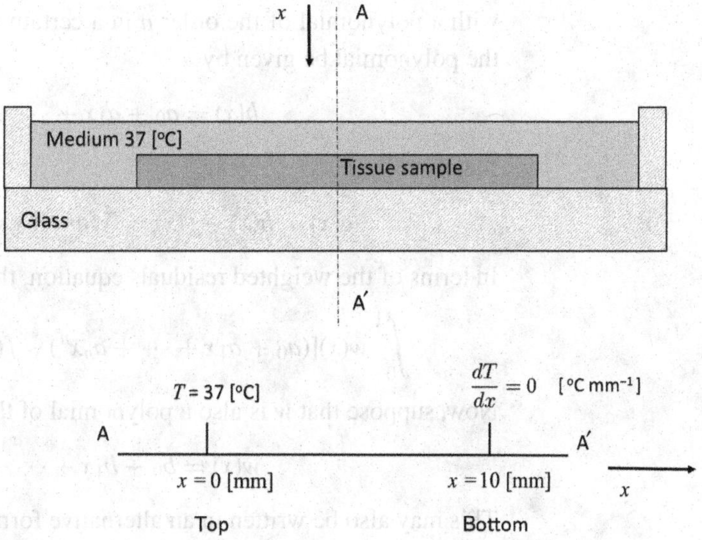

Figure 14.15

Schematic of a tissue sample in medium at 37 [°C] in a glass tray.

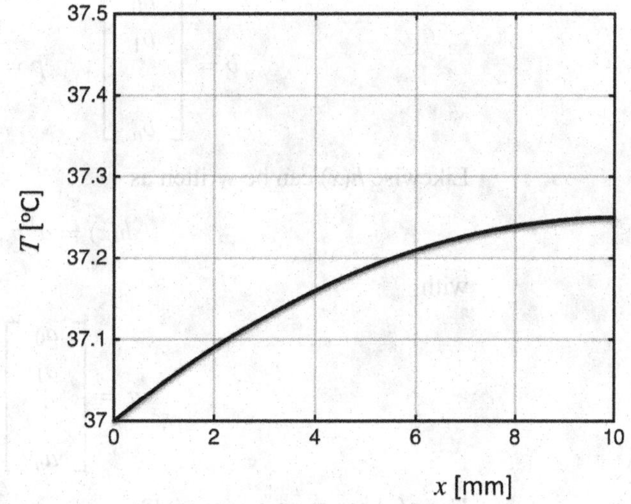

Figure 14.16

Temperature profile in tissue sample as a function of x.

Exercises

14.1 The method of weighted residuals can be used to find approximations of a given function. Let $f(x)$ be a function that one would like to approximate with a polynomial of the order n in a certain domain, say $0 \leq x \leq 1$. Let the polynomial be given by

$$h(x) = a_0 + a_1 x + \cdots + a_n x^n.$$

Ideally

$$g(x) = h(x) - f(x) = 0 \text{ for all } x \text{ with } 0 \leq x \leq 1.$$

In terms of the weighted residuals equation, this may also be expressed as

$$\int_0^1 w(x)[(a_0 + a_1 x + \cdots + a_n x^n) - f(x)]\, dx = 0 \text{ for all } w.$$

Now, suppose that w is also a polynomial of the order n:

$$w(x) = b_0 + b_1 x + \cdots + b_n x^n.$$

This may also be written in an alternative format:

$$w(x) = \underset{\sim}{b}^{\mathrm{T}} \underset{\sim}{p},$$

with

$$\underset{\sim}{b} = \begin{bmatrix} b_0 \\ b_1 \\ \vdots \\ b_n \end{bmatrix}, \quad \underset{\sim}{p} = \begin{bmatrix} 1 \\ x \\ \vdots \\ x^n \end{bmatrix}.$$

Likewise $h(x)$ can be written as

$$h(x) = \underset{\sim}{a}^{\mathrm{T}} \underset{\sim}{p},$$

with

$$\underset{\sim}{a} = \begin{bmatrix} a_0 \\ a_1 \\ \vdots \\ a_n \end{bmatrix}.$$

Use of these expressions yields

$$\int_0^1 w(x)\left[h(x) - f(x)\right] dx$$
$$= \int_0^1 \underset{\sim}{b}^{\mathrm{T}} \underset{\sim}{p} \underset{\sim}{p}^{\mathrm{T}} \underset{\sim}{a}\, dx - \int_0^1 \underset{\sim}{b}^{\mathrm{T}} \underset{\sim}{p} f(x)\, dx = 0.$$

Notice that both $\underset{\sim}{a}$ and $\underset{\sim}{b}$ are arrays with polynomial coefficients independent of x, while $\underset{\sim}{p}$ is an array with known functions of the coordinate x. Therefore, this equation may also be written as

$$\underset{\sim}{b}^{\mathrm{T}} \int_0^1 \underset{\sim}{p}\underset{\sim}{p}^{\mathrm{T}}\, dx\, \underset{\sim}{a} = \underset{\sim}{b}^{\mathrm{T}} \int_0^1 \underset{\sim}{p} f(x)\, dx\,.$$

The integral expression on the left-hand side is a matrix:

$$\underline{K} = \int_0^1 \underset{\sim}{p}\underset{\sim}{p}^{\mathrm{T}}\, dx,$$

while the integral on the right-hand side is a column:

$$\underset{\sim}{f} = \int_0^1 \underset{\sim}{p} f(x)\, dx.$$

The equation must be satisfied for all $\underset{\sim}{b}$; hence, with the use of the above matrix-array notation it follows that:

$$\underline{K}\, \underset{\sim}{a} = \underset{\sim}{f}.$$

(a) Suppose that $w(x)$ and $h(x)$ are both first-order polynomials.

– Show that in that case

$$\underline{K} = \begin{bmatrix} 1 & \frac{1}{2} \\ \frac{1}{2} & \frac{1}{3} \end{bmatrix}.$$

– If $f(x) = 3$, show that

$$\underset{\sim}{f} = \begin{bmatrix} 3 \\ \frac{3}{2} \end{bmatrix}.$$

– Compute the coefficients of the polynomial $h(x)$ collected in $\underset{\sim}{a}$. Explain the results.

(b) If $w(x)$ and $h(x)$ are both polynomials of the order n, show that

$$\underline{K} = \begin{bmatrix} 1 & \frac{1}{2} & \cdots & \frac{1}{n+1} \\ \frac{1}{2} & \frac{1}{3} & \cdots & \frac{1}{n+2} \\ \vdots & \vdots & \vdots & \vdots \\ \frac{1}{n+1} & \frac{1}{n+2} & \cdots & \frac{1}{2n+1} \end{bmatrix}.$$

(c) Let $f(x)$ be such that

$$f(x) = 1 \text{ for } 0 \le x \le 0.5,$$
$$f(x) = 0 \text{ for } 0.5 < x \le 1.$$

- Show that in this case

$$\underset{\sim}{f} = \int_0^1 \underset{\sim}{p} f(x)\, dx = \begin{bmatrix} \frac{1}{2} \\ \frac{1}{2}(\frac{1}{2})^2 \\ \vdots \\ \frac{1}{n+1}(\frac{1}{2})^{n+1} \end{bmatrix}.$$

- Use MATLAB to find the polynomial approximation $h(x)$ of $f(x)$ for $n = 2$, $n = 3$, etc. up to $n = 10$. Plot the original function $f(x)$ as well as the polynomial approximation $h(x)$. Hint: use the function `polyval`. If the MATLAB array `a` represents $\underset{\sim}{a}$ and `n` denotes the order of the polynomial, then to plot the function $h(x)$ you may use
`x=0:0.01:1; plot(x,polyval(a(n+1:-1:1),x))`
- Investigate the 'condition number' of the matrix $\underline{K}$ with increasing n. Hint: use the MATLAB function `cond`. What does this condition number mean and what does this imply with respect to the coefficients of $h(x)$, collected in $\underset{\sim}{a}$?

14.2 Consider the differential equation

$$u + \frac{du}{dx} + \frac{d}{dx}\left(c\frac{du}{dx}\right) + f = 0,$$

on the domain $a \leq x \leq b$.

Derive the weak form of this differential equation, and explain what steps are taken.

14.3 Let $f(x)$ be a function on the domain $0 \leq x \leq 1$. Let $f(x)$ be known at n points, denoted by x_i, homogeneously distributed on the above domain. Hence the distance Δx between two subsequent points equals

$$\Delta x = \frac{1}{n-1}.$$

A polynomial $f_h(x)$ of order $n - 1$ can be constructed through these points, which generally will form an approximation of $f(x)$:

$$f_h(x) = a_0 + a_1 x + \cdots + a_{n-1} x^{n-1}.$$

(a) Show that the coefficients of a_i can be found by solving

$$\begin{bmatrix} 1 & x_1 & x_1^2 & \cdots & x_1^{n-1} \\ 1 & x_2 & x_2^2 & \cdots & x_2^{n-1} \\ 1 & x_3 & x_3^2 & \cdots & x_3^{n-1} \\ \vdots & \vdots & \vdots & \vdots & \vdots \\ 1 & x_n & x_n^2 & \cdots & x_n^{n-1} \end{bmatrix} \begin{bmatrix} a_0 \\ a_1 \\ a_2 \\ \vdots \\ a_{n-1} \end{bmatrix} = \begin{bmatrix} f_1 \\ f_2 \\ f_3 \\ \vdots \\ f_n \end{bmatrix},$$

where $f_i = f(x_i)$.

(b) Use this to find a polynomial approximation for different values of n to the function:

$$f(x) = 1 \text{ for } 0 \leq x \leq 0.5,$$
$$f(x) = 0 \text{ for } 0.5 < x \leq 1.$$

Compare the results with those obtained using the weighted residuals formulation in Exercise 14.1(c). Explain the differences.

14.4 Consider the domain $-1 \leq x \leq 1$. Assume that the function u is known at $x_1 = -1$, $x_2 = 0$ and $x_3 = 1$, say u_1, u_2 and u_3 respectively. The polynomial approximation of u, denoted by u_h, is written as

$$u_\mathrm{h} = a_0 + a_1 x + a_2 x^2.$$

(a) Determine the coefficients a_0, a_1 and a_2 to be expressed as a function of u_1, u_2 and u_3.

(b) If the polynomial u_h is written in terms of the shape functions N_i:

$$u_\mathrm{h} = \sum_{i=1}^{n} N_i(x)\, u_i,$$

then determine N_i, $i = 1, 2, 3$ as a function of x.

(c) Sketch the shape functions N_i.

(d) Is it possible that the shape function $N_i \neq 1$ at $x = x_i$? Explain.

(e) Is it possible that the shape function $N_i \neq 0$ at $x = x_j$ with $j \neq i$? Explain.

14.5 Consider the differential equation

$$\frac{d}{dx}\left(c\frac{du}{dx}\right) = 1$$

on the domain $0 \leq x \leq h_1 + h_2$. At both ends of this domain, u is set to zero. Consider the element distribution as depicted in the figure below.

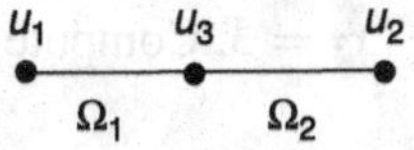

(a) Derive the weak formulation of this problem.

(b) The elements employed have linear shape functions stored in array $\underset{\sim}{N}(x)$. Express these shape functions in terms of x and the element lengths h_1 and h_2.

(c) Show that the coefficient matrix of an element is given by

$$\underset{\sim}{K}_e = \int_{\Omega_e} \frac{d\underset{\sim}{N}}{dx} c \frac{d\underset{\sim}{N}^T}{dx}\, dx.$$

(d) Demonstrate that the coefficient matrix of element e is given by

$$\underset{\sim}{K}_e = \frac{c}{h_e}\begin{bmatrix} 1 & -1 \\ -1 & 1 \end{bmatrix},$$

if c is a constant.

(e) Show that the source term leads to a right-hand-side column on element level given by

$$\underset{\sim}{f}_e = -\frac{h_e}{2}\begin{bmatrix} 1 \\ 1 \end{bmatrix}.$$

(f) After discretization, the resulting set of equations is expressed as

$$\underline{K}\,\underset{\sim}{u} = \underset{\sim}{f}.$$

Define the solution array $\underset{\sim}{u}$ and derive the coefficient matrix $\underline{K}$ and the array $\underset{\sim}{f}$ for the two-element mesh depicted in the figure.

(g) Determine the solution array $\underset{\sim}{u}$.

14.6 Using an isoparametric formulation, the shape functions are defined with respect to a local coordinate system $-1 \leq \xi \leq 1$. Within an element the unknown u_{h} is written as

$$u_{\mathrm{h}} = \sum_{i=1}^{n} N_i(\xi)u_i.$$

(a) What are the above shape functions with respect to the local coordinate system if a linear or quadratic interpolation is used?

(b) How is the derivative of the shape functions

$$\frac{dN_i}{dx}$$

obtained?

(c) Assume a quadratic shape function, and let $x_1 = 0$, $x_2 = 1$, and $x_3 = 3$. Compute

$$\frac{dN_i}{dx} \text{ for } i = 1, 2, 3.$$

(d) Compute from array $\underset{\sim}{h}$ given by

$$\underset{\sim}{h} = \int_0^3 \frac{d\underset{\sim}{N}}{dx}\, x \, dx,$$

the first component using the same quadratic shape functions as above.

14.7 Consider in the code `mlfem_nac` the directory `oneD`. The one-dimensional finite element program `fem1d` solves the diffusion problem:

$$\frac{d}{dx}\left(c\frac{du}{dx}\right) + f = 0,$$

on a domain $a \leq x \leq b$ subject to given boundary conditions for a certain problem. The input data for this program are specified in the m-file `demo_fem1d`, along with the post-processing statements.

(a) Modify the m-file `demo_fem1d` such that five linear elements are used to solve the above differential equation, using the boundary conditions

$$u = 0 \text{ at } x = 0,$$

and

$$p = c\frac{du}{dx} = 1 \text{ at } x = 1,$$

with $c = 1$ and $f = 0$. Compute the finite element method solution (array `sol`) and compare the nodal values with the exact solution.

(b) Use the array `pos` to extract the nodal solutions within the second element from the array `sol`.

(c) Determine the solution for the third node using the array `dest`.

14.8 One of the major problems in the tissue engineering of cartilage is to make thick constructs. Nutrients coming from the surrounding medium have to reach the cells in the middle of the construct by diffusion, but cells close to the edge of the construct may consume so much of the nutrients that there is nothing left for cells in the middle. Consider the experimental set-up as given in the figure below, representing a schematic drawing of a bioreactor for the culture of articular cartilage.

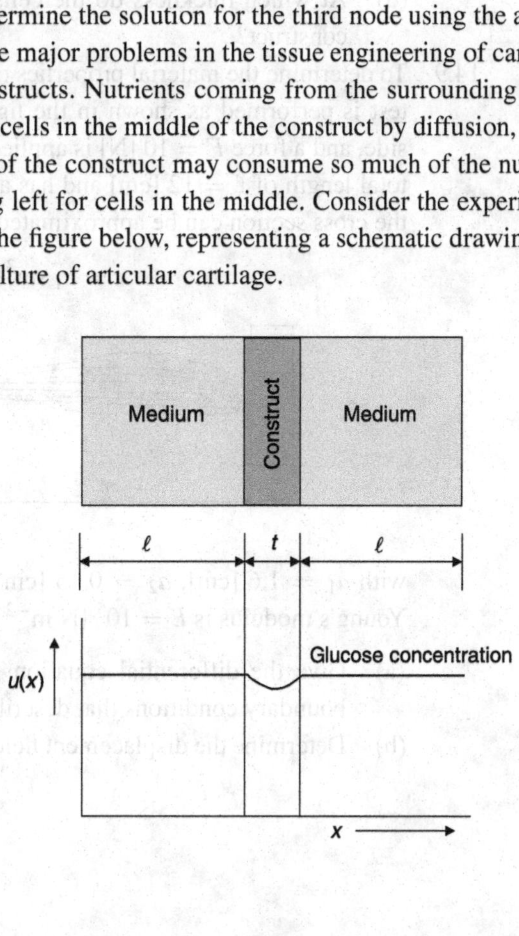

The construct is fixed between two highly permeable membranes allowing free contact with the culture medium. The thickness t is a trade-off between the diffusion coefficient c, the consumption rate of the cells f (neither can be influenced) and the amount of the molecules that can be supplied via the medium, which is usually bound to a maximum.
The diffusion problem can be considered as a one-dimensional problem. The current analysis is meant to determine the glucose concentration $u(x)$, which is an essential nutrient for the cells. The following properties are given: $c = 9.2 \times 10^{-6}[\text{cm}^2\ \text{s}^{-1}]$ and $f = 56 \times 10^{-7}[\text{Mol hour}^{-1}\ \text{cm}^{-3}]$. Low glucose medium is used, concentration = 5×10^{-3} [Mol litre^{-1}].

(a) Assuming the consumption rate f is constant and the medium is refreshed continuously, a stable glucose concentration as a function of the location in the construct will be reached after a while. Give the differential equation and boundary conditions describing this process.
(b) Give the analytical solution of this problem by integrating the differential equation.
(c) Adjust `demo_fem1d` to solve the problem with the finite element method. Solve the problem for $t = 1, 2, 3, 4$ and 5 [mm].
(d) At which thickness do the cells appear to die in the middle of the construct?

14.9 To determine the material properties of a skeletal muscle, a uniaxial tensile test is performed as shown in the figure. The muscle is clamped on one side, and a force $F = 10$ [N] is applied on the other side. The muscle has a total length of $\ell = 12$ [cm] and has a circular cross section. The radius of the cross section can be approximated by:

$$r = a_1 \sin^3(a_2 x + a_3),$$

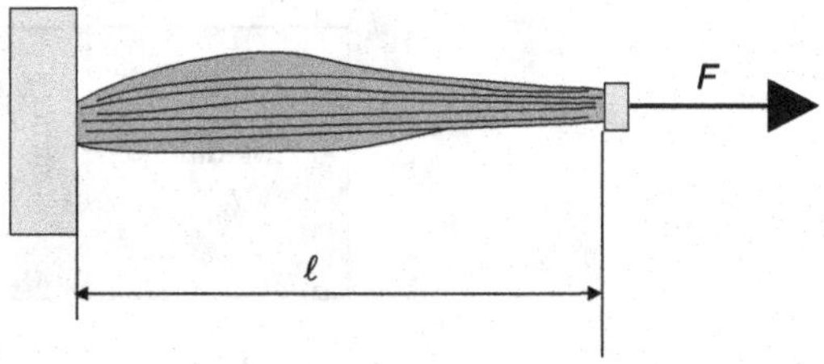

with $a_1 = 1.6$ [cm], $a_2 = 0.15$ [cm^{-1}] and $a_3 = 0.8$ [-]. The estimated Young's modulus is $E = 10^5$ [N m^{-2}].

(a) Give the differential equation for the axial displacement $u(x)$ and boundary conditions that describe the current problem.
(b) Determine the displacement field by adjusting the file `demo_fem1d`.

15 Solution of the One-Dimensional Convection–Diffusion Equation by Means of the Finite Element Method

15.1 Introduction

This chapter extends the formulation of the previous chapter for the one-dimensional diffusion equation to the time-dependent convection–diffusion equation. Although the proper functioning of the human body relies on maintaining a homeostasis or equilibrium in the physiological state of the tissues and organs, it is a dynamic equilibrium. This means that all processes have to respond to changing inputs, which are caused by changes of the environment. The diffusion processes taking place in the body are not constant, but non-stationary, so time has to be included as an independent variable in the diffusion equation. Thus, the instationary diffusion equation becomes a **partial** differential equation.

Convection is the process whereby heat or particles are transported by air or fluid moving from one point to another point. Diffusion could be seen as a process of transport through immobilized fluid or air. When the fluid itself moves, particles in that fluid are dragged along. This is called convection and also plays a major role in biomechanics. An example is the loss of heat because moving air is passing the body. The air next to the body is heated by conduction, moves away and carries off the heat just taken from the body. Another example is a drug that is released at some spot in the circulation and is transported away from that spot by means of the blood flow. In larger blood vessels the prime mechanism of transportation is convection.

15.2 The Convection–Diffusion Equation

Assuming that the source term $f = 0$, the unsteady one-dimensional convection-diffusion equation can be written as

$$\frac{\partial u}{\partial t} + v\frac{\partial u}{\partial x} = \frac{\partial}{\partial x}\left(c\frac{\partial u}{\partial x}\right), \tag{15.1}$$

with u a function of both position x and time t:

$$u = u(x,t). \tag{15.2}$$

The convective velocity is denoted by v and c is the diffusion coefficient. Both v and c are assumed to be constant in the present chapter. Compared with the diffusion equation of the previous chapter, two terms have been added: the time dependency term $\partial u/\partial t$ (inertia term) and the convective term $v\partial u/\partial x$. The convection–diffusion equation holds within a given spatial domain $\Omega = [a,b]$, i.e. with boundaries at $x = a$ and $x = b$, as well as within a time domain, say $S = [0,T]$, and is assumed to be subject to the boundary conditions:

$$u = U \text{ at } \Gamma_u \text{ (located at } x = a), \tag{15.3}$$

and

$$c\frac{\partial u}{\partial x} = P \text{ at } \Gamma_p \text{ (located at } x = b). \tag{15.4}$$

Furthermore, one initial condition on u must be specified, say at $t = 0$:

$$u(x, t = 0) = u_{\text{ini}}(x) \text{ in } \Omega. \tag{15.5}$$

Under certain conditions, the transient character of the solution of the unsteady convection–diffusion equation vanishes. In that case, we deal with a steady convection–diffusion problem, described by the differential equation

$$v\frac{du}{dx} = \frac{d}{dx}\left(c\frac{du}{dx}\right). \tag{15.6}$$

Example 15.1 A typical example of a solution of the **steady** convection–diffusion problem is represented in Fig. 15.1. In this example, the domain spans $0 \leq x \leq 1$, while at $x = 0$ the solution is set to 0 and at $x = 1$ the value $u = 1$ is imposed. That means that, in this example, two essential boundary conditions are used and that there is no natural boundary condition prescribed. The analytical solution for $v \neq 0$ in this case is

$$u = \frac{1}{1 - e^{v/c}}(1 - e^{vx/c}).$$

Clearly, without any convection, i.e. $v = 0$, a spatially linear distribution of u results, while with increasing convective velocity v a boundary layer develops at $x = 1$.

The convection–diffusion equation may be written in a dimensionless form by introducing an appropriate length scale L, a timescale Θ and a reference

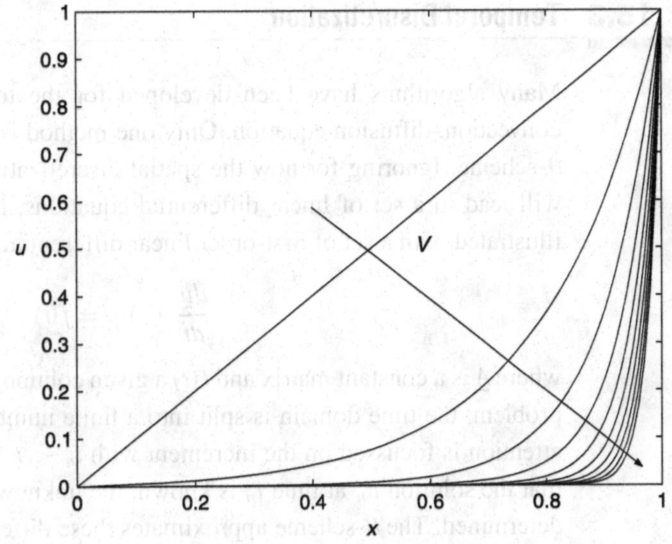

Figure 15.1

Solution to steady one-dimensional convection–diffusion problem.

solution U, for example the reference temperature or concentration. Then, the dimensionless solution u^*, the dimensionless coordinate x^* and the dimensionless time t^* are defined as

$$u^* = \frac{u}{U} \qquad x^* = \frac{x}{L} \qquad t^* = \frac{t}{\Theta}. \tag{15.7}$$

Assuming v and c constant, the dimensionless form may be written as

$$\frac{1}{\text{Fo}} \frac{\partial u^*}{\partial t^*} + \text{Pe}\frac{\partial u^*}{\partial x^*} = \frac{\partial}{\partial x^*}\left(\frac{\partial u^*}{\partial x^*}\right), \tag{15.8}$$

with the Fourier number given by

$$\text{Fo} = \frac{\Theta c}{L^2}, \tag{15.9}$$

and the Peclet number given by

$$\text{Pe} = \frac{vL}{c}. \tag{15.10}$$

The Peclet number reflects the relative importance of convection compared with diffusion. It will be demonstrated that with increasing Peclet number the numerical solution of the convection–diffusion problem becomes more difficult.

In the next section, we will start by studying the time discretization of an non-stationary equation. After that, in Section 15.4, the spatial discretization of the convection–diffusion equation will be discussed.

15.3 Temporal Discretization

Many algorithms have been developed for the temporal discretization of the convection–diffusion equation. Only one method is discussed here: the so-called θ-scheme. Ignoring for now the spatial discretization in the next section, which will lead to a set of linear differential equations, the time discretization is best illustrated with a set of first-order linear differential equations:

$$\frac{d\underset{\sim}{u}}{dt} + \underline{A}\,\underset{\sim}{u} = \underset{\sim}{f}(t), \tag{15.11}$$

where $\underline{A}$ is a constant matrix and $\underset{\sim}{f}(t)$ a given column. To solve this time-dependent problem, the time domain is split into a finite number of time increments. Then, attention is focussed on the increment with $t_n < t < t_{n+1} = t_n + \Delta t$. Assuming that the solution u_n at time t_n is known, the unknown u_{n+1} at time t_{n+1} has to be determined. The θ-scheme approximates these differential equations by

$$\frac{\underset{\sim}{u}_{n+1} - \underset{\sim}{u}_n}{\Delta t} + \theta\underline{A}\,\underset{\sim}{u}_{n+1} + (1-\theta)\underline{A}\,\underset{\sim}{u}_n = \theta\underset{\sim}{f}_{n+1} + (1-\theta)\underset{\sim}{f}_n. \tag{15.12}$$

For $\theta = 0$, this scheme reduces to the **Euler explicit** or **forward Euler** scheme, while for $\theta = 1$ the **Euler implicit** or **backward Euler** scheme results. Both of these schemes are first-order accurate, i.e. $\mathcal{O}(\Delta t)$. This means that the accuracy of the solution is linearly related to the size of the time step Δt. The accuracy improves when the time step becomes smaller. For $\theta = 0.5$, the **Crank–Nicholson** scheme results, which is second-order accurate, i.e. $\mathcal{O}(\Delta t^2)$.

To illustrate the stability properties of the θ-method, consider a single-variable model problem:

$$\frac{du}{dt} + \lambda u = f, \tag{15.13}$$

with $\lambda > 0$. This differential equation has the property that any perturbation to the solution (for example induced by a perturbation of the initial value of u) decays exponentially as a function of time. Assume that $\hat{u}$ satisfies the differential Eq. (15.13) exactly, and let $\tilde{u}$ be a perturbation of $\hat{u}$, hence $u = \hat{u} + \tilde{u}$. Consequently, the perturbation $\tilde{u}$ must obey

$$\frac{d\tilde{u}}{dt} + \lambda\tilde{u} = 0. \tag{15.14}$$

If at $t = 0$, the perturbation equals $\tilde{u} = \tilde{u}_0$, the solution to this equation is

$$\tilde{u} = \mathrm{e}^{-\lambda t}\tilde{u}_0. \tag{15.15}$$

Clearly, if $\lambda > 0$, the perturbation decays exponentially as a function of time.

Application of the θ-scheme to the single variable model problem (15.13) yields

$$\frac{u_{n+1} - u_n}{\Delta t} + \theta\lambda u_{n+1} + (1-\theta)\lambda u_n = \theta f_{n+1} + (1-\theta) f_n. \qquad (15.16)$$

Now, as before, if $\tilde{u}_n$ is a perturbation of $\hat{u}_n$, this perturbation satisfies

$$\frac{\tilde{u}_{n+1} - \tilde{u}_n}{\Delta t} + \theta\lambda\tilde{u}_{n+1} + (1-\theta)\lambda\tilde{u}_n = 0. \qquad (15.17)$$

Clearly, the perturbation at $t = t_{n+1}$ can be expressed as

$$\tilde{u}_{n+1} = \underbrace{\frac{1-(1-\theta)\lambda\Delta t}{1+\theta\lambda\Delta t}}_{A}\tilde{u}_n. \qquad (15.18)$$

The factor A is called the amplification factor. To have a stable integration scheme, the magnitude of $\tilde{u}_{n+1}$ should be smaller than the magnitude of $\tilde{u}_n$, i.e. the perturbation should not grow as time proceeds. Hence, stability requires $|\tilde{u}_{n+1}| \leq |\tilde{u}_n|$, which holds if the amplification factor $|A| \leq 1$.

Figure 15.2 shows the amplification factor A as a function of $\lambda\Delta t$ with θ as a parameter. For $0 \leq \theta < 0.5$ the integration scheme is conditionally stable, meaning that the time step Δt has to be chosen sufficiently small relative to λ. In the multi-variable case, the above corresponds to the requirement that, in the case $0 \leq \theta < 0.5$, $\lambda\Delta t$ should be small compared with the eigenvalues of the matrix $\underline{A}$.

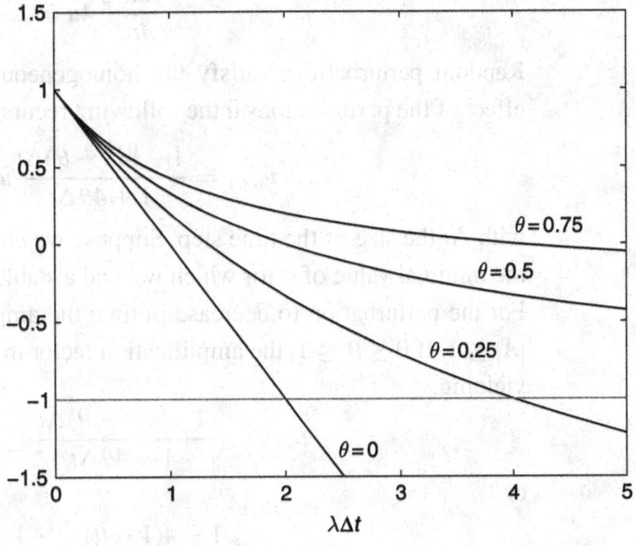

Figure 15.2

Amplification factor A as a function of $\lambda\Delta t$ for various values of θ.

For $0.5 \leq \theta \leq 1$, the scheme is unconditionally stable; hence for any choice of Δt a stable integration process results.

Example 15.2 Consider the following ordinary differential equation, formulated in dimensionless quantities:

$$\frac{du}{dt} + 2u = \frac{1}{4}\sin\left(\frac{1}{5}t\right).$$

The initial condition is $u(0) = 1$. A θ-scheme to find a numerical solution for this equation leads to:

$$u_{n+1} = \frac{u_n - 2(1-\theta)\Delta t u_n + \frac{1}{4}\Delta t\left(\theta \sin(\frac{1}{5}t_{n+1}) + (1-\theta)\sin(\frac{1}{5}t_n)\right)}{1 + 2\theta\Delta t}.$$

By using this scheme it is easy to solve this equation and investigate how it behaves for $\theta = 0$ (explicit scheme), $\theta = 1$ (implicit scheme) and $\theta = 0.5$ (Crank–Nicholson scheme, Cr–Nic). The result is given in Fig. 15.3 for different values of Δt. According to Eq. (15.18), an explicit scheme for this equation should become unstable at $\Delta t = 1$, while the other schemes are stable, which is clearly the case. When we decrease the time step, the explicit scheme also becomes stable and, provided the time step is small enough, all schemes lead to the same solution.

Example 15.3 Consider a problem that is described by the following ordinary differential equation:

$$\frac{du}{dt} + 4u = 2.$$

Random perturbations satisfy the homogeneous part of this equation. For the effect of the perturbations $\tilde{u}$ the following recursive scheme can be derived:

$$\tilde{u}_{n+1} = \frac{1 - 4(1-\theta)\Delta t}{1 + 4\theta\Delta t}\tilde{u}_n = A\,\tilde{u}_n,$$

with Δt the size of the time step. Suppose we choose $\Delta t = 1$. What is in that case the minimal value of θ for which we find a stable solution?

For the perturbation to decrease in time the amplification factor A has to satisfy $|A| \leq 1$. If $0 \leq \theta \leq 1$, the amplification factor in this case is always less than one, yielding:

$$\frac{1 - 4(1-\theta)\Delta t}{1 + 4\theta\Delta t} \geq -1$$

or:

$$1 - 4(1-\theta) \geq -1 - 4\theta$$

which leads to $\theta \geq \frac{1}{4}$.

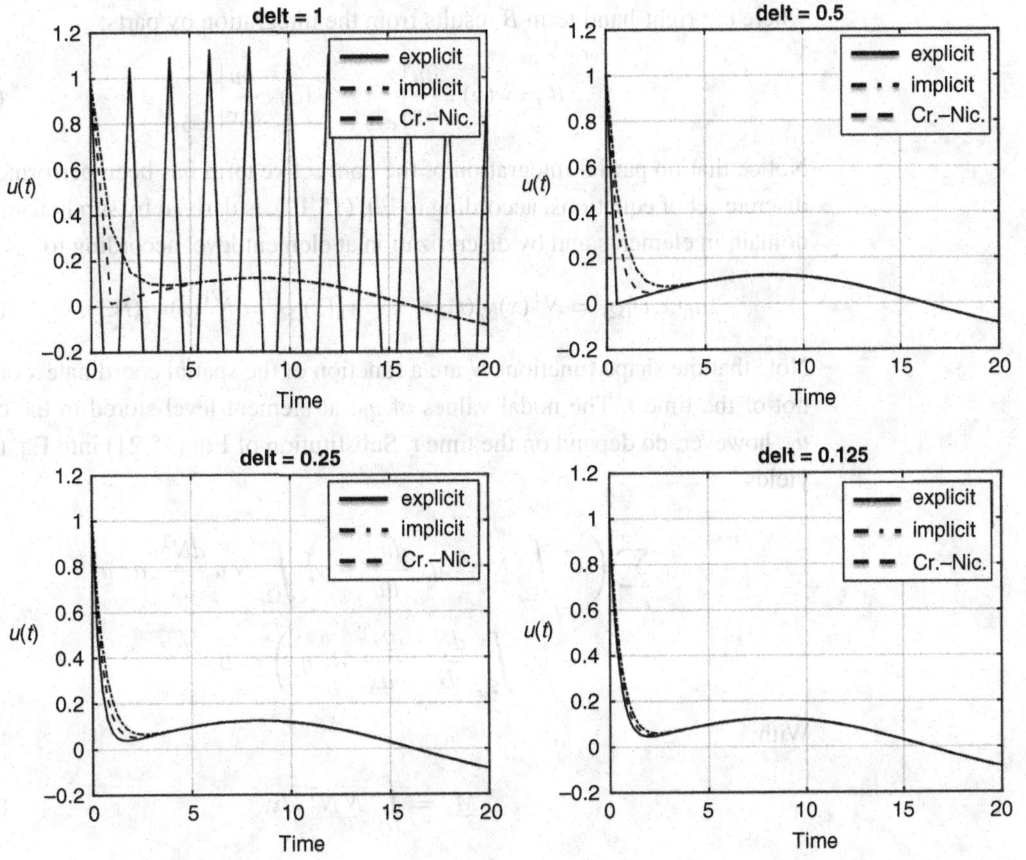

Figure 15.3

Numerical solution for different values of Δt.

15.4 Spatial Discretization

Following a similar derivation to that in the previous chapter, the weak form is obtained by multiplication of Eq. (15.1) with a suitable weighting function w, performing an integration over $\Omega = [a, b]$, followed by an integration by parts:

$$\int_{\Omega} w \frac{\partial u}{\partial t}\, dx + \int_{\Omega} wv \frac{\partial u}{\partial x}\, dx + \int_{\Omega} \frac{dw}{dx} c \frac{\partial u}{\partial x}\, dx = B, \tag{15.19}$$

where the right-hand term B results from the integration by parts:

$$B = w(b)\, c \frac{\partial u}{\partial x}\bigg|_{x=b} - w(a)\, c \frac{\partial u}{\partial x}\bigg|_{x=a}. \tag{15.20}$$

Notice that no partial integration of the convective term has been performed. The discrete set of equations, according to Eq. (15.19), is derived by subdivision of the domain in elements and by discretization at element level according to

$$u_h(x,t)|_{\Omega_e} = \underset{\sim}{N}^T(x)\underset{\sim}{u}_e(t), \qquad w_h(x)|_{\Omega_e} = \underset{\sim}{N}^T(x)\underset{\sim}{w}_e(t). \tag{15.21}$$

Note that the shape functions $\underset{\sim}{N}$ are a function of the spatial coordinate x only and not of the time t. The nodal values of u_h, at element level stored in the column $\underset{\sim}{u}_e$, however, do depend on the time t. Substitution of Eq. (15.21) into Eq. (15.19) yields

$$\sum_{e=1}^{N_{el}} \left(\underset{\sim}{w}_e^T \int_{\Omega_e} \underset{\sim}{N}\,\underset{\sim}{N}^T\, dx\, \frac{d\underset{\sim}{u}_e}{dt} + \underset{\sim}{w}_e^T \int_{\Omega_e} \underset{\sim}{N}\, a \frac{d\underset{\sim}{N}^T}{dx}\, dx\, \underset{\sim}{u}_e \right.$$
$$\left. + \underset{\sim}{w}_e^T \int_{\Omega_e} \frac{d\underset{\sim}{N}}{dx}\, c \frac{d\underset{\sim}{N}^T}{dx}\, dx\, \underset{\sim}{u}_e \right) = B. \tag{15.22}$$

With

$$\underline{M}_e = \int_{\Omega_e} \underset{\sim}{N}\,\underset{\sim}{N}^T\, dx, \tag{15.23}$$

and

$$\underline{K}_e = \int_{\Omega_e} \underset{\sim}{N}\, a \frac{d\underset{\sim}{N}^T}{dx}\, dx + \int_{\Omega_e} \frac{d\underset{\sim}{N}}{dx}\, c \frac{d\underset{\sim}{N}^T}{dx}\, dx, \tag{15.24}$$

Eq. (15.22) can be written as

$$\sum_{e=1}^{N_{el}} \underset{\sim}{w}_e^T \left(\underline{M}_e \frac{d\underset{\sim}{u}_e}{dt} + \underline{K}_e \underset{\sim}{u}_e \right) = B. \tag{15.25}$$

After the usual assembly process, this is written in global quantities:

$$\underset{\sim}{w}^T \left(\underline{M} \frac{d\underset{\sim}{u}}{dt} + \underline{K}\, \underset{\sim}{u} \right) = \underset{\sim}{w}^T \underset{\sim}{f}, \tag{15.26}$$

where $\underset{\sim}{f}$ results from the contribution of B (see Section 14.5). This equation has to be satisfied for all $\underset{\sim}{w}$, hence

$$\underline{M} \frac{d\underset{\sim}{u}}{dt} + \underline{K}\, \underset{\sim}{u} = \underset{\sim}{f}. \tag{15.27}$$

This is a set of first-order differential equations having a similar structure to Eq. (15.11). Therefore, application of the θ-scheme for temporal discretization yields

$$\left(\frac{1}{\Delta t}\underline{M}+\theta\underline{K}\right)\underset{\sim}{u}_{n+1}=\left(\frac{1}{\Delta t}\underline{M}-(1-\theta)\underline{K}\right)\underset{\sim}{u}_{n}+\underset{\sim}{f}^{\theta}, \tag{15.28}$$

with

$$\underset{\sim}{f}^{\theta}=\theta\underset{\sim}{f}_{n+1}+(1-\theta)\underset{\sim}{f}_{n}. \tag{15.29}$$

Clearly, in the steady case, the set of equations, Eqs. (15.27), reduces to

$$\underline{K}\,\underset{\sim}{u}=\underset{\sim}{f}. \tag{15.30}$$

Example 15.4 Consider the steady convection–diffusion problem, according to

$$v\frac{du}{dx}=\frac{d}{dx}\left(c\frac{du}{dx}\right),$$

with the following parameter setting:

$$\begin{aligned}\Omega &= [0\ \ 1]\\ \Gamma_u &: x=0 \text{ and } x=1\\ \Gamma_p &= \emptyset\\ c &= 1\\ u(x=0) &= 0\\ u(x=1) &= 1\end{aligned}$$

The convective velocity v will be varied. For $v = 0$, the diffusion limit, the solution is obvious: u varies linearly in x from $u = 0$ at $x = 0$ to $u = 1$ at $x = 1$. Figures 15.4(a) to (d) show the solutions for $v = 1, 10, 25$ and 100, respectively, using a uniform element distribution with ten linear elements. For $v = 1$ and $v = 10$ the approximate solution u_h (solid line) closely (but not exactly) follows the exact solution (dashed line). However, for $v = 25$ the numerical solution starts to demonstrate an oscillatory behaviour that is more prominent for $v = 100$. Careful analysis of the discrete set of equations shows that the so-called element Peclet number governs this oscillatory behaviour. The element Peclet number is defined as

$$\mathrm{Pe_h}=\frac{vh}{2c},$$

where h is the element length. Above a certain critical value of $\mathrm{Pe_h}$, the solution behaves in an oscillatory fashion. To reduce possible oscillations, the element Peclet number should be reduced. For fixed v and c this can only be achieved

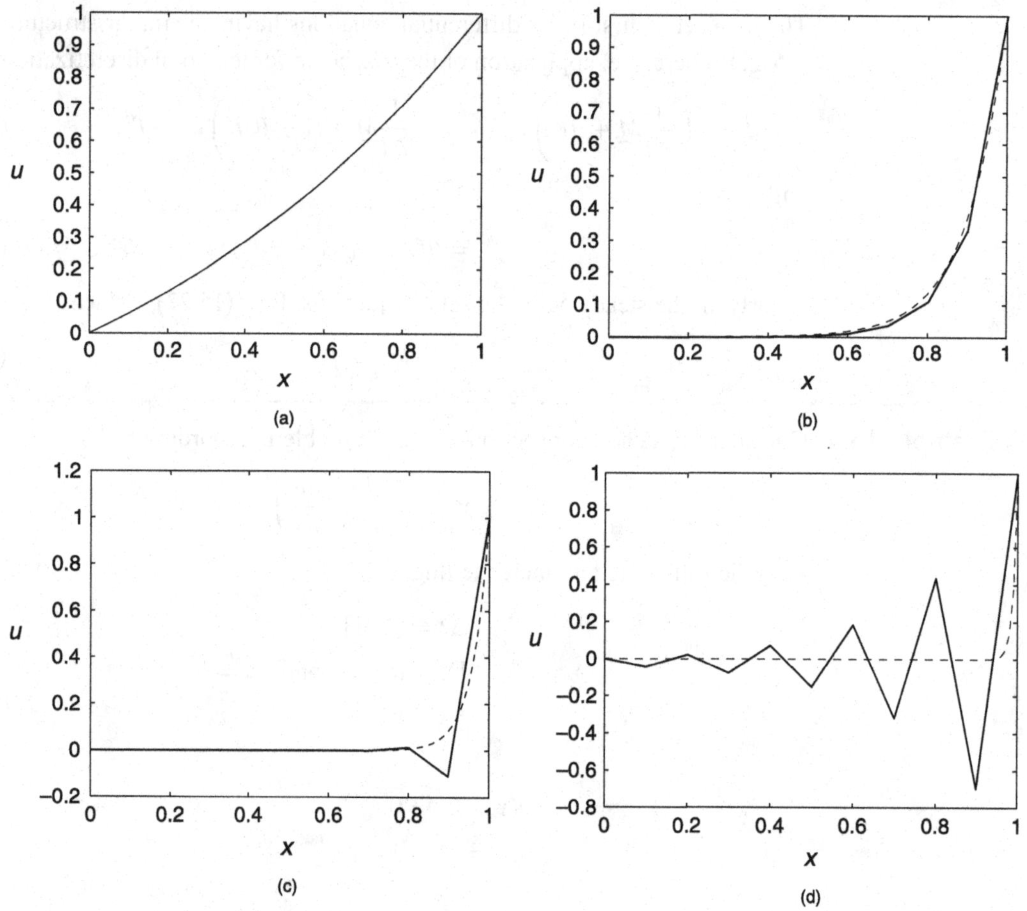

Figure 15.4

Solution of the steady convection–diffusion equation for $\nu = 1, 10, 25$ and 100, respectively using ten linear elements: solid line, approximate solution u_h; dashed line, exact solution u.

by reducing the element size h. For example, doubling the number of elements from 10 to 20 eliminates the oscillations at $\nu = 25$: see Fig. 15.5.

The oscillations that appear in the numerical solution of the steady convection–diffusion equation may be examined as follows. Consider a domain that is subdivided into two linear elements, each having a length equal to h. At one end of the domain the solution is fixed to $u = 0$, while at the other end the solution is set to $u = 1$, or any other arbitrary non-zero value. For constant ν and c, the governing differential equation may be rewritten as

$$\frac{\nu}{c}\frac{du}{dx} - \frac{d^2u}{dx^2} = 0.$$

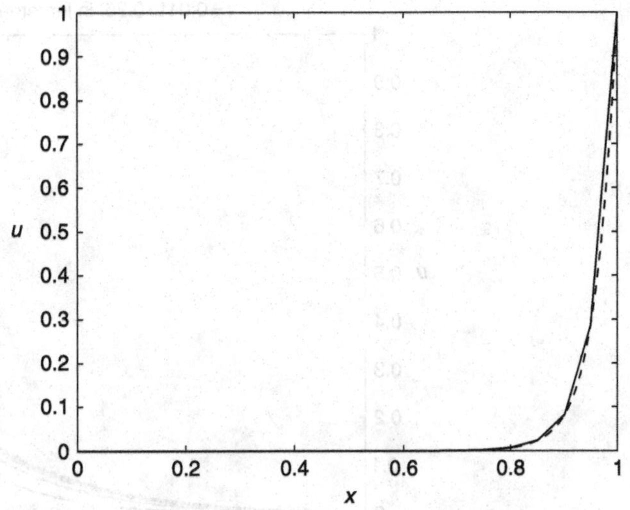

Figure 15.5

Solution of the steady convection–diffusion problem using 20 elements at $v = 25$.

The set of equations that results after discretization is, as usual:

$$\underline{K}\underset{\sim}{u} = \underset{\sim}{f}\,.$$

If only two linear elements of equal length h are used, the coefficient matrix $\underline{K}$ may be written as

$$\underline{K} = \frac{v}{2c}\begin{bmatrix} -1 & 1 & 0 \\ -1 & 0 & 1 \\ 0 & -1 & 1 \end{bmatrix} + \frac{1}{h}\begin{bmatrix} 1 & -1 & 0 \\ -1 & 2 & -1 \\ 0 & -1 & 1 \end{bmatrix},$$

where the first, asymmetric, part corresponds to the convective term and the second, symmetric, part to the diffusion term. In the absence of a source term, the second component of $\underset{\sim}{f}$ is zero. Let u_1 and u_3 be located at the ends of the domain such that $u_1 = 0$ and $u_3 = 1$, then u_2 is obtained from

$$2u_2 = 1 - \frac{vh}{2c}.$$

An oscillation becomes manifest if $u_2 < 0$. To avoid this, the element Peclet number should be smaller than one:

$$\mathrm{Pe_h} = \frac{vh}{2c} < 1.$$

Consequently, at a given convective velocity v and diffusion constant c, the mesh size h can be chosen such that an oscillation-free solution results. In particular for large values of v/c, this may result in very fine meshes. To avoid the

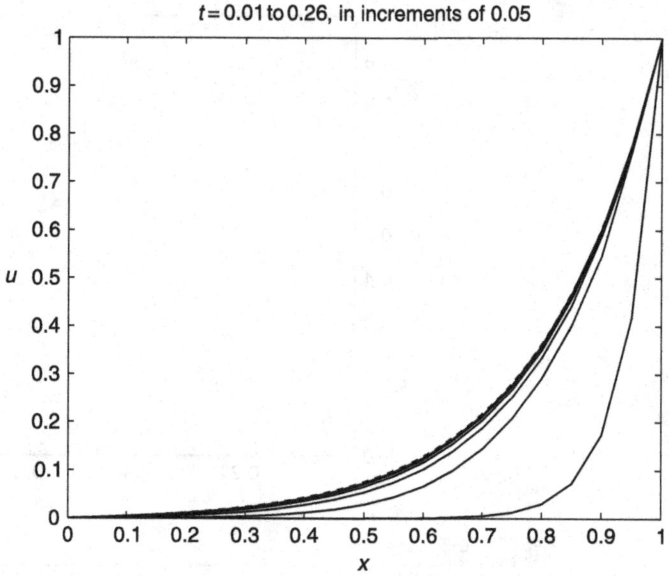

Figure 15.6

Solution of the unsteady convection–diffusion problem using 20 elements at $v = 10$.

use of very fine meshes, an alternative, stabilized formulation has been developed: the so-called SUPG (streamline-upwind/Petrov–Galerkin) formulation [3]. A discussion of this method, however, is beyond the scope of the present book.

Example 15.5 Let us consider the non-stationary convection–diffusion problem. For this problem, the same outline as in the previous example for $v = 10$ is chosen, and we use a uniform distribution of 20 linear elements. The initial condition is $u(x, t = 0) = 0$ throughout the domain. At the first time step, the boundary condition $u(x = 1, t) = 1$ is imposed. The unsteady solution is obtained using a time step of $\Delta t = 0.01$, while $\theta = 0.5$ is selected for the θ-scheme. Figure 15.6 shows the time-evolution of the solution towards the steady–state value (denoted by the dashed line) for $v = 10$.

Exercises

15.1 Consider the domain $\Omega = [0\ 1]$. In this domain, the one-dimensional steady convection–diffusion equation:

$$v\frac{du}{dx} = c\frac{d^2u}{dx^2}$$

holds. As boundary conditions, at $x = 0$, $u = 0$ and at $x = 1$, $u = 1$ are specified.

(a) Prove that the exact solution is given by

$$u = \frac{1}{1 - e^{\frac{v}{c}}}(1 - e^{\frac{v}{c}x}).$$

(b) Verify this by means of the script `demo_fem1dcd`, to solve the one-dimensional convection–diffusion problem, which can be found in the directory `oned` of the programme library `mlfem_nac`. Use five elements and select $c = 1$, while v is varied. Choose $v = 0$, $v = 1$, $v = 10$ and $v = 20$. Explain the results.

(c) According to Section 15.4, the solution is expected to be oscillation-free if the element Peclet number is smaller than 1:

$$\text{Pe}_h = \frac{ah}{2c} < 1.$$

Verify that this is indeed the case.

15.2 Investigate the unsteady convection–diffusion problem:

$$\frac{\partial u}{dt} + v\frac{\partial u}{\partial x} = \frac{\partial}{\partial x}\left(c\frac{\partial u}{\partial x}\right),$$

on the domain $\Omega = [0\ 1]$ subject to the initial condition:

$$u_{\text{ini}}(x, t = 0) = 0,$$

inside the domain Ω and the boundary conditions:

$$u = 0 \text{ at } x = 0, \quad u = 1 \text{ at } x = 1.$$

The θ-scheme for time integration is applied. Modify the m-file `demo_fem1dcd` accordingly. Use ten linear elements.

(a) Choose $v = c = 1$ and solve the problem with different values of θ. Use $\theta = 0.5, \theta = 0.4$ and $\theta = 0.25$. For each problem, start with a time step of 0.01 and increase the time step by 0.01 until a maximum of 0.05. Describe what happens with the solution.

(b) In the steady-state case, what is the maximum value of the convective velocity v such that the solution is oscillation-free for $c = 1$?

(c) Does the numerical solution remain oscillation-free in the unsteady case for $\theta = 0.5$ and $\Delta t = 0.001, 0.01, 0.1$? What happens?

(d) What happens at $\Delta t = 0.001$ if the convective velocity is reduced?

15.3 Investigate the unsteady convection problem:

$$\frac{\partial u}{\partial t} + v\frac{\partial u}{\partial x} = \frac{\partial}{\partial x}\left(c\frac{\partial u}{\partial x}\right)$$

on the spatial domain $\Omega = [0\ \ 1]$ and a temporal domain that spans $t = [0\ 0.5]$. At $x = 0$, the boundary condition $u = 0$ is prescribed, while at $x = 1$, $c\, du/dx = 0$ is selected. The convection-dominated case is investigated, with $v = 1$ and $c = 0.01$. The problem is solved using the θ-scheme with $\theta = 0.5$, a time step $\Delta t = 0.05$ and 40 linear elements.

(a) First, the initial condition:

$$u(x, t = 0) = \sin(2\pi x)$$

is considered. Adapt the `demo_fem1dcd` and the finite element model program accordingly. In particular, make sure that the initial condition is handled properly in `fem1dcd`. (Hint: the initial condition is specified in `sol(:,1)`.) Solve this problem. Plot the solution for all time steps.

(b) Second, let the initial condition be given by

$$u(x, t = 0) = 0 \text{ for } x < 0.25 \text{ and } x > 0.5,$$

while

$$u(x, t = 0) = 1 \text{ for } 0.25 \leq x \leq 0.5.$$

Solve this problem in the same way.

15.4 Many biological materials can be described as a mixture of a porous solid and a fluid, for example the articular cartilage [16], skin [17], intervertebral disk [7], heart tissue [13], etc. A confined compression test can be used to determine the material parameters. A circular specimen is placed in a confining ring. A porous filter supports the solid phase, while fluid from the specimen can be expelled. On top, a tight-fitting indenter is placed to mechanically load the specimen with a constant force. Because of the confining ring, the specimen can only deform in one direction. When a step load is applied to the tissue, at first a pressure will be built up in the fluid. Immediately after loading, the fluid will start moving through the filter, and the solid will gradually take over the load. After some time, the pressure in

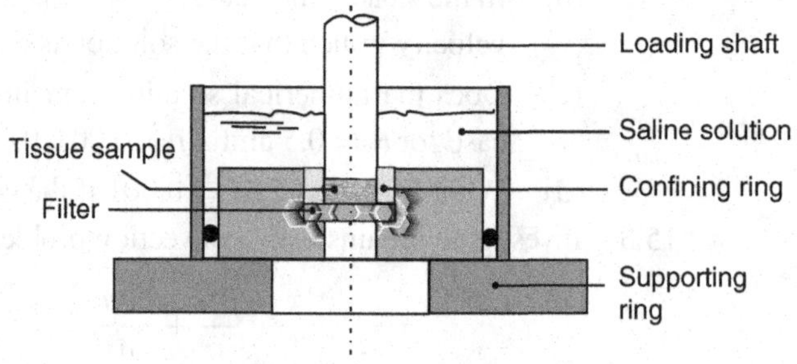

the fluid is zero and all load is taken by the solid. It can be derived from the theory of mixtures that the pressure in the fluid is described by the non-stationary diffusion equation:

$$\frac{\partial p}{\partial t} = HK\frac{\partial^2 p}{\partial^2 x},$$

with K the confined compression modulus (also called aggregate modulus) and H the permeability of the porous solid. The boundary conditions are:

- $p = 0$ at $x = 0$; free drainage at the porous filter
- $\partial p/\partial x = 0$ at $x = 10^{-2}$ [m]; no flow through the surface where the indenter makes contact with the specimen

The initial condition is $p = 1000$ [N m^{-2}] at every point in the specimen at $t = 0$ (except at $x = 0$). This means that, at $t = 0$, all load is carried by the fluid and not by the solid. The aggregate modulus satisfies $K = 10^5$ [N m^{-2}]; the permeability is given by $H = 10^{-14}$ [m^4 N^{-1}s^{-1}].

(a) Adjust `demo_fem1dcd` to calculate the pressure as a function of time for the above given problem. Choose $v = 0$ and take as a time step $\Delta t = 2500$ [s]. Use $\theta = 1$ for the time integration scheme.

(b) Determine the fluid pressure near the contact surface with the indenter and plot this pressure as a function of time.

15.5 Investigate the stationary convection diffusion problem (dimensionless properties):

$$v\frac{du}{dx} - \frac{d}{dx}\left(c\frac{du}{dx}\right) = 0$$

on the spatial domain $\Omega = [0\ 5]$. At $x = 0$ the boundary condition $u(0) = 0$ is prescribed, while at $x = 5$ the boundary $u(5) = 4$ is prescribed. The diffusion constant c is equal to 1. The convection-dominated case is investigated with velocity $v = 8$. The problem is solved on the domain using a mesh of equally distributed elements.

What is the minimal number of elements needed to obtain a stable solution for this problem?

16 Solution of the Three-Dimensional Convection–Diffusion Equation by Means of the Finite Element Method

16.1 Introduction

The two- and three-dimensional convection–diffusion equations play an important role in many applications in biomedical engineering. One typical example is the analysis of the effectiveness of different types of bioreactors for tissue engineering. Tissue engineering is a rapidly evolving interdisciplinary research area aiming at the replacement or restoration of diseased or damaged tissue. In many cases, devices made of artificial materials are only capable of partially restoring the original function of native tissues, and may not last for the full lifetime of a patient. In addition, there is no artificial replacement for a large number of tissues and organs. In tissue engineering, new, autologous tissues are grown. The tissue proliferation and differentiation process is strongly affected by mechanical stimuli and transport of oxygen, minerals, nutrients and growth factors. To optimize bioreactor systems, it is necessary to analyse how these systems behave. The convection–diffusion equation plays an important role in this kind of simulating analysis.

Figure 16.1 shows two different bioreactor configurations, both of which have been used in the past to tissue engineer articular cartilage. The work was especially focussed on glucose, oxygen and lactate, because these metabolites play a major role in the biosynthesis and survival of chondrocyte. Questions ranged from: 'Does significant nutrient depletion occur at the high cells concentrations required for chondrogenesis?' to 'Do increasing transport limitations due to matrix accumulation significantly affect metabolite distributions?' Figure 16.2 shows a typical result for the calculated oxygen distributions in the two bioreactor configurations.

This chapter explains the discretization of the convection-diffusion equation in two or three dimensions. First, the diffusion equation is discussed; thereafter, the convection–diffusion equation is elaborated. The spatial discretization of the weighting function is based on the Galerkin method.

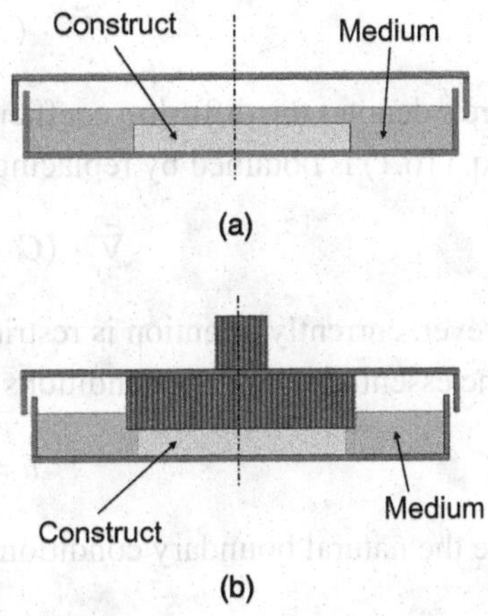

Figure 16.1

Culture configurations. (a) Petri dish; (b) compression set-up. Adapted from [19].

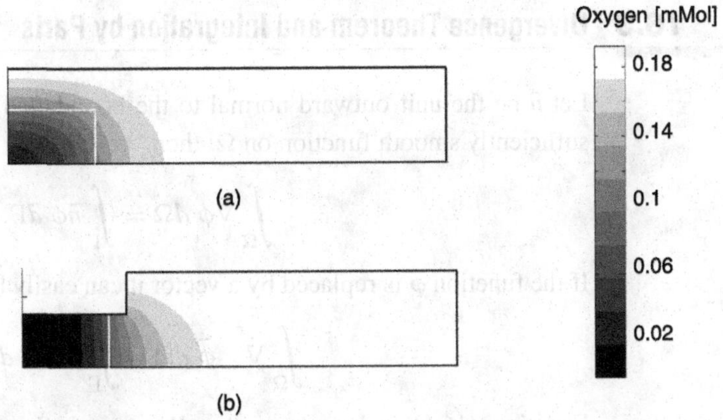

Figure 16.2

Oxygen distribution in the static case at 48 h. Because of axisymmetry, only the right half of the domain cross section is shown. The construct position is indicated with a white line. (a) Petri dish; (b) compression set-up. Adapted from [19].

16.2 Diffusion Equation

Consider a two- or three-dimensional domain Ω with boundary Γ. As in the one-dimensional model problem, the boundary Γ is split into a part Γ_u along which the essential boundary conditions are specified, and a part Γ_p along which the natural boundary conditions may be specified. The generic form of the diffusion equation is given by

$$\vec{\nabla} \cdot (c\vec{\nabla} u) + f = 0, \tag{16.1}$$

where c denotes the diffusion coefficient and f a source term. A more general form of Eq. (16.1) is obtained by replacing the scalar c with a second-order tensor:

$$\vec{\nabla} \cdot (\boldsymbol{C} \cdot \vec{\nabla} u) + f = 0\,. \tag{16.2}$$

However, currently attention is restricted to Eq. (16.1).

The essential boundary conditions along Γ_u read:

$$u = U \text{ at } \Gamma_u, \tag{16.3}$$

while the natural boundary conditions along Γ_p are given by

$$\vec{n} \cdot c\vec{\nabla} u = P \text{ at } \Gamma_p, \tag{16.4}$$

with $\vec{n}$ the unit outward normal vector to the boundary Γ.

16.3 Divergence Theorem and Integration by Parts

Let $\vec{n}$ be the unit outward normal to the boundary Γ of the domain Ω, and ϕ a sufficiently smooth function on Ω; then

$$\int_\Omega \vec{\nabla}\phi \, d\Omega = \int_\Gamma \vec{n}\phi \, d\Gamma. \tag{16.5}$$

If the function ϕ is replaced by a vector it can easily be derived that:

$$\int_\Omega \vec{\nabla} \cdot \vec{\phi} \, d\Omega = \int_\Gamma \vec{n} \cdot \vec{\phi} \, d\Gamma. \tag{16.6}$$

Equation (16.6) is known as the **divergence theorem**. For a proof of these equations, see for example Adams [1].

Let both ϕ and ψ be sufficiently smooth functions on Ω, then

$$\int_\Omega (\vec{\nabla}\phi)\psi \, d\Omega = \int_\Gamma \vec{n}\phi\psi \, d\Gamma - \int_\Omega \phi\vec{\nabla}\psi \, d\Omega. \tag{16.7}$$

This is called integration by parts. To prove this, we must integrate the product rule of differentiation:

$$\vec{\nabla}(\phi\psi) = (\vec{\nabla}\phi)\psi + \phi\vec{\nabla}\psi, \tag{16.8}$$

to obtain

$$\int_\Omega \vec{\nabla}(\phi\psi) \, d\Omega = \int_\Omega (\vec{\nabla}\phi)\psi \, d\Omega + \int_\Omega \phi\vec{\nabla}\psi \, d\Omega\,. \tag{16.9}$$

Subsequently, use the divergence theorem to convert the left-hand side into the boundary integral:

$$\int_{\Omega} \vec{\nabla}(\phi\psi)\, d\Omega = \int_{\Gamma} \vec{n}\phi\psi\, d\Gamma. \tag{16.10}$$

This yields the desired result.

16.4 Weak Form

Following the same steps as in Chapter 14, the differential equation Eq. (16.1) is multiplied by a weighting function w and integrated over the domain Ω:

$$\int_{\Omega} w\left(\vec{\nabla}\cdot(c\vec{\nabla}u) + f\right) d\Omega = 0, \quad \text{for all } w. \tag{16.11}$$

Next, the integration by parts rule according to Eq. (16.7) is used:

$$\int_{\Gamma} w\,\vec{n}\cdot c\vec{\nabla}u\, d\Gamma - \int_{\Omega} \vec{\nabla}w\cdot(c\vec{\nabla}u)\, d\Omega + \int_{\Omega} wf\, d\Omega = 0. \tag{16.12}$$

The boundary integral can be split into two parts, depending on the essential and natural boundary conditions:

$$\int_{\Gamma} w\,\vec{n}\cdot c\vec{\nabla}u\, d\Gamma = \int_{\Gamma_u} w\,\vec{n}\cdot c\vec{\nabla}u\, d\Gamma + \int_{\Gamma_p} wP\, d\Gamma, \tag{16.13}$$

where $c\vec{\nabla}u\cdot\vec{n} = P$ at Γ_p is used. It will be clear that, similar to the derivations in Chapter 14, the first integral on the right-hand side of Eq. (16.13) is unknown, while the second integral offers the possibility to incorporate the natural boundary conditions. For the time being, we keep both integrals together (to limit the complexity of the equations, and rewrite Eq. (16.12) according to:

$$\int_{\Omega} \vec{\nabla}w\cdot(c\vec{\nabla}u)\, d\Omega = \int_{\Omega} wf\, d\Omega + \int_{\Gamma} w\,\vec{n}\cdot c\vec{\nabla}u\, d\Gamma. \tag{16.14}$$

16.5 Galerkin Discretization

Step 1 Introduce a mesh by splitting the domain Ω into a number of non-overlapping elements Ω_e. In a two-dimensional configuration, the elements typically have either a triangular (in this case the mesh is sometimes referred to as a triangulation) or a quadrilateral shape. A typical example of triangulation is given in Fig. 16.3. Each triangle corresponds to an element.

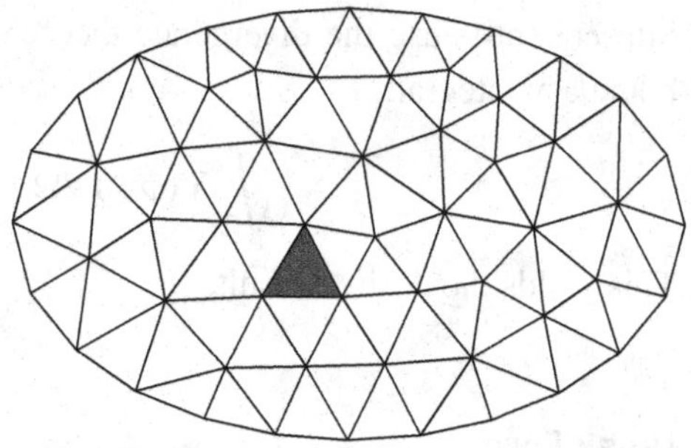

Figure 16.3

Example of a two-dimensional finite element mesh using triangular elements. One element has been highlighted.

The integration over the domain Ω can be performed by a summation of the integrals over each element:

$$\sum_{e=1}^{N_{el}} \int_{\Omega_e} \vec{\nabla} w \cdot (c\vec{\nabla} u)\, d\Omega = \sum_{e=1}^{N_{el}} \left(\int_{\Omega_e} wf\, d\Omega + \int_{\Gamma_e} w\vec{n} \cdot c\vec{\nabla} u\, d\Gamma \right). \tag{16.15}$$

The boundary part Γ_e denotes the intersection of element Ω_e with the boundary Γ, hence $\Gamma_e = \Omega_e \cap \Gamma$. Clearly, not every element will have an intersection with Γ.

Step 2 If a Cartesian coordinate system is used for two-dimensional problems (extension to the three-dimensional case is straightforward), the inner product $\vec{\nabla} w \cdot (c\vec{\nabla} u)$ yields

$$\begin{aligned} \vec{\nabla} w \cdot (c\vec{\nabla} u) &= \left(\frac{\partial w}{\partial x}\vec{e}_x + \frac{\partial w}{\partial y}\vec{e}_y \right) \cdot c \left(\frac{\partial u}{\partial x}\vec{e}_x + \frac{\partial u}{\partial y}\vec{e}_y \right) \\ &= c \left(\frac{\partial w}{\partial x}\frac{\partial u}{\partial x} + \frac{\partial w}{\partial y}\frac{\partial u}{\partial y} \right). \end{aligned} \tag{16.16}$$

Step 3 Introduce a discretization for both the weighting function w and the unknown u, so within each element

$$w_h|_{\Omega_e} = \sum_{i=1}^{n} N_i(x,y) w_{e,i} = \underset{\sim}{N}^{\mathrm{T}}(x,y) \underset{\sim}{w}_e \tag{16.17}$$

$$u_h|_{\Omega_e} = \sum_{i=1}^{n} N_i(x,y) u_{e,i} = \underset{\sim}{N}^{\mathrm{T}}(x,y) \underset{\sim}{u}_e\,. \tag{16.18}$$

Step 4 Substitution of this discretization into Eq. (16.16) yields

$$\begin{aligned}\vec{\nabla} w_{\mathrm{h}} \cdot \left(c\vec{\nabla} u_{\mathrm{h}}\right) &= c\left(\frac{\partial w_{\mathrm{h}}}{\partial x}\frac{\partial u_{\mathrm{h}}}{\partial x} + \frac{\partial w_{\mathrm{h}}}{\partial y}\frac{\partial u_{\mathrm{h}}}{\partial y}\right) \\ &= c\left(\frac{\partial \underset{\sim}{N}^{\mathrm{T}} \underset{\sim}{w}_e}{\partial x}\frac{\partial \underset{\sim}{N}^{\mathrm{T}} \underset{\sim}{u}_e}{\partial x} + \frac{\partial \underset{\sim}{N}^{\mathrm{T}} \underset{\sim}{w}_e}{\partial y}\frac{\partial \underset{\sim}{N}^{\mathrm{T}} \underset{\sim}{u}_e}{\partial y}\right) \\ &= c\,\underset{\sim}{w}_e^{\mathrm{T}}\left(\frac{\partial \underset{\sim}{N}}{\partial x}\frac{\partial \underset{\sim}{N}^{\mathrm{T}}}{\partial x} + \frac{\partial \underset{\sim}{N}}{\partial y}\frac{\partial \underset{\sim}{N}^{\mathrm{T}}}{\partial y}\right)\underset{\sim}{u}_e. \end{aligned} \tag{16.19}$$

Step 5 Using the discretization in Eq. (16.15) yields

$$\begin{aligned}&\sum_{e=1}^{N_{\mathrm{el}}} \underset{\sim}{w}_e^{\mathrm{T}} \int_{\Omega_e} c\left(\frac{\partial \underset{\sim}{N}}{\partial x}\frac{\partial \underset{\sim}{N}^{\mathrm{T}}}{\partial x} + \frac{\partial \underset{\sim}{N}}{\partial y}\frac{\partial \underset{\sim}{N}^{\mathrm{T}}}{\partial y}\right) d\Omega\, \underset{\sim}{u}_e \\ &= \sum_{e=1}^{N_{\mathrm{el}}}\left(\underset{\sim}{w}_e^{\mathrm{T}} \int_{\Omega_e} \underset{\sim}{N} f\, d\Omega + \underset{\sim}{w}_e^{\mathrm{T}} \int_{\Gamma_e} \underset{\sim}{N} \vec{n} \cdot c\vec{\nabla} u\, d\Gamma\right). \end{aligned} \tag{16.20}$$

The element matrix is given by

$$\underline{K}_e = \int_{\Omega_e} c\left(\frac{\partial \underset{\sim}{N}}{\partial x}\frac{\partial \underset{\sim}{N}^{\mathrm{T}}}{\partial x} + \frac{\partial \underset{\sim}{N}}{\partial y}\frac{\partial \underset{\sim}{N}^{\mathrm{T}}}{\partial y}\right) d\Omega, \tag{16.21}$$

and the element column by

$$\underset{\sim}{f}_e = \int_{\Omega_e} \underset{\sim}{N} f\, d\Omega + \int_{\Gamma_e} \underset{\sim}{N} \vec{n} \cdot c\vec{\nabla} u\, d\Gamma. \tag{16.22}$$

Using this notation, Eq. (16.20) may be written as

$$\sum_{e=1}^{N_{\mathrm{el}}} \underset{\sim}{w}_e^{\mathrm{T}} \underline{K}_e\, \underset{\sim}{u}_e = \sum_{e=1}^{N_{el}} \underset{\sim}{w}_e^{\mathrm{T}} \underset{\sim}{f}_e. \tag{16.23}$$

Following a similar procedure to that outlined in Chapter 14, this may be rearranged into

$$\underset{\sim}{w}^{\mathrm{T}} \underline{K}\, \underset{\sim}{u} = \underset{\sim}{w}^{\mathrm{T}} \underset{\sim}{f}, \tag{16.24}$$

which should hold for all $\underset{\sim}{w}$. This finally leads to:

$$\underline{K}\underset{\sim}{u} = \underset{\sim}{f}. \tag{16.25}$$

Similar to the situation described in Section 14.6, the column $\underset{\sim}{u}$ contains an unknown and a known part depending on the essential boundary conditions. In the nodes where essential boundary conditions are prescribed, the associated external loads in the right-hand side integrals are unknown. However, the set equations can be partitioned as has been done in Chapter 14 to arrive at a set that can be solved.

16.6 Convection–Diffusion Equation

Assuming isotropic diffusion, the convection–diffusion equation is given by

$$\frac{\partial u}{\partial t} + \vec{v} \cdot \vec{\nabla} u = \vec{\nabla} \cdot (c\vec{\nabla} u) + f, \tag{16.26}$$

with $\vec{v}$ the convective velocity. This equation should hold on the spatial domain Ω during a certain period of time, say $S = [0, T]$. Initial boundary conditions must be specified:

$$u(\vec{x}, t = 0) = u_{\text{ini}}(\vec{x}) \text{ in } \Omega, \tag{16.27}$$

as well as essential and natural boundary conditions:

$$u = U \text{ at } \Gamma_u \tag{16.28}$$

$$\vec{n} \cdot c\vec{\nabla} u = P \text{ at } \Gamma_p. \tag{16.29}$$

The weak form is obtained analogously to the procedure of Section 16.4, giving

$$\int_\Omega \left(w\frac{\partial u}{\partial t} + w\vec{v} \cdot \vec{\nabla} u + \vec{\nabla} w \cdot (c\vec{\nabla} u) \right) d\Omega = \int_\Omega wf \, d\Omega + \int_\Gamma w\vec{n} \cdot c\vec{\nabla} u \, d\Gamma. \tag{16.30}$$

Spatial discretization is performed in a two-dimensional configuration by introducing

$$w_{\text{h}}|_{\Omega_e} = \underset{\sim}{N}^{\text{T}}(x, y)\underset{\sim}{w}_e, \tag{16.31}$$

and

$$u_{\text{h}}|_{\Omega_e} = \underset{\sim}{N}^{\text{T}}(x, y)\underset{\sim}{u}_e. \tag{16.32}$$

For a particular element Ω_e, the individual integrals of Eq. (16.30) can be converted to:

$$\int_{\Omega_e} w\frac{\partial u}{\partial t} \, d\Omega = \underset{\sim}{w}_e^{\text{T}} \underline{M}_e \frac{d\underset{\sim}{u}_e}{dt}, \tag{16.33}$$

with:

$$\underline{M}_e = \int_{\Omega_e} \underset{\sim}{N}\underset{\sim}{N}^{\text{T}} \, d\Omega. \tag{16.34}$$

Further, by using

$$\vec{v} = v_x \vec{e}_x + v_y \vec{e}_y, \tag{16.35}$$

we can write:

$$\int_{\Omega_e} w\vec{v} \cdot \vec{\nabla} u \, d\Omega = \underset{\sim}{w}_e^{\text{T}} \underline{C}_e \underset{\sim}{u}_e, \tag{16.36}$$

with

$$\underline{C}_e = \int_{\Omega_e} \underset{\sim}{N} \left(v_x \frac{\partial \underset{\sim}{N}^{\mathrm{T}}}{\partial x} + v_y \frac{\partial \underset{\sim}{N}^{\mathrm{T}}}{\partial y} \right) d\Omega. \tag{16.37}$$

The remaining integrals from Eq. (16.30) follow from the previous section. Therefore, Eq. (16.30) may be formulated as

$$\sum_{e=1}^{N_{\mathrm{el}}} \underset{\sim}{w}_e^{\mathrm{T}} \left(\underline{M}_e \frac{d\underset{\sim}{u}_e}{dt} + (\underline{C}_e + \underline{K}_e) \underset{\sim}{u}_e \right) = \sum_{e=1}^{N_{\mathrm{el}}} \underset{\sim}{w}_e^{\mathrm{T}} \underset{\sim}{f}_e, \tag{16.38}$$

which must be satisfied for all admissible weighting values; hence, after the assembly process, the following set of equations results:

$$\underline{M} \frac{d\underset{\sim}{u}}{dt} + (\underline{C} + \underline{K}) \underset{\sim}{u} = \underset{\sim}{f}. \tag{16.39}$$

Application of the θ-scheme for temporal discretization in the time increment $[t_n, t_{n+1}]$ results in:

$$\begin{aligned} &\left(\underline{M} \frac{1}{\Delta t} + \theta \underline{C} + \theta \underline{K} \right) \underset{\sim}{u}_{n+1} \\ &\quad = \left(\underline{M} \frac{1}{\Delta t} - (1-\theta) \underline{C} - (1-\theta) \underline{K} \right) \underset{\sim}{u}_n + \underset{\sim}{f}^{\theta}, \end{aligned} \tag{16.40}$$

with $\Delta t = t_{n+1} - t_n$ and

$$\underset{\sim}{f}^{\theta} = \theta \underset{\sim}{f}_{n+1} + (1-\theta) \underset{\sim}{f}_n. \tag{16.41}$$

16.7 Isoparametric Elements and Numerical Integration

To compute the element matrices $\underline{M}_e$, $\underline{C}_e$ and $\underline{K}_e$ and the element array $\underset{\sim}{f}_e$, the shape functions $\underset{\sim}{N}$ and the shape function derivatives with respect to the coordinates x and y need to be available. To define the shape functions it is convenient to introduce a local coordinate system.

In particular for an arbitrarily shaped quadrilateral element, for instance as depicted in Fig. 16.4(a), it is difficult or impossible to define explicitly the shape functions with respect to the global coordinates x and y. However, defining the shape functions in a unit square domain, as depicted in Fig. 16.4(b), is straightforward. The local coordinate system defined in Fig. 16.4(b) is chosen such that $-1 \leq \xi \leq 1$ and $-1 \leq \eta \leq 1$. Node one, for instance, has local coordinates $\xi = -1$ and $\eta = -1$. Remember that the shape function of a certain node has to be one at the spatial location of this node and zero at the spatial location of

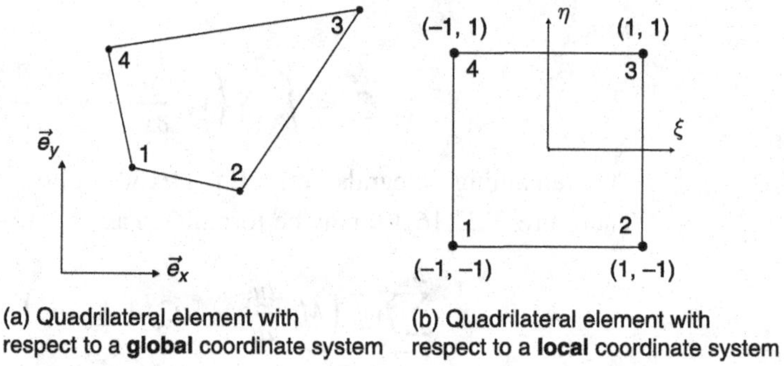

Figure 16.4

Quadrilateral element with respect to global and local coordinate systems.

all other nodes (see Section 14.4). Using the local coordinate system allows an elegant definition of the shape functions of a four-node quadrilateral element:

$$
\begin{aligned}
N_1 &= \frac{1}{4}(1-\xi)(1-\eta) \\
N_2 &= \frac{1}{4}(1+\xi)(1-\eta) \\
N_3 &= \frac{1}{4}(1+\xi)(1+\eta) \\
N_4 &= \frac{1}{4}(1-\xi)(1+\eta).
\end{aligned}
\tag{16.42}
$$

An element having these shape functions is called a **bi-linear** element. Along the edges of the element, the shape functions are linear with respect to either ξ or η. Within the element, however, the shape functions are bi-linear with respect to ξ and η. For instance:

$$
N_1 = \frac{1}{4}(1 - \xi - \eta + \xi\eta). \tag{16.43}
$$

Fig. 16.5 shows N_1 visualized as a contour plot.

The shape function derivatives with respect to the local coordinates ξ and η are easily computed. However, the shape function derivatives with respect to the global coordinates x and y are needed. For this purpose, the concept of **isoparametric elements** is used.

For isoparametric elements, the global coordinates within an element are interpolated based on the nodal coordinates using the shape functions

$$
x|_{\Omega_e} = \underset{\sim}{N}^{\mathrm{T}}(\xi,\eta)\,\underset{\sim}{x}_e, \qquad y|_{\Omega_e} = \underset{\sim}{N}^{\mathrm{T}}(\xi,\eta)\,\underset{\sim}{y}_e, \tag{16.44}
$$

where $\underset{\sim}{x}_e$ and $\underset{\sim}{y}_e$ contain the nodal x- and y-coordinates, respectively. These equations reflect the transformation from the local coordinates (ξ, η) to the global

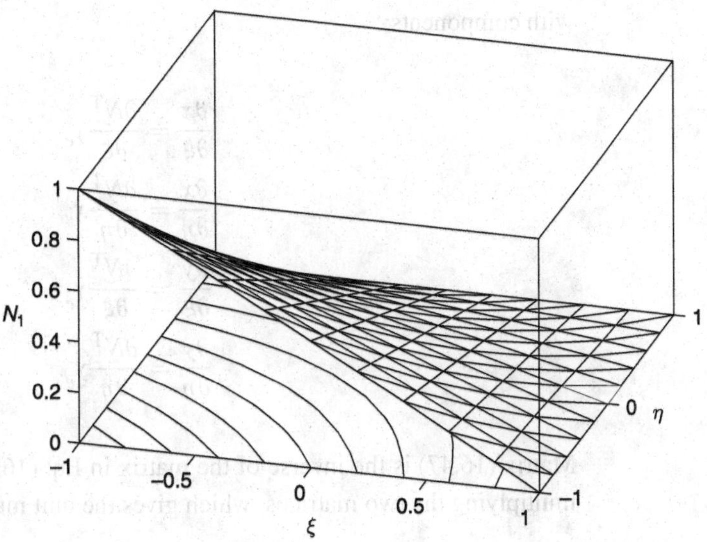

Figure 16.5

Shape function associated with node 1.

coordinates (x, y). The derivatives of N_i with respect to the Cartesian coordinates x and y can be evaluated with the aid of the chain rule:

$$\frac{\partial N_i}{dx} = \frac{\partial N_i}{\partial \xi}\frac{\partial \xi}{\partial x} + \frac{\partial N_i}{\partial \eta}\frac{\partial \eta}{\partial x}$$

$$\frac{\partial N_i}{dy} = \frac{\partial N_i}{\partial \xi}\frac{\partial \xi}{\partial y} + \frac{\partial N_i}{\partial \eta}\frac{\partial \eta}{\partial y}. \tag{16.45}$$

These relations can be rewritten in the following matrix form:

$$\begin{bmatrix} \dfrac{\partial N_i}{\partial x} \\ \dfrac{\partial N_i}{\partial y} \end{bmatrix} = \begin{bmatrix} \dfrac{\partial \xi}{\partial x} & \dfrac{\partial \eta}{\partial x} \\ \dfrac{\partial \xi}{\partial y} & \dfrac{\partial \eta}{\partial y} \end{bmatrix} \begin{bmatrix} \dfrac{\partial N_i}{\partial \xi} \\ \dfrac{\partial N_i}{\partial \eta} \end{bmatrix}. \tag{16.46}$$

The derivatives $\partial N_i/\partial \xi$ and $\partial N_i/\partial \eta$ are readily available, but the terms in the matrix cannot be directly computed since the explicit expressions $\xi = \xi(x, y)$ and $\eta = \eta(x, y)$ are not known. However, owing to the isoparametric formulation, the inverse relations are known, so the following matrix can be calculated easily:

$$\underline{x}_{,\xi} = \begin{bmatrix} \dfrac{\partial x}{\partial \xi} & \dfrac{\partial y}{\partial \xi} \\ \dfrac{\partial x}{\partial \eta} & \dfrac{\partial y}{\partial \eta} \end{bmatrix} \tag{16.47}$$

with components:

$$\begin{aligned}
\frac{\partial x}{\partial \xi} &= \frac{\partial \underset{\sim}{N}^{\mathrm{T}}}{\partial \xi} \underset{\sim}{x}_e \\
\frac{\partial x}{\partial \eta} &= \frac{\partial \underset{\sim}{N}^{\mathrm{T}}}{\partial \eta} \underset{\sim}{x}_e \\
\frac{\partial y}{\partial \xi} &= \frac{\partial \underset{\sim}{N}^{\mathrm{T}}}{\partial \xi} \underset{\sim}{y}_e \\
\frac{\partial y}{\partial \eta} &= \frac{\partial \underset{\sim}{N}^{\mathrm{T}}}{\partial \eta} \underset{\sim}{y}_e .
\end{aligned} \tag{16.48}$$

Matrix (16.47) is the inverse of the matrix in Eq. (16.46) (this can be checked by multiplying the two matrices, which gives the unit matrix). Accordingly

$$\begin{bmatrix} \frac{\partial \xi}{\partial x} & \frac{\partial \eta}{\partial x} \\ \frac{\partial \xi}{\partial y} & \frac{\partial \eta}{\partial y} \end{bmatrix} = (\underline{x}_{,\xi})^{-1} = \frac{1}{j} \begin{bmatrix} \frac{\partial y}{\partial \eta} & -\frac{\partial y}{\partial \xi} \\ -\frac{\partial x}{\partial \eta} & \frac{\partial x}{\partial \xi} \end{bmatrix} \tag{16.49}$$

where

$$j = \det(\underline{x}_{,\xi}) = \frac{\partial x}{\partial \xi}\frac{\partial y}{\partial \eta} - \frac{\partial x}{\partial \eta}\frac{\partial y}{\partial \xi}. \tag{16.50}$$

It is an elaborate process and usually not possible to analytically compute the integrals in the expressions for the element matrices or arrays, so generally they are approximated by numerical integration. Each of the components of the matrices, such as $\underline{K}_e$ etc., that need to be computed consists of integrals of a given function, say $g(x, y)$, over the domain of the element Ω_e. These may be transformed to an integral over the unit square $-1 \leq \xi \leq 1$, $-1 \leq \eta \leq 1$, according to

$$\int_{\Omega_e} f(x, y)\, d\Omega = \int_{-1}^{1} \int_{-1}^{1} f(x(\xi, \eta), y(\xi, \eta)) j(\xi, \eta)\, d\xi\, d\eta, \tag{16.51}$$

with $j(\xi, \eta)$ defined by Eq. (16.50).

The integral over the unit square may be approximated by a numerical integration (quadrature) rule, giving:

$$\int_{-1}^{1}\int_{-1}^{1} f(\xi,\eta)\, j(\xi,\eta)\, d\xi\, d\eta \approx \sum_{i=1}^{n_{\text{int}}} f(\xi_i,\eta_i) j(\xi_i,\eta_i) W_i\,. \tag{16.52}$$

For example, in the case of a two-by-two Gaussian integration rule, the locations of the integration points have ξ, η-coordinates and associated weights:

$$\begin{aligned}
\xi_1 &= \frac{-1}{\sqrt{3}}, & \eta_1 &= \frac{-1}{\sqrt{3}}, & W_1 &= 1\\
\xi_2 &= \frac{1}{\sqrt{3}}, & \eta_2 &= \frac{-1}{\sqrt{3}}, & W_2 &= 1\\
\xi_3 &= \frac{1}{\sqrt{3}}, & \eta_3 &= \frac{1}{\sqrt{3}}, & W_3 &= 1\\
\xi_4 &= \frac{-1}{\sqrt{3}}, & \eta_4 &= \frac{1}{\sqrt{3}}, & W_4 &= 1.
\end{aligned} \tag{16.53}$$

16.8 Example

One of the treatments for coronary occlusions that may lead to an infarct is to put a stent at the location of the occlusion. A problem with this intervention is that the blood vessels often occlude again quite soon after the stent is placed. One solution may be to design a stent that gradually releases drugs to prevent such an occlusion from occurring again: see Fig. 16.6. How these drugs propagate through the vascular tree is a convection–diffusion problem.

Consider the domain as sketched in Fig. 16.7. It represents a section of a long channel. For reasons of simplicity, the three-dimensional problem is modelled as a two-dimensional problem: the relevant fields in the configuration are assumed to be independent of the coordinate perpendicular to the *xy*-plane. A so-called Newtonian fluid (modelling blood in a first approximation) flows through the channel as indicated in the figure.

Let u denote the concentration of a certain drug. Along the entrance of the domain the drug concentration is zero. Along the small part of the wall indicated with a thick line, the drug concentration is prescribed, say $u = 1$. The drug diffuses

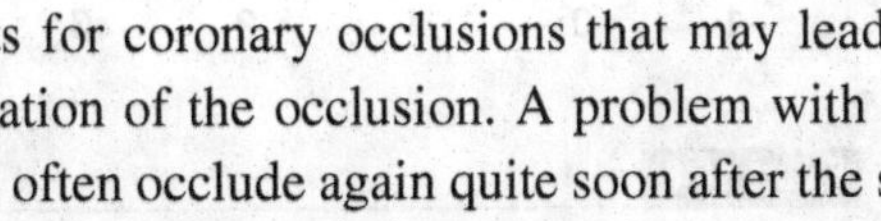

Figure 16.6

Schematic of a stent in a blood vessel.

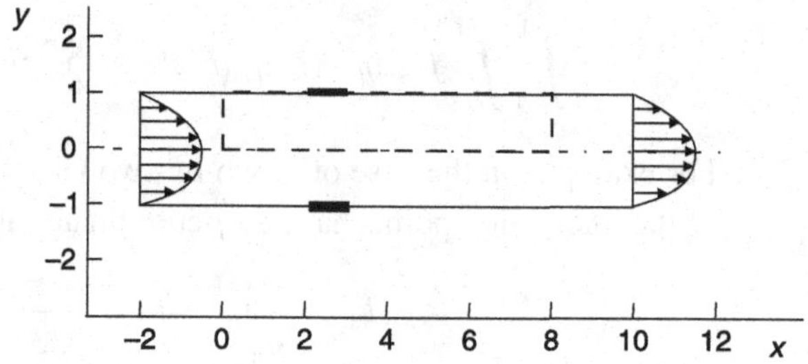

Figure 16.7

Specification of the convection–diffusion problem.

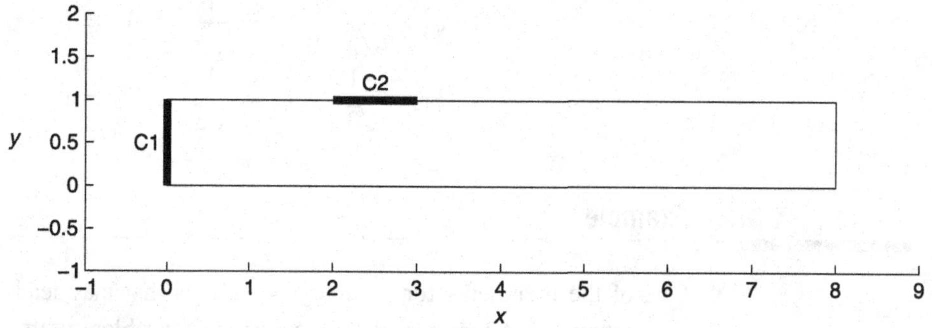

Figure 16.8

Computational domain of the convection–diffusion problem.

into the liquid with a diffusion constant c, but is also convected by the fluid. The aim is to compute the concentration profile in the two-dimensional channel for a number of fluid velocities. The computational domain is indicated by the dashed line, and is further outlined in Fig. 16.8. Because of symmetry, only the top half of the vessel is modelled.

For stationary flow conditions, the velocity field $\vec{v}$ is described by means of a parabolic profile (Poiseuille flow) according to

$$\vec{v} = a(1 - y^2)\vec{e}_x.$$

As mentioned before, along boundary C1 the fluid flows into the domain with a concentration $u = 0$, while along boundary C2 the concentration $u = 1$ is prescribed. Along the remaining parts of the boundary the natural boundary condition:

$$\vec{n} \cdot c\vec{\nabla}u = 0,$$

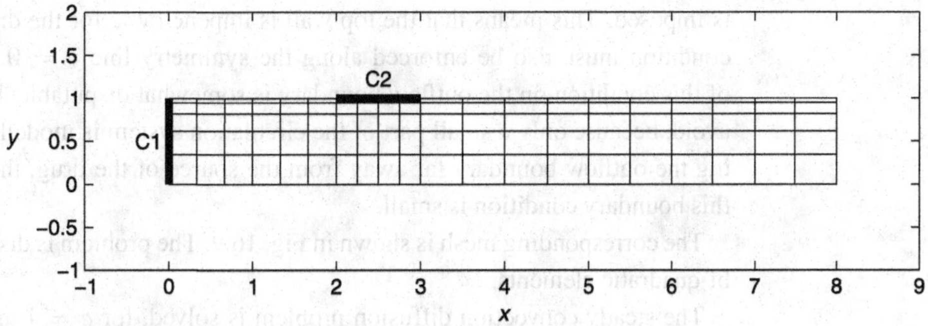

Figure 16.9

Mesh for the convection–diffusion problem.

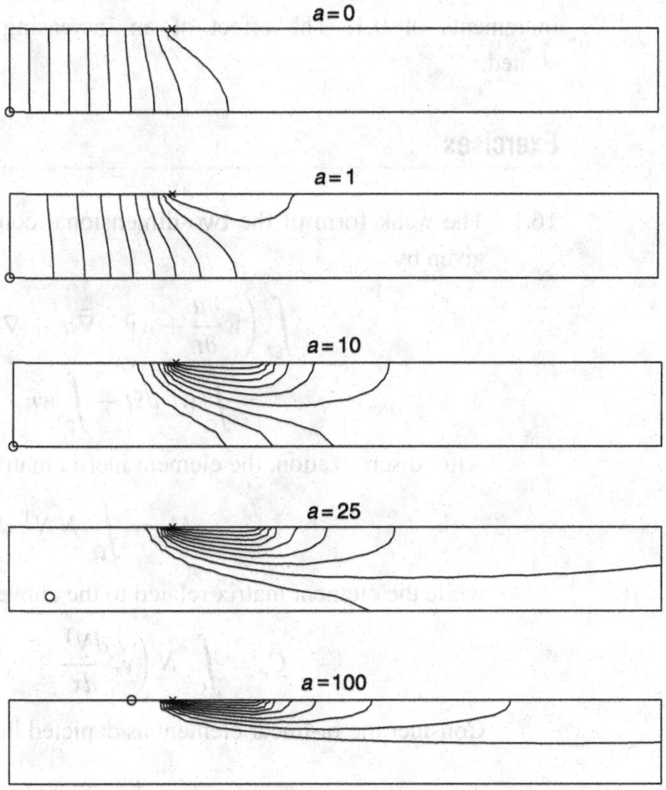

Figure 16.10

Contour lines of constant u values for a range of values of a. With increasing a, the effect of convection increases.

is imposed. This means that the top wall is impenetrable for the drug, while this condition must also be enforced along the symmetry line $y = 0$. Specification of this condition on the outflow boundary is somewhat disputable, but difficult to avoid, because only a small part of the circulation system is modelled. By choosing the outflow boundary far away from the source of the drug, the influence of this boundary condition is small.

The corresponding mesh is shown in Fig. 16.9. The problem is discretized using bi-quadratic elements.

The steady convection diffusion problem is solved, for $c = 1$ and a sequence 0,1, 10, 25, 100 of parameter a. Clearly, with increasing a, the velocity in the x-direction increases proportionally, and hence convection becomes increasingly important. For increasing a, contours of constant u are depicted in Fig. 16.10. In all cases, 10 contour lines are shown ranging from 0.1 to 1 with increments of 0.1. The effect of an increasing velocity is clearly demonstrated.

Exercises

16.1 The weak form of the two-dimensional convection–diffusion equation is given by

$$\int_{\Omega} \left(w\frac{\partial u}{\partial t} + w\vec{v} \cdot \vec{\nabla} u + \vec{\nabla} w \cdot (c\vec{\nabla} u) \right) d\Omega$$
$$= \int_{\Omega} wf \, d\Omega + \int_{\Gamma} w\vec{n} \cdot c\vec{\nabla} u = d\Gamma.$$

After discretization, the element inertia matrix is defined as

$$\underline{M}_e = \int_{\Omega_e} \underset{\sim}{N} \underset{\sim}{N}^{\mathrm{T}} \, d\Omega,$$

while the element matrix related to the convective part is given by

$$\underline{C}_e = \int_{\Omega_e} \underset{\sim}{N} \left(v_x \frac{d\underset{\sim}{N}^{\mathrm{T}}}{dx} + v_y \frac{d\underset{\sim}{N}^{\mathrm{T}}}{dy} \right) d\Omega.$$

Consider the bi-linear element as depicted in the figure below.

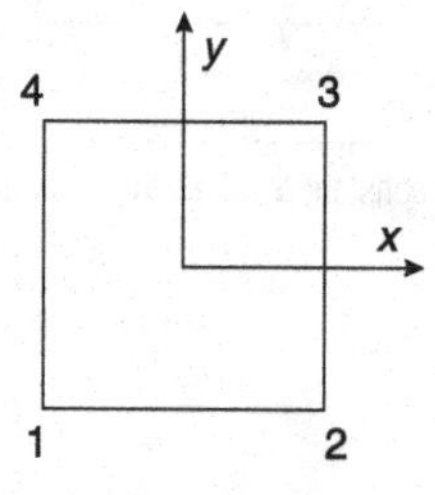

The element spans the spatial domain $-2 \leq x \leq 2$ and $-2 \leq y \leq 2$.

(a) Compute the element inertia matrix $\underline{M}_e$ using a 2×2 Gauss integration rule. It is recommended to use MATLAB for this computation.

(b) Suppose that the location of the integration points coincides with the nodes of the element. What is the element inertia matrix $\underline{M}_e$ in this case?

(c) Compute the matrix $\underline{C}_e$ if $v_x = 1$ and $v_y = 0$, using a 2×2 Gauss integration rule.

(d) Suppose that along the edge of the element located at $x = 2$ a constant flux $q = \vec{n} \cdot c\vec{\nabla} u$ is prescribed. Then compute the column

$$\int_{\Gamma_e} \underset{\sim}{N} q \, d\Gamma,$$

for this edge, contributing to the element column $\underset{\sim}{f}_e$.

16.2 Consider the mesh depicted in the figure below. The solution vector $\underset{\sim}{u}$ contains the nodal solutions in the sequence defined by the node numbers, hence:

$$\underset{\sim}{u}^{\mathrm{T}} = [u_1 \; u_2 \; \cdots \; u_{16}] \, .$$

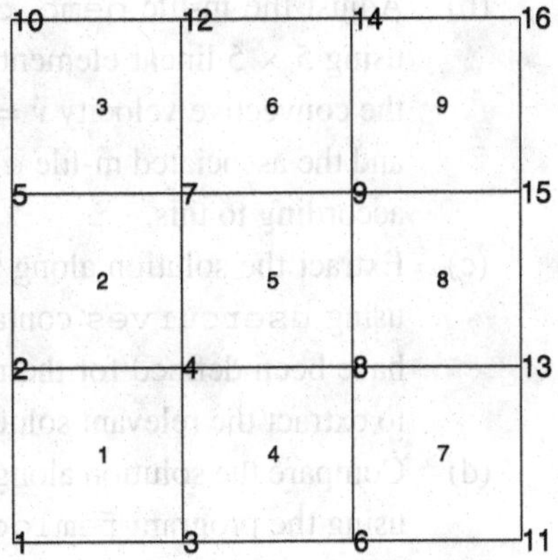

Let the element topology array of the third element be given by `top(3,:)=[5 7 12 10 1 1]` (see software manual).

(a) Is this topology array unique?

(b) What is the `pos` array for this element?

(c) Define the `dest` array.

(d) The solution $\underset{\sim}{u}$ is stored in the array `sol`. How are the nodal solutions of element 8 extracted from `sol` using the array `pos`?

(e) Suppose that the array `nodes` contains the node numbers of the left edge, say `nodes=[1 2 5 10]`. How are the nodal solutions of these nodes extracted from the solution array `sol` via the array `dest`?

(f) Consider the third element whose element topology array is given by `top(3,:)=[5 7 12 10 1 1]`. If the nodal solution array $\underset{\sim}{u}_e$ for this element is given by:

$$\underset{\sim}{u}_e^{\mathrm{T}} = [1\ 4\ 3\ 7],$$

compute the solution u for $\xi = \frac{1}{4}$ and $\eta = \frac{3}{4}$.

(g) Suppose the element topology array is given by `top(3,:)=[12 10 5 7 1 1]`. What is the corresponding element nodal solution array $\underset{\sim}{u}_e$?

16.3 Consider the square domain $0 \leq x \leq 1, 0 \leq y \leq 1$. Along the line $x = 0$ the boundary condition $u = 0$ is imposed, and along the line $x = 1$ the boundary condition $u = 1$ is prescribed. Along the other boundaries the natural boundary condition $\vec{n} \cdot c\vec{\nabla} u = 0$ is specified. Consider the steady convection–diffusion problem on this domain.

(a) Suppose that the convection–diffusion problem represents the temperature equation. What physical meaning does the above natural boundary condition have?

(b) Adjust the m-file `demo_cd` such that the above problem is solved using 5×5 linear elements. Select the diffusion constant $c = 1$ and the convective velocity $\vec{v} = \vec{e}_x$ (remark: see the element m-file `elcd` and the associated m-file `elcd_a`). Specify the structured array `mat` according to this.

(c) Extract the solution along the line $y = 0$. Hint, first select the nodes using `usercurves` containing the nodes along all the curves that have been defined for the mesh generator `crmesh`. Then use `dest` to extract the relevant solution components.

(d) Compare the solution along $y = 0$ with the one-dimensional solution, using the program `fem1dcd`.

16.4 Consider the domain as sketched in the figure. Along the boundary Γ_1 the specification $u = 1$ is chosen, while along Γ_7 $u = 0$ is chosen as a boundary condition. Along all other boundaries, $\vec{n} \cdot c\vec{\nabla} u = 0$ is imposed. Choose the convective velocity $\vec{v}$ such that:

$$\begin{aligned} y < 0: &\quad \vec{v} = \vec{0} \\ y \geq 0: &\quad \vec{v} = y\,\vec{e}_x. \end{aligned}$$

Furthermore, choose $c = 1$ throughout the domain. Model this problem using `femlin_cd`. Beyond which approximate value of x along the boundary Γ_6 does $u > 0.8$ hold?

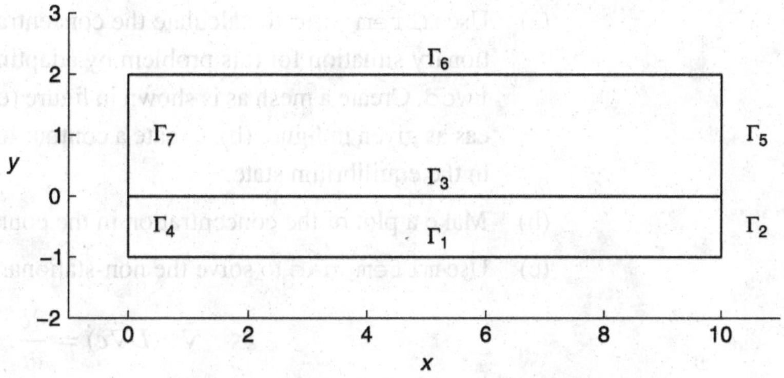

16.5 If the flux $\vec{n} \cdot c\vec{\nabla}u = P$ along Γ_p is prescribed, the integral (element level)

$$\int_{\Gamma_e} \underset{\sim}{N} P \, d\Gamma,$$

needs to be computed. It may be transformed to the local coordinate system $\xi \in [-1 \; 1]$, such that:

$$\int_{\Gamma_e} \underset{\sim}{N} P \, d\Gamma = \int_{-1}^{1} \underset{\sim}{N} P \frac{\partial x}{\partial \xi} \, d\xi,$$

where it is assumed that Γ_e is oriented in the ξ-direction.
If linear elements are chosen, the shape functions along the boundary are given by:

$$\underset{\sim}{N} = \begin{bmatrix} \frac{1}{2}(1-\xi) \\ \frac{1}{2}(1+\xi) \end{bmatrix}.$$

Use an isoparametric element and let L denote the length of Γ_e, then demonstrate that for *constant P*:

$$\int_{\Gamma_e} \underset{\sim}{N} P \, d\Gamma = \frac{LP}{2} \begin{bmatrix} 1 \\ 1 \end{bmatrix}.$$

16.6 We want to study the concentration of a growth factor in a porous container, the grey area in figure (a), with an inclusion, the white area in figure (a), with a high (dimensionless) concentration ($c = 1$). Outside the grey area, we have a (dimensionless) concentration $c = 0$. First we solve a stationary diffusion problem for $c(x, y)$:

$$\vec{\nabla} \cdot (D\vec{\nabla}c) = 0$$

The diffusion constant of the grey area is $D = 1$ [mm^2 s^{-1}]. Because of symmetry, we model only one-quarter of the container. The size of the container is shown in the figure. The dimensions are given in [mm].

(a) Use `mlfem_nac` to calculate the concentration distribution in a stationary situation for this problem by adapting `demo_cd` in directory `twod`. Create a mesh as is shown in figure (c) by defining three subareas as given in figure (b). Create a contour image of the concentration in the equilibrium state.

(b) Make a plot of the concentration in the container along curve 3.

(c) Use `mlfem_nac` to solve the non-stationary diffusion equation:

$$\vec{\nabla} \cdot (D\vec{\nabla}c) = \frac{\partial c}{\partial t}.$$

Assume that the initial condition in the grey part of the container $c(x, y) = 0$ at $t = 0$. Vary the time step and mesh size to find a reliable approximation.

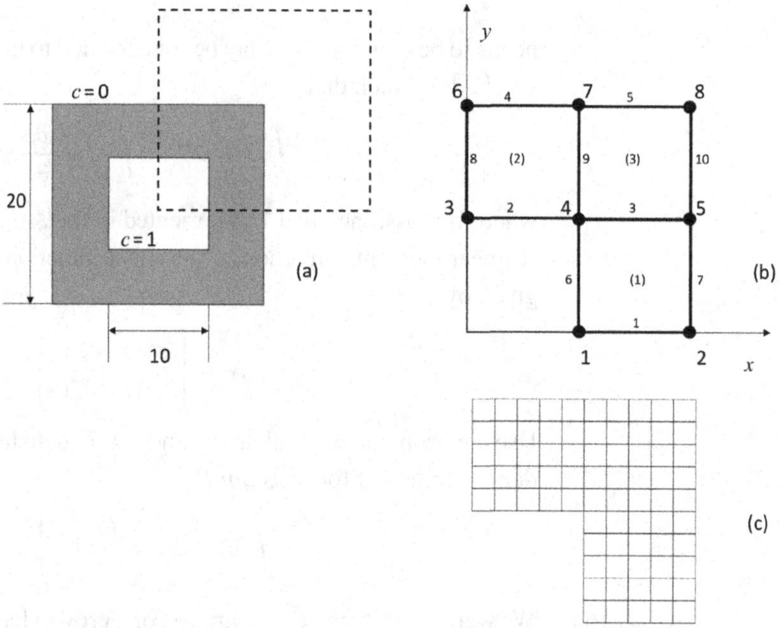

16.7 A researcher is working on a project on trans-epidermal vaccination with microneedles. These microneedles with a length of 200 [μm] can penetrate skin without pain sensation, because they are very short and do not touch the nerves in the skin. Panel (a) shows a typical needle for such a procedure. The needles are equipped with a pH-dependent coating that releases the drug once the needle is placed into the skin [15]. We want to study the resulting drug concentration in the skin. For this we solve the stationary two-dimensional diffusion equation. Because of symmetry, we only consider half of the problem. The userpoints, curves and subareas to create a finite element mesh for this model are drawn in panel (b).

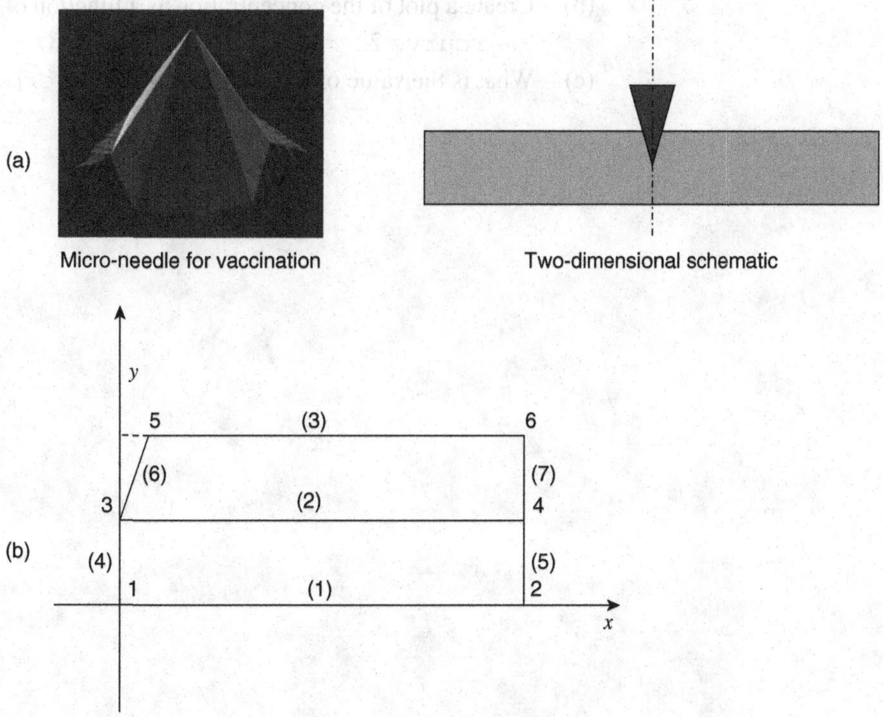

The coordinates of the userpoints are given by:

Point	x [μm]	y [μm]
1	0	0
2	1000	0
3	0	200
4	1000	200
5	40	400
6	1000	400

The diffusion coefficient is 100 [$\mu m^2\ s^{-1}$]. We assume that the (dimensionless) concentration on the needle, `curve` 6, is equal to 1. Along `curve` 1, the blood is transporting the drug away from the skin via the blood stream, so we assume that the concentration equals 0 at this curve. The remaining part of the outer boundary of the domain is assumed to be impenetrable for the drug ($\vec{\nabla} c \cdot \vec{n} = 0$).

(a) Create a finite element mesh of this problem by adjusting the file `demo_cd` in the directory `TwoD` of `mlfem_nac`. Chose 20×5 elements for each subsurface. Use second-order elements.

(b) Create a plot of the concentration as a function of the x-coordinate for `usercurve` 2.

(c) What is the value of the concentration in `userpoint` 4?

17 Shape Functions and Numerical Integration

17.1 Introduction

In the previous chapter, the shape functions N_i were hardly discussed in any detail. The key purpose of this chapter is first to introduce isoparametric shape functions, and second to outline numerical integration of the integrals appearing in the element coefficient matrices and element column. Before this can be done, it is useful to understand the minimum requirements to be imposed on the shape functions. The key question involved is what conditions should at least be satisfied such that the approximate solution of the boundary value problems, dealt with in the previous chapter, generated by a finite element analysis, converges to the exact solution at mesh refinement. The answer is:

(i) The shape functions should be smooth within each element Ω_e, i.e. shape functions are not allowed to be discontinuous within an element.

(ii) The shape functions should be continuous across each element boundary. *This condition does not always have to be satisfied, but this is beyond the scope of the present book.*

(iii) The shape functions should be complete, i.e. at element level the shape functions should enable the representation of uniform gradients of the field variable(s) to be approximated.

Conditions (i) and (ii) allow that the gradients of the shape functions show finite jumps across the element interface. However, smoothness in the element interior assures that all integrals in which gradients of the unknown function, say u, occur can be evaluated. In Fig. 17.1(a), an example is given of an admissible shape function. In this case, the derivative of the shape function is discontinuous over the element boundary, but the jump is finite. In Fig. 17.1(b), the discontinuous shape function at the element boundary leads to an infinite derivative and the integrals in the weighted residual equations can no longer be evaluated.

Completeness When the finite element mesh is refined further and further, at the element level the exact solution becomes more and more linear in the coordinates, and its derivatives approach constant values in each element. To ensure that

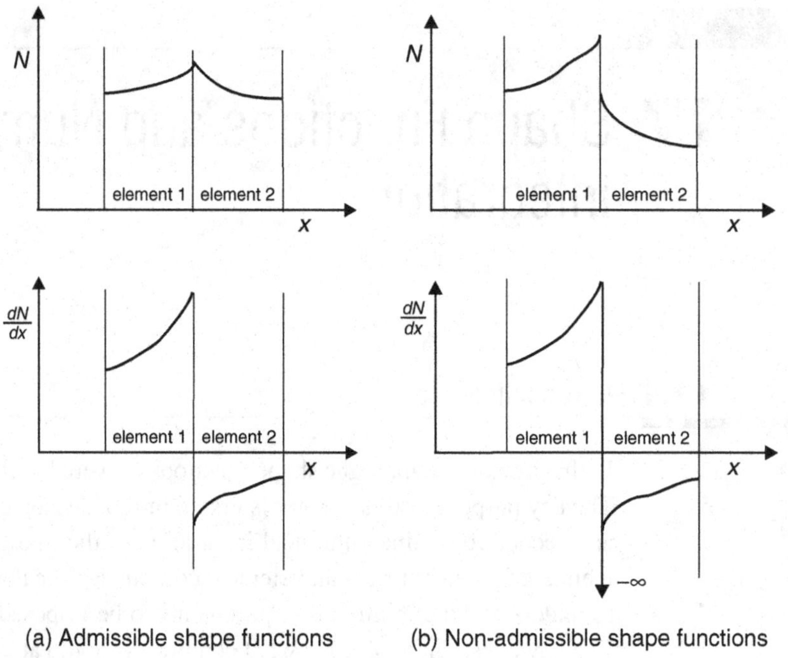

Figure 17.1

Continuity of shape functions.

an adequate approximation can be achieved, the shape functions have to contain all constant and linear functions. Assuming that the exact solution is described by an arbitrary linear polynomial, the element interpolation has to be able to describe this field exactly. In mathematical terms, this means the following. Assume u to be approximated by

$$u_{\mathrm{h}}(\vec{x}) = \sum_{i=1}^{n} N_i(\vec{x}) u_i, \tag{17.1}$$

with $N_i(\vec{x})$ the interpolation functions and u_i the nodal values of $u_{\mathrm{h}}(\vec{x})$. Consider the case where the nodal values u_i are selected to be related to the nodal coordinates x_i, y_i, z_i by

$$u_i = c_0 + c_1 x_i + c_2 y_i + c_3 z_i, \tag{17.2}$$

according to a linear field. Substitution of Eq. (17.2) into (17.1) reveals:

$$u_{\mathrm{h}} = c_0 \sum_{i=1}^{n} N_i(\vec{x}) + c_1 \sum_{i=1}^{n} N_i(\vec{x}) x_i + c_2 \sum_{i=1}^{n} N_i(\vec{x}) y_i + c_3 \sum_{i=1}^{n} N_i(\vec{x}) z_i. \tag{17.3}$$

For the element types described in the current chapter, the coordinates within the elements are interpolated in the following way:

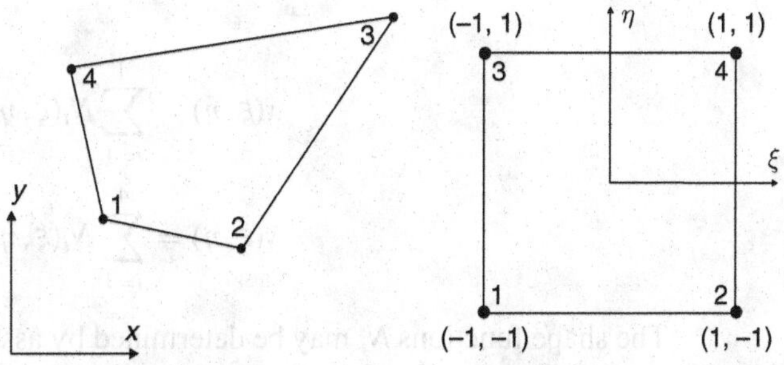

(a) Quadrilateral element with respect to global coordinates

(b) Quadrilateral element with respect to local coordinates

Figure 17.2

Quadrilateral element with respect to a global and a local coordinate system.

$$x = \sum_{i=1}^{n} N_i(\vec{x})x_i$$

$$y = \sum_{i=1}^{n} N_i(\vec{x})y_i$$

$$z = \sum_{i=1}^{n} N_i(\vec{x})z_i. \tag{17.4}$$

Completeness implies that Eq. (17.3) has to lead to

$$u_\mathrm{h}(x, y, z) = c_0 + c_1 x + c_2 y + c_3 z, \tag{17.5}$$

for arbitrary $c_0, \ldots, c_3$. Hence

$$\sum_{i=1}^{n} N_i(\vec{x}) = 1. \tag{17.6}$$

The last equation can be used to check whether the shape functions have been specified in an adequate way.

17.2 Isoparametric, Bi-Linear Quadrilateral Element

Consider a four-noded, straight-edged element $\Omega_e \in \mathbf{R}^2$ as depicted in Fig. 17.2(a). The nodal points are assumed to be numbered counterclockwise from 1 to 4. This element can be mapped onto a square, with local coordinates ξ and η, with $\xi, \eta \in [-1, 1]$. A point with coordinates $x, y \in \Omega_e$ is related to a point $\xi, \eta \in [-1, 1]$ by the following mapping:

$$x(\xi,\eta) = \sum_{i=1}^{4} N_i(\xi,\eta)x_i \tag{17.7}$$

$$y(\xi,\eta) = \sum_{i=1}^{4} N_i(\xi,\eta)y_i. \tag{17.8}$$

The shape functions N_i may be determined by assuming the 'bi-linear' expansion:

$$\begin{aligned} x(\xi,\eta) &= \alpha_0 + \alpha_1\xi + \alpha_2\eta + \alpha_3\xi\eta \\ y(\xi,\eta) &= \beta_0 + \beta_1\xi + \beta_2\eta + \beta_3\xi\eta. \end{aligned} \tag{17.9}$$

The parameters α_j and β_j ($j = 0, 1, 2, 3$) must be calculated such that the relations

$$\begin{aligned} x(\xi_i, \eta_i) &= x_i \\ y(\xi_i, \eta_i) &= y_i, \end{aligned} \tag{17.10}$$

are satisfied, where ξ_i and η_i refer to the local coordinates of the nodes. Applying the restriction (17.10) to Eq. (17.9) leads to

$$\begin{pmatrix} x_1 \\ x_2 \\ x_3 \\ x_4 \end{pmatrix} = \begin{pmatrix} 1 & -1 & -1 & 1 \\ 1 & 1 & -1 & -1 \\ 1 & 1 & 1 & 1 \\ 1 & -1 & 1 & -1 \end{pmatrix} \begin{pmatrix} \alpha_0 \\ \alpha_1 \\ \alpha_2 \\ \alpha_3 \end{pmatrix}. \tag{17.11}$$

From this set of equations, the coefficients α_j can be linearly expressed in terms of the nodal coordinates x_1, x_2, x_3 and x_4. By reorganizing, the eventual expressions for the shape functions become:

$$\begin{aligned} N_1(\xi,\eta) &= \frac{1}{4}(1-\xi)(1-\eta) \\ N_2(\xi,\eta) &= \frac{1}{4}(1+\xi)(1-\eta) \\ N_3(\xi,\eta) &= \frac{1}{4}(1+\xi)(1+\eta) \\ N_4(\xi,\eta) &= \frac{1}{4}(1-\xi)(1+\eta). \end{aligned} \tag{17.12}$$

The element is called *isoparametric* as both the spatial coordinates and the element interpolation function u_h are interpolated with the same shape functions, that is

$$u_\mathrm{h}(\xi,\eta) = \sum_{i=1}^{n} N_i(\xi,\eta)u_i. \tag{17.13}$$

17.3 Linear Triangular Element

There are two ways to arrive at a triangular element. First, it is possible to coalesce two nodes of the quadrilateral element, for instance nodes 3 and 4. This is done by setting $\underset{\sim}{x}_4 = \underset{\sim}{x}_3$ (with $\underset{\sim}{x}_3^{\mathrm{T}} = [x_3, y_3]$ and $\underset{\sim}{x}_4^{\mathrm{T}} = [x_4, y_4]$) and by defining new shape functions $\hat{N}_i$ according to

$$
\begin{aligned}
\underset{\sim}{x} &= \sum_{i=1}^{4} N_i \underset{\sim}{x}_i \\
&= \underbrace{N_1}_{\hat{N}_1} \underset{\sim}{x}_1 + \underbrace{N_2}_{\hat{N}_2} \underset{\sim}{x}_2 + \underbrace{(N_3 + N_4)}_{\hat{N}_3} \underset{\sim}{x}_3 \\
&= \sum_{i=1}^{3} \hat{N}_i \underset{\sim}{x}_i .
\end{aligned}
\qquad (17.14)
$$

Figure 17.3 illustrates this operation.

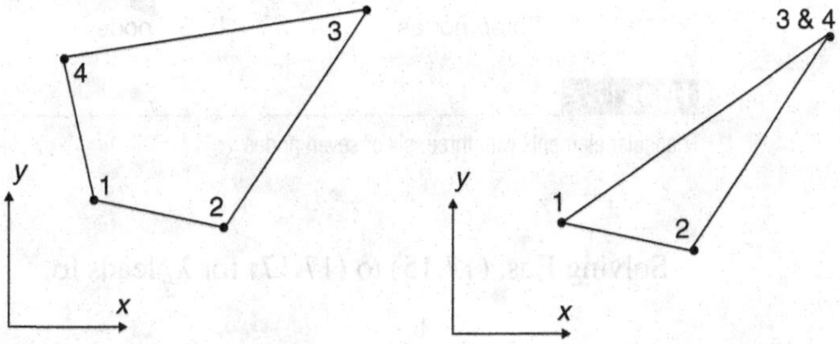

Figure 17.3

Degeneration from quadrilateral element to triangular element.

The second method is based on using so-called triangle coordinates. A convenient set of coordinates $\lambda_1, \lambda_2, \lambda_3$ for a triangle can be defined by means of the following equations:

$$x = \lambda_1 x_1 + \lambda_2 x_2 + \lambda_3 x_3 \qquad (17.15)$$

$$y = \lambda_1 y_1 + \lambda_2 y_2 + \lambda_3 y_3 \qquad (17.16)$$

$$1 = \lambda_1 + \lambda_2 + \lambda_3. \qquad (17.17)$$

To every set $\lambda_1, \lambda_2, \lambda_3 \in [0, 1]$ there corresponds a unique set of Cartesian coordinates x, y (see Fig. 17.4). The triangle coordinates are not independent, but related by Eq. (17.17).

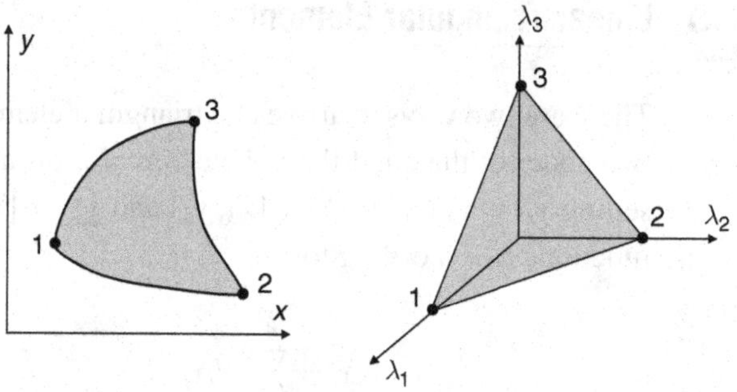

Figure 17.4

The mapping of a triangle from the global coordinate system to the local coordinate system with triangle coordinates.

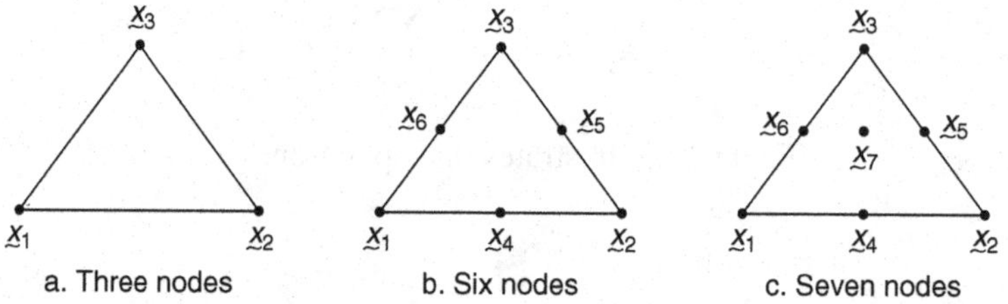

Figure 17.5

Triangular elements with three, six or seven nodes.

Solving Eqs. (17.15) to (17.17) for λ_i leads to

$$\lambda_1 = \frac{1}{\Delta}\left((x_2y_3 - x_3y_2) + (y_2 - y_3)x + (x_3 - x_2)y\right)$$
$$\lambda_2 = \frac{1}{\Delta}\left((x_3y_1 - x_1y_3) + (y_3 - y_1)x + (x_1 - x_3)y\right)$$
$$\lambda_3 = \frac{1}{\Delta}\left((x_1y_2 - x_2y_1) + (y_1 - y_2)x + (x_2 - x_1)y\right), \tag{17.18}$$

where

$$\Delta = (x_3 - x_2)(y_1 - y_2) - (y_2 - y_3)(x_2 - x_1). \tag{17.19}$$

Note that $|\Delta|$ is 2 times the square of the triangle surface. Equations (17.15) to (17.17) show that λ_i can be (linearly) expressed in x and y: $\lambda_i = \lambda_i(x, y)$. The relation satisfies

$$\lambda_i(x_j, y_j) = \delta_{ij}. \tag{17.20}$$

Now, the shape functions for the three-node, six-node and seven-node triangular elements (see Fig. 17.5) are:

(a) Three-node triangular element, linear interpolation:

$$\begin{aligned} N_1 &= \lambda_1 \\ N_2 &= \lambda_2 \\ N_3 &= \lambda_3. \end{aligned} \tag{17.21}$$

(b) Six-node triangular element, quadratic interpolation:

$$\begin{aligned} N_1 &= \lambda_1(2\lambda_1 - 1) \\ N_2 &= \lambda_2(2\lambda_2 - 1) \\ N_3 &= \lambda_3(2\lambda_3 - 1) \\ N_4 &= 4\lambda_1\lambda_2 \\ N_5 &= 4\lambda_2\lambda_3 \\ N_6 &= 4\lambda_1\lambda_3. \end{aligned} \tag{17.22}$$

(c) Seven-node triangular element, bi-quadratic interpolation:

$$\begin{aligned} N_1 &= \lambda_1(2\lambda_1 - 1) + 3\lambda_1\lambda_2\lambda_3 \\ N_2 &= \lambda_2(2\lambda_2 - 1) + 3\lambda_1\lambda_2\lambda_3 \\ N_3 &= \lambda_3(2\lambda_3 - 1) + 3\lambda_1\lambda_2\lambda_3 \\ N_4 &= 4\lambda_1\lambda_2 - 12\lambda_1\lambda_2\lambda_3 \\ N_5 &= 4\lambda_2\lambda_3 - 12\lambda_1\lambda_2\lambda_3 \\ N_6 &= 4\lambda_1\lambda_3 - 12\lambda_1\lambda_2\lambda_3 \\ N_7 &= 27\lambda_1\lambda_2\lambda_3. \end{aligned} \tag{17.23}$$

The factor $\lambda_1\lambda_2\lambda_3$ is called a 'bubble' function giving zero contributions along the boundaries of the element.

There are two major differences between the method used in Section 17.2 and the method with triangle coordinates:

- Determining the derivatives of these shape functions with respect to the global coordinates is not trivial. Consider the derivative of a shape function N_i with respect to x. Applying the chain rule for differentiation would lead to

$$\frac{\partial N_i}{\partial x} = \frac{\partial N_i}{\partial \lambda_1}\frac{\partial \lambda_1}{\partial x} + \frac{\partial N_i}{\partial \lambda_2}\frac{\partial \lambda_2}{\partial x} + \frac{\partial N_i}{\partial \lambda_3}\frac{\partial \lambda_3}{\partial x}. \tag{17.24}$$

  But, by definition, a partial derivative as to one variable implies that the other variables have to be considered as constant. In this case, a partial derivative with respect to λ_1 means that, when this derivative is determined, λ_2 and λ_3 have to be considered constant. However, the λ_i values are related by Eq. (17.17). Only two variables can be considered as independent. A solution for this dilemma is to eliminate λ_3 from the shape functions by using:

$$\lambda_3 = 1 - \lambda_1 - \lambda_2. \tag{17.25}$$

A more obvious way of solving the problem is to substitute Eq. (17.18) into the equations for the shape functions, thus eliminating all triangular coordinates, and directly determine $\partial N_i/\partial x$ and $\partial N_i/\partial y$.

- The second issue is the difference in integration limits which have to correspond with a triangle. For the square element in Section 17.2 the domain for integration is simple, meaning that the surface integral can be split into two successive single integrals with ξ and η as variables. The integration limits are -1 to $+1$. For the triangle, this is more complicated, and the limits of integration now involve the coordinate itself. This item will be discussed shortly in Section 17.5.

17.4 Lagrangian and Serendipity Elements

In principle, higher-order elements are more accurate than the linear ones discussed so far. However, the computation of element coefficient matrices and element arrays is more expensive for higher-order elements, and the cost-effectiveness depends on the particular problem investigated (cost-effectiveness in the sense that there is a trade-off between the accuracy, using a smaller number of higher-order elements, versus using more linear elements).

Higher-order elements can systematically be derived from Lagrange polynomials. A (one-dimensional) set of Lagrange polynomials on an element with domain $\xi_1 \leq \xi \leq \xi_n$ is defined by

$$l_a^{n-1}(\xi) = \frac{\prod_{b=1,b\neq a}^{n}(\xi - \xi_b)}{\prod_{b=1,b\neq a}^{n}(\xi_a - \xi_b)} \tag{17.26}$$

$$= \frac{(\xi - \xi_1)\ldots(\xi - \xi_{a-1})(\xi - \xi_{a+1})\ldots(\xi - \xi_n)}{(\xi_a - \xi_1)\ldots(\xi_a - \xi_{a-1})(\xi_a - \xi_{a+1})\ldots(\xi_a - \xi_n)},$$

with n the number of nodes of the element and with $a = 1, 2, \ldots, n$ referring to a node number. Notice that the above polynomial is of the order $(n-1)$.

For instance, first-order (linear) polynomials are found for $n = 2$, hence

$$l_1^1 = \frac{\xi - \xi_2}{\xi_1 - \xi_2}, \tag{17.27}$$

and:

$$l_2^1 = \frac{\xi - \xi_1}{\xi_2 - \xi_1}. \tag{17.28}$$

Example 17.1 A one-dimensional element has four nodes located as shown in Fig 17.6, unequally distributed.

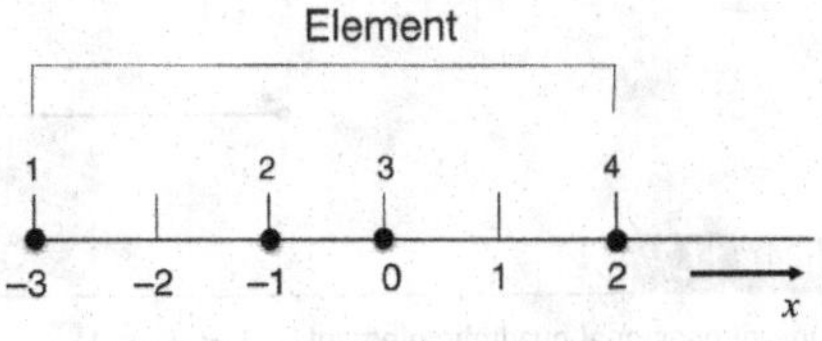

Figure 17.6

One-dimensional element with four nodes.

We want to determine the four shape functions $N_i(x)$, with $i = 1, 2, 3, 4$ for this element using Lagrange polynomials. Because we have four nodes, $n = 4$. When we start with node 1 we can apply Eq. (17.26) with $a = 1$:

$$N_1(x) = \frac{(x+1)(x-0)(x-2)}{(-3+1)(-3-0)(-3-2)} = -\frac{1}{30}(x+1)x(x-2).$$

For node 2 we find:

$$N_2(x) = \frac{(x+3)(x-0)(x-2)}{(-1+3)(-1-0)(-1-2)} = \frac{1}{6}(x+3)x(x-2),$$

and for nodes 3 and 4:

$$N_3 = \frac{(x+3)(x+1)(x-2)}{(0+3)(0+1)(0-2)} = -\frac{1}{6}(x+3)(x+1)(x-2)$$
$$N_4 = \frac{(x+3)(x+1)(x-0)}{(2+3)(2+1)(2-0)} = \frac{1}{30}(x+3)(x+1)x.$$

17.4.1 Lagrangian Elements

For Lagrangian elements, the full Lagrangian polynomials are used in ξ and η direction (and in the ζ-direction in the case of three-dimensional elements). The shape functions of an element of order $(n-1)$ in one dimension are chosen as

$$N_a = l_a^{n-1}. \tag{17.29}$$

The quadratic shape function associated with node 1 of the element depicted in Fig. 17.7 (with $\xi_1 = -1$, $\xi_2 = 0$, $\xi_3 = 1$) satisfies

$$N_1(\xi) = l_1^2(\xi) = \frac{(\xi-\xi_2)(\xi-\xi_3)}{(\xi_1-\xi_2)(\xi_1-\xi_3)} = \frac{1}{2}\xi(\xi-1). \tag{17.30}$$

Likewise, it follows that

$$N_2(\xi) = -(\xi+1)(\xi-1) = 1-\xi^2, \tag{17.31}$$

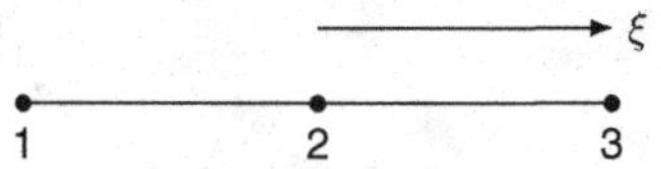

Figure 17.7

A one-dimensional quadratic element, $-1 \leq \xi \leq 1$.

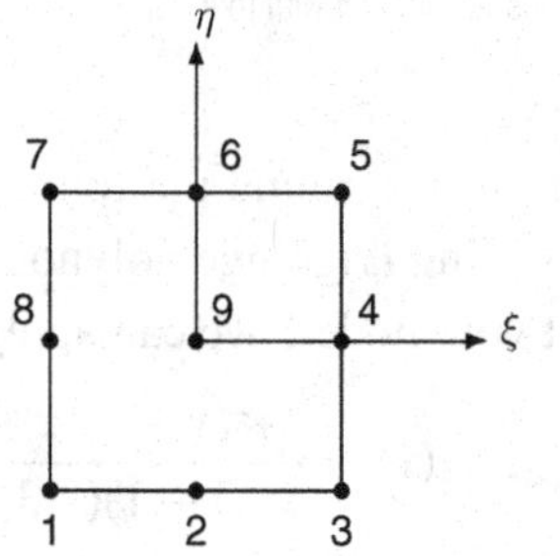

Figure 17.8

A two-dimensional quadratic element, $-1 \leq \xi \leq 1, -1 \leq \eta \leq 1$.

and

$$N_3(\xi) = \frac{1}{2}\xi(\xi + 1). \tag{17.32}$$

In two dimensions, the shape functions for the nine-node element as visualized in Fig. 17.8 are formed by multiplication of two Lagrangian polynomials, leading to:

$$\begin{aligned}
N_1(\xi, \eta) &= \frac{1}{4}\xi(\xi - 1)\eta(\eta - 1) \\
N_2(\xi, \eta) &= -\frac{1}{2}(\xi + 1)(\xi - 1)\eta(\eta - 1) \\
N_3(\xi, \eta) &= -\frac{1}{4}\xi(\xi + 1)\eta(\eta - 1) \\
N_4(\xi, \eta) &= -\frac{1}{2}\xi(\xi + 1)(\eta + 1)(\eta - 1) \\
N_5(\xi, \eta) &= \frac{1}{4}\xi(\xi + 1)(\eta + 1)\eta \\
N_6(\xi, \eta) &= -\frac{1}{2}(\xi + 1)(\xi - 1)(\eta + 1)\eta \\
N_7(\xi, \eta) &= -\frac{1}{4}\xi(\xi - 1)(\eta + 1)\eta \\
N_8(\xi, \eta) &= -\frac{1}{2}\xi(\xi - 1)(\eta + 1)(\eta - 1) \\
N_9(\xi, \eta) &= (\xi + 1)(\xi - 1)(\eta + 1)(\eta - 1).
\end{aligned} \tag{17.33}$$

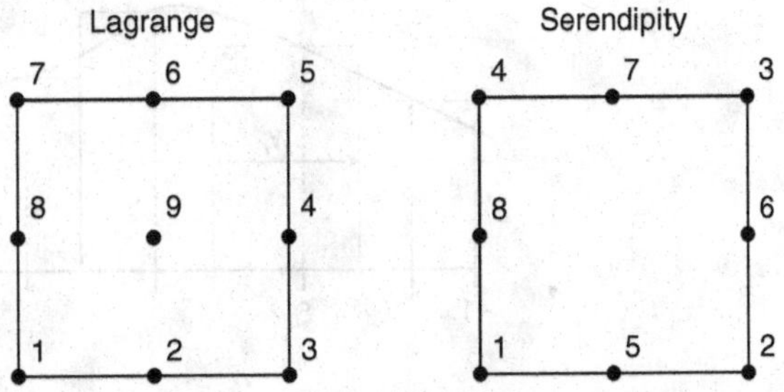

Figure 17.9

Example of a Serendipity element compared to the 'equal order' Lagrangian element.

17.4.2 Serendipity Elements

For serendipity elements, no internal nodes are used. Consider a 'quadratic element', as depicted in Fig. 17.9 (right). The shape functions of the corner nodes are defined by

$$\begin{aligned} N_1 &= \frac{1}{4}(1-\xi)(1-\eta)(-\xi-\eta-1) \\ N_2 &= \frac{1}{4}(1+\xi)(1-\eta)(+\xi-\eta-1) \\ N_3 &= \frac{1}{4}(1+\xi)(1+\eta)(+\xi+\eta-1) \\ N_4 &= \frac{1}{4}(1-\xi)(1+\eta)(-\xi+\eta-1), \end{aligned} \tag{17.34}$$

while the shape functions for the mid-side nodes read

$$\begin{aligned} N_5 &= \frac{1}{2}(1-\xi^2)(1-\eta) \\ N_6 &= \frac{1}{2}(1+\xi)(1-\eta^2) \\ N_7 &= \frac{1}{2}(1-\xi^2)(1+\eta) \\ N_8 &= \frac{1}{2}(1-\xi)(1-\eta^2). \end{aligned} \tag{17.35}$$

Other examples can be found in Zienkiewicz [23] and Hughes [12].

17.5 Numerical Integration

Let $f : \Omega_e \mapsto \mathbf{R}$ be some function, and assume that the integral:

$$\int_{\Omega_e} f(x)\, dx, \tag{17.36}$$

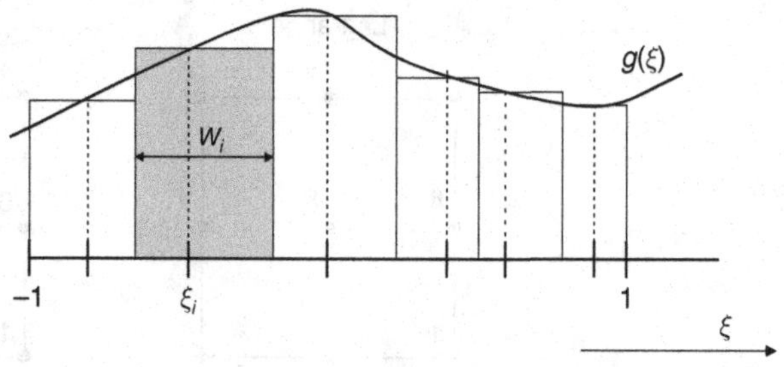

Figure 17.10

Numerical integration of a function $g(\xi)$.

over the domain Ω_e of an element is to be computed. In finite element computations, there is a mapping from the x-space to the ξ-space, such that (see Section 16.7, on isoparametric elements)

$$\int_{\Omega_e} f(x)\, dx = \int_{-1}^{1} \underbrace{f(\xi)\frac{dx}{d\xi}(\xi)}_{g(\xi)}\, d\xi. \tag{17.37}$$

This integral can be approximated with a numerical integration rule:

$$\int_{-1}^{1} g(\xi)\, d\xi \approx \sum_{i=1}^{n_{\text{int}}} g(\xi_i)W_i, \tag{17.38}$$

where ξ_i denotes the location of an integration point and W_i the associated weight factor. The number of integration points is given by n_{int}.

In Fig. 17.10, an interpretation is given of the above numerical integration rule. At a discrete number of points ξ_i within the interval $\xi \in [-1, +1]$ the function value $g(\xi_i)$ is evaluated. Related to each point ξ_i a rectangle is defined with height $g(\xi_i)$ and width W_i. Note that it is not necessary that the point ξ_i is located on the symmetry line of the rectangle. By adding up the surfaces $g(\xi_i)W_i$ of all rectangles, an approximation is obtained of the total surface underneath the function, which is the integral. It is clear that the weight factor W_i in Eq. (17.38) can be interpreted as the width of the interval around ξ_i.

The integration rule that is mostly used is the **Gaussian quadrature**. In that case, the locations of the integration points and weight factors are chosen so as to obtain optimal accuracy for polynomial expressions of $g(\xi)$. In Table 17.1 the location of the integration points and the associated weight factors are given up to $n_{\text{int}} = 3$. For two-dimensional problems, the above generalizes to

$$\begin{aligned}\int_{\Omega_e} f(x,y)\, d\Omega &= \int_{-1}^{1}\int_{-1}^{1} f(x(\xi,\eta), y(\xi,\eta))j(\xi,\eta)\, d\xi\, d\eta \\ &= \int_{-1}^{1}\int_{-1}^{1} g(\xi,\eta) d\xi\, d\eta,\end{aligned} \tag{17.39}$$

Table 17.1 Gaussian quadrature up to $n_{\text{int}} = 3$.

n_{int}	ξ_i	W_i
1	$\xi_1 = 0$	$W_1 = 2$
2	$\xi_1 = \frac{-1}{\sqrt{3}},\ \xi_2 = \frac{1}{\sqrt{3}}$	$W_1 = W_2 = 1$
3	$\xi_1 = -\sqrt{\frac{3}{5}},\ \xi_2 = 0,\ \xi_3 = \sqrt{\frac{3}{5}}$	$W_1 = W_3 = \frac{5}{9},\ W_2 = \frac{8}{9}$

Table 17.2 Integration points in the square and associated weight factors.

n_{int}	Point	Location of the integration points		W_i
		ξ	η	
9	1	−0.774 596 669 2	−0.774 596 669 2	0.308 642 004 7
	2	0.774 596 669 2	−0.774 596 669 2	0.308 642 004 7
	3	0.774 596 669 2	0.774 596 669 2	0.308 642 004 7
	4	−0.774 596 669 2	0.774 596 669 2	0.308 642 004 7
	5	0	−0.774 596 669 2	0.493 827 181 8
	6	0.774 596 669 2	0	0.493 827 181 8
	7	0	0.774 596 669 2	0.493 827 181 8
	8	−0.774 596 669 2	0	0.493 827 181 8
	9	0	0	0.790 123 468 6

with $j(\xi, \eta)$ according to Eq. (16.50) and

$$\int_{-1}^{1}\int_{-1}^{1} g(\xi,\eta)\, d\xi\, d\eta \approx \sum_{i=1}^{n_{\text{int}}}\sum_{j=1}^{n_{\text{int}}} g(\xi_i,\eta_j) W_i W_j. \tag{17.40}$$

The above integration scheme can be elaborated for the nine-node rectangular Lagrangian element in Fig. 17.11 using Table 17.2.

As was already remarked in Section 17.3 integration over a triangular domain is not trivial. For a triangular element that is formed by degeneration from a quadrilateral element the integration can be performed in the same way, with the same integration points, as given above.

In the case that triangular coordinates are used, the evaluation of the integrals is far from trivial. If the triangle coordinates λ_1 and λ_2 are maintained by eliminating the coordinate λ_3, these can represented in a rectangular coordinate system as given in Fig. 17.12. It can easily be seen that the surface integral of a function $\phi(\lambda_1, \lambda_2)$ can be written as

Table 17.3 Integration points in the triangles.

n_{int}	Point	Location of the integration points			Weight factors
		λ_1	λ_2	λ_3	W_i
3	1	0.5	0.5	0	0.166 67
	2	0	0.5	0.5	0.166 67
	3	0.5	0	0.5	0.166 67
7	1	0.333 333 333 3	0.333 333 333 3	0.333 333 333 3	0.112 50
	2	0.059 715 871 7	0.470 142 064 1	0.470 142 064 1	0.006 62
	3	0.470 142 064 1	0.059 715 871 7	0.470 142 064 1	0.006 62
	4	0.470 142 064 1	0.470 142 064 1	0.059 715 871 7	0.006 62
	5	0.797 426 985 3	0.101 286 507 3	0.101 286 507 3	0.006 30
	6	0.101 286 507 3	0.797 426 985 3	0.101 286 507 3	0.006 30
	7	0.101 286 507 3	0.101 286 507 3	0.797 426 985 3	0.006 30

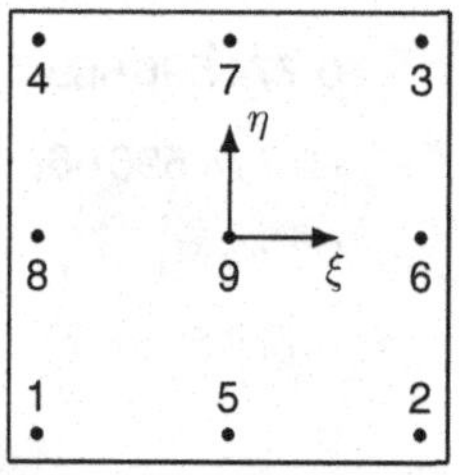

Figure 17.11

Position of the integration points in the square.

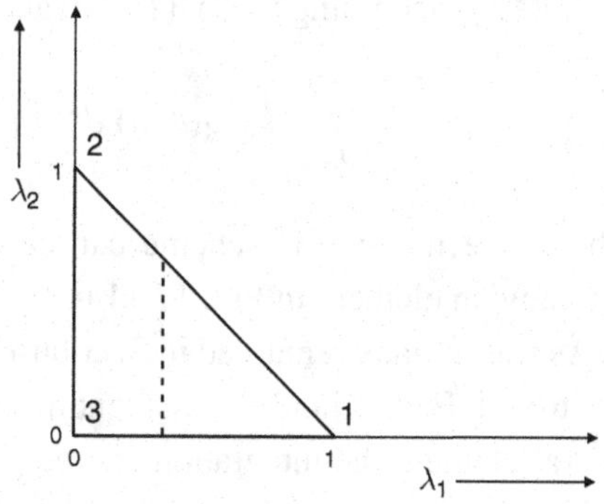

Figure 17.12

Mapping of triangle in a λ_1, λ_2-coordinate system.

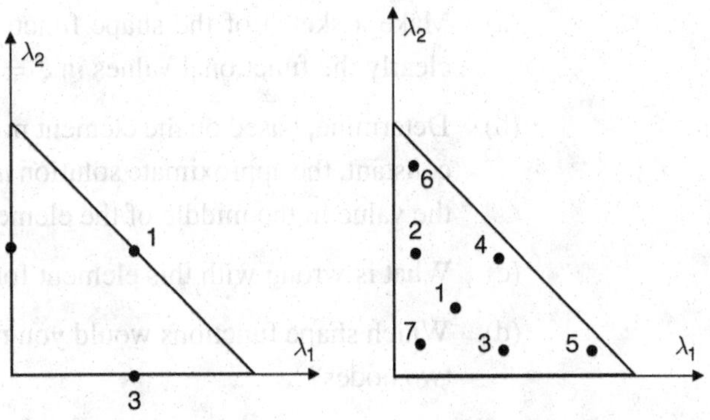

Figure 17.13
Positions of the integration points in the triangles.

$$\int_0^1 \int_0^{1-\lambda_1} \phi(\lambda_1, \lambda_2) d\lambda_2 d\lambda_1. \qquad (17.41)$$

Despite a problem with a variable integral limit, it appears possible to derive numerical integration rules for triangles. In Table 17.3 and Fig. 17.13, the position of integration points in triangular coordinates as well as weight factors are given for two triangular elements.

Exercises

17.1 We consider a very special one-dimensional isoparametric element with two nodes. The element is given in its isoparametric configuration (local coordinates) in the figure below

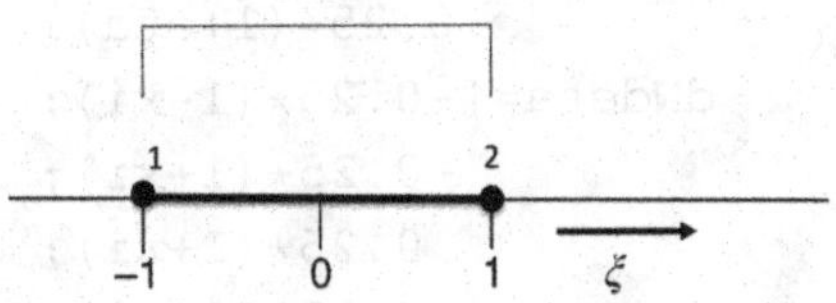

The shape functions for the nodes 1 and 2 are different from the commonly used linear functions and are given by:

$$\underset{\sim}{N}(\xi) = \begin{bmatrix} 1 - \frac{1}{4}(\xi + 1)^2 \\ 1 - \frac{1}{4}(\xi - 1)^2 \end{bmatrix}.$$

(a) Make a sketch of the shape functions on the interval (−1,1). Show clearly the functional values in $\xi = -1, 0$ and 1.

(b) Determine, based on the element nodal values $\underset{\sim}{u} = [u_0 \; u_0]^{\mathrm{T}}$ with u_0 a constant, the approximate solution $u^{\mathrm{h}}(\xi)$ as a function of ξ. Determine the value in the middle of the element for $\xi = 0$.

(c) What is wrong with this element formulation and why?

(d) Which shape functions would you normally use for this element with two nodes?

(e) Determine, based on the element nodal values $\underset{\sim}{u} = [u_0, u_0]^{\mathrm{T}}$, the approximate solution $u^{\mathrm{h}}(\xi)$ as a function of ξ, using the normal linear shape functions.

17.2 A MATLAB script to calculate the shape functions and the derivatives with respect to the isoparametric coordinates of four-noded isoparametric bi-linear elements in some point of the elements may consist of the following code:

```
% Shape = program to calculate shape functions
xi= ;
eta= ;
x=[ , , , ]';
y=[ , , , ]';
N=[0.25*(1-xi)*(1-eta);
   0.25*(1+xi)*(1-eta);
   0.25*(1+xi)*(1+eta);
   0.25*(1-xi)*(1+eta)]
dNdxi=[-0.25*(1-eta);
        0.25*(1-eta);
        0.25*(1+eta);
       -0.25*(1+eta)]
dNdeta=[-0.25*(1-xi);
        -0.25*(1+xi);
         0.25*(1+xi);
         0.25*(1-xi)]
```

To study some of the properties of the shape functions, the script will be completed by considering the elements in the following figure and extended accordingly. In this figure, the local node numbering has also been indicated.

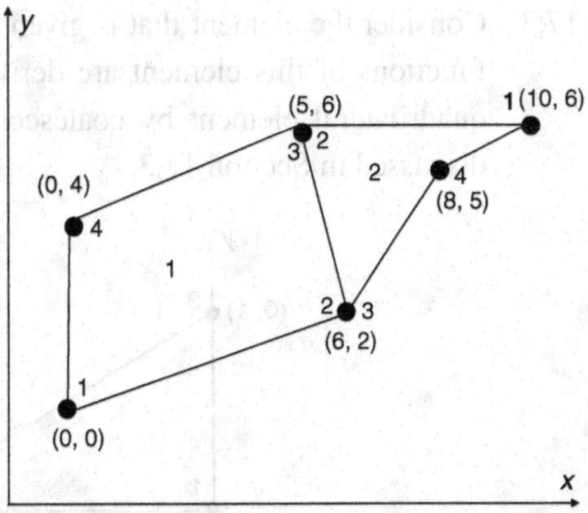

(a) Use the program to test whether

$$\sum_{i=1}^{n} N_i = 1,$$

and

$$\sum_{i=1}^{n} \frac{\partial N_i}{\partial \xi} = \sum_{i=1}^{n} \frac{\partial N_i}{\partial \eta} = 0,$$

for a number of combinations of (ξ, η).

(b) Extend the program to calculate the Jacobian matrix:

$$\underline{x}_{,\xi} = \begin{pmatrix} \frac{\partial x}{\partial \xi} & \frac{\partial y}{\partial \xi} \\ \frac{\partial x}{\partial \eta} & \frac{\partial y}{\partial \eta} \end{pmatrix},$$

and the Jacobian determinant $j = \det(\underline{x}_{,\xi})$.

(c) Determine the Jacobian determinant at the following points:

$$(\xi_1, \eta_1) = (0, 0)$$

$$(\xi_2, \eta_2) = (0.5, 0.5)$$

$$(\xi_3, \eta_3) = (1, 0)$$

$$(\xi_4, \eta_4) = (1, -1)$$

for both elements which are shown in the figure. What can you conclude from this?

17.3 Consider the element that is given in the figure below. The element shape functions of this element are derived after degeneration of a four-noded quadrilateral element by coalescence of two nodes in the same way as discussed in Section 17.3.

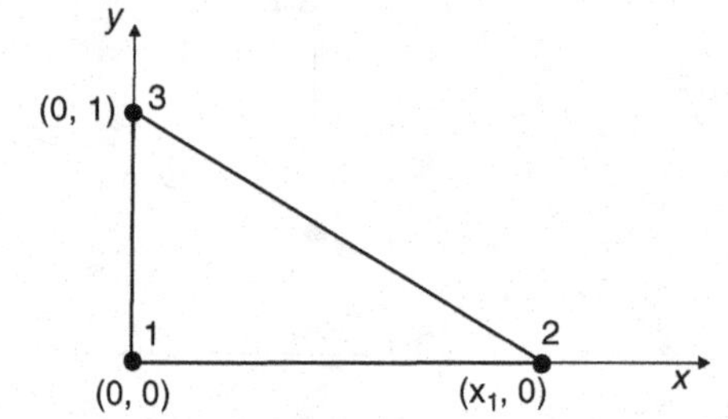

(a) Give analytical expressions for the shape functions $N_i(\xi, \eta)$.
(b) Compute the Jacobian determinant for $\xi = \eta = 0$.
(c) Plot the result as a function of x_1 on the interval $[-2, +2]$ and comment on the result.

17.4 Consider the one-dimensional four-noded element as given in the figure.

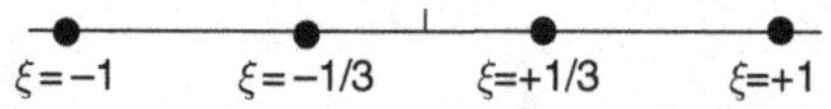

Use Lagrange polynomials to derive the element shape functions.

17.5 After solving the stationary convection–diffusion equation for the four-noded quadrilateral element e, the following element solution vector has been found:

$$\underset{\sim}{u}_e = \begin{bmatrix} 0.4 \\ 0.1 \\ 0.3 \\ 1 \end{bmatrix}.$$

(a) Use the script from Exercise 17.1 as the basis to calculate the u_h at the point $(\xi, \eta) = (0.5, 0.5)$.
(b) The nodal coordinates of the element are given:

$$\text{coord} = \begin{bmatrix} 5 & 2 \\ 15 & 1 \\ 10 & 4 \\ 4 & 4 \end{bmatrix}.$$

Calculate $\partial u_h/\partial x$ and $\partial u_h/\partial y$ for $(\xi, \eta) = (0.5, 0.5)$.

17.6 Consider the quadratic six-node triangular element in the figure.

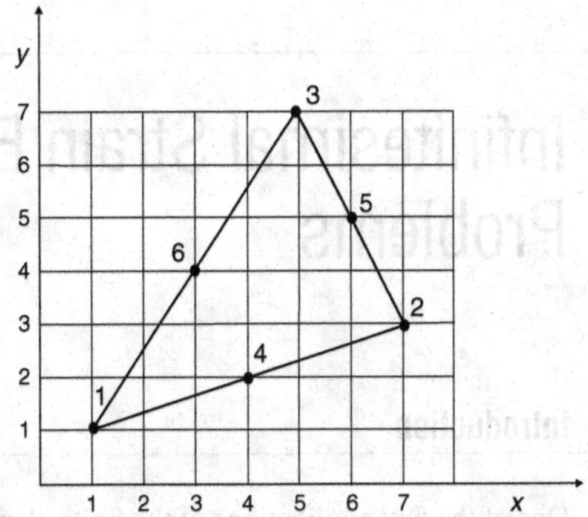

The solution vector for this element after a computation is given as

$$u_e = \begin{bmatrix} 2 \\ 5 \\ 3 \\ 2 \\ 0 \\ 2 \end{bmatrix}.$$

(a) Write a MATLAB program to calculate the solution u_h at point $(x, y) = (4, 4)$.

(b) Give expressions for $\partial u_\mathrm{h}/\partial x$ and $\partial u_\mathrm{h}/\partial y$ as a function of x and y.

18 Infinitesimal Strain Elasticity Problems

18.1 Introduction

One of the first applications of the finite element method in biomechanics has been the analysis of the mechanical behaviour of bone [2]. In particular, the impact of prosthesis implants has been extensively investigated. An example of a finite element mesh used to analyse the mechanical stresses and strain in a human femur is given in Fig. 18.1(a). In Fig. 18.1(b) the femur head is replaced by a prosthesis. In this case, the prosthesis has different mechanical properties compared to the bone, leading to high stress concentrations at some points in the bone and stress shielding (lower stresses than normal) in other parts. This normally leads to a remodelling process in the bone that has to be accounted for when new prostheses are designed. The purpose of this chapter is to introduce the finite element theory that forms the basis of these analyses.

18.2 Linear Elasticity

Neglecting inertia, the momentum equation, Eq. (11.9), may be written as

$$\vec{\nabla} \cdot \boldsymbol{\sigma} + \vec{f} = \vec{0}, \tag{18.1}$$

where $\vec{\nabla}$ denotes the gradient operator, $\boldsymbol{\sigma}$ the Cauchy stress tensor and $\vec{f} = \rho \vec{q}$ a given distributed volume load. This equation should hold at each position within the domain of interest Ω, having boundary Γ, and must be supplemented with suitable boundary conditions. Either the displacement field $\vec{u}$ is specified along Γ_u:

$$\vec{u} = \vec{u}_0 \quad \text{at} \quad \Gamma_u, \tag{18.2}$$

or the external load along Γ_p is specified:

$$\boldsymbol{\sigma} \cdot \vec{n} = \vec{p} \quad \text{at} \quad \Gamma_p. \tag{18.3}$$

The vector $\vec{n}$ denotes the unit outward normal at the boundary Γ.

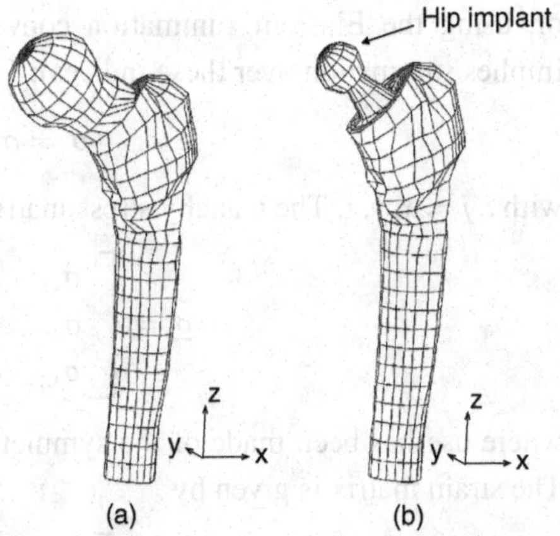

Figure 18.1

(a) Mesh to analyse stress and strain distributions in a human femur; (b) a finite element mesh of a femur, where the femur head has been replaced by a prosthesis. (Image courtesy of Bert van Rietbergen.)

If isotropic, linearly elastic material behaviour according to Hooke's law is assumed (see Section 12.2), the Cauchy stress tensor is related to the infinitesimal strain tensor $\boldsymbol{\varepsilon}$ by:

$$\boldsymbol{\sigma} = K\,\mathrm{tr}(\boldsymbol{\varepsilon})\boldsymbol{I} + 2G\boldsymbol{\varepsilon}^{\mathrm{d}}, \tag{18.4}$$

where K is the compression modulus and G the shear modulus. The infinitesimal strain tensor is given by:

$$\boldsymbol{\varepsilon} = \frac{1}{2}(\vec{\nabla}\vec{u} + (\vec{\nabla}\vec{u})^{\mathrm{T}}), \tag{18.5}$$

where $\vec{u}$ denotes the displacement field. Furthermore, $\mathrm{tr}(\boldsymbol{\varepsilon})$ is the trace and $\boldsymbol{\varepsilon}^{\mathrm{d}}$ the deviatoric part of the infinitesimal strain tensor. The shear modulus G and the compression modulus K may be expressed in terms of Young's modulus E and Poisson's ratio ν via

$$G = \frac{E}{2(1+\nu)} \tag{18.6}$$

$$K = \frac{E}{3(1-2\nu)}. \tag{18.7}$$

Now let us introduce the Cartesian basis $\{\vec{e}_x, \vec{e}_y, \vec{e}_z\}$. With respect to this basis, the Cauchy stress tensor may be written as

$$\boldsymbol{\sigma} = \sigma_{xx}\vec{e}_x\vec{e}_x + \sigma_{xy}\vec{e}_x\vec{e}_y + \cdots + \sigma_{zz}\vec{e}_z\vec{e}_z, \tag{18.8}$$

or, using the Einstein summation convention (the presence of double indices implies summation over these indices):

$$\boldsymbol{\sigma} = \sigma_{ij}\vec{e}_i\vec{e}_j, \tag{18.9}$$

with $i,j = x,y,z$. The Cauchy stress matrix is given by

$$\underline{\sigma} = \begin{bmatrix} \sigma_{xx} & \sigma_{xy} & \sigma_{xz} \\ \sigma_{xy} & \sigma_{yy} & \sigma_{yz} \\ \sigma_{xz} & \sigma_{yz} & \sigma_{zz} \end{bmatrix}, \tag{18.10}$$

where use has been made of the symmetry of the Cauchy stress tensor $\boldsymbol{\sigma}^{\mathrm{T}} = \boldsymbol{\sigma}$. The strain matrix is given by

$$\underline{\varepsilon} = \begin{bmatrix} \varepsilon_{xx} & \varepsilon_{xy} & \varepsilon_{xz} \\ \varepsilon_{xy} & \varepsilon_{yy} & \varepsilon_{yz} \\ \varepsilon_{xz} & \varepsilon_{yz} & \varepsilon_{zz} \end{bmatrix}. \tag{18.11}$$

The trace of the strain tensor is given by

$$\mathrm{tr}(\boldsymbol{\varepsilon}) = \varepsilon_{xx} + \varepsilon_{yy} + \varepsilon_{zz}, \tag{18.12}$$

and the deviatoric part of the strain matrix is given by

$$\underline{\varepsilon}^{\mathrm{d}} = \begin{bmatrix} \frac{2}{3}\varepsilon_{xx} - \frac{1}{3}(\varepsilon_{yy}+\varepsilon_{zz}) & \varepsilon_{xy} & \varepsilon_{xz} \\ \varepsilon_{xy} & \frac{2}{3}\varepsilon_{yy} - \frac{1}{3}(\varepsilon_{xx}+\varepsilon_{zz}) & \varepsilon_{yz} \\ \varepsilon_{xz} & \varepsilon_{yz} & \frac{2}{3}\varepsilon_{zz} - \frac{1}{3}(\varepsilon_{xx}+\varepsilon_{yy}) \end{bmatrix}. \tag{18.13}$$

18.3 Weak Formulation

As before, the weak formulation based on the method of weighted residuals is obtained by pre-multiplying Eq. (18.1) with a weight function. Since Eq. (18.1) is a vector equation, the weighting function is chosen to be a vector field as well: $\vec{w}$. Taking the inner product of the weight function $\vec{w}$ with the momentum Eq. (18.1) yields a scalar expression, which upon integration over the domain Ω yields

$$\int_{\Omega} \vec{w} \cdot (\vec{\nabla} \cdot \boldsymbol{\sigma} + \vec{f})\, d\Omega = 0. \tag{18.14}$$

This integral equation must hold for all weighting functions $\vec{w}$. By application of the product rule, it can be shown that

$$\vec{\nabla} \cdot (\boldsymbol{\sigma} \cdot \vec{w}) = (\vec{\nabla} \cdot \boldsymbol{\sigma}) \cdot \vec{w} + (\vec{\nabla}\vec{w})^{\mathrm{T}} : \boldsymbol{\sigma}. \tag{18.15}$$

In Eq. (18.15) the double dot product $(\vec{\nabla}\vec{w})^{\mathrm{T}} : \boldsymbol{\sigma}$ is used. The definition of the double dot product of two tensors $\boldsymbol{A}$ and $\boldsymbol{B}$ is:

$$\boldsymbol{A} : \boldsymbol{B} = \mathrm{tr}(\boldsymbol{A} \cdot \boldsymbol{B}) = A_{ij}B_{ji}, \tag{18.16}$$

where the Einstein summation convention has been used. (Note that in the mechanics literature the definition $\boldsymbol{A} : \boldsymbol{B} = \operatorname{tr}(\boldsymbol{A} \cdot \boldsymbol{B}^{\mathrm{T}}) = A_{ij}B_{ij}$ is also quite common.)

With respect to a Cartesian basis, and using index notation, it is straightforward to prove Eq. (18.15). First of all, notice that

$$\boldsymbol{\sigma} \cdot \vec{w} = \sigma_{ij} w_j \vec{e}_i. \tag{18.17}$$

Consequently

$$\vec{\nabla} \cdot (\boldsymbol{\sigma} \cdot \vec{w}) = \frac{\partial}{\partial x_i}(\sigma_{ij} w_j). \tag{18.18}$$

Application of the product rule of differentiation yields

$$\begin{aligned} \vec{\nabla} \cdot (\boldsymbol{\sigma} \cdot \vec{w}) &= \frac{\partial}{\partial x_i}(\sigma_{ij} w_j) \\ &= \frac{\partial \sigma_{ij}}{\partial x_i} w_j + \sigma_{ij} \frac{\partial w_j}{\partial x_i}. \end{aligned} \tag{18.19}$$

With the identifications

$$(\vec{\nabla} \cdot \boldsymbol{\sigma}) \cdot \vec{w} = \frac{\partial \sigma_{ij}}{\partial x_i} w_j \, , \tag{18.20}$$

and

$$(\vec{\nabla} \vec{w})^{\mathrm{T}} : \boldsymbol{\sigma} = \frac{\partial w_j}{\partial x_i} \sigma_{ij}, \tag{18.21}$$

the product rule according to Eq. (18.15) is obtained.

Use of this result in Eq. (18.14) yields

$$\int_{\Omega} \vec{\nabla} \cdot (\boldsymbol{\sigma} \cdot \vec{w}) \, d\Omega - \int_{\Omega} (\vec{\nabla} \vec{w})^{\mathrm{T}} : \boldsymbol{\sigma} \, d\Omega + \int_{\Omega} \vec{w} \cdot \vec{f} \, d\Omega = 0. \tag{18.22}$$

The first integral may be transformed using the divergence theorem Eq. (16.5). This yields

$$\int_{\Omega} (\vec{\nabla} \vec{w})^{\mathrm{T}} : \boldsymbol{\sigma} \, d\Omega = \int_{\Gamma} \vec{w} \cdot (\boldsymbol{\sigma} \cdot \vec{n}) \, d\Gamma + \int_{\Omega} \vec{w} \cdot \vec{f} \, d\Omega \, . \tag{18.23}$$

18.4 Galerkin Discretization

Within the context of the finite element method, the domain Ω is split into a number of non-overlapping subdomains (elements) Ω_e, such that this integral equation is rewritten as

$$\sum_{e=1}^{N_{\text{el}}} \int_{\Omega_e} (\vec{\nabla}\vec{w})^{\text{T}} : \boldsymbol{\sigma}\, d\Omega = \sum_{e=1}^{N_{\text{el}}} \left(\int_{\Gamma_e} \vec{w} \cdot (\boldsymbol{\sigma} \cdot \vec{n})\, d\Gamma + \int_{\Omega_e} \vec{w} \cdot \vec{f}\, d\Omega \right). \tag{18.24}$$

Clearly, the above $\int_{\Gamma_e} (.)\, d\Gamma$ denotes the integral along those element boundaries that coincide with the domain boundary Γ ($\Gamma_e = \Omega_e \cap \Gamma$). As an example of how to proceed based on Eq. (18.24), a **plane strain** problem is considered.

Step 1 The integrand of the integral in the left-hand side of Eq. (18.24) is elaborated first. If the double inner product of a symmetric tensor $\boldsymbol{A}$ and a skew-symmetric tensor $\boldsymbol{B}$ is taken, then

$$\boldsymbol{A} : \boldsymbol{B} = 0 \text{ if } \boldsymbol{A}^{\text{T}} = \boldsymbol{A} \text{ and } \boldsymbol{B}^{\text{T}} = -\boldsymbol{B}\,. \tag{18.25}$$

This can easily be proved. Using the properties of the double inner product it follows that

$$\boldsymbol{A} : \boldsymbol{B} = \boldsymbol{A}^{\text{T}} : \boldsymbol{B}^{\text{T}} = -\boldsymbol{A} : \boldsymbol{B}\,. \tag{18.26}$$

This equality can only hold if $\boldsymbol{A} : \boldsymbol{B} = 0$.

The dyadic product $(\vec{\nabla}\vec{w})^{\text{T}}$ appearing in $(\vec{\nabla}\vec{w})^{\text{T}} : \boldsymbol{\sigma}$ may be split into a symmetric and a skew-symmetric part:

$$(\vec{\nabla}\vec{w})^{\text{T}} = \frac{1}{2}\left[(\vec{\nabla}\vec{w}) + (\vec{\nabla}\vec{w})^{\text{T}}\right] - \frac{1}{2}\left[(\vec{\nabla}\vec{w}) - (\vec{\nabla}\vec{w})^{\text{T}}\right]. \tag{18.27}$$

Because of the symmetry of the Cauchy stress tensor, it follows that

$$(\vec{\nabla}\vec{w})^{\text{T}} : \boldsymbol{\sigma} = \frac{1}{2}\left[(\vec{\nabla}\vec{w}) + (\vec{\nabla}\vec{w})^{\text{T}}\right] : \boldsymbol{\sigma}. \tag{18.28}$$

Notice that the expression $\frac{1}{2}\left[(\vec{\nabla}\vec{w}) + (\vec{\nabla}\vec{w})^{\text{T}}\right]$ has a similar form to the infinitesimal strain tensor $\boldsymbol{\varepsilon}$ defined as

$$\boldsymbol{\varepsilon} = \frac{1}{2}\left[(\vec{\nabla}\vec{u}) + (\vec{\nabla}\vec{u})^{\text{T}}\right], \tag{18.29}$$

where $\vec{u}$ denotes the displacement field. This motivates the abbreviation:

$$\boldsymbol{\varepsilon}^{w} = \frac{1}{2}\left[(\vec{\nabla}\vec{w}) + (\vec{\nabla}\vec{w})^{\text{T}}\right]. \tag{18.30}$$

Consequently, the symmetry of the Cauchy stress tensor allows the double inner product $(\vec{\nabla}\vec{w})^{\text{T}} : \boldsymbol{\sigma}$ to be rewritten as

$$(\vec{\nabla}\vec{w})^{\text{T}} : \boldsymbol{\sigma} = \boldsymbol{\varepsilon}^{w} : \boldsymbol{\sigma}. \tag{18.31}$$

Step 2 To elaborate further, it is convenient to introduce a vector basis. Here, a Cartesian vector basis $\{\vec{e}_x, \vec{e}_y, \vec{e}_z\}$ is chosen. In the plane strain case it is assumed that $\varepsilon_{xz} = \varepsilon_{yz} = \varepsilon_{zz} = 0$ while the displacement field is written as

$$\vec{u} = u_x(x,y)\vec{e}_x + u_y(x,y)\vec{e}_y. \tag{18.32}$$

Likewise, the weighting function $\vec{w}$ is written as

$$\vec{w} = w_x(x,y)\vec{e}_x + w_y(x,y)\vec{e}_y. \tag{18.33}$$

In the plane strain case, the matrices associated with the tensors $\boldsymbol{\varepsilon}$ and $\boldsymbol{\varepsilon}^w$ are given by

$$\underline{\varepsilon} = \begin{bmatrix} \varepsilon_{xx} & \varepsilon_{xy} & 0 \\ \varepsilon_{xy} & \varepsilon_{yy} & 0 \\ 0 & 0 & 0 \end{bmatrix}, \quad \underline{\varepsilon}^w = \begin{bmatrix} \varepsilon^w_{xx} & \varepsilon^w_{xy} & 0 \\ \varepsilon^w_{xy} & \varepsilon^w_{yy} & 0 \\ 0 & 0 & 0 \end{bmatrix}. \tag{18.34}$$

Consequently, the inner product $\boldsymbol{\varepsilon}^w : \boldsymbol{\sigma}$ equals

$$\boldsymbol{\varepsilon}^w : \boldsymbol{\sigma} = \varepsilon^w_{xx}\sigma_{xx} + \varepsilon^w_{yy}\sigma_{yy} + 2\varepsilon^w_{xy}\sigma_{xy}. \tag{18.35}$$

Notice the factor 2 in front of the last product on the right-hand side owing to the symmetry of both $\boldsymbol{\varepsilon}^w$ and $\boldsymbol{\sigma}$. It is convenient to gather the relevant components of $\underline{\varepsilon}^w$, $\underline{\varepsilon}$ (for future purposes) and $\underline{\sigma}$ in a column:

$$(\underset{\sim}{\varepsilon}^w)^{\mathrm{T}} = \begin{bmatrix} \varepsilon^w_{xx} & \varepsilon^w_{yy} & 2\varepsilon^w_{xy} \end{bmatrix}$$

$$\underset{\sim}{\varepsilon}^{\mathrm{T}} = \begin{bmatrix} \varepsilon_{xx} & \varepsilon_{yy} & 2\varepsilon_{xy} \end{bmatrix}, \tag{18.36}$$

(notice the 2 in front of ε^w_{xy} and ε_{xy}) and

$$\underset{\sim}{\sigma}^{\mathrm{T}} = \begin{bmatrix} \sigma_{xx} & \sigma_{yy} & \sigma_{xy} \end{bmatrix}. \tag{18.37}$$

This allows the inner product $\boldsymbol{\varepsilon}^w : \boldsymbol{\sigma}$ to be written as

$$\boldsymbol{\varepsilon}^w : \boldsymbol{\sigma} = (\underset{\sim}{\varepsilon}^w)^{\mathrm{T}}\underset{\sim}{\sigma}. \tag{18.38}$$

Step 3 The constitutive equation according to Eq. (18.4) may be recast in the form

$$\underset{\sim}{\sigma} = \underline{H}\,\underset{\sim}{\varepsilon}. \tag{18.39}$$

Dealing with the isotropic Hooke's law and plane strain conditions, and after introduction of Eqs. (18.12) and (18.13) into Eq. (18.4), the matrix $\underline{H}$ can be written as

$$\underline{H} = K\begin{bmatrix} 1 & 1 & 0 \\ 1 & 1 & 0 \\ 0 & 0 & 0 \end{bmatrix} + \frac{G}{3}\begin{bmatrix} 4 & -2 & 0 \\ -2 & 4 & 0 \\ 0 & 0 & 3 \end{bmatrix}. \tag{18.40}$$

Consequently

$$\boldsymbol{\varepsilon}^w : \boldsymbol{\sigma} = (\underset{\sim}{\varepsilon}^w)^{\mathrm{T}}\underline{H}\,\underset{\sim}{\varepsilon}\,. \tag{18.41}$$

Step 4 With $\vec{u} = u_x\vec{e}_x + u_y\vec{e}_y$ and $\vec{x} = x\vec{e}_x + y\vec{e}_y$, the strain components may be written as

$$\underset{\sim}{\varepsilon} = \begin{bmatrix} \varepsilon_{xx} \\ \varepsilon_{yy} \\ 2\varepsilon_{xy} \end{bmatrix} = \begin{bmatrix} \frac{\partial u_x}{\partial x} \\ \frac{\partial u_y}{\partial y} \\ \frac{\partial u_x}{\partial y} + \frac{\partial u_y}{\partial x} \end{bmatrix}. \tag{18.42}$$

This is frequently rewritten as

$$\underset{\sim}{\varepsilon} = \underline{\hat{B}}\, \underset{\sim}{u}, \tag{18.43}$$

with $\underline{\hat{B}}$ an operator defined by

$$\underline{\hat{B}} = \begin{bmatrix} \frac{\partial}{\partial x} & 0 \\ 0 & \frac{\partial}{\partial y} \\ \frac{\partial}{\partial y} & \frac{\partial}{\partial x} \end{bmatrix}, \tag{18.44}$$

while $\underset{\sim}{u}$ represents the displacement field:

$$\underset{\sim}{u} = \begin{bmatrix} u_x \\ u_y \end{bmatrix}. \tag{18.45}$$

Step 5 Within each element, the displacement field is interpolated according to

$$\begin{aligned} u_x|_{\Omega_e} &= \sum_{i=1}^{n} N_i(x,y)u_{xei} = \underset{\sim}{N}^{\mathrm{T}} \underset{\sim}{u}_{xe} \\ u_y\big|_{\Omega_e} &= \sum_{i=1}^{n} N_i(x,y)u_{yei} = \underset{\sim}{N}^{\mathrm{T}} \underset{\sim}{u}_{ye}, \end{aligned} \tag{18.46}$$

where $\underset{\sim}{u}_{xe}$ and $\underset{\sim}{u}_{ye}$ denote the nodal displacements of element Ω_e in the x- and y-directions, respectively. Using this discretization, the strain within an element can be written as

$$\underset{\sim}{\varepsilon} = \sum_{i=1}^{n} \begin{bmatrix} \frac{\partial N_i}{\partial x} u_{xei} \\ \frac{\partial N_i}{\partial y} u_{yei} \\ \frac{\partial N_i}{\partial y} u_{xei} + \frac{\partial N_i}{\partial x} u_{yei} \end{bmatrix}. \tag{18.47}$$

It is customary, and convenient, to gather all the nodal displacements u_{xei} and u_{yei} into one column, indicated by $\underset{\sim}{u}_e$, according to

$$\underset{\sim}{u}_e = \begin{bmatrix} \left.\begin{matrix} u_{xe1} \\ u_{ye1} \end{matrix}\right\} \text{node } 1 \\ \\ \left.\begin{matrix} u_{xe2} \\ u_{ye2} \end{matrix}\right\} \text{node } 2 \\ \vdots \\ \left.\begin{matrix} u_{xen} \\ u_{yen} \end{matrix}\right\} \text{node } n \end{bmatrix}. \tag{18.48}$$

Using this definition, the strain column for an element e can be rewritten as

$$\underset{\sim}{\varepsilon} = \underline{B}\, \underset{\sim}{u}_e, \tag{18.49}$$

with $\underline{B}$ the so-called **strain displacement matrix**:

$$\underline{B} = \begin{bmatrix} \frac{\partial N_1}{\partial x} & 0 & \frac{\partial N_2}{\partial x} & 0 & \cdots & \frac{\partial N_n}{\partial x} & 0 \\ 0 & \frac{\partial N_1}{\partial y} & 0 & \frac{\partial N_2}{\partial y} & \cdots & 0 & \frac{\partial N_n}{\partial y} \\ \frac{\partial N_1}{\partial y} & \frac{\partial N_1}{\partial x} & \frac{\partial N_2}{\partial y} & \frac{\partial N_2}{\partial x} & \cdots & \frac{\partial N_n}{\partial y} & \frac{\partial N_n}{\partial x} \end{bmatrix}. \tag{18.50}$$

Clearly, a similar expression holds for $\underset{\sim}{\varepsilon}^w$. So, patching everything together, the double inner product $\boldsymbol{\varepsilon}^w : \boldsymbol{\sigma}$ may be written as:

$$\begin{aligned} \boldsymbol{\varepsilon}^w : \boldsymbol{\sigma} &= (\underset{\sim}{\varepsilon}^w)^{\mathrm{T}} \underset{\sim}{\sigma} \\ &= (\underset{\sim}{\varepsilon}^w)^{\mathrm{T}} \underline{H}\, \underset{\sim}{\varepsilon} \\ &= \underset{\sim}{w}_e^{\mathrm{T}} \underline{B}^{\mathrm{T}} \underline{H}\, \underline{B}\, \underset{\sim}{u}_e, \end{aligned} \tag{18.51}$$

where $\underset{\sim}{w}_e$ stores the components of the weighting vector $\vec{w}$ structured in the same way as $\underset{\sim}{u}_e$. This result can be exploited to elaborate the left-hand side of Eq. (18.24):

$$\int_{\Omega_e} (\vec{\nabla}\vec{w})^{\mathrm{T}} : \boldsymbol{\sigma}\, d\Omega = \underset{\sim}{w}_e^{\mathrm{T}} \int_{\Omega_e} \underline{B}^{\mathrm{T}} \underline{H}\, \underline{B}\, d\Omega\, \underset{\sim}{u}_e. \tag{18.52}$$

The element coefficient matrix, or stiffness matrix $\underline{K}_e$, is defined as

$$\underline{K}_e = \int_{\Omega_e} \underline{B}^{\mathrm{T}} \underline{H}\, \underline{B}\, d\Omega. \tag{18.53}$$

Step 6 Writing the force per unit volume vector $\vec{f}$ as

$$\vec{f} = f_x \vec{e}_x + f_y \vec{e}_y, \tag{18.54}$$

and the weighting function $\vec{w}$ within element Ω_e as

$$\vec{w}|_{\Omega_e} = \underset{\sim}{N}^{\mathrm{T}} \underset{\sim}{w}_{xe} \vec{e}_x + \underset{\sim}{N}^{\mathrm{T}} \underset{\sim}{w}_{ye} \vec{e}_y, \tag{18.55}$$

the second integral on the right-hand side of Eq. (18.24) can be written as

$$\int_{\Omega_e} \vec{w} \cdot \vec{f}\, d\Omega = \int_{\Omega_e} (\underset{\sim}{N}^{\mathrm{T}} \underset{\sim}{w}_{xe} f_x + \underset{\sim}{N}^{\mathrm{T}} \underset{\sim}{w}_{ye} f_y)\, d\Omega$$

$$= \underset{\sim}{w}_e^{\mathrm{T}} \underset{\sim}{f}_e^{\mathrm{v}}, \tag{18.56}$$

where

$$\underset{\sim}{f}_e^{\mathrm{v}} = \int_{\Omega_e} \begin{bmatrix} N_1 f_{ex} \\ N_1 f_{ey} \\ \\ N_2 f_{ex} \\ N_2 f_{ey} \\ \vdots \\ N_n f_{ex} \\ N_n f_{ey} \end{bmatrix} d\Omega. \tag{18.57}$$

The first integral on the right-hand side of Eq. (18.24) may, if applicable, be elaborated in exactly the same manner. This results in

$$\int_{\Gamma_e} \vec{w} \cdot (\boldsymbol{\sigma} \cdot \vec{n})\, d\Gamma = \underset{\sim}{w}_e^{\mathrm{T}} \underset{\sim}{f}_e^{\mathrm{p}}, \tag{18.58}$$

where $\underset{\sim}{f}_e^{\mathrm{p}}$ is structured analogously to $\underset{\sim}{f}_e^{\mathrm{v}}$. The contribution of the right-hand side triggers the abbreviation:

$$\underset{\sim}{f}_e = \underset{\sim}{f}_e^{\mathrm{v}} + \underset{\sim}{f}_e^{\mathrm{p}}, \tag{18.59}$$

often referred to as the element load contribution.

Step 7 Putting all the pieces together, the discrete weak formulation of Eq. (18.24) is written as:

$$\sum_{e=1}^{N_{\mathrm{el}}} \underset{\sim}{w}_e^{\mathrm{T}} \underline{K}_e \underset{\sim}{u}_e = \sum_{e=1}^{N_{\mathrm{el}}} \underset{\sim}{w}_e^{\mathrm{T}} \underset{\sim}{f}_e. \tag{18.60}$$

Performing an equivalent assembling procedure, as outlined in Chapter 14, the following result may be obtained:

$$\underset{\sim}{w}^{\mathrm{T}} \underline{K}\, \underset{\sim}{u} = \underset{\sim}{w}^{\mathrm{T}} \underset{\sim}{f}, \tag{18.61}$$

where $\underset{\sim}{w}$ and $\underset{\sim}{u}$ contain the global nodal weighting factors and displacements, respectively, and $\underline{K}$ is the global stiffness matrix. This equation should hold for all weighting factors, and thus

$$\underline{K}\,\underset{\sim}{u} = \underset{\sim}{f}. \tag{18.62}$$

18.5 Solution

As outlined in Chapter 14, the nodal displacements $\underset{\sim}{u}$ may be partitioned into two groups. The first displacement group consists of components of $\underset{\sim}{u}$ that are prescribed: $\underset{\sim}{u}_{\mathrm{p}}$. The remaining nodal displacements, which are initially unknown, are gathered in $\underset{\sim}{u}_{\mathrm{u}}$. Hence:

$$\underset{\sim}{u} = \begin{bmatrix} \underset{\sim}{u}_{\mathrm{u}} \\ \underset{\sim}{u}_{\mathrm{p}} \end{bmatrix}. \tag{18.63}$$

In a similar fashion, the stiffness matrix $\underline{K}$ and the load vector $\underset{\sim}{f}$ are partitioned. As a result, Eq. (18.62) can be written as:

$$\begin{bmatrix} \underline{K}_{\mathrm{uu}} & \underline{K}_{\mathrm{up}} \\ \underline{K}_{\mathrm{pu}} & \underline{K}_{\mathrm{pp}} \end{bmatrix} \begin{bmatrix} \underset{\sim}{u}_{\mathrm{u}} \\ \underset{\sim}{u}_{\mathrm{p}} \end{bmatrix} = \begin{bmatrix} \underset{\sim}{f}_{\mathrm{u}} \\ \underset{\sim}{f}_{\mathrm{p}} \end{bmatrix}. \tag{18.64}$$

The force column $\underset{\sim}{f}$ is split into two parts: $\underset{\sim}{f}_{\mathrm{u}}$ and $\underset{\sim}{f}_{\mathrm{p}}$. The column $\underset{\sim}{f}_{\mathrm{u}}$ is known, since it stores the external loads applied to the body. The column $\underset{\sim}{f}_{\mathrm{p}}$, on the other hand, is not known, since no external load may be applied to points at which the displacement is prescribed. The following set of equations results:

$$\underline{K}_{\mathrm{uu}}\underset{\sim}{u}_{\mathrm{u}} = \underset{\sim}{f}_{\mathrm{u}} - \underline{K}_{\mathrm{up}}\underset{\sim}{u}_{\mathrm{p}}, \tag{18.65}$$

$$\underset{\sim}{f}_{\mathrm{p}} = \underline{K}_{\mathrm{pu}}\underset{\sim}{u}_{\mathrm{u}} + \underline{K}_{\mathrm{pp}}\underset{\sim}{u}_{\mathrm{p}}. \tag{18.66}$$

The first equation is used to calculate the unknown displacements $\underset{\sim}{u}_{\mathrm{u}}$. The result is substituted into the second equation to calculate the unknown forces $\underset{\sim}{f}_{\mathrm{p}}$.

Example 18.1 Consider the bending of a beam subjected to a concentrated force (Fig. 18.2). Let the beam be clamped at $x = 0$ and the point load F be applied at $x = L$. It is interesting to investigate the response of the bar for different kinds of elements using a similar element distribution. Four different elements are tested: the linear and the quadratic triangular element, and the bi-linear and bi-quadratic quadrilateral element. The meshes for the linear triangular and the bi-linear quadrilateral element are shown in Fig. 18.3. The displacement in vertical direction at $x = L$ can be computed using standard beam theory, giving:

$$u_L = \frac{FL^3}{3EI}, \tag{18.67}$$

Table 18.1 Comparison of the relative accuracy of different element types for the beam bending case.

Element type	u_c/u_L
Linear triangle	0.231
Bi-linear quadrilateral	0.697
Quadratic triangle	0.999
Bi-quadratic quadrilateral	1.003

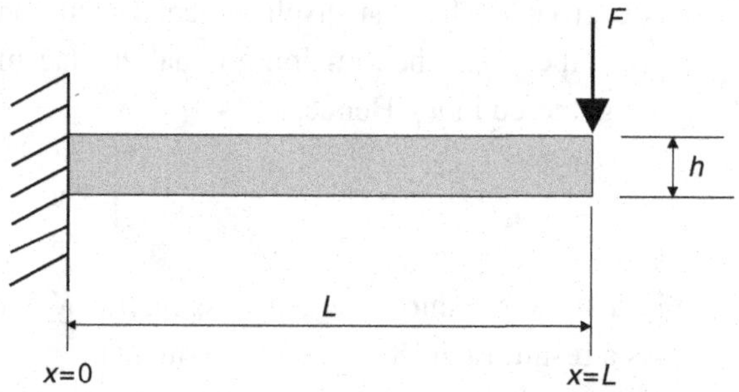

Figure 18.2

A beam clamped on one side and loaded with a vertical concentrated force at the other side.

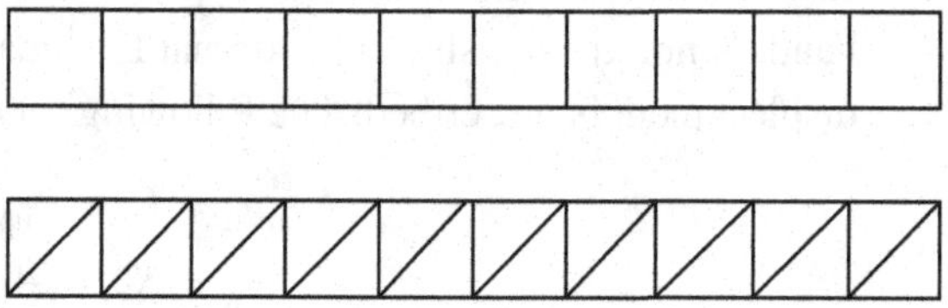

Figure 18.3

Triangular and quadrilateral element distribution for the beam problem.

with

$$I = \frac{bh^3}{12}, \tag{18.68}$$

where b is the width of the beam, h the height of the beam and E the Young's modulus of the material. The ratio h/L of height over length equal to 0.1 is chosen. Using the meshes depicted in Fig. 18.3, the results of Table 18.1 are obtained, which presents the ratio of u_L and the computed displacement u_c at $x = L$.

It is clear that the displacement of the beam obtained by using a mesh of linear triangles is much too small. The poor performance of the linear triangle can easily be understood; because of the linear interpolation of the displacement field $\vec{u}$, the associated strains computed from

$$\underset{\sim}{\varepsilon} = \underset{\sim}{B}\,\underset{\sim}{u} \qquad (18.69)$$

are constant per element. The bi-linear quadrilateral element, on the other hand, is clearly enhanced. A typical shape function, for example that of the first node in an element, is given by (with respect to the local coordinate system)

$$N_1 = \frac{1}{4}(1-\xi)(1-\eta). \qquad (18.70)$$

Hence

$$N_1 = \frac{1}{4}(1-\xi-\eta+\xi\eta), \qquad (18.71)$$

which means that an additional non-linear term is present in the shape functions. Therefore a *linear* variation of the stress field within an element is represented.

Two remarks have to be made at this point:

- The numerical analysis in this example was based on plane stress theory, whereas in the rest of this chapter the equations were elaborated for a plane strain problem. How this elaboration is done for a plane stress problem is discussed in Exercise 18.1.
- Strictly speaking, a concentrated force in one node of the mesh is not correct. Mesh refinement in this case would lead to an infinite displacement of the node where the force is acting. This can be avoided by applying a distributed load over the height of the beam.

Example 18.2 The bending deformation described in the previous example is often used in biological experiments to quantify the forces exerted by soft tissues on their surroundings. In Fig. 18.4, an experimental set-up is shown where tissues are cultured around silicone microposts that bend when forces are exerted on them by the tissue. When researchers develop computational models to predict and understand

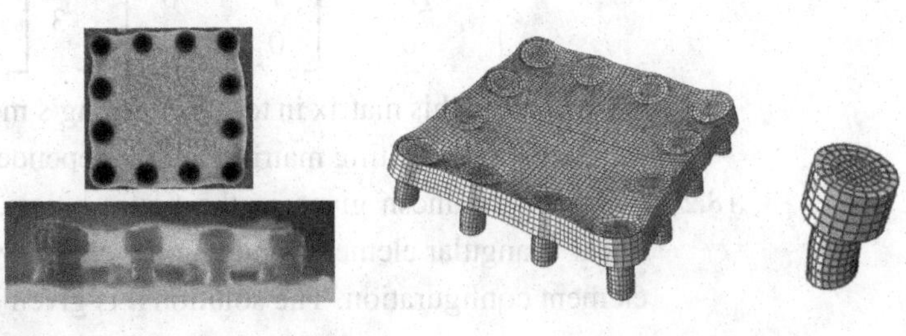

Figure 18.4

(a) Example of an experiment in which a tissue has been cultured around silicone microposts, top view and side view (image courtesy of Jasper Foolen). (b) Finite element model of the complete experimental configuration. (c) Finite element model of a single micropost.

how much force is exerted on the posts, they have to ensure that the element choice and distribution is adequate for describing the correct bending properties of the posts. If linear triangles were used, for example, then the force exerted by the tissue in the model would be overestimated if a similar degree of bending is to be modelled compared to the experimental observations.

Exercises

18.1 For Hooke's law, the Cauchy stress tensor $\boldsymbol{\sigma}$ is related to the infinitesimal strain tensor $\boldsymbol{\varepsilon}$ via

$$\boldsymbol{\sigma} = K\,\mathrm{tr}(\boldsymbol{\varepsilon})\boldsymbol{I} + 2G\boldsymbol{\varepsilon}^{\mathrm{d}}.$$

(a) What is tr($\boldsymbol{\varepsilon}$) for the plane strain case?
(b) What is tr($\boldsymbol{\varepsilon}$) for the plane stress case?
(c) What are the non-zero components of $\boldsymbol{\varepsilon}$ and $\boldsymbol{\sigma}$ for the plane stress case?
(d) What are the non-zero components of $\boldsymbol{\varepsilon}$ and $\boldsymbol{\sigma}$ for the plane strain case?
(e) Consider the plane stress case. Let

$$\underset{\sim}{\sigma} = \begin{bmatrix} \sigma_{xx} \\ \sigma_{yy} \\ \sigma_{xy} \end{bmatrix} \quad \underset{\sim}{\epsilon} = \begin{bmatrix} \varepsilon_{xx} \\ \varepsilon_{yy} \\ 2\varepsilon_{xy} \end{bmatrix}.$$

What is the related $\underline{H}$ matrix for this case, using

$$\underset{\sim}{\sigma} = \underline{H}\,\underset{\sim}{\varepsilon}?$$

(f) For the plane strain case, the matrix $\underline{H}$ is given by

$$\underline{H} = K\begin{bmatrix} 1 & 1 & 0 \\ 1 & 1 & 0 \\ 0 & 0 & 0 \end{bmatrix} + \frac{G}{3}\begin{bmatrix} 4 & -2 & 0 \\ -2 & 4 & 0 \\ 0 & 0 & 3 \end{bmatrix}.$$

Rewrite this matrix in terms of Young's modulus E and Poisson's ratio ν. Is the resulting matrix linearly dependent on E?

18.2 Consider the mesh given in the figure below. The mesh consists of two linear triangular elements and a linear elasticity formulation applies to this element configuration. The solution $\underset{\sim}{u}$ is given by

$$\underset{\sim}{u}^{\mathrm{T}} = [u_1\ w_1\ u_2\ w_2\ u_3\ w_3\ u_4\ w_4],$$

where u and v denote the displacements in the x- and y-directions, respectively.

(a) What is a possible `top` array of this element configuration assuming equal material and type identifiers for both elements?

(b) What is the `pos` array, for this element configuration?

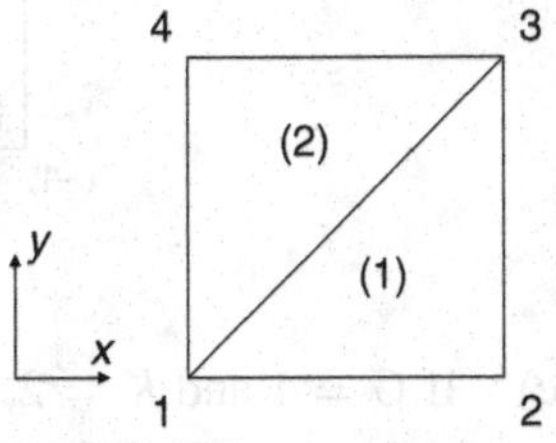

(c) What is the `dest` array?

(d) Based on the `pos` array, the non-zero entries of the stiffness matrix can be identified. Visualize the non-zero entries of the stiffness matrix.

(e) Suppose that the boundary nodes in the array `usercurves` are stored as

```
usercurves=[
  1 2
  2 3
  3 4
  4 1
  1 3];
```

The solution $\underset{\sim}{u}$ is stored in the array `sol`. How are the displacements in the x-direction extracted from the `sol` array along the last `usercurve`?

(f) Let the solution array `sol` and the global stiffness matrix `q` be given. Suppose that both displacements at the nodes 1 and 2 are suppressed. Compute the reaction forces in these nodes. Compute the total reaction force in the y-direction along the boundary containing the nodes 1 and 2.

18.3 Consider the bi-linear element of the figure below. It covers exactly the domain $-1 \leq x \leq 1$ and $-1 \leq y \leq 1$. Assume that the plane strain condition applies.

(a) Compute the strain displacement matrix $\underline{B}$ for this element in point $(x, y) = (\frac{1}{2}, \frac{1}{2})$.

(b) Let the nodal displacements for this element be given by

$$\underset{\sim}{u}_e^{\mathrm{T}} = \begin{bmatrix} 0 & 0 & 0 & 0 & 1 & 0 & 1 & 0 \end{bmatrix}.$$

What are the strains in $(x, y) = (\frac{1}{2}, \frac{1}{2})$?

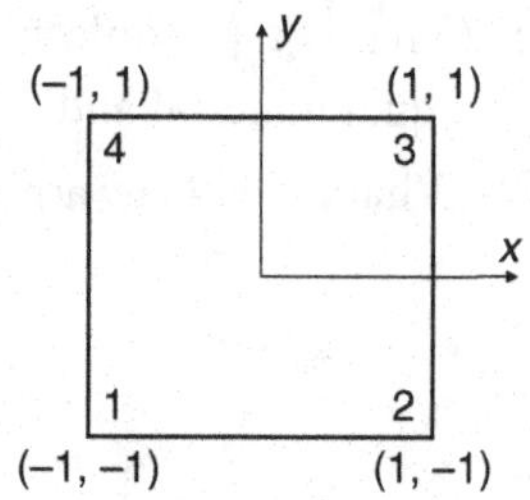

(c) If $G = 1$ and $K = 2$, what are the stress components σ_{xx}, σ_{yy} and σ_{xy} at $(x, y) = (0, 0)$? What is the value of σ_{zz}?

18.4 Given the solution vector `sol` and the `pos` array,

(a) Extract the solution vector for a given element `ielem`.

(b) Compute the strain at an integration point.

(c) Compute the stress at an integration point.

18.5 The m-file `demo_bend` in the directory `twode` analyses the pure bending of a single element (for quadrilaterals) or two elements (for triangles). The analysis is based on plane stress theory. The geometry is a simple square domain of dimensions 1×1. Along the left edge the displacements in the x-direction are set to zero, while at the lower left corner the displacement in the y-direction is set to zero to prevent rigid body motions. The two nodes at the right edge are loaded with a force F of opposite sign, to represent a pure bending moment. Investigate the stress field for various elements: linear triangle, bi-linear quadrilateral, and their quadratic equivalents. Explain the observed differences. To make these choices, use `itype` and `norder`:

```
itype = 1 : quadrilateral element
itype = 20 : triangular element
norder = 1 :(bi-)linear element
norder = 2 :(bi-)quadratic element
```

18.6 In a shearing test, a rectangular piece of material is clamped between a top and bottom plate, as schematically represented in the figure. This experiment is generally set up to represent the so-called ‘simple-shear’ configuration. In the simple-shear configuration, the strain tensor is given as

$$\boldsymbol{\varepsilon} = \varepsilon_{xy}\vec{e}_x\vec{e}_y + \varepsilon_{xy}\vec{e}_y\vec{e}_x$$

using the symmetry of the strain tensor. As a consequence, the stress–strain relation according to Hooke’s law reduces to

$$\sigma_{xy} = 2G\varepsilon_{xy}.$$

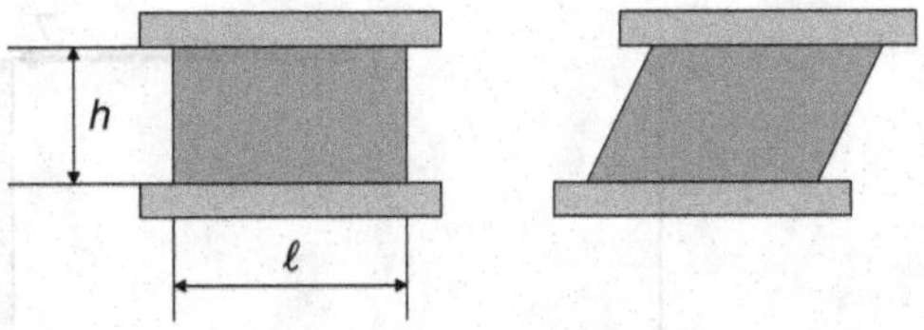

Hence, measuring the clamp forces and the shear displacement provides a direct means to identify the shear modulus G. However, the 'simple-shear' state is difficult to realize experimentally, since the configuration of the figure does not exactly represent the simple-shear case. This may be analysed using the m-file `demo_shear`.

(a) Analyse the shear and the simple-shear case using this m-file. What is the difference in boundary conditions for these two cases?
(b) Why is the `simple_shear=0` case not equal to the exact simple-shear case?
(c) What ratio ℓ/h is required to measure the modulus G within 10% accuracy? How many elements did you use to obtain this result? Explain the way in which the m-file computes the modulus G.

18.7 In an experimental set-up, the bending of a beam is used to measure forces. To be able to design the set-up with suitable dimensions and known mechanical properties, the researcher performs a number of simulations using the model of the sensor shown in the figure below. The mechanical load is applied at `userpoint 7`. By measuring the bending of the beam, the load can be determined.

Create a finite element model in `mlfem_nac`. The dimensions of the sensor are given in the figure in [mm]. Assume the sensor can be modelled with two-dimensional plane stress elements. Use bi-linear quadrilateral elements. Subarea I is modelled with 10×5 elements. Subarea II with 3×5 elements and subarea III with 3×10 elements. The displacements of the boundary at $y = 0$ are suppressed. At `userpoint` 7, a force is prescribed in x-direction with magnitude 0.01 [N]. The material comprising the sensor can be modelled as linear elastic with Young's modulus $E = 100$ [Nmm^{-2}], the Poisson's ratio equals 0.3 [-].

Determine the horizontal displacement of the beam at `userpoint` 7.

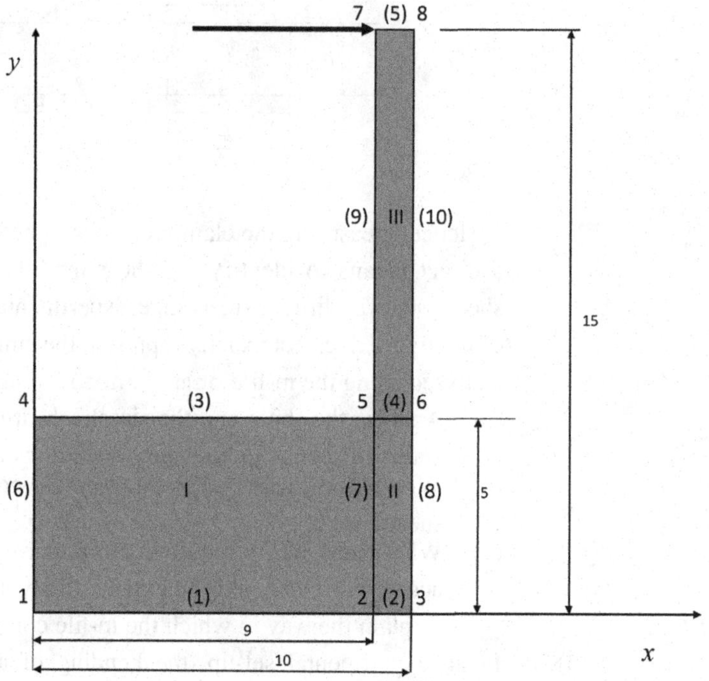
7 (5) 8
y
(9) III (10)
15
4 (3) 5 (4) 6
(6) I (7) II (8)
5
1 (1) 2 (2) 3
9
10
x

References

[1] Adams, R. A. (2003) *Calculus: A Complete Course* (Addison, Wesley, Longman).

[2] Brekelmans, W. A. M., Poort, H. W. and Slooff, T. J. J. H. (1972) A new method to analyse the mechanical behaviour of skeletal parts. *Acta Orthop. Scand.* **43**, 301–317.

[3] Brooks, A. N. and Hughes, T. J. R. (1990) Streamline upwind/Petrov–Galerkin formulations for convection dominated flows with particular emphasis on the incompressible Navier–Stokes equations. *Computer Methods in Applied Mechanics and Engineering Archive* Special Edition, 199–259.

[4] Carslaw, H. S. and Jaeger, J. C. (1980) *Conduction of Heat in Solids* (Clarendon Press).

[5] Cacciola, G. R. C. (1998) *Design, Simulation and Manufacturing of Fibre Reinforced Polymer Heart Valves.* Ph.D. thesis, Eindhoven University of Technology.

[6] Delfino, A., Stergiopoulos, N., Moore, J. E. and Meister, J. J. (1997) Residual strain effects on the stress field in thick wall finite element model of the human carotid bifurcation. *J. Biomech.* **30**, 777–786.

[7] Frijns, A. J. H., Huyghe, J. M. R. J. and Janssen, J. D. (1997) A validation of the quadriphasic mixture theory for intervertebral disc tissue. *Int. J. Eng. Sci.* **35**, 1419–1429.

[8] Fung, Y. C. (1990) *Biomechanics: Motion, Flow, Stress, and Growth* (Springer-Verlag).

[9] Fung, Y. C. (1993) *Biomechanics: Mechanical Properties of Living Tissues*, 2nd Edition (Springer-Verlag).

[10] Gerhardt, L. C., Schmidt, J., Sanz-Herrera, J. A. *et al.* (2012) A novel method for visualising and quantifying through-plane skin layer deformations. *J. Mech. Behav. Biomed. Mat.* **14**, 199–207.

[11] Hill, A. V. (1938) The heat of shortening and the dynamic constants in muscle. *Proc. Roy. Soc. Lond.* **126**, 136–165.

[12] Hughes, T. J. R. (1987) *The Finite Element Method* (Prentice Hall).

[13] Huyghe, J. M. R. J., Arts, T. and Campen, D. H. van (1992) Porous medium finite element model of the beating left ventricle. *Am. J. Physiol.* **262**, H1256–H1267.

[14] Huxley, A. F. (1957). Muscle structure and theory of contraction. *Prog. Biochem. Biophys. Chem.* **7**, 255–318.

[15] Maaden, K. van der, Sekerdag, E., Schipper, P. *et al.* (2015) Layer-by-layer assembly of inactivated poliovirus and N-trimethyl chitosan on pH-sensitive microneedles for dermal vaccination. *Langmuir* **31**, 8654–8660.

[16] Mow, V. C., Kuei, S. C. and Lai, W. M. (1980) Biphasic creep and stress relaxation of articular cartilage in compression. *J. Biomech. Eng.* **102**, 73–84.

[17] Oomens, C. W. J., Campen, D. H. van and Grootenboer, H.J. (1985) A mixture approach to the mechanics of skin. *J. Biomech.* **20**, 877–885.

[18] Oomens, C. W. J., Maenhout, M., Oijen, C. H. van, Drost, M. R. and Baaijens, F. P. T. (2003) Finite element modelling of contracting skeletal muscle. *Phil. Trans. R. Soc. Lond. B* **358**, 1453–1460.

[19] Sengers, B. G., Oomens, C. W. J., Donkelaar, C. C. van and Baaijens, F. P. T. (2005) A computational study of culture conditions and nutrient supply in cartilage tissue engineering. *Biotechnol. Prog.* **21**, 1252–1261.

[20] Sengers, B. G. (2005) *Modeling the Development of Tissue Engineered Cartilage.* Ph.D. thesis, Eindhoven University of Technology.

[21] Vlimmeren, M. A. A. van, Driessen-Mol, A., Oomens, C. W. J. and Baaijens, F. P. T. (2011) An in vitro model system to quantify stress generation, compaction and retraction in engineered heart valve tissue. *Tissue Eng. Part C* **17**, 983–991.

[22] Ward, I. M., and Hardley, D. W. (1993) *Mechanical Properties of Solid Polymers* (John Wiley & Sons).

[23] Zienkiewicz, O. C. (1989) *The Finite Element Method,* 4th Edition (McGraw-Hill).

Index

Almansi–Euler strain tensor, 217
anisotropy, 383
assembly process, 303, 305

basis
  Cartesian, 5
  orthogonal, 5
  orthonormal, 5
boundary conditions, 328
  essential, 122, 293
  natural, 123, 293
boundary value problem, 123
Bubnov–Galerkin, 302
bulk modulus, 237

cantilever beam, 47
Cartesian basis, 134
Cauchy–Green tensor
  left, 211
  right, 210
commutative, 2
completeness, 363
compression modulus, 237, 382
configuration, material, 183
confined compression, 265
constitutive model, 56, 235
convection, 189
convection–diffusion equation, 327, 348
  in 3-D, 342
convective
  contribution, 189
  velocity, 189
convolution integral, 97
coordinates, material, 183
Couette flow, 285
Coulomb friction, 278
Crank–Nicholson scheme, 330
creep, 96
creep function, 97
cross product, 3

Darcy's law, 264
deformation
  gradient tensor, 206
  matrix, 192
  tensor, 193, 206
degeneration, 367
differential equation, partial, 327
diffusion coefficient, 264
diffusion equation, 292, 342
divergence theorem, 344
dot product, 2
dyadic product, 4

eigenvalue, 172
eigenvector, 12, 172
elastic behaviour, 235
element
  bi-linear, 365
  isoparametric, 365
  Lagrangian, 370
  quadrilateral, 365
  Serendipity, 370
  triangular, 367
element column, 347
element matrix, 347
element Peclet number, 335
elongational rate, 82
equilibrium equations, 162
Eulerian description, 185

fibre, 56
  elastic, 56
  non-linear, 58
Fick's law, 264
Finger tensor, 217
force
  decomposition, 22
  normal, 23
  parallel, 22
  vector, 16, 17
force equilibrium, 117, 118
Fourier number, 329
free body diagram, 42, 157
friction coefficient, 279

Galerkin method, 300, 345, 385
Gaussian integration, 310, 373
geometrically non-linear, 65, 272
Green–Lagrange strain tensor, 214

harmonic excitation, 98
Heaviside function, 85
homogenization, 133, 139
Hooke's law, 236, 383
hydrostatic pressure, 176

initial condition, 328
inner product, 2, 20
integration by parts, 344
integration points, 311, 373
integration scheme
  backward Euler, 330
  Crank–Nicholson, 330
  explicit, 330
  forward Euler, 330
  implicit, 330
internal mechanical energy, 231
isochoric deformation, 219
isoparametric, 308, 349, 366
isotropy, 236, 383

Kelvin–Voigt model, 96
kinetic energy, 230

Lagrangian description, 185
line-of-action, 16
linear elastic stress–strain relation, 121
linear elasticity, 382
linear strain tensor, 216
local coordinate system, 349
loss modulus, 99

matrices
  `pos`, 313
  `top`, 312
Maxwell model, 92, 93
muscle, contraction, 60
myofibrils, 59

Navier–Stokes equation, 282
Newton's law, 18
Newtonian fluid, 262
non-Newtonian fluid, 263
numerical integration, 310, 349, 373

Peclet number, 329
permeability, 264
Poiseuille flow, 284
polynomial interpolation, 297
polynomials, Lagrangian, 370
proportionality, 85

relaxation, 95
relaxation function, 89, 97
relaxation time, 93
retardation time, 96

scalar multiplication, 19
shape functions, 297
shear modulus, 237
snap-through behaviour, 66
spatial discretization, 333
spin matrix, 150
static equilibrium, 39
statically determinate, 43
statically indeterminate, 43
stent, 353
storage modulus, 99
strain ε, 120
streamline upwind scheme, 338
stress
  deviatoric, 176
  equivalent, 177
  hydrostatic, 176
  principal, 172
  tensor, 164
  Tresca, 177
  vector, 155
  von Mises, 177
stress σ, 119
stretch ratio, 210
superposition, 85

temporal discretization, 330
tensor
  definition, 4
  deformation rate, 220
  determinant, 12
  deviatoric, 12
  invariant, 171
  inverse, 11
  product, 4
  rotation velocity or spin, 220
  trace, 12
theta-scheme, 330
time derivative
  material, 185
  spatial, 185
transfer function, 103
triple product, 4
Trouton's law, 263

vector addition, 19
vector basis, 4
vector product, 3
viscosity, 262
viscous behaviour, 81

weak form, 296, 345, 348
weak formulation, 384
weighted residuals, 295
weighting function, 295

Young's modulus, 382

Printed in the United States
By Bookmasters